HEALTH RISKS OF RADON AND OTHER INTERNALLY DEPOSITED ALPHA-EMITTERS

BEIR IV

HEALTH RISKS OF RADON AND OTHER INTERNALLY DEPOSITED ALPHA-EMITTERS

BEIR IV

Committee on the Biological Effects
of Ionizing Radiations
Board on Radiation Effects Research
Commission on Life Sciences
National Research Council

NATIONAL ACADEMY PRESS
Washington, D.C. 1988

National Academy Press • 2101 Constitution Avenue, N.W. • Washington, D. C. 20418

NOTICE: The project that is the subject of this report was approved by the Governing Board of the National Research Council, whose members are drawn from the councils of the National Academy of Sciences, the National Academy of Engineering, and the Institute of Medicine. The members of the committee responsible for the report were chosen for their special competences and with regard for appropriate balance.

This report has been reviewed by a group other than the authors according to procedures approved by a Report Review Committee consisting of members of the National Academy of Sciences, the National Academy of Engineering, and the Institute of Medicine.

The National Academy of Sciences is a private, nonprofit, self-perpetuating society of distinguished scholars engaged in scientific and engineering research, dedicated to the furtherance of science and technology and to their use for the general welfare. Upon the authority of the charter granted to it by the Congress in 1863, the Academy has a mandate that requires it to advise the federal government on scientific and technical matters. Dr. Frank Press is president of the National Academy of Sciences.

The National Academy of Engineering was established in 1964, under the charter of the National Academy of Sciences, as a parallel organisation of outstanding engineers. It is autonomous in its administration and in the selection of its members, sharing with the National Academy of Sciences the responsibility for advising the federal government. The National Academy of Engineering also sponsors engineering programs aimed at meeting national needs, encourages education and research, and recognizes the superior achievements of engineers. Dr. Robert M. White is president of the National Academy of Engineering.

The Institute of Medicine was established in 1970 by the National Academy of Sciences to secure the services of eminent members of appropriate professions in the examination of policy matters pertaining to the health of the public. The Institute acts under the responsibility given to the National Academy of Sciences by its congressional charter to be an adviser to the federal government and, upon its own initiative, to identify issues of medical care, research, and education. Dr. Samuel O. Thier is president of the Institute of Medicine.

The National Research Council was organized by the National Academy of Sciences in 1916 to associate the broad community of science and technology with the Academy's purposes of furthering knowledge and advising the federal government. Functioning in accordance with general policies determined by the Academy, the Council has become the principal operating agency of both the National Academy of Sciences and the National Academy of Engineering in providing services to the government, the public, and the scientific and engineering communities. The Council is administered jointly by both Academies and the Institute of Medicine. Dr. Frank Press and Dr. Robert M. White are chairman and vice chairman, respectively, of the National Research Council.

The study summarized in this report was supported by the U.S. Environmental Protection Agency and the U.S. Nuclear Regulatory Commission under Contract No. 68-02-3895.

Library of Congress Cataloging-in-Publication Data
Health risks of radon and other internally deposited alpha-emitters.

This report was supported by the U.S. Environmental Protection Agency and the U.S. Nuclear Regulatory Commission under Contract No. 68-02-3895.
Includes bibliographies and index.
1. Alpha rays—Health aspects. 2. Radioisotopes in the body—Health aspects. 3. Radon—Health aspects. 4. Health risk assessment. I. National Research Council (U.S.). Committee on the Biological Effects of Ionizing Radiations. II. United States. Environmental Protection Agency. III. U.S. Nuclear Regulatory Commission. [DNLM: 1. Probability. 2. Radiation Effects. 3. Radioactive Pollutants—adverse effects. 4. Radon—adverse effects. 5. Risk WN 620 H4345]
RC95.A42H43 1987 616.9′897 87-31280
ISBN 0-309-03789-1, soft cover; ISBN 0-309-03797-2, hard cover

FRONTISPIECE: *Radium-dial painters.*

Printed in the United States of America

First Printing, January 1988
Second Printing, May 1988

COMMITTEE ON THE BIOLOGICAL EFFECTS OF IONIZING RADIATIONS

Jacob I. Fabrikant (*Chairman*), Donner Laboratory, University of California, Berkeley, California
William J. Bair (*Vice-Chairman*), Battelle Pacific Northwest Laboratories, Richland, Washington
Michael A. Bender, Brookhaven National Laboratory, Upton, New York
Carmia G. Borek, College of Physicians and Surgeons, Columbia University, New York, New York
Peter G. Groer, Center for Epidemiologic Research, Oak Ridge Associated Universities, Oak Ridge, Tennessee
Jay H. Lubin, National Cancer Institute, Bethesda, Maryland
David C. F. Muir, McMaster University, Hamilton, Ontario, Canada
Donald A. Pierce, Department of Statistics, Oregon State University, Corvallis, Oregon
William C. Roesch, Richland, Washington
Jonathan M. Samet, University of New Mexico Cancer Center, Albuquerque, New Mexico
Robert A. Schlenker, Argonne National Laboratory, Argonne, Illinois
Richard J. Waxweiler, Centers for Disease Control, Atlanta, Georgia

National Research Council Staff

William H Ellett, Project Director, Board on Radiation Effects Research, Commission on Life Sciences
Raymond D. Cooper, Senior Program Officer, Board on Radiation Effects Research, Commission on Life Sciences

Sponsors' Project Officers

Neal Nelson, U.S. Environmental Protection Agency
Shlomo Yaniv, U.S. Nuclear Regulatory Commission

Preface

BACKGROUND

In June 1984, the Environmental Protection Agency (EPA) and the Nuclear Regulatory Commission (NRC) asked the National Academy of Sciences to submit a proposal in response to EPA Solicitation DU-84-C165 for a study of the "biological effects of internally deposited alpha-emitting radionuclides and their decay products." The proposal constituted an extension of the work of the National Research Council's Committees on the Biological Effects of Ionizing Radiations (BEIR), which began in the early 1970s and most recently culminated in the report *The Effects on Populations of Exposure to Low Levels of Ionizing Radiation: 1980.* That report, the so-called BEIR III report, dealt mainly with the effects of radiation of low linear energy transfer (low LET), primarily external x rays and gamma rays.

At the time of the BEIR III deliberations, the human and animal studies on high-LET radiation effects were limited, and epidemiological surveys were only beginning to provide reliable data on potential health effects. The reported epidemiological and laboratory animal studies pointed to a need to extend the series of BEIR reports, to appraise the state of scientific knowledge concerning the biological and health effects of alpha radiation (internally deposited alpha-emitting radionuclides and their decay products). This will enable government officials and the public to make decisions about the potential community and workplace health hazards associated with exposure to internal alpha-emitters, such as those from indoor radon and uranium mining.

The task before the current BEIR committee was specified in detail in the contract agreement between the National Academy of Sciences and the EPA and NRC signed on October 1, 1984.

CHARGE TO THE COMMITTEE

In response to the EPA and NRC request, the Committee on the Biological Effects of Ionizing Radiations was established within the National Research Council's Commission on Life Sciences. This committee, the fourth in a series originally established in 1969, was asked for a comprehensive assessment of available knowledge of the risks associated with internally deposited alpha-emitters. Radiobiological and animal data were to be reviewed, but relevant epidemiological data were also to be used to the greatest possible extent in estimating the risks.

The first phase of the study was to be a review of current knowledge of the somatic and genetic effects of internal alpha-emitters, including clinical and epidemiological evidence of human effects, results of animal studies, alpha-particle damage at the cellular level, metabolic pathways for internal alpha-emitters, dosimetry and microdosimetry of alpha-emitters deposited in specific tissues, and the possible chemical toxicity of low-specific-activity alpha-emitters. The committee was also asked to review the evidence of dependence of the biological effects on age, sex, route of entry, dose, dose rate, physical and chemical properties of the radioactive materials, and similar factors.

During the second phase of the study, the committee was requested to suggest methods for estimating the risks to human health, with their related uncertainties, associated with internally deposited alpha-emitters and then to apply the methods to the principal alpha-emitters in the environment. This phase was to include the provision of formulas and coefficients to estimate individual and population risks associated with single and chronic exposure to internal alpha-emitters and, where appropriate, threshold formulas and coefficients for nonstochastic effects. This information was to be applied to estimating numbers of genetic effects, risks to unborn children, and risks of carcinogenic effects. The committee was asked to describe the metabolic models they used and provide examples of the methods to be used in applying their risk estimates to exposed populations. Finally, the committee was asked to discuss the uncertainty in their risk estimates and provide recommendations for further research based on the limitations in the data available for assessing the risks which the committee identified.

The third phase of the committee's work was preparation and submission of a report covering the results and findings of the first two phases.

The committee's review and evaluation of the current epidemiological and basic research involved not only an assessment of the relevant research data and their analyses in the scientific literature, but also an independent evaluation and analysis of relevant epidemiological data considered essential to the committee's charge. The committee critically reviewed the scientific literature on the biological and health effects of internally deposited alpha-emitting radionuclides, relying wherever possible on original scientific publications and on current data that were generously provided by investigators in the United States, Canada, Western Europe, and Japan.

ORGANIZATION OF THE STUDY

To carry out the charge, the NRC appointed a committee of scientists experienced in radiation epidemiology, radiobiology, genetics, biostatistics, metabolism and pharmacokinetics, pathology, radiation dosimetry, inhalation physics, chemistry, biology, radiology and nuclear medicine, and mathematical modeling and risk assessment. The study was conducted under the general guidance of the Board on Radiation Effects Research of the Commission on Life Sciences.

To facilitate its work and to augment its expertise so as to encompass a wider spectrum of scientific subjects, the committee solicited specific contributions from a number of scientific experts other than its own members. These experts participated in the committee's deliberations throughout the course of its work.

The committee held eight meetings over a period of 24 months—six in Washington, D.C., one in Berkeley, California, and one in Woods Hole, Massachusetts. The second meeting, on May 15, 1985, included a public meeting, at which open discussion and contributions from interested scientists and the public at large were invited. Several additional meetings of subgroups of the committee were held, to plan and outline specific work assignments.

The committee organized its work according to the main objectives of the charge and divided the study into the following main categories:

- Genetic, teratogenic, and fetal effects of internally deposited alpha-emitting radionuclides.
- Carcinogenic and other health effects of radon, radium, thorium, polonium, uranium, and the transuranic radionuclides.
- The scientific basis and mechanisms underlying the biological and health effects, including the relevant physics and dosimetry, radiobiology, anatomy and physiology, and method of risk analysis.

The structure, composition, and expertise of the committee, including its invited participants, permitted considerable overlapping of assignments among the different categories, ensuring the interaction of scientific disciplines.

The committee also conducted two informal workshops that focused on radon. These workshops were designed to review with a number of investigators the current scientific knowledge with respect to uranium-miner epidemiology, lung modeling and dosimetry, and risk estimation.

JACOB I. FABRIKANT
Chairman
Committee on the Biological Effects of Ionizing Radiations

Acknowledgments

During this study the committee was aided by many experts from the scientific community. Scientific data, advice, and help in the preparation of the text were freely offered, and the committee wishes to acknowledge this very important assistance. Special thanks are due to Richard Hornung, Geoffrey Howe, Jan Muller, and their respective government authorities in the United States and Canada, and Edward Radford for providing their data tapes on the followup of miners exposed to radon progeny for reanalysis by this committee. The resulting combined data base of radon effects in uranium miners is the largest that has been analyzed.

The preparation of the report required expertise in many disciplines. The committee gratefully acknowledges the help it received from Victor P. Bond, Anthony L. Brooks, Fred Cross, Carter Denniston, William H. DuMouchel, Marvin Goldman, Douglas Grahn, Webster Jee, Robert E. Rowland, Charles L. Sanders, Melvin R. Sikov, Newell Stannard, and McDonald E. Wrenn. These invited participants provided invaluable aid in the preparation and review of specific chapters; they are not responsible for errors that might have crept in to the final report. In addition, several other scientists, including Bernard Cohen, Richard Cuddihy, Leonard Hamilton, and Charles Land, provided useful information on lung dosimetry, epidemiological studies, and risk modeling. The analyses of the radon-exposed cohorts presented in this report were made possible by the use of the AMFIT program developed by Dale Preston and his colleagues at the Radiation Effects Research Foundation, Hiroshima.

We wish to acknowledge also the help and support of Stephen L. Brown and William L. Lappenbusch, formerly of the National Research Council, who were instrumental in getting this study started. Special appreciation is extended to Colette A. Carmi and Doris E. Taylor for handling the myriad administrative details associated with this committee's work and for preparing the many drafts of the report. Their patience and good cheer helped the committee over the innumerable difficulties that inevitably arise during the course of its work.

Contents

PRINCIPAL TABLES

HEALTH RISKS OF RADON AND OTHER INTERNALLY DEPOSITED ALPHA-EMITTERS

BEIR IV

1
Overview

INTRODUCTION

This report addresses demonstrated and potential health effects of exposure of human populations to internally deposited alpha-emitting radionuclides and their decay products. It emphasizes carcinogenic effects and, where possible, presents quantitative risk estimates for cancer induction. The largest part of the report deals with health effects of exposure to radon and its progeny, primarily because of a need to characterize the lung-cancer risk associated with exposure to radon and its short-lived daughters in indoor domestic environments. The report also addresses health effects of exposure to other groups of radionuclides and their progeny that emit alpha particles—the isotopes of polonium, radium, thorium, uranium, and the transuranic elements.

Several alpha-emitting radionuclides occur naturally in our environment; others are produced for industrial, military, and medical applications. Recent attention has focused on the alpha-emitting radioisotopes because of their presence in drinking water, in indoor air in buildings, and in mines and because of their potential release into the environment from the nuclear fuel cycle (including radioactive waste disposal) and from accidents during space exploration. The radionuclides of concern are mainly radon-222 and radium-226 and their alpha-emitting daughter products and the transuranic elements plutonium-238 and -239.

Alpha-emitting radionuclides can be absorbed into the tissues of the body and irradiate adjacent cells after inhalation or ingestion, after entry through a wound in the skin, or after injection for diagnostic or therapeutic purposes. Radiation effects depend not only on the physical properties of emitted radiation, but also on the physiology and biochemistry of the exposed person and the physical and chemical characteristics of the radionuclides, which control their deposition, transport, metabolism, excretion, and reuse in the body. The health effects of radiation in humans include cancer induction, genetic disease, teratogenesis (induction of developmental abnormalities), and degenerative changes. The most important target tissues for cancer induction and degenerative changes are the respiratory tract, bone, liver, and the reticuloendothelium system.

Both natural and man-made alpha-emitting radionuclides in our environment can pose a risk to human health, but the natural sources currently make the largest contribution to human exposure. Among the natural sources, inhaled radon and radon decay products indoors are the largest contributors to population exposure and might be responsible for a large number of lung-cancer deaths each year.[3] That has led to recommendations, now being implemented, for national studies to assess the magnitude of the problem, for adopting remedial action levels of radon progeny in the indoor environment, and for introducing mitigation procedures to take effect at or below such levels to reduce population exposures from this source.[8,9]

For estimation of risks associated with exposure to the alpha-emitting radionuclides, the most important human populations examined are the underground miners who are exposed to widely differing concentrations of radon-222 progeny,[3] the American radium-dial workers who ingested various amounts of long-lived radium-226 and radium-228,[6] the German patients who received injections of short-lived radium-224[11] with different activities, and the German patients who received injections of graded volumes of Thorotrast (colloidal thorium-232 dioxide).[10] Human data on cancer induction by alpha-particle irradiation are sparse, but preliminary risk estimates have been calculated for some sites and tissues—lung, bone, head sinus and mastoid, and liver.

All of these epidemiological surveys are presently in progress, none is completed, and the person-years of follow-up are still relatively small, so that the lifetime carcinogenic risks of alpha-radiation exposure remain uncertain. Sufficient human data are not available for assessing the late health effects of the transuranic elements, e.g., plutonium-239; and here it has been necessary to estimate risks from these internally deposited alpha-emitters in humans by simplified mathematical and dosimetric models[1] or from comparison of effects with other radionuclides, where both direct experimental observation in laboratory animals and knowledge of radiation effects in humans are available. Complications arise in evaluating such comparisons because of such factors as different time patterns of deposition and resorption of the various radionuclides, e.g., radium vis à vis plutonium in bone.[2]

This report attempts to respond to a broad range of scientific questions related to current public health issues. Not all the questions can be addressed directly. There is considerable variation in the amount of data on each radionuclide from epidemiological studies and animal investigations. Epidemiological data are available on some alpha-emitting radionuclides, such as radon and its daughters, radium, and thorium. Little human information is available, however, on the transuranic elements, so dependence must be placed on animal experiments. As in all experimental animal studies, the extent to which the results can be extrapolated to humans and the confidence that can be placed on such extrapolation are uncertain. Even when human data were available, the committee has tried to rely on its own studies using newly developed methods for the analysis of occupational cohort data rather than relying solely on published information. The committee has also used novel statistical methods to analyze interspecies comparisons of the risks associated with different radionuclides when human data were insufficient. The committee recognizes that these analyses are preliminary and that large uncertainties are inherent in such extrapolations. Nevertheless, the committee believes that the methods introduced here will help to point the way to more detailed comparisons as additional data from epidemiological and animal studies become available.

This report consists of eight chapters and eight appendixes. The remainder of this chapter presents a summary of the committee's findings and its recommendations for future research. The next six chapters review the epidemiological and experimental evidence of the biological and health effects of the internally deposited alpha-emitting radionuclides and their daughter products. Chapter 8 summarizes the scientific evidence on genetic and fetal effects. The eight appendixes provide much of the scientific basis for the committee's conclusions, dealing primarily with radon and its progeny and with molecular and cellular radiobiology. Throughout the committee's deliberations, the sources of uncertainty that should be recognized in connection with radiation risk estimation are discussed; they are particularly important with regard to the effects of radon and its progeny.

The committee found it necessary, because of constraints on time and resources, to narrow its charge to an examination only of alpha-emitting radionuclides known to induce health effects in exposed human populations and to concentrate its efforts on specific subjects in each case. The committee's focus and efforts were strongly influenced by the need to address the health effects of inhaled radon progeny, because of the concern over lung-cancer risk associated with increased indoor concentrations of radon. When results of epidemiological surveys were available (e.g., on radon, radium, and thorium), analysis of human data was preferred to analysis of laboratory animal data (e.g., on polonium, uranium, and the transuranic elements) for quantitative human risk estimation.

As in earlier reports from the Committee on the Biological Effects of Ionizing Radiations, the so-called BEIR reports, the committee cautions that the risk estimates derived from epidemiological and experimental animal data should not be considered precise. They are derived from analyses of incomplete data and involve numerous uncertainties. The risk estimates presented here will change as new information and analytical methods become available.

Finally, the committee notes that it assumes no responsibility to address the subject of regulatory guidance on exposure levels or societal cost-benefit issues that involve the radionuclides of concern. Clearly, such issues are beyond the scope of the committee's task and beyond its expertise.

SUMMARY OF FINDINGS

Most primordial radionuclides are isotopes of heavy elements and belong to the three radioactive series headed by uranium-238, thorium-232, and uranium-235. These contribute significantly to the general population collective dose equivalent. The relevant radionuclides in the body include the isotopes of uranium, radium, radon, polonium, bismuth, and lead; these enter the body by inhalation or by ingestion of food and water and only rarely through wounds in the skin. They follow normal chemical metabolism, and the concentrations of the long-lived radionuclides are usually maintained at equilibrium or increase slowly with age. The shorter-lived radionuclides disappear by decay, but might be continually replenished by renewed intake.

The annual dose equivalent to the bronchial epithelium from inhaled alpha-emitting radionuclides and their progeny approaches 2,500 mrem/yr (25 mSv/yr),[3] due almost entirely to the radon progeny polonium-218–polonium-214 pair. The important tissue is the bronchial epithelium, which is the site of most lung cancers thought to be induced by radiation. The major contributors are the short-lived decay products of radon, measurements of which show an apparent log-normal distribution of concentrations in indoor air. For smokers, the additional exposure to the lungs from naturally occurring radionuclides in tobacco increases the dose to the bronchial epithelium.[4] For other soft tissues, bone surfaces, and bone marrow, the largest contributors to the dose equivalent from the alpha-emitters are the lead-210–polonium-210 pair in bone. Exposure of the general and worker populations from man-made or enhanced sources comes primarily from consumer products (e.g., tobacco), the nuclear fuel cycle, and emissions from government and industrial facilities, including those from mineral extraction. In the past, enhanced materials produced for medical applications, such as colloidal thorium dioxide, were injected or instilled directly into body tissues and resulted in high doses to some organs.

RADON

The evaluation of the lung-cancer risk associated with radon and its progeny has been the most challenging task of the committee. Numerous studies of underground miners exposed to radon daughters in the air of mines have shown an increased risk of lung cancer in comparison with nonexposed populations. Laboratory animals exposed to radon daughters also develop lung cancer. The abundant epidemiological and experimental data have established the carcinogenicity of radon progeny. Those observations are of considerable importance, because uranium, from which radon and its progeny arise, is ubiquitous in the earth's crust, and radon in indoor environments can reach relatively high concentrations. Although the carcinogenicity of radon daughters is established and the hazards of exposure during mining are well recognized, the hazards of exposure in other environments have not yet been adequately quantified. Risk estimates of the health effects of long-term exposures at relatively low levels are required, to address the potential health effects of radon and radon daughters in homes and to refine estimates of the risk in occupational environments.

Two approaches are being used to characterize the lung-cancer risks associated with radon-daughter exposure: mathematical representations of the respiratory tract that model radiation doses to target cells and epidemiological investigation of exposed populations, mainly underground miners. The dosimetric approach used by other investigators and committees provides an estimate of lung-cancer risk related to radon-daughter exposure that is based specifically on modeling of the dose to target cells. The various dosimetric models all require assumptions, some of which are not subject to direct verification, as to breathing rates; the deposition of radon daughters in the respiratory tract; and the type, nature, and location of the target cells for cancer induction. Accordingly, the committee chose not to use dosimetric models for calculating lung-cancer risk estimates in this report. However, the results of dose models were used to extrapolate lung-cancer risks derived from the epidemiological studies of underground miners to the general population in indoor environments. The lung-cancer risk estimates for radon-daughter exposure derived by the committee in this report are based solely on epidemiological evidence.

The committee preferred a direct epidemiological approach, because the studies of radon-daughter-exposed miners provided a direct assessment of human health effects. Although each of the epidemiological studies that the committee assessed has limitations, the approach of a combined analysis of major data sets permitted a comprehensive assessment of the health risks associated with radon-daughter exposure and of other factors that influence the risk, such as age and time since exposure. In analyzing the data, the committee used a descriptive analytical approach, rather than statistical methods based on conceptual models of carcinogenesis. The committee obtained data from four of the principal studies of radon-exposed miners (Ontario uranium miners, Saskatchewan uranium miners, Swedish metal miners, and Colorado Plateau uranium miners) and developed risk models for lung cancer based on analyses of these data. By means of statistical regression techniques appropriate for survival-time data, the committee found that the probability of dying of lung cancer at age a in the combined cohorts was best described by the following expression:

$$r(a) = r_0(a)[1 + 0.025\gamma(a)(W_1 + 0.5W_2)], \qquad (1\text{-}1)$$

where $r(a)$ is the lung-cancer mortality rate at age a; $r_0(a)$ is the baseline lung-cancer mortality rate in the 1980–1984 U.S. population; $\gamma(a)$ is 1.2 for ages less than 55 yr, 1.0 for ages 55–64 yr, and 0.4 for age 65 yr or greater; W_1 is the cumulative radiation exposure, in WLM,* from 5 to 15 yr before age a; and W_2 is the cumulative exposure, in WLM, 15 yr or more before age a.

In this model, the excess relative risk varies with time since exposure, rather than remaining constant, and depends on age at risk; the expression, therefore, is a departure from most previous risk models, which have assumed that the relative risk is constant over both age and time. In the committee's modified relative-risk model, radon exposures more distant in time have a smaller impact on the age-specific excess relative risk than more recent exposures. Moreover, the age-specific excess relative risk is higher for younger persons and declines at higher ages. The committee's analysis did not assume a priori that analysis based on the relative risk was necessarily more appropriate than alternatives, such as analysis based on absolute risk. However, an absolute-risk model would have involved a complex power function of age. Since it requires fewer variables, the relative-risk form adopted by the committee provides a simpler description of observed lung-cancer risks in the miner cohorts.

Recognition that radon and its daughter products can accumulate to high concentrations in homes has led to concern about the potential lung-cancer risk associated with indoor domestic exposure. Although such risks can be estimated with the mathematical expression in Equation 1-1 for excess relative risks, it must be recognized that the committee's model is based on occupational exposure data. Several assumptions are required to transfer risk estimates from an occupational setting to the indoor domestic environment. Accordingly, the committee assumed that the epidemiological findings in the underground miners could be extended across the entire life span, that cigarette smoking and exposure to radon daughters interact multiplicatively, that exposure to radon progeny increases the risk of lung cancer in proportion to the sex-specific ambient risk of lung cancer associated with other causes, and that, to a reasonable

*Working level month (WLM) is a unit of exposure to radon progeny. It is defined in Chapter 2 and in the Glossary. The current occupational limit is 4 WLM/yr.

TABLE 1-1 Comparisons of Estimates of Lifetime Risk of Lung-Cancer Mortality due to a Lifetime Exposure to Radon Progeny

Study	Excess Lifetime Lung-Cancer Mortality (deaths/10^6 person WLM)
BEIR IV (1987, this report)	350[a]
NCRP[3] (1984)	130
BEIR III[5] (1980)	730
UNSCEAR[7] (1977)	200–450

[a]See Chapter 2 of this report.

approximation, a WLM yields an equivalent dose to the bronchial epithelium in both occupational and environmental settings. This last assumption is tentative, as it is based on very limited information. The committee concluded that more complete specifications of aerosol characteristics in mines and homes and the relevant physiological parameters are needed to permit quantitative assessment of the comparative dosimetry of radon daughters in the occupational and environmental settings.

On the basis of the estimates of excess relative risks per WLM of exposure to radon progeny derived from analysis of the four miner cohorts examined and the assumptions outlined above, the committee projected lung-cancer risks for U.S. males and females. The committee's risk projections estimate the ratio of lifetime risks relative to baseline risks, the probability of lung-cancer mortality, and average years of life lost for various exposure rates and durations of exposure. The report includes tables for estimating risks conditional on survival to a particular age and for smokers and nonsmokers of either sex.

The risk projections cover exposure situations of current public-health concern. Lifetime exposure to 1 WLM/yr is estimated to increase the number of deaths due to lung cancer by a factor of about 1.5 over the current rate for both males and females in a population having the current prevalence of cigarette smoking. Occupational exposure to 4 WLM/year from ages 20 to 40 is projected to increase male lung-cancer deaths by a factor of 1.6 over the current rate in this age cohort in the general population. In all these cases, most of the increased risk is in smokers in whom the risk is 10 or more times greater than that in nonsmokers.

Comparisons of estimates of the lifetime risk of lung-cancer mortality due to a lifetime exposure to radon progeny in terms of WLM made by this and other scientific committees yield the data presented in Table 1-1.

The BEIR IV (this report) committee's modified relative-risk model differs from the others, in that it incorporates dependence of the relative risk of lung-cancer mortality on both time since exposure and age at risk. Unlike the modified relative-risk model developed by this committee, risk estimates by the 1980 BEIR III committee[5] were based on the assumption of an additive risk of lung-cancer mortality due to exposure to radon progeny that increased with age.

Users must be aware of the uncertainties that affect the estimates of the lung-cancer risk due to exposure to radon progeny given in this report. The uncertainties include sampling variation in the primary data, random and possibly systematic errors in the original data on exposure and lung-cancer occurrence, inappropriate statistical models for analysis or misspecification of the components of the models, and incorrect description of the interaction between radon-daughter exposure and cigarette smoking. In addition, the actual computed lifetime risk and expected life-shortening depend on the age-specific disease rates in the referent population—in the committee's examples, the 1980–1984 U.S. population mortality rates. Projections based on a different referent population would be expected to differ, although the ratios of lifetime risks and years of life lost to baseline values are believed to be more stable across populations.

In its review and analysis, the committee found gaps in information related to some aspects of radiation carcinogenesis by radon daughters. The cells of the respiratory tract that give rise to radon-daughter-associated lung cancer are still not known. A unique link between radon-daughter exposure and small-cell carcinoma of the lung was not found; in the studies of underground miners, this histological type occurred in greatest excess, but other cancer-cell types were also increased.

Review of the literature and the committee's own analyses of the relevant data did not lead to a conclusive description of the interaction between radon daughters and cigarette smoking for the induction of lung cancer. Several data sets were analyzed, and although the committee chose a multiplicative interaction for its risk projections on a relative-risk scale, it recognizes that a submultiplicative model is also consistent with the data analyzed. Neither an additive nor a subadditive model appears consistent with these data.

Health effects of exposure to radon daughters other than respiratory cancer are also of concern, but the data are sparse and associations are weak. Reductions in lung function in some uranium miners cannot be attributed directly to radon-daughter exposure.

The data on increased occurrence of chromosomal aberrations in lymphocytes and on adverse reproductive outcomes in uranium miners are inconclusive.

Research in the United States and other countries has provided data on concentrations of radon and radon progeny in homes. The studies have also described the sources of radon and determinants of its concentration. A few exploratory epidemiological investigations of the lung-cancer risk associated with radon-daughter exposure in homes have been carried out, but the study populations have been small and the results inconclusive. The committee judged these exploratory studies to be inadequate for the purposes of risk estimation. Its risk projections for the general population are therefore based on the studies of miners. The committee concluded that estimates of lung-cancer risks based on studies on miners can be used to estimate the potential lung-cancer risk associated with increased concentrations of indoor radon; however, the estimates derived are imprecise. The committee recognizes that the differences between risks in mining and domestic environments and the interaction between smoking and exposure to radon progeny remain incompletely resolved.

POLONIUM

Polonium isotopes occur in nature; they appear in tissues as a result of ingestion in foods, inhalation of tobacco smoke, and decay of lead-210 deposited in bone. Polonium-214 and polonium-218 are short-lived daughters of radon-222 and contribute a large fraction of the radiation dose from inhaled radon. Extensive work with animals, primarily with polonium-210, has indicated that it does not localize appreciably in bone, in contrast with many other alpha-emitters; it concentrates instead in the reticuloendothelial system, in kidney, and in blood cells. Its effects at higher doses resemble those of generalized whole-body radiation and involve all major organ systems. At lower doses, soft-tissue tumors, nephrosclerosis, hypertension, cataracts, generalized atrophy of the lymphoid system, and nonspecific life-span shortening occur.

In laboratory animal experiments, the relative toxicity of polonium-210 is a function of duration of exposure and dose. At high doses, it is much more toxic than uranium, plutonium, radium, or the transplutonic elements. Because of its shorter half-life and its toxicity at longer times and lower doses, it is comparable with plutonium-239, i.e., about 5 times as effective as radium-226; at very low doses and very long times, its effectiveness approaches that of radium-226.

Experimental studies in humans and accidental exposures have indicated that metabolism in the human body is similar to that in laboratory animals. Only a few cases of effects in humans due directly to exposure to polonium-210 have been documented, so carcinogenic risk associated with exposure to polonium cannot be estimated directly. Risks can be estimated indirectly from the experience with other internally deposited alpha-particle emitters.

RADIUM

The main sources of information on the health effects of radium deposited in human tissues are the U.S. cases of occupational exposure (mostly in dial painters and radium chemists) and medical exposure to radium-226 and radium-228 and the German cases of repeated injection of radium-224 into patients for treatment of ankylosing spondylitis in adult life or tuberculosis in childhood. Malignant effects are almost exclusively the induction of skeletal tumors and of carcinomas in the paranasal sinuses and mastoid air cells. The evidence of induction of leukemia is weak, except at doses far greater than those in occupational, environmental, or therapeutic exposures currently encountered.

The dose-response data on bone sarcomas are characterized by low-dose regions of zero observed risk. Depending on which isotope of radium is being considered, a variety of dose-response relationships are consistent with the human data—linear, dose squared, linear with correction for dose protraction, dose-squared exponential, linear-quadratic exponential, 1 minus an exponential, and threshold.[6] In the dose range in which bone tumors have occurred, the lifetime risk associated with radium-224 is estimated to be about 2×10^{-2} excess bone sarcomas per person Gy (200 per million person-rad) when a linear function is assumed and an apparent increase in risk with dose protraction is taken into account. However, analyses that take into account competing risks lead to the rejection of a linear dose response on statistical grounds, and the best fit to the data on children and adults is found to be linear-quadratic exponential. The lifetime probability of excess bone cancer induction per person Gy to bone is then estimated to be approximately $(0.0085D + 0.0017D^2)$ $\exp - 0.025D$ after an average skeletal dose of less than 1 Gy and a 25-yr expression period. Tumors are distributed over time, with their frequency diminishing with a half-life of about 4 yr after a minimum latent period of 5 yr. In the low-dose range in which no tumors have been observed, the uncertainty in the risk estimates for radium-224 increases monotonically with decreasing dose.

For radium-226 and radium-228 bone sarcoma induction, a number of dose-response functions provide statistically acceptable fits to the data within the range of doses where tumors have been observed. All of these functions predict approximately the same risk for a given exposure, but not at lower doses where no tumors have been observed. Below the dose range in which tumors have been observed, the uncertainty in the estimate of risk based on extrapolation of the dose-response function increases monotonically with decreasing exposure. Bone sarcomas induced by radium-226 and radium-228 have appeared 7 yr after first exposure and continued to appear throughout life. The time to tumor appearance apparently increases with decreasing dose and dose rate. Below an average skeletal dose of about 0.8 Gy, the chance of developing bone cancer from radium-226 and radium-228 during a normal lifetime is extremely small—possibly zero.

Carcinomas in the paranasal sinuses and mastoid air cells are observed after exposure to radium-226 or to radium-226 in combination with radium-228, but have not yet been observed among persons exposed to radium-224. The working hypothesis in most analyses of the data is that radionuclides other than radium-226 are ineffective for the induction of these carcinomas (although such carcinomas have occurred at a statistically significant frequency in dogs exposed to other radium isotopes and to the actinides). The tumors occurred as early as 10 yr after exposure and continued to occur throughout life. A linear dose-response relationship describes the data either as a function of average skeletal dose or of radium-226 intake. In terms of systemic intake, the risk coefficient for these carcinomas is estimated as 16 excess cancers per million person-yr at risk per μCi of intake. Causation is thought to be associated partly with the generation of radon-222 by radium-226 decay and later irradiation of the sinus and mastoid epithelial tissues by radon-222 and its progeny.

The cells at risk of bone cancer induction appear to be proliferating osteogenic cells or their precursors at bone surfaces. Identification of cell type and location is complicated by the diversity of cells that lie within the range of alpha particles emitted from bone surfaces. In the mastoids, the cells at risk for carcinoma appear to be the epithelial cells in the squamous or cuboidal epithelium of the lining mucosa. In the paranasal sinuses, where the epithelial structure is more complex, the location and identity of the cell at risk are less certain.

THORIUM

Thorium-232 is a primordial, long-lived, alpha-emitting radionuclide; its decay series can be considered as consisting of two steps: the formation of radium-224 by successive decays from thorium-232 and then the decay of radium-224 and its daughters to stable lead. The alpha-emitters from radium-224 are biologically the most important in the dosimetry concerned with the radioactive properties of the thorium series. Colloidal [^{232}Th]thorium dioxide (Thorotrast) was used widely as a contrast medium in diagnostic radiology from 1928 to 1955. Intravascularly injected Thorotrast aggregates tend to be incorporated into the tissues of the reticuloendothelial system, mainly the liver, the bone marrow, and the lymph nodes. The radioactive daughter products can escape into the bloodstream and thus reach the bone and bone marrow; the important bone-seeking daughter products are radium-224, radium-228, and thorium-228. Aggregates in the liver, bone, and bone marrow are often taken up by macrophages that are mobile, thereby distributing the radiation in relation to the reticuloendothelial, hematopoietic, and endosteal cells. The radiation dosimetry is therefore complex and can be further complicated by the colloidal and elemental chemical and physical characteristics.

Epidemiological surveys of Thorotrast patients are in progress in Germany, Denmark, and Portugal; additional studies are being carried out in Japan and the United States. Approximately 4,000 patients are being followed. A typical injection of 25 ml of Thorotrast would result in an average liver dose rate of 25 rads/yr (0.25 Gy/yr) and an average endosteal bone dose rate of about 16 rads/yr (0.16 Gy/yr). The late effects of Thorotrast incorporated in the body are primarily the induction of liver cancers, bone sarcomas, and myeloproliferative disorders, including leukemias. Liver cancers appear in excess in all epidemiological studies. Hemangioendotheliomas in the liver occur uniquely after Thorotrast is intravascularly administered; it has been described as a Thorotrast-specific liver cancer.

Risk estimates for thorium-232-induced liver cancer, bone cancer, and leukemia have been calculated on the basis of Thorotrast patients who received injections of colloidal [^{232}Th]thorium dioxide and its progeny. For liver cancer, a lifetime risk is estimated to be about 3×10^{-2} per person-Gy (300 excess liver cancers per million person-rad), where the alpha radiation dose is to the liver. For bone sarcomas, the lifetime risk is estimated to be about $(0.55–1.2) \times 10^{-2}$ excess bone sarcomas per person-Gy (55–120 per million person-rad), where the dose is to the skeleton without bone marrow. For leukemia, a lifetime risk of about $(0.5–0.6) \times 10^{-2}$ per person-Gy (50–60 excess leukemia cases per million person-rad) is estimated. Those estimates are uncertain because of the nonuniform deposition of thorium in the tissues (which results in high local tissue doses), the chemical nature of thorium, the wasted radiation dose in necrotic and fibrotic tissues (particularly in the liver), and the incomplete follow-up in the epidemiological studies.

URANIUM

Natural uranium is of low specific activity and consists mainly of uranium-238 (over 99% by weight) with smaller amounts of uranium-235 and -234. The latter radionuclides have shorter half-lives than uranium-238 and account for about 50% of the radioactivity in natural uranium. Uranium is ubiquitous in rocks and soil and is a trace element in foods, particularly crops or cereals, and in drinking water. Wide geographical differences have been noted. Gastrointestinal absorption from food or water is the principal source of internally deposited uranium in the general population. It is stored mainly in bone, where it has a uniform distribution. Inhalation of aerosols containing uranium is a hazard of industrial exposure, and this uranium might consist of depleted or enriched uranium. The distribution and retention of uranium in the body after inhalation of an aerosol depends critically on the aerodynamic size of the particles and on their solubility in biological fluids. Inhalation of insoluble compounds is associated with uranium retention in lung tissue and hilar lymph glands.

Uranium compounds may induce detrimental health effects due to both chemical toxicity and alpha-radiation damage. Animal experiments have demonstrated a specific toxic effect of uranium on the kidney, but with little evidence of toxic effects on other organs. There are considerable interspecies differences in sensitivity, possibly owing to differences in the acidity of urine. The dog is thought to be the animal model with greatest similarity to humans. Uranium of high specific activity (uranium-232 and -233) can cause bone sarcomas in mice, and massive doses of uranium oxide have produced lung fibrosis and lung cancer in the primates, dogs, and rodents. This is interpreted as resulting from alpha-particle irradiation of the lung.

Epidemiological surveys of uranium millers and miners occupationally exposed to dusts containing natural uranium at relatively high concentrations have not yielded convincing evidence of serious renal damage nor of increased rates of malignant tumors. Those studies had limited power to detect increased rates of disease, and confounding factors obscured the interpretations. Emphasis has therefore been on animal data concerning renal damage after exposure to uranium and on data on animals and humans exposed to other alpha-emitting elements, such as radium-226.

Observations on animals exposed to high-specific-activity uranium suggest that a small excess of bone sarcomas in human populations could result from naturally occurring uranium, but that the magnitude of the excess depends on which mathematical model is chosen. If the dose-response relationship is quadratic, virtually no effect is expected at environmental natural uranium concentrations. If a linear dose-response relationship is chosen, it has been estimated that ingestion in water or food at an environmental rate of 1 pCi/day could be associated with a lifetime risk of 1.5 bone sarcomas per million persons. That may be contrasted with about 750 naturally occurring bone sarcomas per million persons in the United States. It is concluded, on the basis of present evidence, that the general population risk associated with natural uranium is very low and might be negligible. Higher risks could be associated with higher uranium concentrations in local water supplies.

TRANSURANIC ELEMENTS

Transuranic elements are members of the actinide series beyond uranium; all are artificially produced in nuclear reactors, accelerators, and explosions of nuclear weapons and several include alpha-emitting radioisotopes with very long half-lives. Neptunium, plutonium, americium, and curium are the most abundant and the most extensively used. The transuranic elements are not readily absorbed through the skin or from the gastrointestinal tract. Because of the short range of alpha radiation in tissues, these elements are not of potential health concern unless they enter the body and deposit in tissues through wounds or the respiratory tract. Inhalation of airborne particles into the respiratory tract and subsequent deposition probably represents the most common pathway by which transuranic elements might enter the body to cause alpha irradiation of human tissues and eventual health effects. Following deposition in the lungs, inhaled aerosol particles are quickly phagocytized by alveolar macrophages, and may be transported from the lungs, depending on solubility; the target tissues include primarily the lungs, liver, bone, bone marrow, and lymph nodes.

Insoluble transuranic compounds, primarily plutonium oxide, are retained in the lungs and thoracic lymph nodes. Other plutonium compounds are more mobile when taken into the body through the respiratory tract or through wounds and deposited primarily in the liver, and bone. Distribution within tissues tends to be diffuse initially, but the compounds often accumulate or form aggregates within cells. Only under conditions of very high deposition would there be more than a few percent of the total cells exposed to alpha radiation. Nevertheless, an association exists between cancers of the lung, bone, and liver and deposition of transuranic elements in these tissues in several animal species under experimental conditions. Inhalation of large amounts of transuranic compounds, e.g., plutonium oxide particles, in experimental rodents and dogs results in radiation pneumonitis, pulmonary fibrosis, and lung cancer. Inhaled plutonium compounds can also cause an increase in the incidence of bone tumors but this has not been observed in experimental animals that inhaled highly insoluble $^{239}PuO_2$ particles. Alpha particles from plutonium are considerably more mutagenic and carcinogenic than are x rays; the experimental animal data in rats and dogs are extensive. In the absence of sufficient human surveys to calculate risk estimates for cancer induction, the animal data, together with data on radium-224 and radium-226 in humans, provide a basis for cancer risk estimation.

Human exposures occur primarily among occupationally exposed workers in nuclear facilities. The United States Transuranium Registry and other studies involving several thousand workers who have been accidentally exposed, predominately to low levels of transuranic elements, have shown that plutonium tends to concentrate in the tracheobronchial lymph nodes, with smaller amounts accumulating in the lungs, liver, and bone. The most extensive epidemiologic study of plutonium workers found that mortality experience for the entire cohort was less than that expected based on U.S. mortality rates.

The only significant excess risk was for benign and unspecified neoplasms. The analysis showed no elevated risks for cancer in tissues with the highest concentrations of plutonium, namely, lung, liver, and bone. The human data and the alpha-radiation dosimetry alone are, at present, inadequate to provide direct calculation of cancer-risk coefficients in the radiosensitive organs and tissues.

Although cancer-risk estimates have been derived from the animal studies, extrapolation of these numerical values to humans introduces uncertainties and technical difficulties. The experimental animal data are quite extensive, and the committee has applied Bayesian components of variance models to 15 data sets for bone sarcoma induction in humans and laboratory animals. The analysis yields, for plutonium deposition in human bone, a lifetime risk estimate of 3×10^{-2} per person-Gy (300 excess bone-cancer deaths per million person-rad) to bone. This is consistent with risk estimates based on data from laboratory animals.

Genetic and Fetal Effects

The genetic disorders that can arise in the progeny of persons exposed to alpha radiation are of the same classes as those arising after exposure to low linear energy transfer (LET) radiation: single-gene autosomal dominant and X-linked disorders, irregularly inherited disorders, recessive disorders, and chromosomal aberrations. Estimates of genetic risk have been made by the BEIR III committee[5] based on the current incidence of hereditary disorders and their estimates of the dose of low-LET radiation required to double the mutational frequency. That information was combined with relative biological effectiveness (RBE) values for alpha irradiation derived from plutonium-239 experiments in mice—specifically, RBEs of 2.5 for mutations and 15 for chromosomal aberrations—to estimate the risk due to internally deposited alpha-particle emitters. Numerical estimates of the incidence of genetic effects over a 150-yr span (five generations) were made for continuous average population gonadal doses of 0.01 Gy (1 rad) per 30 yr reproductive generation, 0.33-mGy alpha dose/yr. For a stable population of about 1 million persons, nearly 200 dominant, X-linked, and translocation genetic effects would accumulate over 150 yr.

Although alpha-emitting radionuclides can be transmitted across the placenta and incorporated in the body of the developing fetus, only the alpha decays that occur during intrauterine life can cause teratogenesis. The teratogenic effects are closely related to the stage of embryonic development at which the radiation dose is received; preimplantation is the stage of specific teratogenic effects that can occur only during specific, relatively brief periods during intrauterine development. Data on radiation effects on the developing embryo and fetus in humans are sparse, and risk estimates must be based mainly on experimental animal data.

Most of the alpha-emitting radionuclides demonstrate low fetal accretion in laboratory animals, although they vary widely in fetoplacental distribution. Developmental studies on internal alpha-emitters have included radon and its daughters, radium, polonium, uranium, and the transuranic elements. Almost all the teratogenic effects are considered to be due to cell killing. RBE values for cell killing by alpha particles exceed 10, but could be higher for very low dose rates. However, because alpha irradiation is delivered chronically, most of the total dose accumulated during gestation is not effective—only that received during the sensitive interval is effective.

RECOMMENDATIONS FOR FURTHER RESEARCH

RADON

- The committee's model for estimating the lung-cancer risks due to radon exposures is based on the application of multivariate statistical procedures to the data from four major epidemiological surveys of underground miners. Several current underground-miner surveys could provide a more extensive data base with increased person-years of follow-up and help to refine lung-cancer risk coefficients; provide more information on the interaction between smoking and radon exposure; and, with improved dosimetry, narrow the uncertainties in the application of lung-cancer risk data derived from miners to the estimation of risk in the general population. Collecting and reporting smoking data on these miners should be an essential part of the study design.
- The committee recommends continued epidemiological study, with parallel multivariate analysis, of the temporal expression of lung cancer in underground miners exposed to radon progeny.
- The present need to apply lung-cancer risk projections from surveys of underground miners to estimate risk to the general population associated with indoor radon introduces uncertainties and technical difficulties. The domestic environment has not been characterized adequately in terms of the variables affecting the dose and

risk related to radon progeny. Variations in indoor radon concentrations, alterations of aerosol characteristics, and impacts of smoking-related risk factors suggest that health consequences of indoor radon exposure require more epidemiological study and basic research. Further studies of dosimetric modeling in the indoor environment and in mines are necessary to determine the comparability of risk per WLM in domestic environments and underground mines.

- The committee recommends continuation of epidemiological studies of lung cancer and other health outcomes resulting from indoor radon exposure; such studies must have sufficient statistical power to quantify any significant differences between the risks in environmental and occupational settings.

Polonium

- The committee recommends that studies continue to evaluate the role of polonium from tobacco smoke in the production of lung cancer, including bronchial and lung dosimetry, identification and characterization of target cells, and the role of cofactors and mechanisms of the carcinogenic response.
- The induction of nonstochastic health effects, both acute and long term, particularly in the renal, cardiovascular, and reproductive systems, requires further study.
- The committee recommends that the effects of small exposures to polonium on the pathophysiological response in some organs and tissues deserve continued study in laboratory animals.

Radium

- The bone-cancer risk appears to have been completely expressed in the populations exposed to radium-224 in the 1940s and to have been nearly completely expressed in the populations exposed to radium-226 and radium-228 before 1930. Further analysis of these data should involve reevaluation of the dosimetry. More quantitative information is required for the evaluation of the magnitude of the dosimetric uncertainties and their impact on uncertainties of quantitative risk estimation.
- The committee recommends that the bone-cancer risk data from the two studies be integrated and analyzed with newer statistical methods.

- The committee recommends that the follow-up studies of the lower-dose radium-224 patients exposed since the 1940s now in progress in Germany and of similar groups of radium-226 and radium-228 patients continue.
- The discovery of bone cancer or sinus/mastoid cancer after exposure at the lower doses and the additional person-years of follow-up should substantially reduce the uncertainties of risk estimation related to the low doses.
- The committee recommends that research should continue on the identification of the cells at risk of bone-cancer induction; on cell behavior over time, including where the cells are in the radiation field at various stages of their life cycles; on modifying factors, such as the formation of fibrotic layers that might reduce the radiation that the cells receive; and on the time course and distribution of radioactivity in bone.
- The sinus and mastoid carcinomas in persons exposed to radium-226 and radium-228 are produced largely by the action of radon-222 and its daughters; continued study might offer insights into the effects of occupational and environmental radon. The dosimetry of the mastoid air cell system is much simpler than that of the bronchial tree; the mastoid mucosa might be the only respiratory tissues whose epithelial structure is simple enough to permit accurate dose estimation.
- The committee recommends that the dosimetry of the mastoids should be examined as completely as possible, so that the risk per unit of epithelial tissue dose and per unit of cell dose can be determined accurately; this might improve the understanding and estimation of the carcinogenic risk in the epithelium of the lower respiratory tract.

THORIUM

- The carcinogenic risk estimates related to thorium-232 depend primarily on studies of patients who received Thorotrast. These studies are incomplete, and except for those of the German patients, they have little statistical power to establish with precision the types of diseases produced, their influence on carcinogenic risks at low doses, the effects of dose and dose rate, and the chemical effects of the colloidal heavy metal, particularly in the liver.
- The committee recommends that the data be obtained from all five principal epidemiological studies of Thorotrast-exposed pa-

tients, to develop risk models for liver and other cancers from original multivariate analyses.

- The dosimetry of Thorotrast and thorium radionuclides in target organs is poorly understood. The radiation effects depend on the physical properties of the emitted radiation and on the physical and chemical characteristics of the radionuclide and its aggregation, movement, and deposition.
- The committee recommends further study of the dosimetry of thorium radioisotopes at the cellular level in the target organs or tissues; these processes are central to an understanding of the biological effects, notably in liver and bone.

Uranium

- The committee recommends that experimental studies of the nephrotoxic effects of uranium should be continued to determine the threshold concentration of uranium that is associated with substantial renal tubular damage and the animal and metabolic models most appropriate for predicting human effects.
- The committee recommends that cross-sectional and longitudinal epidemiological investigations of occupational exposure to natural uranium be vigorously pursued.
- Assessment of renal function and other health outcomes should be examined and correlated with environmental measurements designed specifically to estimate individual exposures. Studies of mortality and morbidity might be warranted if stable populations of sufficient size can be identified in areas with high concentrations of uranium in drinking water or food.
- The committee recommends that the mechanism of uranium deposition and redistribution in bone should be further investigated, so that the potential carcinogenic effect of natural uranium can be more reliably predicted from the results obtained with enriched uranium or with other alpha-emitters, such as radium-226 and radium-228 decay chains.

Transuranic Elements

- While no health effects have been associated with such human exposures, the results of experimental animal studies suggest that effects may eventually be observed in the highest-exposed worker populations. Such studies should emphasize the importance of thorough postmortem examinations of deceased persons whose deaths may be related to transuranic element exposures to confirm the cause of death and the possible presence of other lesions and to obtain tissue samples for radiochemical analysis and cellular-molecular biological studies.

- The committee recommends the continuation of current epidemiological studies of worker populations exposed to transuranic elements.
- Analysis of data from life span experimental animal studies using an epidemiological approach will ensure maximum use of this invaluable large data base for extrapolating the results of animal studies to humans. This should have a high priority because it is unlikely that these expensive life-span studies, particularly the dog experiments, will be repeated.
- The committee recommends that current life-span studies with dogs be completed and reported in a manner that will ensure that the maximum information is obtained.
- It is important to expand the effort to correlate the available human and experimental animal data on the deposition, translocation, metabolism, clearance, and excretion of transuranic elements. Considerably more work is required with respect to biokinetics and to the development of models that can be applied to the practice of radiation protection, including bioassay procedures and assessment of exposures. Additional research is needed to correlate the gross and microscopic distribution of transuranic elements within tissues and the site of tumor formation to ensure relevant dosimetry.
- The application of the powerful new tools of modern biology to multilevel studies (molecular, cellular, tissue, organ, animal, and human) will lead to improved understanding of the interactions of alpha radiation with biological targets from transuranic elements deposited in various tissues. Such studies have the potential for detecting potential harmful biological effects at low radiation doses, identifying persons of special risk to radiation injury, determining whether certain diseases are attributable to transuranic exposures, and directing therapeutic measures to sites of injury.
- The committee recommends that the Bayesian methods for interspecies extrapolation be developed further and applied to the determination of other risk factors in humans.

GENETIC EFFECTS

- The committee recommends that investigations continue on the retention and cellular distribution of the alpha-emitting radionuclides in appropriate chemical forms in the ovaries and testes of selected primates suitable as surrogates for humans.

REFERENCES

1. International Commission on Radiological Protection (ICRP). 1979. Limits of Intake of Radionuclides by Workers. ICRP Publication 30. Oxford: Pergamon.
2. Mays, C. W., G. N. Taylor, and R. D. Lloyd. 1986. Toxicity ratios: Their use and abuse in predicting the risk from induced cancer. Pp. 299–310 in Life Span Radiation Studies in Animals: What Can They Tell Us?, R. C. Thompson and J. A. Mahaffey, eds. CONF-830951. Springfield, Va.: National Technical Information Service.
3. National Council on Radiation Protection and Measurements (NCRP). 1984. Evaluation of Occupational and Environmental Exposures to Radon and Radon Daughters in the United States. NCRP Report 78. Bethesda, Md.: National Council on Radiation Protection and Measurements.
4. National Council on Radiation Protection and Measurements (NCRP). Exposures of the Population in the United States to Ionizing Radiation. NCRP draft report. Bethesda, Md.: National Council on Radiation Protection and Measurements.
5. National Research Council, Committee on the Biological Effects of Ionizing Radiation (BEIR). 1980. The Effects on Populations of Exposure to Low Levels of Ionizing Radiation. Washington D.C.: National Academy Press. 524 pp.
6. Rowland, R. E., A. F. Stehney, and H. F. Lucas. 1983. Dose-response relationships for radium-induced bone sarcomas. Health Phys. 44(Suppl. 1):15–31.
7. United Nations Scientific committee on the Effects of Atomic Radiation (UNSCEAR). 1977. Sources and Effects of Ionizing Radiation. Report E.77.IX.1. New York: United Nations. 725 pp.
8. U.S. Environmental Protection Agency. 1986. Radon Reduction Methods: A Homeowner's Guide. OPA-86-005. Washington D.C.: U.S. Environmental Protection Agency.
9. U.S. Environmental Protection Agency and U.S. Department of Health and Human Services. 1986. A Citizens Guide to Radon. OPA-86-004. Washington, D.C: .U.S. Government Printing Office.
10. van Kaick, G., H. Muth, A. Kaul, H. Wesch, H. Immich, D. Liebermann, D. Lorenz, W. J. Lorenz, H. Luhrs, K. E. Scheer, G. Wagner, and K. Wegener. 1986. Report on the German Thorotrast study. Strahlentherapie (80 Suppl.):114–118.
11. Wick, R. R., D. Chmelevsky, and W. Gössner. 1986. ^{224}Ra: Risk to bone and haematopoietic tissue in ankylosing spondylitis patients. Strahlentherapie (80 Suppl.):38–44.

2
Radon

INTRODUCTION

Of the several isotopes of radon, radon-222 has the most important impact on human health (see the box entitled "Isotopes of Radon"). An inert gas at temperatures above −61.8°C, radon-222 is a naturally occurring decay product of radium-226, the fifth daughter of uranium-238. Both uranium-238 and radium-226 are present in most soils and rocks in widely varied concentrations.[21] As radon forms from the decay of radium-226, it can leave the soil or rock and enter the surrounding air or water. Radon gas thus becomes ubiquitous, and its concentration is increased by the presence of a rich source and by low ventilation in the vicinity of a source. As illustrated in Figure 2.1, radon decays with a half-life of 3.82 days into a series of solid, short-lived radioisotopes collectively referred to as *radon daughters* or *progeny* (Figure 2-1) (see Annex 2B). Two of these daughters, polonium-218 and polonium-214, emit alpha particles, which, when emission occurs in the lung, can damage the cells lining the airways. The resulting biological changes can ultimately lead to lung cancer.

Underground mining was the first occupation associated with an increased risk of lung cancer. Uranium ores contain particularly high concentrations of radium, and radon-daughter exposure has been associated with lung cancer in uranium miners. Miners of other types of ore can also be placed at risk by the combination of a sufficiently strong source of radon and inadequate ventilation.

ISOTOPES OF RADON

The committee's discussion of radon is limited to radon-222, the most common isotope. Other radioisotopes of radon—radon-219 (actinon) and radon-220 (thoron)—occur naturally and have alpha-emitting decay products. Actinon has an extremely short half-life (3.9 s). Accordingly, concentrations of actinon and its daughters are extremely low, and decay of actinon contributes little to human exposure. Because of its short half-life (56 s), the concentration of thoron is also usually low. Dosimetric considerations suggest that the dose to the tracheobronchial epithelium from thoron progeny is, for an equal concentration of inhaled alpha energy, less by a factor of 3 than that due to the progeny of radon-222.[26] The potential for lung cancer due to inhalation of thoron cannot be addressed directly, because the available epidemiological data are based almost exclusively on exposures to radon-222 and its daughters.

Radon progeny are also present in the air of dwellings. Their source is the underlying soil, but building materials, water used routinely in the building, and utility natural gas also contribute. The concentration of radon progeny in dwellings is highly variable and depends mainly on the pressure in the house and on the ventilation.

Because of their wide distribution, radon daughters are a major source of exposure to radioactivity for the general public, as well as for special occupational groups. The estimated dose to the bronchial epithelium from radon daughters far exceeds that to any other organ from natural background radiation.[20] The recent recognition that some homes have high concentrations of radon has focused concern on the potential lung-cancer risk associated with environmental radon. Measured concentrations of radon in homes in the United States appear to follow a log-normal distribution.[24]

In addressing the risks associated with radon exposure, the committee responsible for this report considered the extensive information accumulated during nearly a century of research on radon. Epidemiological studies have described the risks associated with radon-daughter exposure of underground miners; animal studies have provided complementary data; and experimental and theoretical research has provided insights into radon-daughter carcinogenesis. The

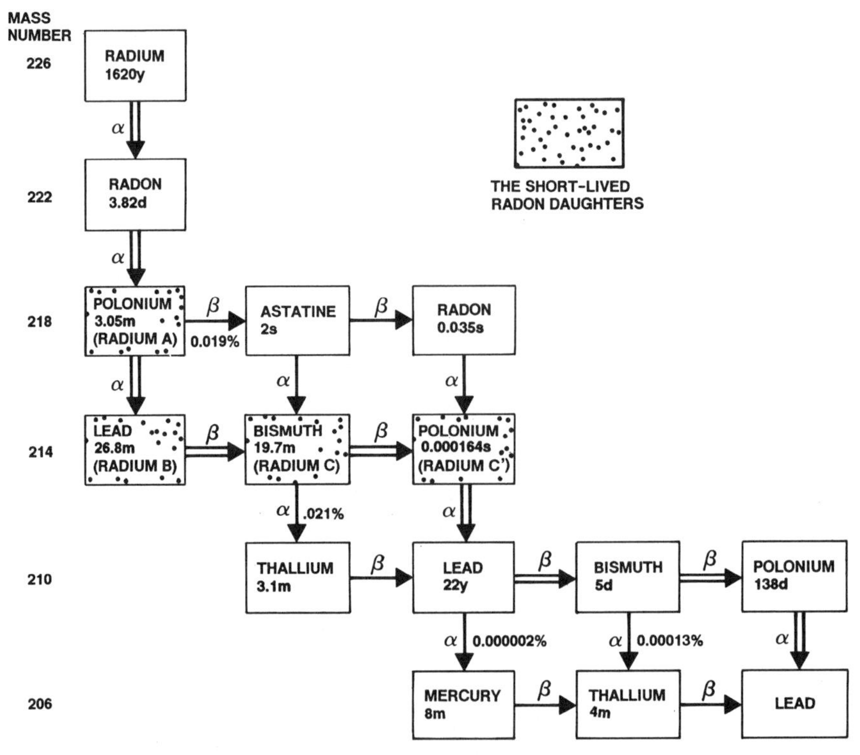

FIGURE 2.1 The radon decay chain. An arrow pointing downward indicates a decay by alpha-particle emission; an arrow pointing to the right indicates a decay by beta-particle emission. The historical symbols for the nuclides are in parentheses below the modern symbols. Most of the decays take place along the unbranched chain marked by the double arrows. The negligible percentage of the decays going along the single arrows is shown at critical points. The end of the chain, lead-206, is stable, not radioactive. Half-lives are shown for each isotope with s = seconds, m = minutes, d = days, and y = years.

research, described in detail in the appendixes of this report, is briefly summarized below.

DOSIMETRY

By convention, the concentration of radon daughters is measured in working levels (WL), and cumulative exposures over time are measured in working-level months (WLM) (see box entitled "Special Quantities and Units for Radon Exposures"). As described in Annex 2B, the relationship between exposure, measured as WLM,

SPECIAL QUANTITIES AND UNITS FOR RADON EXPOSURES

The ***working level*** (WL) is defined as any combination of the short-lived radon daughters in 1 liter of air that results in the ultimate release of 1.3×10^5 MeV of potential alpha energy. As detailed in Annex 2B, this is approximately the amount of alpha energy emitted by the short-half-life daughters in equilibrium with 100 pCi of radon. Exposure of a miner to this concentration for a working month of 170 h (or twice this concentration for half as long, etc.) is defined as a ***working-level month*** (WLM). Note that the cumulative exposure in WLM is the sum of the products of radon-progeny concentrations and the times of exposure. For historical reasons, time is quantified into blocks of 170 h when the concentration is expressed in WL. This can lead to confusion in domestic environments, because a 30-day month is 720 h. Exposure to 1 WL for 720 h results in a cumulative exposure of 4.235 WLM. Home occupancy for 12 h/day at 1 WL would result in a cumulative exposure of about 2.12 WLM per month of occupancy.

and dose to target cells and tissues in the respiratory tract is extremely complex and depends on both biological and nonbiological factors.[20] Because of differences in the circumstances of exposure, it cannot be assumed a priori that exposure to 1 WLM in a home and to 1 WLM in a mine will result in the same dose of alpha radiation to cells in the target tissues of the respiratory tract. Thus, an understanding of the dosimetry of radon daughters in the respiratory tract is essential for extrapolating risk estimates derived from epidemiological studies of miners to the general population in indoor domestic environments. Factors influencing the dosimetry of radon daughters include physical characteristics of the inhaled air, breathing patterns, and the biological characteristics of the lung (Table 2.1).

Radon daughters are initially formed as condensation nuclei. Although most of these attach to aerosols immediately after formation, a variable proportion remain unattached and are referred to as the unattached fraction. This fraction is an important determinant of the dose received by target cells; as the unattached fraction increases, the dose also increases because of the efficient deposition of

TABLE 2-1 Factors Influencing the Dose to Target Cells in the Respiratory Tract from Radon Exposure

Characteristics of Inhaled Air
Fraction of daughters unattached to particles
Aerosol characteristics
Equilibrium of radon with its daughters
Breathing Pattern
Tidal volume
Respiratory frequency
Nose or mouth breathing
Characteristics of Lung
Bronchial morphometry
Mucociliary clearance rate
Mucus thickness
Location of target cells

the unattached daughters in the airways. The particle size distribution in the inhaled air also influences the dose to the airways, because particles of different sizes deposit preferentially in different generations of the lung airways. The specific mixture of radon daughters also affects the dose to target cells, but to a smaller extent.

The amount of radon daughters inhaled varies directly with the minute ventilation, i.e., the total volume of air inhaled in each minute. The deposition of radon daughters within the lungs, however, does not depend in a simple fashion on the minute ventilation, but varies with the flow rates in each airway generation. The flow rates vary with both tidal volume and breathing frequency. The proportions of oral and nasal breathing also influence the relationship between exposure and dose. A substantial proportion of the unattached radon daughters deposits in the nose with nasal breathing, whereas it is likely that a smaller fraction deposits in the mouth with oral breathing.

Characteristics of the lung also influence the relationship between exposure and dose (Table 2-1). The sizes and branching patterns of the airways affect depositions and can differ between children and adults and between males and females. The rate of mucociliary clearance and the thickness of the mucus layer in the airways also enter into dose calculations, as does the location of the target cells in the bronchial epithelium. As outlined in Part 3 of Appendix VII, smoking and presumably other pollutants modulate these factors. The effect of the physical and biological factors outlined in Table 2-1 on the dosimetry of radon daughters can be estimated by computer modeling (see Annex 2B). The committee used the results of such models to provide guidance on estimating the risk of lung cancer due to radon in indoor environments.

HUMAN AND ANIMAL STUDIES

The association of radon-daughter exposure with human lung cancer has been the subject of extensive epidemiological studies of underground miners. The lung-cancer hazard faced by underground miners was first recognized by Harting and Hesse in 1879[7] on the basis of their autopsy observations of European miners. Excess lung-cancer occurrence has been found in uranium miners in the United States, Czechoslovakia, France, and Canada and in other underground miners exposed to radon daughters, including Newfoundland fluorspar miners, Swedish metal miners, British iron and tin miners, French iron miners, Chinese tin miners, and American metal miners (see Appendix IV). Epidemiological studies of these mining groups have shown increasing lung-cancer risk as cumulative exposure to radon daughters increases and have provided some insights into the combined effects of cigarette smoking and radon-daughter exposure (see Appendix VII).

Exposures of animals to radon and its daughters have confirmed that exposure to radon daughters causes lung cancer (see Appendix III). Animal experiments have also provided data on exposure-response relationships and on the modifying effects of exposure rate and the physical characteristics of the inhaled radon (see Appendix III). Animal models have proved less useful for studying the interaction of radon daughters with cigarette smoking because of the difficulties of replicating smoking patterns in humans with animals.

The committee also considered relevant information from the extensive literature on the biology and epidemiology of lung cancer. This malignancy, although relatively uncommon at the start of the twentieth century, has become the leading cause of cancer death in the United States.[35] Most lung cancers are caused by cigarette smoking; only 5–10% of the total cases occur in lifelong nonsmokers.[35,38] In cigarette smokers, the risk of developing lung cancer increases with the number of cigarettes smoked daily and with the number of years of smoking.[3,35] The risk of lung cancer for a smoker is some 10 times higher than that for a nonsmoker, and up to 20 times higher for heavy smokers. Because cigarette smoking predominates as the cause of lung cancer, the committee needed to address separately the risks of radon-daughter exposure for smokers and for nonsmokers.

The committee's analyses of the interaction between radiation and smoking are discussed in Part 2 of Appendix VII.

The committee was faced with the challenge of using the epidemiological and experimental evidence in combination with its understanding of the dosimetry of radon daughters to address the topical issue of exposure in domestic environments. This followed from the charge to the committee to develop risk coefficients applicable to exposure to radon in homes. The rationale for this charge emerges from recognition of the potential for radon in domestic environments as a public-health hazard. The urgency of completing this task has increased as more data have become available on indoor radon in the United States. Radon concentrations in American homes have not been systematically surveyed, but available data indicate that some homes have levels approaching or greater than the control levels in underground mines.[25] Radon daughters are also a hazard for underground miners. Thus, the committee was also charged with addressing the risks for occupational exposure to underground miners.

Other expert groups and individual investigators have derived risk estimates for radon-daughter exposure (see Appendix VII). The approaches have been diverse and based variously on the epidemiological data, on extensive dosimetric modeling, and on informal expert judgment. The resulting risk estimates have a wide range. The committee did not select any of the published lung-cancer risk estimates associated with exposure to radon and its progeny as appropriate for meeting its charge.

THE COMMITTEE'S APPROACH TO ESTIMATION OF LUNG-CANCER RISK

Evaluation of the lung-cancer risk associated with radon daughters was the most challenging task faced by the committee. Numerous studies of underground miners exposed to radon daughters have shown an increased risk of lung cancer, in comparison with unexposed populations. Animal studies have confirmed this risk, and the development of multicomponent dosimetric models has provided an understanding of factors influencing the carcinogenic potential of exposure. However, the human data are for occupational exposure in an underground environment and do not address directly the risks at the generally lower levels of exposure that are typically of concern in the home.

Two approaches that have been used previously to characterize the risks associated with radon-daughter exposure were considered by the committee: dosimetric models of the respiratory tract and statistical models applied to one or more of the epidemiological data sets. The dosimetric approach provides an estimate of risk that is based on modeling the dose to target cells in the respiratory tract. Diverse dosimetric models have been developed; all require assumptions (some not subject to direct verification) concerning the deposition of radon daughters in the lung and the nature and location of the target cells for cancer induction. Additional assumptions concerning the carcinogenic potential of alpha radiation in the respiratory tract are also required. The committee preferred an epidemiological approach that provides risk coefficients based directly on the substantial body of available human data. While the committee did not use dosimetric models for calculating the lung-cancer risk coefficients, it found such models useful in applying its risk model, derived from studies of underground miners, to the general population.

Rather than basing its risk estimates solely on review of published reports, the committee obtained and analyzed the original exposure and follow-up data from four of the most important epidemiological studies of underground miners (see the next section). Each of the studies has limitations, but a combined analysis of these major data sets permitted a comprehensive assessment of the risk associated with radon-daughter exposure and other factors influencing this risk. In analyzing the epidemiological data, the committee used a descriptive approach, rather than methods based on models of carcinogenesis. The committee's analytical approach was appropriate for meeting its charge with a minimum of assumptions as to the underlying mechanisms of cancer initiation and promotion. Although a few epidemiological studies of lung-cancer risk associated with indoor domestic exposure to radon have been reported, these studies have been preliminary and are inadequate for the purpose of risk estimation. In the future, however, epidemiological studies of indoor exposure may serve as a basis for lung-cancer risk estimates.

THE COMMITTEE'S ANALYSIS OF THE RISK OF LUNG CANCER ASSOCIATED WITH EXPOSURE TO RADON PROGENY

The committee's risk estimates for radon-daughter exposures are based largely on its own reanalysis of the four principal data sets on the epidemiological follow-up of underground miners. The committee obtained data based on two Canadian uranium-miner cohorts, Eldorado Beaverlodge[10] and Ontario;[17–19] on Swedish iron miners, Malmberget;[29] and on Colorado Plateau uranium miners.[8,9,12,13,37] (see box entitled "Characteristics of the Four Underground-Miner Groups Analyzed by the Committee"). The committee attempted to obtain comparable data from the studies of Czechoslovakian uranium miners, but was unsuccessful. For the first three of the cohorts, we obtained data on individual miners; for the Colorado cohort, we were able to obtain only detailed summaries of the type described below. Some of the miners in the Colorado cohort were occupationally exposed to radon progeny prior to their employment in uranium mines. Although exposures have been estimated for this earlier period,[13] their accuracy is uncertain and they were not considered in the committee's analysis. Cigarette-smoking information on all exposed subjects was available only for the Colorado cohort, so the committee's primary dose-response analysis does not include this factor. However, the committee's analyses of the combined effects of smoking and radon exposure on the Colorado cohort, described in Appendix VII, support the results presented here (and in Annex 2A).

Recent developments in statistical methods have provided better approaches for analyzing data from occupational-cohort studies than were available when these data were first analyzed. With these multivariate methods for analysis of the data, it is possible to examine systematically many aspects of risk estimation that have been the most uncertain in the past, particularly the temporal patterns of excess risk. By analyzing the combined data from the four cohort populations, it was the committee's intent to gain a clearer understanding of appropriate models for describing the risk associated with exposure and to obtain a more meaningful comparison of the risks in these primary cohorts.

The committee first carried out separate but parallel analyses of the four cohorts to gain a clearer understanding of the determinants of risk within each. The committee then carried out a formal analysis of the combined cohort data to obtain better estimates of the effects that seemed important and consistent in the separate analyses. This approach led to the development of a relative-risk time-since-exposure model, based on the combined data, which is more complex than ordinarily used for estimating radiation risks. However, it is the simplest mathematical expression that adequately

Characteristics of the Four Underground-Miner Groups Analyzed by the Committee

Cohort	No. of Workers Followed	Average Duration of Followup (yr)	Average Age at End of Followup (yr)	Average Duration of Exposure (yr)	Average Cumulative Exposure (WLM)	No. of Lung Cancer Deaths
Eldorado, Beaverlodge[10]	1,580[a]	14	45	—	—	65
	6,847	14	43	3.2	22	
Ontario[18]	570[a]	16	52	—	—	87
	11,076	19	50	3.7	37	
Malmberget[29]	1,292	21	67	20	98	51
Colorado[8,14] (all exposed)	3,347	25	57	8	822	256
Colorado (≤2,000 WLM)	2,975	25	57	7	509	157

[a]These rows of data indicate surface workers not exposed to radon from the mines.

NOTE: Appendix IV refers to a thorough review of the four studies of underground miners that the committee used as the basis for its analysis: uranium miners on the Colorado plateau, uranium miners in Ontario, uranium miners employed at the Eldorado Mine at Beaverlodge, Saskatchewan, and iron miners at Malmberget, Sweden. The followup experience of these groups totals over 400,000 person-years and includes 459 lung-cancer deaths. There are important differences among the four studies, including duration of exposure and followup, exposure rate, and degree of uncertainty and potential biases in estimated exposures. These factors are described in Appendix IV and were examined to the extent possible in the committee's analysis.

describes the level and temporal pattern of risk in the four cohorts. This section provides a summary of the committee's approach and results. Detailed discussion of our statistical models and methods and of their application to these cohort data is given in Annex 2A.

Undoubtedly, many factors influence the occurrence of lung cancer in miners exposed to radon daughters. In carrying out its own original analyses of the data on the cohorts, the committee focused on the following potential risk factors: cumulative exposure, duration of exposure, age at which risk is being evaluated, age at first exposure, time since cessation of exposure, and time since each part of the exposure. In the one cohort for which it was possible to do so (the Colorado Plateau uranium miners), the effects of smoking were also evaluated. The original investigators of these data relied primarily on calculation of standardized mortality ratios (SMRs) by exposure category, a method that provides a useful but limited analysis of the data. In addition to a thorough investigation of effects of the above risk factors, a substantial part of the analysis was in terms of comparisons purely within the cohort data, as opposed to comparison with data on external populations, as in analyses based on SMRs.

STATISTICAL MODELS AND METHODS

The committee's general approach was to examine how the age-specific relative risk depends on the variables of interest. This was done by making a cross-classification of numbers of lung-cancer deaths and person-years at risk, by categories of these variables, and then fitting models to the rates given by the ratio of deaths to person-years in such a tabular cross-classification. The committee fitted regression models with a Poisson probability model for the number of deaths in each cell of the table, where the expected value was taken as the product of the person-years at risk for the cell and a cancer rate given by a parametric model. For the case of purely internal cohort comparisons, not relying on external rates, this is a grouped-data analog of the widely used Cox relative-risk regression method.[2] For the case of comparison with external rates, it is a generalization of standard SMR methods that provides more detailed examination of the relative risk. (A very useful reference for these methods is the survey paper by Breslow.[1])

The parametric models for this analysis were expressed in terms of the excess relative risk, that is, the ratio of the excess risk to the background age-specific risk. The choice of analyzing the relative risk in this way, rather than as absolute excess risk, was made because of the fairly general observation in these and other cohort studies that the excess risk increases markedly with age, in a similar fashion as the background risk. Modeling of relative risks is somewhat simpler than modeling of absolute excess risks (see Annex 2A). The committee did not assume that the relative risk was constant in age or time since exposure—an assumption that would correspond to the usual relative-risk model—but rather used statistical methods appropriate for examining how the relative risk in the four cohorts actually depends on these (and other) factors.

Much of the committee's analysis was based on models for relative risk of the general mathematical form:

$$r(a) = r_0(a)[1 + \beta\gamma(\nu)W], \tag{2-1}$$

where $r(a)$ is the lung-cancer mortality rate for a given age and calendar period, $r_0(a)$ is the background or baseline risk of lung-cancer mortality in the population, W is the cumulative exposure in WLM at 5 yr before age a, β is a basic slope of the dose-response relation, and $\gamma(\nu)$ represents modifying effects of variables (ν) other than cumulative exposure. The 5-yr lag period used in evaluating cumulative dose represents the assumption that increments in exposure have no substantial effect on the risk of lung-cancer mortality for at least 5 yr. The variables that $\gamma(\nu)$ might depend on are those itemized earlier in this section, including age in particular, so that the variation of relative risk with age could be investigated. This model was first used to carry out separate parallel analyses of the four miner cohorts. The analyses were made both by comparison with external population rates and by comparisons restricted to the cohorts (internal comparisons).

For the external comparisons, population rates were scaled to make them consistent with the experience of the cohort at zero and low exposures. This is essentially the same approach as the commonly used procedure of estimating an intercept in the regression of SMRs on dose. Thus, the external rates were used only, in effect, to incorporate knowledge of trends in age and calendar time, as opposed to absolute levels of risk. Such a procedure is well accepted in epidemiological analyses as being necessary to allow for differences between the cohorts and the comparison population other than the occupational exposure under analysis.[1]

It is widely held that linear models are adequate for extrapolation to low doses of high linear energy transfer (LET) radiations.[23,33] The relative risk in this model is linear in cumulative exposure. The only statistically significant evidence against linearity in any of the committee's analyses of these four cohorts was in the data for the Colorado uranium miners, where risk per unit exposure in WLM decreased at doses above approximately 2,000 WLM. Consequently, these very high exposure levels were excluded from our primary analysis (as discussed in more detail in Annex 2A).

Applying models of the form of Equation 2-1 to the four cohorts, the committee did not find consistently significant effects on relative risk due to age at first exposure or duration of exposure after adjusting for cumulative exposure, an analytical procedure equivalent to investigating an effect due to the average rate of exposure. That is, the term $\gamma(\nu)$ did not depend significantly and consistently on either of these factors. This type of analysis did reveal, however, consistent and significant effects of age, a, at which risk is being evaluated (age at risk) and of time since cessation of exposure. The excess relative risk, for a given cumulative exposure, decreased substantially with an increase in each of these variables.

For prolonged exposures, as is the case for much of the cohorts' experience, a risk model that incorporates time since cessation of exposure was felt to overemphasize the date of the last recorded exposure, in contrast with the full exposure history. The following alternative form, which is called a time-since-exposure (TSE) model, described the same features of the data and seemed more plausible to the committee. Let d_1, d_2, and d_3 be the WLM received by a miner in three time intervals prior to a specified age at risk. The three intervals are the 5th through the 9th yr (5–10), the 10th through the 14th yr (10–15), and 15 yr or more. A time-since-exposure effect was introduced by replacing the (lagged) cumulative exposure W in Equation 2-1 by an effective exposure at any given age at risk $\theta_1 d_1 + \theta_2 d_2 + \theta_3 d_3$ so that the parameters θ_i allow the effect of exposures to depend on how long ago they occurred. Since this effective exposure will always be multiplied by an estimated parameter, β, the parameter θ_1 can be taken arbitrarily as 1.0 (i.e., absorbed into β), so that θ_2 and θ_3 represent the effect of exposures relative to those in the time-interval window 5–10 years ago. Thus, a model incorporating an effect of time since exposure can be expressed as:

$$r(a) = r_0(a)[1 + \beta(d_1 + \theta_2 d_2 + \theta_3 d_3)]. \qquad (2\text{-}2)$$

Using this model, consistent effects of time since exposure were found in analyses of the separate cohorts.

Based on these analyses, using Equations 2-1 and 2-2 for the individual cohorts, it was thus determined that effects of age at risk and time since exposure should be examined in the combined cohort data. Our aims in this analysis included (1) formal assessment of the consistency of these effects over the cohorts, (2) more precise estimation of the parameters based on the larger sample provided by the combined cohorts, and (3) examination of the joint effects of these two variables, to determine the extent to which they are independent effects or manifestations of the same temporal trends. The model was taken as

$$r(a) = r_0(a)[1 + \beta\gamma(a)(d_1 + \theta_2 d_2 + \theta_3 d_3)], \qquad (2\text{-}3)$$

where $\gamma(a)$ represents the effect of age at risk. Here, though not explicit in the notation, both the background rate $r_0(a)$ and the risk coefficient β were allowed to depend on the cohort in most of these analyses. This is an important feature of such analyses, because it avoids the biases that can result from confounding the effects under investigation with unexplained variations in the general levels of risks between cohorts, for example, the variation introduced by systematic differences in estimation of exposure. In addition, background rates used in the analyses depended on calendar time. Such details are discussed in Annex 2A.

THE COMMITTEE'S TSE MODEL

The committee's analyses show consistent results among the miner cohorts, in spite of the uncertainties in exposure assessment and the varying cohort experiences with respect to magnitude and duration of exposure. Detailed support of the procedures and conclusions is provided in Annex 2A. In particular, the decline in excess relative risk with both age at risk evaluation and time since exposure, with these two factors acting largely independently of one another, was reasonably consistent over the cohorts. The estimated general levels of risk among the cohorts, although differing substantially, were in fact as close to one another as could be expected on the grounds of ordinary statistical variation, even if all the exposures had been estimated perfectly. Neither age at first exposure nor duration (or rate) of exposure had an estimable effect on the relative risk in the combined analysis of the four cohorts. The analyses based on internal and external comparisons were in remarkably good agreement.

The committee's analysis for time since exposure did not show a clear difference in risks due to exposures in the first two of the three intervals considered, i.e., 5–10 and 10–15 yr before to the age at risk. Thus, the parameter θ_2 in Equation 2-3 was taken as 1.0; that resulted effectively in two categories for time since exposure: 5–15 and $\geq$15 yr. The variation in relative risk with age at risk is represented by three values, i.e., intervals less than 55, 55–64, and 65 yr of age or more. These abrupt changes in risk with age and time since exposure only approximate the actual pattern of variation, which is presumably gradual. Statistical methods could be developed for fitting less-abrupt models, but such models would not alter the committee's findings in a substantive way. Application of the model involves a cumulation of risk in age and time, and this cumulative risk varies more smoothly.

As a general description of the pattern of risk for the miner cohorts, the committee recommends the following relative-risk, TSE model for $r(a)$, the age-specific lung-cancer mortality rate:

$$r(a) = r_0(a)[1 + 0.025\gamma(a)(W_1 + 0.5W_2)], \qquad (2\text{-}4)$$

where $r_0(a)$ is the age-specific background lung-cancer mortality rate; $\gamma(a)$ is 1.2 when age a is less than 55 yr, 1.0 when a is 55–64 yr, and 0.4 when a is 65 yr or more; W_1 is WLM incurred between 5 and 15 yr before this age; and W_2 is WLM incurred 15 yr or more before this age. Note that $r(a)$ is the lung-cancer mortality rate from all causative agents, not just that due to radon exposure alone.

This TSE model is applied as follows. First, exposures are separated into the two intervals as indicated above for each year in the period of interest. Then the total annual risk for the person's age is calculated from Equation 2-4 with an appropriate background age-specific risk $r_0(a)$. $r(a)$ is multiplied by the chance of surviving all causes of death to that age, including the increased risk due to exposure, and these products are summed over the ages in the desired period, as outlined in Annex 2A, Part 2. The choice of an appropriate age-specific background rate for this calculation involves proper treatment of smoking, sex, and calendar time. These issues and procedures are discussed in the section "Projecting Risks Associated with Radon-Daughter Exposure," where tables are provided for a number of ages and exposure situations.

As discussed in the section, "Sources of Uncertainty in the Committee's Data Analysis," many difficult issues must be considered in extrapolating the results of these analyses to the populations exposed to radon daughters in homes. The most direct interpretation of the committee's model is as a summary of the total lung cancer experienced by the four cohorts of underground miners. It is not to be taken as a true model, even for that setting, because there are undoubtedly some factors influencing risk that are not accounted for by such a simple model. The committee's recommendation of the use of the TSE model is based on its inclusion of all the effects that are consistent and statistically significant among the four cohorts.

The committee is cognizant of the current widespread preference for constant-relative-risk models (i.e., constant in age) for radiation-induced epithelial cancers. The TSE model is a departure from this preference with regard to the dependence of relative risk both on age and on time since exposure. In particular, the Ad Hoc Working Group to Develop Radioepidemiological Tables of the National Institutes of Health[22] presented important evidence for constant-relative-risk models, but considered mainly acute exposure to low-LET radiation. A model with relative risk constant in cumulative exposure was also the recommendation of Thomas et al.[32] and Thomas and McNeill,[31] who considered largely the same studies as the Committee did, but without access to unpublished details of the data. On the other hand, Harley and Pasternack[6] and other investigators,[9,14,30] have come to conclusions similar to the committee's with regard to an effect of time since exposure. We took an appropriately conservative attitude toward departing from the constant-relative-risk model, but concluded that the evidence from the four cohorts' experience supporting a nonconstant relative risk was too strong to be ignored. Most of the evidence for constant relative risk is from acute external exposures to low-LET radiation, e.g., the case of the Japanese atomic-bomb survivors; that situation is biologically and dosimetrically different from the chronic exposure to inhaled alpha-emitters sustained by underground miners. Although the committee considered the radiobiological implications of the results of its analyses, particularly the observed decrease in risk with time since exposure,

TABLE 2-2 Lung-Cancer Risk Coefficients, β, with a Constant-Relative-Risk Model

Cohort	β Excess Relative Risk[a] per WLM	95% Confidence Interval
Ontario	0.014	0.006–0.033
Eldorado	0.026	0.013–0.060
Malmberget	0.014	0.003–0.089
Colorado	0.006	0.003–0.013
All cohorts together	0.0134[b]	0.008–0.023

[a]Based on internal comparison (see text).
[b]Differences between internal and external comparisons are negligible, e.g., all cohorts together, external comparison, excess relative risk = 0.015/WLM.

it decided that speculation on possible underlying biological mechanisms was unwarranted, because the subject obviously deserves further laboratory and clinical investigation.

To provide a clear comparison with other published reports, it seemed important for the committee to provide the estimated lung-cancer risks in these miner cohorts under a constant-relative-risk model, inasmuch as the most common method of analysis is in terms of dose-group-specific SMRs, which corresponds to an implicit assumption of constant relative risk. In this case, the estimates correspond to average risk over the follow-up period.

Estimates of β, the slopes of the dose-response curves, in the model for relative risk (Equation 2-1), with $\gamma(\nu)$ omitted, are given in Table 2-2.

The cohort-specific risk estimates in Table 2-2 are statistically significantly different at only a marginal level ($\chi^2 = 6.6$ on 3 degrees of freedom; $P = 0.09$). The confidence intervals represent only basic sampling variation, assuming, for example, no errors in exposure assessment, and ascertainment of death from lung cancer. The confidence interval for all cohorts together (Table 2-2) requires especially careful interpretation, because it relates only to the statistical variation in the estimates. This is discussed further below.

THE STATISTICAL UNCERTAINTY OF THE COMMITTEE'S ANALYSES

The following discussion concerns only basic statistical issues in assessment of the errors in the above estimates. More specific matters, including attention to the quality of the data for specific cohorts and the uncertainty inherent in applying these risk estimates to exposure situations other than underground mining, are discussed in the next section.

Although the statistical calculations yield formal standard errors for the parameter estimates in the TSE model (Equation 2-4), errors in estimation of age-specific risks might not be of primary interest. Rather, one would like to assess errors in the use of the TSE model to predict overall risks for a given exposure history. Moreover, it is not a simple matter, even if one were interested in age-specific risks, to combine appropriately the errors in estimation of all the parameters involved. Finally, and most importantly, formal standard errors from statistical calculations do not take into account errors in dosimetry, inadequacy of the form of the model, and other factors discussed in the next section. The first two of those difficulties are met to a useful approximation by relying, at least informally, on the errors in estimation of the average (over age and time since exposure) relative risks calculated in Table 2-2. Although those risks were computed for a constant-relative-risk model, they can also be interpreted as a kind of average relative risk over the exposure history and period of observation. That is, if a model of the nature of Equation 2-4 were used to compute total lung-cancer risk over an exposure history and follow-up not substantially different from the experiences of these miners, then the errors of estimation indicated in Table 2-2 would be indicative of the error in estimating the total risk. Only the last line of Table 2-2 is directly relevant to the actual use of Equation 2-4, since it is based on the combined cohorts.

The procedure suggested above yields a multiplicative standard error of 30% for the calculated total lung-cancer risk for the TSE model as defined in Equation 2-4. When an estimate has a multiplicative standard error of 30%, the true value lies, with 67% confidence, in an interval given by multiplying and dividing the estimate by 1.30. The 67% confidence level corresponds to the commonly used method of expressing uncertainty as ± 1 standard error; in this context, however, a multiplicative factor is more appropriate than an additive one. The uncertainty at a 95% level of confidence, corresponding to ± 2 standard errors, would, in this instance, be represented by multiplication and division by 1.7 (approximately the square of 1.3; cf. Table 2-2).

These confidence intervals reflect only the uncertainty due to basic sampling variation and not that due to systematic errors in dosimetry and other nonsampling errors. It is important to reconsider the apparent variation in the cohort-specific average relative risks given above, because that is the primary evidence regarding systematic errors in dosimetry and other factors. The range seems wide, so results from one or another cohort have often been discounted. However, the sampling errors represented by the confidence intervals in Table 2-2 are large enough to account for the variation among the estimates. The logarithms of the estimates have approximately normal distributions. Four such estimates of precisely the same quantity, with (logarithmic) standard errors corresponding to the individual confidence intervals shown above, would vary, with reasonable chance, like the results for the individual cohorts in Table 2-2.

Nevertheless, it is certain that the 95% confidence intervals indicated in Table 2-2 do not reflect all sources of uncertainty, for example, the inevitable errors in exposure assessment and inadequacies of any model, like that given by Equation 2-4. Although it is difficult to assess how much more uncertainty to allow for other factors, the committee gives some guidance in the next section.

SOURCES OF UNCERTAINTY IN THE COMMITTEE'S DATA ANALYSIS

Those who use the committee's TSE model to estimate the lung-cancer risks associated with radon-daughter exposure must consider the sources of uncertainty that affect the results. Uncertainty related to the committee's analyses of the cohort data is discussed in this section; uncertainty related to the projection of risk to other groups is discussed in the next section.

The risk coefficients were derived from analyses of four data sets; random or systematic errors in the original data might have biased the risk coefficients in the recommended TSE model. Several statistical models were evaluated in the development of the TSE model. Use of a model that was inappropriate from a biological perspective or one in which the components were not specified correctly could also have influenced the estimated risk coefficients. The risk coefficients themselves have an associated degree of imprecision due to sampling variation. The imprecision that results from sampling variation can be readily quantified, but other sources of variation cannot be estimated in a quantitative fashion. Therefore, the committee chose not to combine the various uncertainties into a single numerical value.

ERRORS IN EXPOSURE ESTIMATES

In assessing the lung-cancer risks associated with exposure to radon daughters, the committee has based its quantitative conclusions on parallel analyses of original data from four epidemiological studies of underground miners. Two sources of potential error in the exposure estimates in these studies must be addressed: random error in the assignment of the exposure of individual subjects and nonrandom or systematic errors in the exposure estimates. Random misclassification is inherent in exposure assessment for an underground miner. Working levels fluctuate rapidly in time and from place to place within a mine; this variation cannot be captured by the sparse exposure data, based on measurements in work areas, that have been available for epidemiological investigations. The committee notes that any random error in the assignment of exposures reduces the magnitude of the estimated risk. Systematic errors can also affect the exposure estimates. Systematic errors refer to biases in the exposure estimates that cause either upward or downward deviations. Upward bias would tend to reduce risk estimates, whereas downward bias would tend to increase them. The committee could not directly assess the magnitude of either the random or the systematic errors in the exposure estimates of the four epidemiological studies it analyzed. It assumes that random error is present. With regard to systematic error, comprehensive descriptions of the techniques used to estimate exposure have been provided for each study (see Appendix IV). To the extent that the original investigators were aware of potential bias, their concerns have been described.

The committee recognized that the magnitude of error in the exposure estimates might differ among the studies. The extent and sources of information on exposure vary among the studies. In the two Canadian investigations (Ontario and Saskatchewan), WLM for recent years had been calculated on the basis of periodic sampling of mine air and records maintained for each miner by the mining companies. Exposures for earlier years were reconstructed retrospectively from the available data. In the investigation of Swedish metal miners at Malmberget, all exposures were retrospectively estimated on the basis of ventilation records and more recent measurements. The U.S. Public Health Service combined data from industry and state regulatory agencies with its own measurements to estimate exposures of the Colorado Plateau miners.

The exposure variable that was entered into the model represented only occupational exposure to radon daughters. Data on environmental exposures were not available, but those exposures can reasonably be assumed to have been much lower than the occupational exposures of most of the underground miners. Moreover, as mentioned above, population rates were scaled to take account of the zero-dose and low-dose occupational experience of the miner cohorts. Furthermore, the findings of the committee's analyses were not changed significantly whether internally or externally controlled approaches were used. The committee found no reason to assume that environmental exposures correlated with the categories of occupational exposure used in its analyses. The similarity of the internal and external analyses suggests that the committee's inability to control for environmental exposure did not bias the results.

MISCLASSIFICATION OF DISEASE

The disease outcome considered in these analyses is death from lung cancer. By misclassification of disease, the committee refers to the incorrect attribution of death to lung cancer when some other cause was responsible and to the incorrect attribution of a lung-cancer death to some other cause. As for exposure, misclassification of disease may be systematic or random. The validity of the listed cause of death on death certificates has been evaluated by several investigators and found to be generally satisfactory for lung cancer. Percy et al.[27] assessed the accuracy of death certificates for cancer in eight areas of the United States during 1970 and 1971. The clinical diagnosis of lung cancer was confirmed by 95% of death certificates. An earlier study in the United States also showed good concordance for lung cancer.[16] Even though some lung cancers may not be diagnosed during life, the committee does not consider random misclassification of disease to be an important source of uncertainty. The quality of death certificates could not be assessed for Sweden and Canada, but standards of medical care in those countries are high, and the confirmation rates should be similar to that in the United States.

Differential misclassification is of some concern. The diagnosis of lung cancer in uranium miners might be more vigorously pursued by clinicians who are aware of the association between lung cancer and underground mining. For the committee's analyses, based on internal comparisons, that type of bias seems unimportant.

SMOKING

The committee's risk model is based on analyses of cohorts in which most of the members smoked. The estimates of values of the parameters for time since exposure and age in Equation 2-4 might be more applicable to smokers than to nonsmokers, who were poorly represented in the lung-cancer mortality data. In the next section, "Projecting Risks Associated with Radon-Daughter Exposure," the committee shows how to apply the TSE model to groups in which smoking is less common. If data on individual smoking habits among all four miner cohorts were available, the committee's analyses could have been controlled for smoking as a cofactor in the risk of lung cancer. But such information is available only for the Colorado miners, and Appendix VII presents the results of analyses for this cohort that consider combinations of smoking and levels of cumulative exposure. The results are consistent with the analyses of the Colorado cohort in Annex 2A, which underlies the TSE model, in which smoking status was unspecified.

The committee recognizes that smoking is the most important risk factor for lung cancer and, as described in Appendix VII, has evaluated the interaction between smoking and radon-daughter exposure as comprehensively as possible. The committee has accepted that smoking and radon exposure combine in a fashion that is multiplicative (or nearly so) on the relative-risk scale. The combined effect of smoking and radon exposure is unlikely to be consistent with additivity. Unless smoking status was correlated with cumulative exposure, which seems unlikely, the cohort analyses on which the TSE model is based are not confounded by smoking.

MODELING UNCERTAINTY

Any simple mathematical model, such as the committee's TSE model, can provide only a crude approximation of complex biological processes, such as the evolution of excess lung-cancer risk after radon-daughter exposure. In developing the committee's model, compromise was necessary between simple models that represent only general features of average risk among the miners and more complex models that involve too many parameters for estimation. In making the compromise, the committee recognized that some potential determinants of risk were not incorporated into the final model. Therefore, limitations of available data and complexity of computational approaches necessarily constrained the development of the final model that was derived. For example, exposure rate was not included, even though different exposure rates could lead to the same final cumulative exposure. Age at exposure might also plausibly affect relative risk. However, the limited variation of this factor in the miner data prevented an adequate assessment of its effect.

The magnitude of errors due to the choice of model depends substantially on the extent to which one is extrapolating from exposure and follow-up experiences similar to those of the miner cohorts. For nonextrapolative inferences, the choice of model does not have a large effect; in fact, the constant-relative-risk model and the TSE model give similar lung-cancer estimates.

The final model also involves extrapolations of risk in time and age beyond the range of currently available data. Substantial extrapolations of the observed patterns of relative risk are made in applying the TSE model to lifetime exposures, as described in the next section.

The committee has received and considered the suggestion that the decline in lung-cancer risk with time since exposure reflects a temporal pattern of smoking cessation in the various cohorts. Because the effect of smoking on lung-cancer risk is large, this potential source of bias merited serious consideration. First, the committee's analyses have been standardized by calendar year, so the observed decline in risk must be over time, independent of calendar year. Moreover, the estimate of the declining risk with time since exposure is based on the estimated effect of exposure, obtained by comparing miners at different levels of cumulative exposure. Smoking cessation would explain this effect, only if the pattern of cessation has been associated with both cumulative exposure level and time since exposure.

In arriving at a final risk model, the committee combined the data on four miner cohorts. In essence, that weighted each cohort in proportion to the number of person-years it contributed to the combined data. Whereas risk patterns were generally consistent when the cohorts were analyzed separately, the magnitude of the effects varied, although in most cases the variation was not greater than would be expected by chance. If one cohort were better or worse than the others, the unweighted averages of the effects of age and time since exposure might not be optimal. In the view of the committee, however, there was no a priori reason to suspect such an occurrence, nor were any substantial discrepancies detected during analyses. The committee did not believe that a subjective weighting of the cohorts was appropriate.

In summary, a number of sources of uncertainty may substantially affect the committee's risk projections; the magnitude of uncertainty associated with each of these sources cannot be readily quantified. Accordingly, the committee acknowledges that the total uncertainty in its risk estimates is large. Readers of this report must take this into account in interpreting the numerical risk estimates presented in the tables below. An illustrative calculation may help to emphasize this point. As was noted in the section, "The Committee's Analysis of the Risk of Lung Cancer Associated with Exposure to Radon Progeny" above, the multiplicative standard error due to sampling variation alone is about 30%. It is of some value to consider what would be the joint effect on the committee's risk estimates of the six sources of uncertainty considered, if each had this degree of variation. In multiplicative terms, the result would be approximately a twofold uncertainty at a 67% confidence level, i.e., $1.3^{\sqrt{6}}$, and hence an uncertainty of about 4 at a 95% confidence level. This example is based on representing the uncertainty effects as independent additive terms on the log scale. It illustrates how total uncertainty may be increased by the contribution of multiple individual sources of uncertainty.

The uncertainty in the committee's risk projections varies with the population being considered. For example, the risk projections for males are more certain than those for females, and those for smokers are more certain than those for nonsmokers. In general, however, it would be overly optimistic to interpret the estimates as more precise than the illustration above suggests.

PROJECTING RISKS ASSOCIATED WITH RADON-DAUGHTER EXPOSURE.

The TSE model (Equation 2-4) provides the basis for the committee's assessment of the risk of lung cancer associated with radon-daughter exposure. The application of this model to the general population requires a number of epidemiological techniques as well as assumptions concerning the population at risk. The methods used to project risks are described in Part 2 of Annex 2A. Briefly, the committee applied the risk coefficients derived from Equation 2-4 for various patterns of exposure and age at risk to U.S. rates for lung-cancer mortality. A life-table method was used to calculate the lifetime risks of lung-cancer mortality and the years of life lost—two measures used by the committee to evaluate the lifetime consequences of exposure to radon daughters.

It must be recognized that the risk model was developed with data on four groups of miners; the subjects were male, a limited age span and duration of exposure were covered, smoking was not explicitly considered in the analyses, and follow-up has not yet extended across the subjects' lifetimes. Of necessity, the committee needed to make assumptions concerning, among other factors, the effect of gender, the effect of age at exposure, the interaction between cigarette smoking and radon daughters, and the lifetime expression of risk. Furthermore, the committee extrapolated a risk model based on the mining environment to exposure in the general domestic indoor environment. This section first reviews each of these sources of uncertainty and then presents the lung-cancer risk projections for exposures in mining and indoor environments.

UNCERTAINTIES IN THE RISK PROJECTIONS

Gender

The committee's model is based only on data on males; with the exception of the small studies of indoor exposure, epidemiological data on females exposed to radon daughters are unavailable. Accordingly, the committee had to decide how the risk model should be applied to females.

The committee made risk projections based on the assumption that the relative risk was the same for males and females. This approach was adopted because it seemed biologically appropriate and consistent with the apparent lack of sex-specific effects for cigarette smoking. Exposure to radon progeny was considered to act in concert with other agents to increase the pool of target cells intermediate in the carcinogenic process or to increase the rate of cell transformation.[15] The committee could identify no biological rationale for considering that sex influences the development of radon-related carcinoma of the lung. For cigarette smoking, the predominant cause of lung cancer, little investigation has addressed the effects of sex directly. However, the relative risks appear to be similar in males and females when quantitative aspects of smoking habits are adjusted for.[4,11]

As an alternative, the committee considered that the excess risk associated with radon-daughter exposure is additive with background rates for males and females, as suggested by recent analyses of the Japanese atomic-bomb survivor data.[28] The contrasting models are presented in a simplified form below. Suppose that $\phi(g)$ represents sex effects, with $\phi(1)$ for females and $\phi(0)$ for males. Among atomic-bomb survivors, a multiplicative model for lung-cancer risk in age (a), dose (d), and sex (g) of the form:

$$r(a, d, g) = r_0(a)[1 + \phi(g)](1 + \beta d) \tag{2-5}$$

requires that β depend on sex. Alternatively, for the atomic-bomb survivors, the additive model:

$$r(a, d, g) = r_0(a)[1 + \phi(g) + \beta d] \tag{2-6}$$

provides an adequate fit without having β depend on sex. Note that the baseline lung-cancer rates for no radon exposure are $r_0(a)$ for males and $r_0(a)(1 + \phi)$ for females. With Equation 2-6, the risk for females, $g = 1$, can be written:

$$r(a, d, g) = r_0(a, 1) + r_0(a)\beta d. \tag{2-7}$$

The first term, $r_0(a,1)$, is the female baseline rate, and the second term, $r_0(a)\beta d$, is the absolute excess for males. Thus, this analysis shows that, among the atomic-bomb survivors, the absolute radiogenic excess is similar in males and females.

Because lung-cancer data on females exposed to radon daughters are not available, the committee cannot formally support its preference for a sex-specific relative-risk model. The alternative suggested by analysis of the atomic-bomb survivor data, described above, might not be biologically relevant for exposure to radon progeny, but the committee presents the results of risk estimates based on this alternative approach below, so that the consequences of its choice of risk models regarding sex can be appreciated.

Age at Exposure

In its analysis of the miner data, the committee did not find an effect of age at first exposure, after controlling for other correlates of age. Exposure at an early age, particularly before the age of 20 yr, might have greater effects than exposure at later ages. The relative risks of radiation-induced cancers are increased in Japanese atomic-bomb survivors exposed before the age of 20 yr. Dosimetric models for radon daughters also project greater risks associated with exposure during childhood.[20] However, the committee had only very limited data on young miners and none on persons exposed during childhood.

In making its risk projections, the committee assumed that age at exposure did not affect the risk associated with radon-daughter exposure. Although that assumption is supported by the analyses of the miner data, the committee recognizes the limitations of these data, as well as the conflicting conclusions that arise from dosimetric considerations. Compared with other sources of uncertainty, the assumption might not have major consequence for lifetime projections. The background lung-cancer rates are virtually nil in the young, their duration of exposure is relatively brief, and the time since exposure is long.

Cigarette Smoking

As discussed in Appendixes V and VII, cigarette smoking causes widespread and extensive changes in the lungs and is the predominant cause of lung cancer. It was thus essential for the committee to provide risk projections for both smokers and nonsmokers. This required the assumption of a model of the interaction between cigarette smoking and exposure to radon daughters. Unfortunately, most of the data sets available to the committee provided little or no information on several potentially important aspects of smoking. Data on tobacco use were available only for the Colorado Plateau and New Mexico uranium miners (see Appendix VII). Such potentially important variables as time since cessation of tobacco use and inhalation practices were unknown. Those limitations and the relatively small number of nonsmoking subjects precluded precise description of the interaction between smoking and radon-daughter exposure. On the basis of its review of the relevant evidence, the committee selected a multiplicative model of the interaction of the two exposures, smoking and exposure to radon progeny, even though several studies provided evidence of a submultiplicative interaction (see Appendix VII). Largely on the strength of the Colorado Plateau data, a multiplicative interaction was considered appropriate. However, the committee did find that a submultiplicative model provided the best fit to the Colorado Plateau data among the class of mixture models considered in its analysis. The committee did not find this analysis sufficiently compelling to abandon the more parsimonious multiplicative model in favor of any specific submultiplicative alternative. The selection of a multiplicative model for risk projections was consistent with the committee's analysis of the miner data, which implicitly assumed a multiplicative interaction between radon-daughter exposure and factors determining background lung-cancer rates.

If the committee's assumption of a multiplicative interaction is incorrect, the most likely alternative would be in the direction of a submultiplicative interaction. As discussed below, if a submultiplicative model is correct, use of a multiplicative model would overestimate the risk associated with exposure to radon progeny for smokers and, more substantially, underestimate the risk for nonsmokers.

Temporal Expression of Risk

The committee used data from a limited period of observation of underground miners to develop its TSE model. Only the Swedish miner study provided lifetime observation of a substantial proportion of cohort members; for others, the follow-up intervals were much shorter. Thus, use of any model for risk projection requires assumptions concerning the temporal expression of risk that cannot yet be verified.

In the TSE model (Equation 2-4), W_2 is the cumulative exposure 15 yr or more before the age at risk. The model implies that the effects of radon progeny decline, but not to zero, regardless of the numbers of years since exposure. Data on the four cohorts were insufficient to evaluate risk associated with exposures that occurred more than some 30 or 40 yr in the past. Although in principle it would increase the risk associated with W_2, the committee's model could arbitrarily be modified to incorporate a return to baseline risk by stipulating that W_2 is exposure 15–40 yr before the age at risk. This ad hoc modification would reduce the estimated lifetime risk and years of life lost. Whether the risk of radon-induced effects associated with a given exposure does indeed eventually return to the background value, and if so, the time required for this to occur cannot now be established. A decline in relative risk has not yet been observed among the Japanese atomic-bomb survivors.

A constant-relative-risk model is often used for risk projection. In contrast with the committee's TSE model, a constant relative risk applied to the background rates does not vary with age or time since exposure. A comparison of risk projections made with the TSE model and a constant-relative-risk model is provided below.

Extrapolation from a Mining Environment to the Indoor Environment

As reviewed above in the introduction to this chapter and in Annex 2B, several factors that affect the dosimetry of radon daughters differ between homes and mines, and these differences must be taken into account when the TSE model, developed from data on miners, is used to project risks associated with the indoor domestic environment. Dosimetric models offer a quantitative framework for assessing the differences between estimated doses in mines and in homes, although little information is available on some of the relevant factors in the two environments.

In addressing this source of uncertainty, the committee relied on the results of others who have constructed dosimetric models and described the effects of varying the values of parameters that differ in mines and homes: breathing pattern, ventilation pattern, fraction of unattached radon daughters in inhaled air, particle size distribution in inhaled air, and equilibrium of radon with its daughters. The characteristics and structures of the various models are convergent in providing guidance to the committee on extrapolation from miners to people living in homes. As illustrated in Annex 2B, within a range of uncertainty, the ratio of exposure-to-dose relationships in homes and mines is quantitatively similar. Thus, the committee has assumed, in the risk projections described below, that exposure to 1 WLM in a home and exposure to 1 WLM in a mine have equivalent potency in causing lung cancer. In situations where, as outlined in Annex 2B, detailed information on the domestic environment allows a more exact estimate of relative potency, it would be appropriate to scale the committee's risk estimates.

MEASURES OF RISK

The risk of lung-cancer mortality associated with long-term exposure to radon daughters is a function of the duration of exposure and, because of competing risks of death from other causes, of age as well. Part 2 of Annex 2A provides analytical descriptions of the measures that were used to evaluate the impact of exposure to radon progeny. As outlined there, the lifetime probability of lung-cancer mortality in a nonexposed person is R_0, and the lifetime probability of lung-cancer mortality in an exposed person is R_e. Note that R_e includes R_0. The ratio of lifetime risks R_e/R_0 is the proportional increased lifetime risk associated with exposure and is, for reasons outlined below, the preferred measure of increased risk. Another measure of risk is the number of years of life lost because of exposure, $L_0 - L_e$, where L_0 is expected years of life at birth for a nonexposed subject, and L_e is expected years for an exposed subject. Note that $L_0 - L_e$ is the average for a population of exposed persons. The number of years of life lost by those who actually die of lung cancer is usually much greater—typically, about 15 yr for males and about 18 yr for females.

In calculating R_e/R_0, R_e, and $L_0 - L_e$ for various magnitudes of annual exposure, the committee has used the male and female 1980–1984 U.S. mortality rates, R_0, listed in Table 2-3. Mortality rates based on a specific period in the past can only approximate future rates in a population whose total mortality rates vary over time. Moreover, the age patterns and the relative importance of lung cancer as a cause of death are expected to change as smoking becomes less prevalent. Estimates of lifetime lung-cancer risk, R_e, and expected lifetime, L_e, based on the TSE model (Equation 2-4), will vary with calendar time. However, the ratios of these quantities to R_0 and L_0, respectively, are remarkably stable. In particular, comparisons of R_e/R_0 in males and females show only small differences, even though the sex-specific background rates yield lung-cancer risks almost 3 times higher in males than females and life expectancy almost 7 yr less in males. Thus, the ratio of lifetime risks shown in the tables below and the ratio of total life expectancy L_e/L_0 are likely to be good approximations in populations in which the age-specific mortality patterns are similar to those in the current U.S. population.

The committee did not adjust the rates in Table 2-3 to reflect lung-cancer rates in a population that is not exposed to ambient radon progeny. As demonstrated in the section "Summary and Recommendations" at the end of this chapter, this correction would be small, compared with the uncertainty in the estimated risks. However, cigarette use has such a large effect on lung-cancer rates that it is necessary to adjust the rates when estimating the risk to nonsmokers. Age-specific lung-cancer rates for nonsmokers, $r_n(a)$, were calculated from population mortality rates as follows. Let r_0 be the U.S. lung-cancer mortality rate; the rate in nonsmokers is:

$$r_n(a) = r_0/[P + RR(1 - P)], \tag{2-8}$$

TABLE 2-3 All Causes and Lung-Cancer Annual Mortality Rates per Thousand by Sex for U.S. Population, 1980–1984

	Mortality Rate			
	All Causes		Lung Cancer	
Age (yr)	Males	Females	Males	Females
0–4	3.6552	2.8986	0.0006	0.0004
5–9	0.3500	0.2556	0.0001	0.0004
10–14	0.3833	0.2289	0.0003	0.0003
15–19	1.4139	0.5307	0.0003	0.0002
20–24	2.0347	0.6192	0.0014	0.0002
25–29	1.9616	0.6822	0.0039	0.0020
30–34	1.9596	0.8448	0.0124	0.0078
35–39	2.4513	1.2435	0.0579	0.0329
40–44	3.6426	2.0084	0.1925	0.1042
45–49	5.8409	3.2123	0.4956	0.2366
50–54	9.4292	4.9875	1.0133	0.4306
55–59	14.6070	7.4910	1.7076	0.6134
60–64	22.3096	11.4394	2.6387	0.8402
65–69	33.9381	17.1748	3.6452	1.0091
70–74	50.7750	26.7248	4.5769	1.0484
75–79	74.7904	42.5674	4.9701	0.9871
80–84	112.4224	72.5938	4.7390	0.8706
85+	187.9436	147.4246	3.6801	0.9192

where P is the proportion of nonsmokers, and RR is the relative risk—the risk for smokers relative to that for nonsmokers.

Smoking characteristics and other lung-cancer risk factors vary within segments of the population characterized by, for example, age, sex, birth cohort, and educational level. Because those risk factors affect current and future baseline lung-cancer rates, they also influence excess risks and risk projections associated with radon exposure. The committee simplified lifetime risk estimates by considering only cigarette use and by assuming a steady-state pattern of tobacco consumption. It further assumed that all smokers begin their habit at the same age and smoke for the remainder of their lives and that smoking-induced lung cancer has a 10-yr latent period. In 1970, approximately 48% of the male and 36% of the female adult population were smokers and had begun to smoke at the age of approximately 18.[36] The relative risk for smokers compared with that for nonsmokers is approximately 12 for males and 10 for females.[34] Therefore, age-specific lung-cancer rates for male nonsmokers are obtained from total lung-cancer rates for males by using $r_n(a) = r_0(a)$ for $a < 28$

and $r_n(a) = r_0(a)/[0.52 + 12(0.48)]$ for $a \geq 28$. Similarly, for women age 28 or over, $r_n(a) = r_0(a)/[0.64 + 10(0.36)]$. In each case, $r_0(a)$ is the sex-specific lung-cancer mortality rate in Table 2-3.

On the basis of this model, the lifetime risks of lung cancer and life expectancies for males are 0.0112 and 70.5 yr, respectively, for nonsmokers and 0.123 and 69.0 yr, respectively, for smokers. For females, the values are 0.0060 and 76.7 yr, respectively, for nonsmokers and 0.058 and 75.9 yr, respectively, for smokers. The increased mortality rate among smokers caused by diseases other than lung cancer is not considered here. Ginevan and Mills[5] have discussed this issue in their analysis of the combined risk of smoking and radon exposure. The committee did not specify risks for specific categories of cigarette consumption. It considered that such detailed projections go beyond the limits imposed by the available data.

Table 2-4 compares the ratio of lifetime risks, R_e/R_0, R_e, and years of life lost in nonsmoking and smoking males and females. With the committee's multiplicative model for the combined effects of smoking and exposure, the lung-cancer risk to smokers associated with exposure to radon progeny is substantially greater than the risk to nonsmokers. As illustrated in Table VII-10 of Appendix VII for the Colorado cohort, this difference becomes smaller for a submultiplicative interaction, with the risk per unit exposure increasing for nonsmokers and decreasing slightly for smokers. Therefore, use of a submultiplicative model could have a substantial impact on the estimated risk to nonsmokers. As shown in Appendix VII, a submultiplicative model is not incompatible with the available evidence.

In some exposure scenarios of interest, smoking status is unknown and the U.S. population rates in Table 2-3 are directly applicable. Figures 2-2, 2-3, and 2-4, respectively, show the ratio of lifetime risks, lung-cancer risk, and years of life lost for males and for females by exposure rate in WLM per year, based on the TSE model (Equation 2-4) when exposure is sustained throughout life. The risk ratios and lifetime risk of lung cancer increase sharply with exposure rate for both males and females. The doubling exposure rate is 2.0 WLM/yr for males and 1.8 WLM/yr for females. Because smoking patterns are not modeled here, the lung-cancer risks for a general population shown in Figures 2-2 to 2-4 and in the other figures and tables later in this chapter can be considered as averaged over smoking status. Under this approximation, lifetime lung-cancer risk and expected years of life for nonexposed persons are 0.067 and

TABLE 2-4 Ratio of Lifetime Risks[a] (R_e/R_0), Lifetime Risk of Lung-Cancer Mortality (R_e), and Years of Life Lost ($L_0 - L_e$)[b] for Lifetime Exposure at Various Rates of Annual Exposure[c]

Exposure Rate	Males						Females					
	Nonsmokers			Smokers			Nonsmokers			Smokers		
(WLM/yr)	R_e/R_0	R_e	$L_0 - L_e$	R_e/R_0	R_e	$L_0 - L_e$	R_e/R_0	R_e	$L_0 - L_e$	R_e/R_0	R_e	$L_0 - L_e$
0	1.0	0.0112	0	1.0	0.123	1.50	1.0	0.00602	0	1.0	0.0582	0.809
0.1	1.09	0.0122	0.0124	1.08	0.133	1.63	1.09	0.00657	0.00751	1.09	0.0634	0.881
0.2	1.19	0.0133	0.0246	1.16	0.143	1.76	1.18	0.00712	0.0149	1.18	0.0684	0.953
0.3	1.28	0.0143	0.0368	1.24	0.153	1.88	1.27	0.00768	0.0223	1.26	0.0735	1.02
0.4	1.37	0.0153	0.0490	1.32	0.162	2.01	1.37	0.00823	0.0298	1.35	0.0785	1.09
0.5	1.46	0.0163	0.0612	1.40	0.172	2.13	1.46	0.00878	0.0372	1.43	0.0835	1.17
0.6	1.55	0.0174	0.0735	1.47	0.181	2.26	1.55	0.00933	0.0446	1.52	0.0885	1.24
0.8	1.73	0.0194	0.0978	1.62	0.199	2.50	1.73	0.0104	0.0595	1.69	0.0983	1.38
1.0	1.91	0.0214	0.122	1.76	0.216	2.73	1.91	0.0115	0.0744	1.86	0.108	1.52
1.5	2.37	0.0265	0.183	2.09	0.257	3.30	2.37	0.0143	0.112	2.27	0.132	1.86
2.0	2.81	0.0315	0.243	2.39	0.294	3.84	2.82	0.0170	0.148	2.65	0.154	2.20
2.5	3.25	0.0364	0.303	2.67	0.328	4.35	3.27	0.0197	0.186	3.02	0.176	2.53
3.0	3.69	0.0413	0.363	2.92	0.359	4.83	3.72	0.0224	0.222	3.40	0.198	2.86
3.5	4.13	0.0462	0.422	3.15	0.387	5.29	4.17	0.0251	0.259	3.75	0.218	3.18
4.0	4.56	0.0511	0.482	3.36	0.413	5.72	4.62	0.0278	0.296	4.09	0.238	3.50
4.5	4.99	0.0559	0.540	3.55	0.437	6.14	5.06	0.0305	0.333	4.42	0.257	3.81
5.0	5.41	0.0606	0.599	3.74	0.460	6.53	5.50	0.0331	0.369	4.74	0.276	4.16
10.0	9.48	0.1062	1.170	4.96	0.610	9.63	9.56	0.0593	0.733	7.39	0.430	6.89

[a]Relative to persons of the same sex and smoking status.
[b]L_0 is the average lifetime of nonsmokers of the same sex.
[c]Estimated with the committee's TSE model (Equation 2-4), and a multiplicative interaction between smoking and exposure to radon progeny.

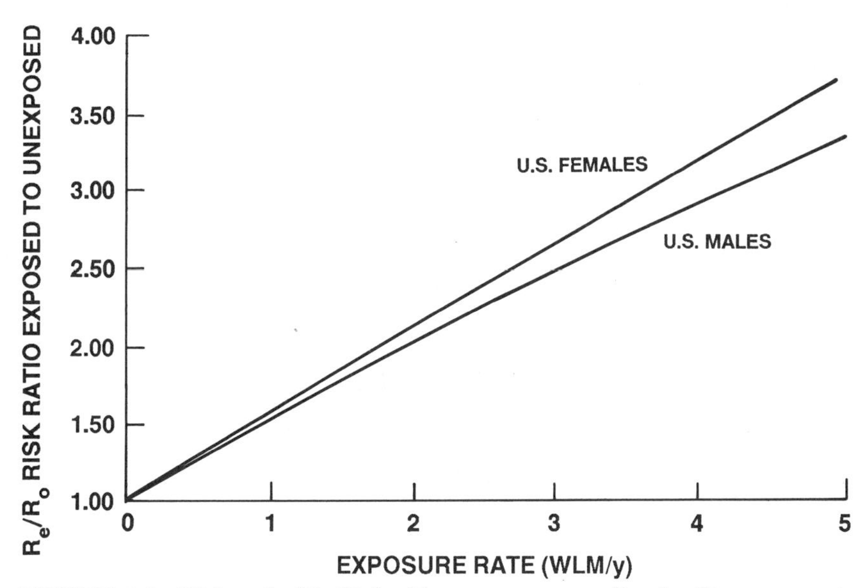

FIGURE 2-2 Risk ratio (R_e/R_0) of lung-cancer mortality for life exposure to radon progeny at constant rates of annual exposure.

69.7 yr, respectively, for males and 0.025 and 76.4 yr, respectively, for females.

Comparison with a Constant-Relative-Risk Model

For comparison with the TSE model defined by Equation 2-4 and other published reports, the committee also analyzed population risks under the constant-relative-risk model,

$$r(a) = r_0(a)(1 + \beta W), \tag{2-9}$$

where W is the cumulative exposure, in WLM, to age at risk, a, minus 5 yr (cf. Equation 2-1). The committee's estimate of the coefficient β for this case is $\beta = 0.0134$ (Table 2-2). Because both relative-risk models were developed with the same data, this model (Equation 2-9) can be viewed as a summary over the four cohorts, although it must be emphasized that important departures from the constant-relative-risk model were identified. Within common constraints of age at risk, doses, and follow-up, as in the miner cohorts, the TSE model and Equation 2-9 would generate similar lung-cancer risk estimates. However, because the recommended TSE model specifies a decline in risk with age and time since exposure and this constant-relative-risk

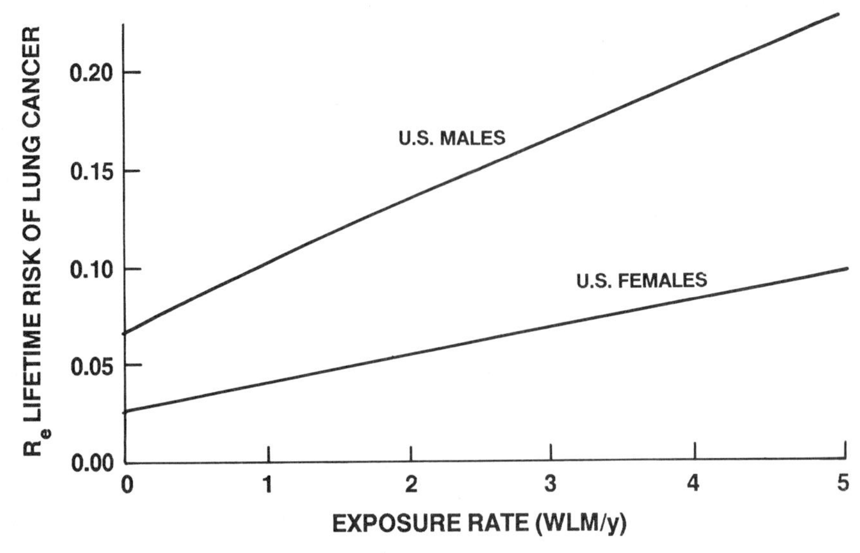

FIGURE 2-3 Lifetime risk of lung-cancer mortality (R_e) for lifetime exposure to radon progeny at constant rates of annual exposure.

model (Equation 2-9) postulates a continued and constant effect of an exposure, estimates of lifetime risk and life lost outside the range of the cohort data, for example, over a lifetime, can be quite different. Figure 2-5 compares lifetime risks for lifetime exposure of males and females in the general population under these two models. It is seen that the choice of model substantially influences the ultimate assessment of effect, with the constant-relative-risk model predicting greater radon effects. For the constant-relative-risk model (Equation 2-9), the risk doubles at an exposure rate of 1.30 WLM/yr for males and 1.20 WLM/yr for females—which is considerably less than the risks estimated above with Equation 2-4.

Comparison of Lung-Cancer Risk Projections for Females

As noted above, an alternative to a simple multiplicative model for lung cancer in females can be developed that is based on the experience of the Japanese atomic-bomb survivors (Equation 2-7). Figure 2-6 compares lifetime risk estimates for females with the two approaches. As would be expected, the predicted lifetime risk of lung cancer for females is substantially larger if the radon-progeny-associated excess for males is applied to the background rates for

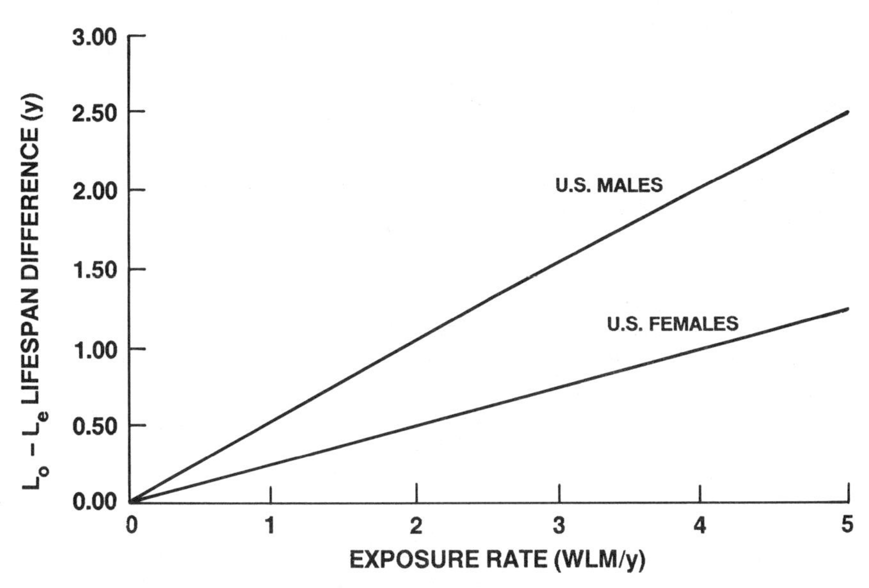

FIGURE 2-4 Average years of life lost (L_0-L_e) due to lifetime exposure to radon progeny at constant rates of annual exposure.

females. The doubling dose rate is 0.60 WLM/yr with this model approach, compared with 1.90 WLM/yr if Equation 2-4 is applied to the background rates for females. The committee prefers the relative-risk TSE model, but these calculations illustrate the potential magnitude of the difference, if the alternative model were correct.

INTERVAL EXPOSURE TO RADON DAUGHTERS

Exposure to radon daughters usually does not take place at a constant rate over a lifetime. A variable pattern is exemplified by occupational exposure; a miner's exposure starts at first employment underground and continues until termination of employment or reduction of exposure. Radon-progeny exposure in the home is another example; it can be sustained until it is detected and remedial action is taken. More complex dose-fractionation schemes can be addressed by means of Equation 2-4 and life-table calculations, similar to those outlined in Part 2 of Annex 2A.

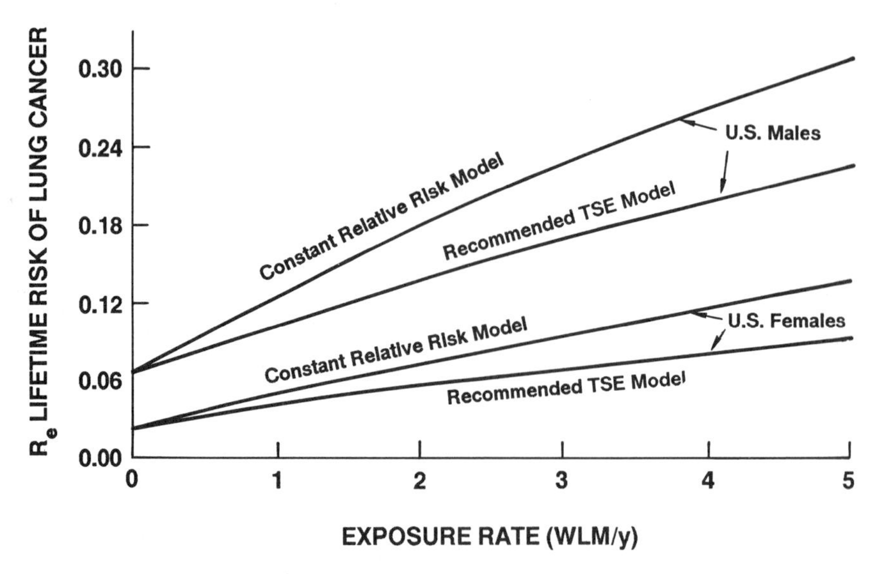

FIGURE 2-5 Comparison of lifetime risks associated with lifetime exposure to radon progeny, as estimated by a constant-relative-risk model and the TSE model recommended by this committee.

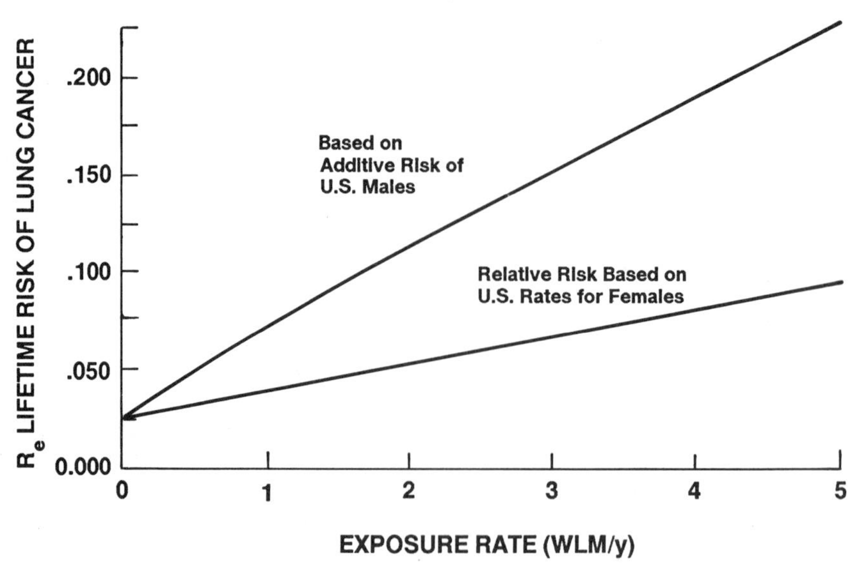

FIGURE 2-6 Lifetime risk to U.S. females for lifetime exposure to radon progeny. Comparison of models for interaction between sex and exposure (see text).

Tables 2-5 to 2-7 for males and Tables 2-8 to 2-10 for females show, respectively, for various annual exposure rates, the ratio of lifetime risks R_e/R_0 of lung-cancer mortality, the lifetime risk R_e, and years of life lost $(L_0 - L_e)$ by age exposure started and age exposure ends. All of these measures of risk increase with duration of exposure and exposure rate. For exposures of fixed duration (along diagonals), effects are similar within a given exposure rate below 1.0 WLM/yr. At higher dose rates, the excess risks increase and then decline with age for a fixed duration of exposure. Tables 2-5 to 2-10 are applicable to populations whose smoking status is unspecified. Additional tables for exposures during specified age intervals applicable to smokers and nonsmokers of each sex are given in Appendix VII.

CONDITIONAL LIFETIME EFFECTS—EXPOSURE FROM BIRTH AND FROM AGE KNOWN ALIVE

Lifetime risk of lung cancer and expected years of life lost are related to a specified exposure profile for persons who are followed from birth. Those measures do not, however, provide information about lung-cancer risk after survival to some specified age, for example, in uranium miners who are free of disease at retirement or homeowners who are currently healthy but want to assess the consequences of lowering radon concentrations in their homes. In those examples, survival up to some age has taken place without the development of lung cancer, and future risk is at issue. Total life expectancy when a person is known to be alive at a given age is greater than that at birth, because the rigors of the intervening years have been survived. Tables 2-11 and 2-12 compare risk measures computed conditionally on survival to ages 20, 40, 60, and 80 yr for males and females, respectively, in a general population. The column headed 0 gives baseline values without information on age. Exposure rates, in WLM per year, are assumed constant throughout life. Conditional lung-cancer risk associated with radon-progeny exposure, R_e, increases with age until middle age, when it falls, owing to the increased risk of death from competing causes. The expected life span increase, of course, as age increases.

Tables 2-11 and 2-12 also show the effects on the lifetime risk of lung-cancer mortality, ratio of lifetime risks, and years of life lost if exposure were to cease at the conditional age known to be alive. The health benefits of eliminating exposure can be substantial into

TABLE 2-5 Ratio of Lifetime Risks (R_e/R_0) for Males by Age Started and Age Exposure Ends[a]

Age (yr) Started	Age (yr) Exposure Ends 10	20	30	40	50	60	70	80	110
				Exposure Rate = 0.10 WLM/yr					
0	1.008	1.015	1.023	1.031	1.040	1.047	1.051	1.052	1.052
10		1.008	1.015	1.023	1.032	1.039	1.043	1.045	1.045
20			1.008	1.016	1.025	1.032	1.036	1.037	1.037
30				1.008	1.017	1.024	1.028	1.029	1.030
40					1.009	1.016	1.020	1.021	1.022
50						1.007	1.011	1.012	1.013
60							1.004	1.005	1.005
				Exposure Rate = 0.20 WLM/yr					
0	1.015	1.030	1.046	1.062	1.079	1.094	1.101	1.104	1.105
10		1.015	1.030	1.047	1.064	1.079	1.086	1.089	1.090
20			1.015	1.032	1.049	1.064	1.071	1.074	1.074
30				1.016	1.034	1.048	1.056	1.059	1.059
40					1.018	1.032	1.040	1.043	1.043
50						1.014	1.022	1.025	1.025
60							1.008	1.010	1.011
				Exposure Rate = 0.50 WLM/yr					
0	1.038	1.076	1.113	1.154	1.197	1.233	1.251	1.258	1.259
10		1.038	1.076	1.116	1.160	1.196	1.214	1.221	1.222
20			1.038	1.079	1.123	1.158	1.177	1.184	1.185
30				1.041	1.085	1.121	1.139	1.146	1.147
40					1.044	1.080	1.099	1.106	1.107
50						1.036	1.055	1.062	1.063
60							1.019	1.026	1.027

				Exposure Rate = 1.00 WLM/yr					
0	1.076	1.151	1.226	1.305	1.391	1.460	1.496	1.509	1.512
10		1.076	1.151	1.231	1.318	1.388	1.424	1.437	1.440
20			1.076	1.157	1.244	1.314	1.351	1.364	1.367
30				1.081	1.169	1.240	1.276	1.290	1.293
40					1.088	1.160	1.197	1.211	1.213
50						1.072	1.109	1.123	1.126
60							1.038	1.052	1.054
				Exposure Rate = 4.00 WLM/yr					
0	1.299	1.589	1.874	2.169	2.480	2.724	2.845	2.889	2.896
10		1.299	1.592	1.896	2.216	2.467	2.592	2.637	2.645
20			1.302	1.614	1.943	2.202	2.331	2.379	2.387
30				1.321	1.660	1.927	2.061	2.110	2.118
40					1.349	1.624	1.763	1.814	1.823
50						1.285	1.429	1.482	1.492
60							1.149	1.204	1.214
				Exposure Rate = 10.00 WLM/yr					
0	1.731	2.412	3.052	3.687	4.326	4.799	5.018	5.092	5.103
10		1.732	2.419	3.100	3.785	4.294	4.532	4.613	4.626
20			1.739	2.469	3.204	3.752	4.011	4.099	4.114
30				1.785	2.574	3.165	3.446	3.543	3.559
40					1.852	2.491	2.797	2.905	2.923
50						1.697	2.034	2.153	2.174
60							1.367	1.498	1.520
				Exposure Rate = 20.00 WLM/yr					
0	2.412	3.636	4.709	5.705	6.634	7.260	7.518	7.595	7.606
10		2.412	3.648	4.790	5.854	6.576	6.879	6.972	6.985
20			2.426	3.738	4.959	5.794	6.150	6.261	6.277
30				2.513	3.919	4.886	5.306	5.440	5.459
40					2.635	3.766	4.266	4.428	4.452
50						2.344	2.947	3.147	3.178
60							1.714	1.954	1.993

[a]Estimated with the committee's TSE model. R_e includes R_0, the baseline risk for males in the 1980–1984 U.S. population, 0.067; the expected lifetime of males is 69.7 yr.

TABLE 2-6 Lifetime Risks (R_e) for Males by Age Started and Age Exposure Ends[a]

	Age (yr) Exposure Ends								
Age (yr) Started	10	20	30	40	50	60	70	80	110
				Exposure Rate = 0.10 WLM/yr					
0	0.068	0.068	0.069	0.069	0.070	0.071	0.071	0.071	0.071
10		0.068	0.068	0.069	0.070	0.070	0.070	0.070	0.070
20			0.068	0.068	0.069	0.069	0.070	0.070	0.070
30				0.068	0.068	0.069	0.069	0.069	0.069
40					0.068	0.068	0.069	0.069	0.069
50						0.068	0.068	0.068	0.068
60							0.068	0.068	0.068
				Exposure Rate = 0.20 WLM/yr					
0	0.068	0.069	0.070	0.072	0.073	0.074	0.074	0.074	0.074
10		0.068	0.069	0.070	0.072	0.073	0.073	0.073	0.073
20			0.068	0.069	0.071	0.072	0.072	0.072	0.072
30				0.068	0.070	0.071	0.071	0.071	0.071
40					0.069	0.070	0.070	0.070	0.070
50						0.068	0.069	0.069	0.069
60							0.068	0.068	0.068
				Exposure Rate = 0.50 WLM/yr					
0	0.070	0.072	0.075	0.078	0.081	0.083	0.084	0.085	0.085
10		0.070	0.072	0.075	0.078	0.081	0.082	0.082	0.082
20			0.070	0.073	0.076	0.078	0.079	0.080	0.080
30				0.070	0.073	0.075	0.077	0.077	0.077
40					0.070	0.073	0.074	0.074	0.075
50						0.070	0.071	0.072	0.072
60							0.069	0.069	0.069

				Exposure Rate = 1.00 WLM/yr					
0	0.072	0.077	0.083	0.088	0.094	0.098	0.101	0.102	0.102
10		0.072	0.078	0.083	0.089	0.093	0.096	0.097	0.097
20			0.072	0.078	0.084	0.089	0.091	0.092	0.092
30				0.073	0.079	0.083	0.086	0.087	0.087
40					0.073	0.078	0.081	0.082	0.082
50						0.072	0.075	0.076	0.076
60							0.070	0.071	0.071
				Exposure Rate = 4.00 WLM/yr					
0	0.087	0.107	0.126	0.146	0.167	0.183	0.192	0.195	0.195
10		0.087	0.107	0.128	0.149	0.166	0.175	0.178	0.178
20			0.088	0.109	0.131	0.148	0.157	0.160	0.161
30				0.089	0.112	0.130	0.139	0.142	0.143
40					0.091	0.109	0.119	0.122	0.123
50						0.087	0.096	0.100	0.100
60							0.077	0.081	0.082
				Exposure Rate = 10.00 WLM/yr					
0	0.117	0.162	0.206	0.248	0.291	0.323	0.338	0.343	0.344
10		0.117	0.163	0.209	0.255	0.289	0.305	0.311	0.312
20			0.117	0.166	0.216	0.253	0.270	0.276	0.277
30				0.120	0.173	0.213	0.232	0.239	0.240
40					0.125	0.168	0.188	0.196	0.197
50						0.114	0.137	0.145	0.146
60							0.092	0.101	0.102
				Exposure Rate = 20.00 WLM/yr					
0	0.162	0.245	0.317	0.384	0.447	0.489	0.506	0.512	0.512
10		0.162	0.246	0.323	0.394	0.443	0.463	0.470	0.470
20			0.163	0.252	0.334	0.390	0.414	0.422	0.423
30				0.169	0.264	0.329	0.357	0.366	0.368
40					0.177	0.254	0.287	0.298	0.300
50						0.158	0.198	0.212	0.214
60							0.115	0.132	0.134

[a]Estimated with the committee's TSE model. R_e includes R_0, the baseline risk for males in the 1981–1984 U.S. population, 0.067; the expected lifetime of males is 69.7 yr.

TABLE 2-7 Years of Life Lost ($L_0 - L_e$) for Males by Age Started and Age Exposure Ends[a]

Age (yr) Started	Age (yr) Exposure Ends								
	10	20	30	40	50	60	70	80	110
				Exposure Rate = 0.10 WLM/yr					
0	0.01	0.02	0.03	0.04	0.04	0.05	0.05	0.05	0.05
10		0.01	0.02	0.03	0.04	0.04	0.05	0.05	0.05
20			0.01	0.02	0.03	0.03	0.04	0.04	0.04
30				0.01	0.02	0.03	0.03	0.03	0.03
40					0.01	0.02	0.02	0.02	0.02
50						0.01	0.01	0.01	0.01
60							0.00	0.00	0.00
				Exposure Rate = 0.20 WLM/yr					
0	0.02	0.03	0.05	0.07	0.09	0.10	0.11	0.11	0.11
10		0.02	0.03	0.05	0.07	0.09	0.09	0.09	0.09
20			0.02	0.04	0.06	0.07	0.07	0.07	0.07
30				0.02	0.04	0.05	0.06	0.06	0.06
40					0.02	0.03	0.04	0.04	0.04
50						0.01	0.02	0.02	0.02
60							0.00	0.01	0.01
				Exposure Rate = 0.50 WLM/yr					
0	0.04	0.08	0.13	0.17	0.22	0.26	0.27	0.27	0.27
10		0.04	0.09	0.13	0.18	0.21	0.22	0.23	0.23
20			0.04	0.09	0.14	0.17	0.18	0.18	0.18
30				0.05	0.10	0.13	0.14	0.14	0.14
40					0.05	0.08	0.09	0.09	0.09
50						0.03	0.04	0.05	0.05
60							0.01	0.01	0.01

				Exposure Rate = 1.00 WLM/yr					
0	0.08	0.17	0.25	0.35	0.44	0.51	0.53	0.53	0.53
10		0.08	0.17	0.26	0.36	0.42	0.44	0.45	0.45
20			0.09	0.18	0.28	0.34	0.36	0.37	0.37
30				0.09	0.19	0.26	0.28	0.28	0.28
40					0.10	0.16	0.18	0.19	0.19
50						0.06	0.09	0.09	0.09
60							0.02	0.03	0.03
				Exposure Rate = 4.00 WLM/yr					
0	0.33	0.66	0.99	1.35	1.71	1.93	2.00	2.02	2.02
10		0.33	0.67	1.03	1.40	1.63	1.71	1.72	1.72
20			0.34	0.71	1.09	1.32	1.40	1.42	1.41
30				0.38	0.76	1.00	1.08	1.10	1.09
40					0.39	0.64	0.72	0.74	0.74
50						0.26	0.34	0.36	0.35
60							0.09	0.10	0.10
				Exposure Rate = 10.00 WLM/yr					
0	0.82	1.61	2.37	3.18	3.96	4.42	4.56	4.58	4.58
10		0.82	1.63	2.47	3.29	3.78	3.93	3.96	3.95
20			0.84	1.73	2.59	3.11	3.26	3.30	3.29
30				0.93	1.84	2.39	2.55	2.59	2.59
40					0.96	1.55	1.73	1.77	1.76
50						0.63	0.83	0.87	0.86
60							0.21	0.26	0.25
				Exposure Rate = 20.00 WLM/yr					
0	1.61	3.07	4.42	5.79	7.04	7.70	7.87	7.90	7.90
10		1.61	3.10	4.60	5.97	6.71	6.91	6.95	6.95
20			1.64	3.28	4.80	5.63	5.86	5.90	5.90
30				1.81	3.48	4.42	4.68	4.73	4.72
40					1.86	2.93	3.23	3.29	3.29
50						1.23	1.58	1.66	1.65
60							0.42	0.50	0.49

[a]Estimated with the committee's TSE model. R_e includes R_0, the baseline risk for males in the 1980–1984 U.S. population, 0.067; the expected lifetime of males is 69.7 yr.

TABLE 2-8 Ratio of Lifetime Risks (R_e/R_0) for Females by Age Started and Age Exposure Ends[a]

	Age (yr) Exposure Ends								
Age (yr) Started	10	20	30	40	50	60	70	80	110
				Exposure Rate = 0.10 WLM/yr					
0	1.008	1.017	1.025	1.034	1.043	1.051	1.054	1.056	1.056
10		1.008	1.017	1.026	1.035	1.042	1.046	1.048	1.048
20			1.008	1.017	1.027	1.034	1.038	1.039	1.040
30				1.009	1.019	1.026	1.029	1.031	1.032
40					1.010	1.017	1.020	1.022	1.023
50						1.007	1.011	1.012	1.013
60							1.004	1.005	1.006
				Exposure Rate = 0.20 WLM/yr					
0	1.017	1.033	1.050	1.068	1.087	1.101	1.108	1.112	1.113
10		1.017	1.033	1.051	1.070	1.085	1.092	1.095	1.096
20			1.017	1.035	1.054	1.068	1.076	1.079	1.080
30				1.018	1.037	1.052	1.059	1.062	1.063
40					1.019	1.034	1.041	1.044	1.045
50						1.014	1.022	1.025	1.026
60							1.007	1.010	1.011
				Exposure Rate = 0.50 WLM/yr					
0	1.041	1.082	1.124	1.169	1.217	1.253	1.271	1.278	1.281
10		1.041	1.083	1.128	1.176	1.212	1.230	1.238	1.240
20			1.042	1.087	1.135	1.171	1.188	1.196	1.199
30				1.045	1.093	1.129	1.147	1.155	1.157
40					1.048	1.084	1.102	1.110	1.112
50						1.036	1.054	1.062	1.065
60							1.018	1.026	1.029

				Exposure Rate = 1.00 WLM/yr					
0	1.082	1.165	1.248	1.338	1.432	1.504	1.539	1.555	1.560
10		1.083	1.166	1.256	1.351	1.422	1.458	1.473	1.479
20			1.084	1.174	1.269	1.341	1.376	1.392	1.397
30				1.090	1.186	1.258	1.293	1.309	1.314
40					1.096	1.168	1.203	1.219	1.224
50						1.072	1.108	1.124	1.129
60							1.036	1.052	1.057
				Exposure Rate = 4.00 WLM/yr					
0	1.329	1.654	1.982	2.331	2.697	2.971	3.104	3.163	3.183
10		1.329	1.659	2.013	2.382	2.658	2.793	2.853	2.873
20			1.334	1.690	2.063	2.342	2.478	2.539	2.559
30				1.360	1.737	2.019	2.156	2.217	2.238
40					1.381	1.666	1.805	1.867	1.888
50						1.288	1.429	1.492	1.513
60							1.142	1.206	1.227
				Exposure Rate = 10.00 WLM/yr					
0	1.816	2.612	3.401	4.232	5.086	5.714	6.014	6.146	6.190
10		1.816	2.625	3.475	4.350	4.994	5.303	5.439	5.483
20			1.828	2.700	3.596	4.257	4.573	4.713	4.759
30				1.893	2.812	3.489	3.814	3.958	4.006
40					1.944	2.639	2.974	3.122	3.171
50						1.716	2.061	2.214	2.265
60							1.353	1.510	1.563
				Exposure Rate = 20.00 WLM/yr					
0	2.611	4.148	5.633	7.157	8.683	9.773	10.281	10.498	10.567
10		2.613	4.172	5.771	7.372	8.518	9.053	9.283	9.356
20			2.637	4.315	5.996	7.200	7.763	8.007	8.085
30				2.763	4.528	5.795	6.389	6.647	6.730
40					2.861	4.198	4.827	5.101	5.190
50						2.416	3.085	3.377	3.473
60							1.702	2.010	2.111

[a]Estimated with the committee's TSE model. R_e includes R_0, the baseline risk for females in the 1980–1984 U.S. population, 0.025; the expected lifetime of females is 76.4 yr.

TABLE 2-9 Lifetime Risks (R_e) for Females by Age Started and Age Exposure Ends[a]

	Age (yr) Exposure Ends								
Age (yr) Started	10	20	30	40	50	60	70	80	110
				Exposure Rate = 0.10 WLM/yr					
0	0.025	0.026	0.026	0.026	0.026	0.026	0.027	0.027	0.027
10		0.025	0.026	0.026	0.026	0.026	0.026	0.026	0.026
20			0.025	0.026	0.026	0.026	0.026	0.026	0.026
30				0.025	0.026	0.026	0.026	0.026	0.026
40					0.025	0.026	0.026	0.026	0.026
50						0.025	0.025	0.026	0.026
60							0.025	0.025	0.025
				Exposure Rate = 0.20 WLM/yr					
0	0.026	0.026	0.026	0.027	0.027	0.028	0.028	0.028	0.028
10		0.026	0.026	0.027	0.027	0.027	0.028	0.028	0.028
20			0.026	0.026	0.027	0.027	0.027	0.027	0.027
30				0.026	0.026	0.027	0.027	0.027	0.027
40					0.026	0.026	0.026	0.026	0.026
50						0.026	0.026	0.026	0.026
60							0.025	0.025	0.025
				Exposure Rate = 0.50 WLM/yr					
0	0.026	0.027	0.028	0.029	0.031	0.032	0.032	0.032	0.032
10		0.026	0.027	0.028	0.030	0.031	0.031	0.031	0.031
20			0.026	0.027	0.029	0.030	0.030	0.030	0.030
30				0.026	0.028	0.028	0.029	0.029	0.029
40					0.026	0.027	0.028	0.028	0.028
50						0.026	0.027	0.027	0.027
60							0.026	0.026	0.026

				Exposure Rate = 1.00 WLM/yr					
0	0.027	0.029	0.031	0.034	0.036	0.038	0.039	0.039	0.039
10		0.027	0.029	0.032	0.034	0.036	0.037	0.037	0.037
20			0.027	0.030	0.032	0.034	0.035	0.035	0.035
30				0.027	0.030	0.032	0.033	0.033	0.033
40					0.028	0.029	0.030	0.031	0.031
50						0.027	0.028	0.028	0.028
60							0.026	0.027	0.027
				Exposure Rate = 4.00 WLM/yr					
0	0.033	0.042	0.050	0.059	0.068	0.075	0.078	0.080	0.080
10		0.034	0.042	0.051	0.060	0.067	0.070	0.072	0.072
20			0.034	0.043	0.052	0.059	0.062	0.064	0.065
30				0.034	0.044	0.051	0.054	0.056	0.056
40					0.035	0.042	0.045	0.047	0.048
50						0.032	0.036	0.038	0.038
60							0.029	0.030	0.031
				Exposure Rate = 10.00 WLM/yr					
0	0.046	0.066	0.086	0.107	0.128	0.144	0.152	0.155	0.156
10		0.046	0.066	0.088	0.110	0.126	0.134	0.137	0.138
20			0.046	0.068	0.091	0.107	0.115	0.119	0.120
30				0.048	0.071	0.088	0.096	0.100	0.101
40					0.049	0.067	0.075	0.079	0.080
50						0.043	0.052	0.056	0.057
60							0.034	0.038	0.039
				Exposure Rate = 20.00 WLM/yr					
0	0.066	0.105	0.142	0.180	0.219	0.246	0.259	0.265	0.266
10		0.066	0.105	0.145	0.186	0.215	0.228	0.234	0.236
20			0.066	0.109	0.151	0.182	0.196	0.202	0.204
30				0.070	0.114	0.146	0.161	0.168	0.170
40					0.072	0.106	0.122	0.129	0.131
50						0.061	0.078	0.085	0.088
60							0.043	0.051	0.053

[a]Estimated with the committee's TSE model. R_e includes R_0, the baseline risk for females in the 1980–1984 U.S. population, 0.025; the expected lifetime of females is 76.4 yr.

TABLE 2-10 Years of Life Lost ($L_0 - L_e$) for Females by Age Started and Age Exposure Ends[a]

Age (yr) Started	Age (yr) Exposure Ends								
	10	20	30	40	50	60	70	80	110
				Exposure Rate = 0.10 WLM/yr					
0	0.00	0.01	0.01	0.02	0.02	0.02	0.03	0.03	0.03
10		0.00	0.01	0.01	0.02	0.02	0.02	0.02	0.02
20			0.00	0.01	0.01	0.02	0.02	0.02	0.02
30				0.00	0.01	0.01	0.01	0.01	0.01
40					0.00	0.01	0.01	0.01	0.01
50						0.00	0.00	0.00	0.00
60							0.00	0.00	0.00
				Exposure Rate = 0.20 WLM/yr					
0	0.01	0.02	0.02	0.03	0.04	0.05	0.05	0.05	0.05
10		0.01	0.02	0.03	0.04	0.04	0.04	0.04	0.04
20			0.01	0.02	0.03	0.03	0.03	0.03	0.03
30				0.01	0.02	0.02	0.03	0.03	0.03
40					0.01	0.01	0.02	0.02	0.02
50						0.01	0.01	0.01	0.01
60							0.00	0.00	0.00
				Exposure Rate = 0.50 WLM/yr					
0	0.02	0.04	0.06	0.09	0.11	0.12	0.13	0.13	0.13
10		0.02	0.04	0.07	0.09	0.10	0.11	0.11	0.11
20			0.02	0.04	0.07	0.08	0.09	0.09	0.09
30				0.02	0.05	0.06	0.06	0.07	0.06
40					0.02	0.04	0.04	0.04	0.04
50						0.01	0.02	0.02	0.02
60							0.00	0.00	0.00

				Exposure Rate = 1.00 WLM/yr					
0	0.04	0.08	0.12	0.17	0.22	0.24	0.25	0.25	0.25
10		0.04	0.08	0.13	0.18	0.20	0.21	0.21	0.21
20			0.04	0.09	0.14	0.16	0.17	0.17	0.17
30				0.05	0.09	0.12	0.13	0.13	0.13
40					0.05	0.07	0.08	0.08	0.08
50						0.03	0.04	0.04	0.03
60							0.01	0.01	0.01
				Exposure Rate = 4.00 WLM/yr					
0	0.16	0.33	0.49	0.68	0.86	0.96	0.99	1.00	0.99
10		0.16	0.33	0.52	0.70	0.81	0.83	0.84	0.83
20			0.17	0.36	0.54	0.65	0.68	0.68	0.67
30				0.19	0.37	0.48	0.51	0.52	0.51
40					0.19	0.30	0.33	0.33	0.32
50						0.11	0.14	0.15	0.14
60							0.03	0.04	0.03
				Exposure Rate = 10.00 WLM/yr					
0	0.41	0.81	1.21	1.66	2.09	2.34	2.40	2.41	2.40
10		0.41	0.82	1.27	1.71	1.96	2.03	2.04	2.03
20			0.42	0.88	1.33	1.58	1.65	1.66	1.65
30				0.47	0.92	1.18	1.26	1.27	1.25
40					0.46	0.73	0.80	0.82	0.80
50						0.27	0.35	0.36	0.34
60							0.08	0.09	0.07
				Exposure Rate = 20.00 WLM/yr					
0	0.80	1.58	2.36	3.20	4.00	4.45	4.56	4.58	4.56
10		0.81	1.61	2.48	3.30	3.76	3.89	3.91	3.89
20			0.83	1.73	2.58	3.06	3.19	3.21	3.19
30				0.93	1.81	2.31	2.44	2.46	2.44
40					0.92	1.44	1.58	1.60	1.57
50						0.54	0.69	0.71	0.68
60							0.15	0.18	0.14

[a]Estimated with the committee's TSE model. R_e includes R_0, the baseline risk for females in the 1980–1984 U.S. population, 0.025; the expected lifetime of females is 76.4 yr.

TABLE 2-11 Measures of Effects of Radon-Daughter Exposure in Males Conditional on Known Survival to a Specified Age

Measure[a]	Exposure Rate (WLM/yr)	Age Known Alive (yr) 0	20	40	60	80
R_0	0.00	0.067	0.069	.072	0.067	0.029
R_e^b	0.20	0.074	0.077	0.079	0.073	0.032
	0.50	0.085	0.087	0.091	0.082	0.036
	1.00	0.102	0.105	0.109	0.097	0.042
	5.00	0.223	0.230	0.238	0.206	0.091
	10.00	0.344	0.354	0.367	0.319	0.146
	20.00	0.512	0.527	0.547	0.487	0.241
R_e^b/R_e^c	0.10	1.052	1.043	1.027	1.014	1.000
	0.20	1.105	1.085	1.039	1.014	1.000
	0.50	1.259	1.160	1.096	1.025	1.029
	1.00	1.512	1.313	1.160	1.054	1.024
	5.00	3.312	1.917	1.360	1.102	1.034
	10.00	5.103	2.120	1.390	1.104	1.028
	20.00	7.606	2.091	1.337	1.087	1.026
L_0 (yr)	0.00	69.72	71.56	73.33	77.14	86.26
L_e^b (yr)	0.10	69.67	71.51	73.28	77.11	86.25
	0.20	69.61	71.45	73.22	77.07	86.25
	0.50	69.45	71.29	73.05	76.97	86.24
	1.00	69.19	71.02	72.77	76.80	86.22
	5.00	67.24	69.01	70.71	75.52	86.04
	10.00	65.14	66.85	68.49	74.12	85.82
	20.00	61.82	63.43	64.99	71.82	85.38
$L_e^c - L_e^b$ (yr)	0.10	0.05	0.04	0.02	0.00	0.00
	0.20	0.11	0.08	0.04	0.01	0.00
	0.50	0.27	0.19	0.10	0.02	0.00
	1.00	0.53	0.37	0.20	0.03	0.01
	5.00	2.48	1.71	0.88	0.14	0.01
	10.00	4.58	3.06	1.51	0.23	0.02
	20.00	7.90	4.98	2.27	0.33	0.01

NOTE: Exposure rate assumed constant starting at birth.

[a]The 1980–1984 population average of smokers and nonsmokers (see text).

[b]Based on exposure sustained over lifetime.

[c]Based on exposure stopped at age known alive.

TABLE 2-12 Measures of Effects of Radon-Daughter Exposure in Females Conditional on Known Survival to a Specified Age

Measure[a]	Exposure Rate (WLM/yr)	Age Known Alive (yr)				
		0	20	40	60	80
R_0	0.00	0.025	0.026	0.026	0.021	0.008
R_e^b	0.10	0.027	0.027	0.027	0.022	0.008
	0.20	0.028	0.029	0.029	0.023	0.009
	0.50	0.032	0.033	0.033	0.026	0.010
	1.00	0.039	0.040	0.040	0.031	0.011
	5.00	0.093	0.095	0.096	0.071	0.026
	10.00	0.156	0.159	0.160	0.118	0.043
	20.00	0.266	0.272	0.274	0.204	0.076
R_e^b/R_e^c	0.10	1.056	1.038	1.000	1.000	1.000
	0.20	1.113	1.074	1.036	1.000	1.000
	0.50	1.281	1.179	1.100	1.040	1.000
	1.00	1.560	1.333	1.143	1.033	1.000
	5.00	3.706	2.021	1.391	1.127	1.083
	10.00	6.190	2.373	1.468	1.146	1.049
	20.00	10.567	2.542	1.481	1.146	1.055
L_0 (yr)	0.00	76.45	77.85	78.64	81.07	86.87
L_e^b (yr)	0.10	76.42	77.83	78.62	81.06	86.87
	0.20	76.40	77.80	78.59	81.04	86.87
	0.50	76.32	77.72	78.52	81.01	86.87
	1.00	76.20	77.60	78.39	80.95	86.88
	5.00	75.22	76.60	77.39	80.46	86.88
	10.00	74.05	75.40	76.21	79.88	86.88
	20.00	71.88	73.20	74.01	78.76	86.86
$L_e^c - L_e^b$ (yr)	0.10	0.03	0.01	0.01	0.00	0.00
	0.20	0.05	0.03	0.02	0.01	0.00
	0.50	0.13	0.09	0.04	0.00	0.00
	1.00	0.25	0.17	0.08	0.00	0.01
	5.00	1.23	0.84	0.41	0.04	0.02
	10.00	2.40	1.63	0.77	0.08	0.03
	20.00	4.56	3.04	1.42	0.15	0.05

NOTE: Exposure rate assumed constant starting at birth.

[a]The 1980–1984 population average of smokers and nonsmokers (see text).

[b]Based on exposure sustained over lifetime.

[c]Based on exposure stopped at age known alive.

TABLE 2-13 Lifetime Risk of Lung-Cancer Mortality due to Lifetime Exposure to Radon Progeny

Study	Cancer Deaths per 10^6 Person WLM
BEIR IV (this report)	350
BEIR III[23]	730
UNSCEAR[33]	200–450
NCRP[20]	130

The estimated lifetime risk for this committee in Table 2-13 was calculated as follows. From the calculations used to prepare Tables 2-5 to 2-10, R_e, the lifetime risk for males at 0.1 WLM/yr, is 0.07087; the excess risk, $R_e - R_0$, is 0.07087 − 0.06734 = 0.00353. The mean cumulative lifetime exposure to a male population having an average life span of 69.7 yr at 0.1 WLM/yr is 6.97 WLM. Therefore, the excess lifetime lung-cancer mortality for males per WLM is 0.00353/6.97 = 5.06×10^{-4}cases per WLM of lifetime exposure. Similarly, for females at 0.1 WLM/yr, $R_e - R_0$ is 0.02663 − 0.02521 = 0.00142, and the average lifetime exposure is 7.64 WLM. The excess lifetime lung-cancer mortality for females per WLM is 0.00142/7.64 = 1.86×10^{-4}/WLM. Summing for a population of 500,000 males and 500,000 females yields $253 + 93 \simeq 350$ lung-cancer deaths per 10^6 person-WLM of lifetime exposure. The estimated lifetime risk attributed to the NCRP was calculated as follows, from Table 10.2 in NCRP Report 78[20]: "Lifetime lung cancer risk under environmental conditions per WLM per year." The lifetime risk for lifetime exposure from age 1 is 9.1×10^{-3}, assuming an average life span in 1976 of 70 yr, yields $9.1 \times 10^{-3}/70$ WLM = 1.3×10^{-4} cases per 1 WLM or 130 per 10^6 person-WLM of lifetime exposure.

the middle ages. For persons over the age of 60, the reduction in lung-cancer risk is minimal, unless exposures are quite high.

COMPARISON WITH RISK ESTIMATES MADE BY OTHERS

In Appendix VIII, the committee reviews risk estimates made by other scientific groups, including those made by the National

Research Council in 1980 (the so-called BEIR III report).[23] Comparisons between all studies are not possible because of large differences in the populations assumed to be at risk, for example, duration of exposure and smoking prevalence. Where there is some comparability, risk estimates for lifetime exposure are shown in Table 2-13. In calculating these results, we assumed equal numbers of males and females initially at risk. Other details and example calculations are given in the box following Table 2-13.

The committee's estimate is near the middle of range of risks listed in Table 2-13. It is about three times larger than that made by the National Council on Radiation Protection and Measurements (NCRP) in 1984[20] and about half of that made by the BEIR III committee in 1980.[23] While both of these groups assumed an additive model for the risk due to radon progeny, the BEIR III committee based its projections on an increasing excess risk with age, while the NCRP projection is based on a diminishing excess risk with time since exposure. This committee's estimate is based on a modified relative risk model that takes into account the reduced risk at age 65 or greater and the smaller effectiveness of exposures occurring 15 yr or more in the past that were identified in the miner cohort data (Annex 2A). Without these two modifying factors, the committee's estimate of the lifetime risk due to lifetime exposure would be about the same as that made by the BEIR III committee.

SUMMARY AND RECOMMENDATIONS

Radon and its daughter products are ubiquitous in indoor environments. Underground miners, exposed to radon daughters in a mine's air, have an increased risk of lung cancer that has been demonstrated in numerous populations. Animals exposed to radon daughters also develop lung cancer. Thus, the committee found abundant experimental and epidemiological data to support the fact that radon daughters are carcinogenic. However, the evidence was less conclusive on the quantitative risks of radon-daughter exposure. Therefore, the committee developed its own risk estimates based on analysis of four of the principal data sets related to lung-cancer occurrence in underground miners. The analysis indicated that the risk of lung-cancer mortality due to radon-daughter exposure was explained best by a model in which the excess relative risk is directly proportional to the cumulative exposure, but modified by the age at which the risk occurs and the length of time since exposure.

Recognition that radon and its progeny can accumulate to high concentration levels in homes has led to concern about the lung-cancer risk associated with domestic exposure. Those risks can be estimated with the committee's model based on occupational exposure, but several assumptions are required. For the purpose of risk estimation the committee assumed that the findings in the miners could be extended across the entire life span, that 1 WLM yields an equivalent dose to the respiratory tract in both occupational and environmental settings, that cigarette smoking and radon-daughter exposure interact multiplicatively, and that the sex-specific baseline risk of lung cancer is increased multiplicatively by radon daughters for males and females. The committee judged that its assumptions were supported by available evidence, although some alternatives are possible.

Tables 2-4 to 2-12 and Tables VII-12 to VII-23 in Appendix VII can be used to describe the risks associated with exposure patterns of current concern. For example, the current standard for radon-daughter exposure in underground mines limits the annual total to 4 WLM. From Table 2-6, the average lifetime risk of lung-cancer mortality for males of unspecified smoking status sustaining 4 WLM annually from age 20 through age 50 is 0.131. That is about twice the baseline risk for all males, 0.067. If the miners are smokers, lifetime risk in this exposure situation (from Table VII-13 in Appendix VII) is 0.226, compared with 0.123 for unexposed male smokers.

Exposure to radon progeny in indoor environments is also a present concern. Consider female smokers exposed at 1.0 WLM annually from age 20 through age 60. From Tables VII-19 in Appendix VII, the lifetime lung-cancer risk associated with that exposure is 0.087, about 1.5 times the risk for unexposed female smokers.

Definitive data on average magnitudes of radon exposures in indoor environments are not available. The U.S. Environmental Protection Agency is, however, planning a national survey that will take into account types of housing and other demographic variables on a regional basis. In its Report 78,[20] *Evaluation of Occupational and Environmental Exposures to Radon and Radon Daughters in the United States*, NCRP assumed an average environmental exposure of 0.2 WLM/yr, and the committee follows that example to estimate roughly the effects of environmental exposures at average ambient levels. On the basis of 1980–1984 U.S. mortality rates, the lung-cancer mortality risk for males of unspecified smoking status is 0.067. From Table 2-6, the lifetime lung-cancer mortality rate associated with lifetime exposure at 0.2 WLM/yr is 0.074—an increase of about 10%. That illustrates why the committee chose not to modify the baseline lung-cancer mortality rate that it used to allow for the fraction of lung-cancer deaths due to ambient radon. Such corrections are small, compared with other sources of uncertainty in the risk estimates.

The committee found substantial gaps in information related to many aspects of respiratory carcinogenesis by radon daughters. As described in Appendix VI, a unique link of radon-daughter exposure to small-cell cancer of the lung was not found. In the studies of miners, that histological type was in greatest excess, but other types were also increased. Review of the literature and the committee's own analyses in Appendix VII did not lead to a firm conclusion on the form of the interaction between radon daughters and cigarette smoking; the data were consistent with a submultiplicative interaction, as well as a multiplicative interaction. Simple additivity was not considered to be compatible with the data that the committee analyzed.

With regard to health outcomes other than respiratory malignancy, reviewed in Appendix V, the committee generally found only scant information. Studies of several populations of miners suggested that uranium mining reduced lung function, but the effect could not be attributed directly to exposure to radon daughters. Studies of miners and other populations exposed to radon daughters showed an increased occurrence of cytogenetic abnormalities of uncertain biological significance. The committee did not find the data on adverse reproductive outcomes associated with uranium mining to be cohesive or conclusive (see Appendix V).

The committee identified several subjects that need additional investigation with regard to respiratory carcinogenesis:

- Studies of underground miners exposed to radon progeny should be extended to cover the full lifetimes of the cohort members, and additional information on their smoking habits should be obtained.
- Analyses of results of current epidemiological studies and new studies should address the temporal expression of the lung-cancer risk associated with radon-daughter exposure.
- Additional information on the interaction between radon daughters and cigarette smoking with regard to the induction of lung cancer is needed. Both animal data and epidemiological investigations are relevant.

- The relationship between WLM and dose to the respiratory tract can differ in the occupational and environmental settings. The dosimetry of radon daughters in various settings should be examined further, to provide a better basis for describing the risk associated with environmental exposures.
- Only a few studies of lung cancer associated with environmental exposure (indoor radon) have been carried out. These studies have had small numbers of subjects and have based exposure assessment on surrogates or limited measurements. Additional studies of the risks associated with environmental radon-daughter exposures are needed, but they should address the comparative risks related to a given exposure in occupational and environmental settings, and not solely the question of whether environmental exposure causes lung cancer. Studies of environmental radon should be designed with sufficient statistical power to quantitate potential biologically significant differences in the risks of exposure in occupational and environmental settings.
- The lung-cancer histological types associated with radon-daughter exposure need further investigation.
- Many questions remain on health outcomes other than respiratory malignancy. The committee encourages investigation related to renal disease, nonmalignant respiratory disease, and reproductive outcomes.

REFERENCES

1. Breslow, N. E. 1985. Cohort analysis in epidemiology. Chapter 6 in A Celebration of Statistics: The ISI Centenary Volume, A. C. Atkinson and S. E. Fineberg, eds. New York: Springer-Verlag.
2. Cox, D. R. 1972. Regression models and life tables (with discussion). J. R. Stat. Soc. Ser. B 34:187–220.
3. Doll, R., and R. Peto. 1978. Cigarette-smoking and bronchial carcinoma: Dose and time relationships among regular smokers and lifelong non-smokers. J. Epidemiol. Commun. Health 32:303–313.
4. Doll, R., R. Gray, B. Hafner, and R. Peto. 1980. Mortality in relation to smoking: 22 year's observation in female British doctors. Br. Med. J. 280:967–971.
5. Ginevan, M. E., and W. A. Mills. 1986. Assessing the risks of Rn exposure: The influence of cigarette smoking. Health Phys. 51:163–174.
6. Harley, N. H., and B. S. Pasternack. 1981. A model for predicting lung cancer risks induced by environmental levels of radon daughters. Health Phys. 40:307–316.
7. Harting, F. H., and W. Hesse. 1879. Der Lungenkrebs, die Bergkrankheit in den Schneeberger Gruben. Vierteljahrsschr. f. Gerichtl. Med. u. Offentl. Gensundheitswesen 30:296–309; 31:102–132, 313-337.

8. Hornung, R. W., and T. J. Meinhardt. 1987. Quantitative risk assessment of lung cancer in U.S. uranium miners. Health Phys. 52P:417–430.
9. Hornung, R. W., and S. Samuels. 1981. Survivorship models for lung cancer mortality in uranium miners—is cumulative dose an appropriate measure of exposure? Pp. 363–368 in International Conference, Radiation Hazards in Mining: Control, Measurement, and Medical Aspects, M. Gomez, ed. New York: Society of Mining Engineers of the American Institute of Mining, Metallurgical, and Petroleum Engineers, Inc.
10. Howe, G. R., R. C. Nair, H. G. Hewcombe, A. B. Miller, and J. D. Abbatt. 1986. Lung cancer mortality (1950-1980) in relation to radon daughter exposure in a cohort of workers at the Eldorado Beaverlodge uranium mine. J. Natl. Cancer Inst. 77(2):357–362.
11. Lubin, J. H., W. J. Blot, F. Bervino, R. Flamant, C. R. Gillis, M. Kunze, D. Schmahl, and G. Visco. 1984. Patterns of lung cancer risk according to type of cigarette smoked. Int. J. Cancer 33:569–576.
12. Lundin, F. D., Jr., J. W. Lloyd, E. A. Smith, V. E. Archer, and D. A. Holaday. 1969. Mortality of uranium miners in relation to radiation exposure, hardrock mining and cigarette smoking—1950 through September 1967. Health Phys. 16:571–578.
13. Lundin, F. D., Jr., J. K. Wagoner, and V. E. Archer. 1971. Radon Daughter Exposure and Respiratory Cancer, Quantitative and Temporal Aspects. Joint Monograph No. 1. Washington, D.C.: U.S. Public Health Service.
14. Lundin, F. D., Jr., V. E. Archer, and J. K. Wagoner. 1979. An exposure-time-response model for lung cancer mortality in uranium miners—effects of radiation exposure, age, and cigarette smoking. Pp. 243–264 in Proceedings of the Work Group at the Second Conference of the Society for Industrial and Applied Mathematics, N. E. Breslow and A. S. Whittemore, eds. Philadelphia, Pa.: Society for Industrial and Applied Mathematics.
15. Moolgavkar, S. H., and A. G. Knudsen. 1981. Mutation and cancer: A model for human carcinogenesis. J. Natl. Cancer Inst. 66:1037–1052.
16. Moriyama, I. W., W. S. Baum, W. M. Haenszel, and B. F. Mattison. 1958. Inquiry into diagnostic evidence supporting medical certifications of death. Am. J. Public Health 48:1376–1387.
17. Muller, J., W. C. Wheeler, J. F. Gentleman, G. Suranyi, and R. Kusiak. 1985. Pp. 359–362 in The Ontario Miners Mortality Study, General Outline and Progress Report: Radiation Hazards in Mining, M. Gomez, ed. New York: Society of Mining Engineers of the American Institute of Mining, Metallurgical, and Petroleum Engineers.
18. Muller, J., W. C. Wheeler, J. F. Gentleman, G. Suranyi, and R. A. Kusiak. 1983. Study of Mortality of Ontario Miners, 1955–1977, Part I. Toronto: Ontario Ministry of Labour, Ontario Workers Compensation Board.
19. Muller, J., W. C. Wheeler, J. F. Gentlemen, G. Suranyi, and R. A. Kusiak. 1985. Study of mortality of Ontario miners. Pp. 335–343 in Occupational Radiation Safety in Mining, Proceedings of the International Conference, H. Stocker, ed. Toronto: Canadian Nuclear Association.
20. National Council on Radiation Protection and Measurements (NCRP). 1984. Evaluation of Occupational and Environmental Exposures to Radon and Radon Daughters in the United States. NCRP Report 78. Washington, D.C.: National Council on Radiation Protection and Measurements.

21. National Council on Radiation Protection and Measurements (NCRP). 1984. Exposures from the Uranium Series with Emphasis on Radon and Its Daughters. NCRP Report 77. Washington, D.C.: National Council on Radiation Protection and Measurements.
22. National Institutes of Health. 1985. Report of the NIH Ad Hoc Working Group to Develop Radioepidemiological Tables. Publication 85-2748. Washington, D.C.: U.S. Government Printing Office.
23. National Research Council, Committee on the Biological Effects of Ionizing Radiation (BEIR). 1980. The Effects on Populations of Exposure to Low Levels of Ionizing Radiation. Washington, D.C.: National Academy Press. 524 pp.
24. Nero, A. V. 1985. Indoor concentrations of radon-222 and its daughters: Sources, range, and environmental influences. Pp. 43–77 in Indoor Air and Human Health, Proceedings of the Seventh Life Sciences Symposium, R. B. Gammage and S. V. Kaye, eds. October 29-31, 1984, Knoxville, Tenn. Chelsea, Mich.: Lewis Publishers, Inc.
25. Nero, A. V., M. B. Schwehr, W. W. Nazaroff, and K. L. Revzan. 1986. Indoor environment program. Lawrence Berkeley Laboratory, University of California, Berkeley. Science 234(4779):992–997.
26. Nuclear Energy Agency, Group of Experts of the Organisation for Economic Co-operation and Development. 1983. Dosimetry Aspects of Exposure to Radon and Thoron Daughter Products. Paris: Nuclear Energy Agency, Organisation for Economic Co-operation and Devleopment.
27. Percy, C., E. Stanck, and L. Gloeclder. 1981. Accuracy of cancer death certificates and its effects on cancer mortality statistics. Am. J. Public Health 71:242–250.
28. Pierce, D. A., and D. L. Preston. 1985. Analysis of cancer mortality in the A-bomb survivor cohort. Pp. 11–13 in Proceedings of the 45th Session of the International Statistical Institute, Vol. 3, No. 3. Amsterdam: International Statistical Institute.
29. Radford, E. P., and K. G. St. Clair Renard. 1984. Lung cancer in Swedish iron miners exposed to low doses of radon daughters. N. Engl. J. Med. 310(23):1485–1494.
30. Thomas, D. C. 1981. General relative risk models for survival time in matched case control analysis. Biometrics 37:673–686.
31. Thomas, D. C., and K. G. McNeill. 1982. Risk estimates for the Health Effects of Alpha Radiation. Info-0081. Ottawa, Canada: Atomic Energy Control Board.
32. Thomas, D. C., K. G. McNeill, and C. Dougherty. 1985. Estimates of lifetime lung cancer risks resulting from Rn progeny exposures. Health Phys. 49:825–846.
33. United Nations Scientific Committee on the Effects of Atomic Radiation (UNSCEAR). 1977. Sources and Effects of Ionizing Radiation. Report E.77.IX.1. New York: United Nations. 725 pp.
34. U.S. Department of Health, Education, and Welfare, U.S. Public Health Service, Office of Smoking and Health. 1971. The Health Consequences of Smoking. A Report to the Surgeon General. DHEW (HSM) 71-7513. Washington, D.C.: U.S. Government Printing Office.
35. U.S. Department of Health and Human Services, U.S. Public Health Service, Office of Smoking and Health. 1982. The Health Consequences of

Smoking: Cancer. A Report to the Surgeon General. Washington, D.C.: U.S. Government Printing Office.

36. U.S. Department of Health and Human Services, U.S. Public Health Service, Office of Smoking and Health. 1985. The Health Consequences of Smoking: Cancer and Chronic Lung Disease in the Workplace. A Report to the Surgeon General, Department of Health and Human Services. PHS 85-50207. Washington, D.C.: U.S. Government Printing Office.
37. Waxweiler, R. J., R. J. Roscoe, V. E. Archer, M. J. Thun, J. K. Wagoner, and F. E. Lundin, Jr. 1981. Mortality follow-up through 1977 of the white underground uranium miners cohort examined by the United States Public Health Service. Pp. 823–830 in International Conference, Radiation Hazards in Mining: Control, Measurement, and Medical Aspects, M. Gomez, ed. New York: Society of Mining Engineers of the American Institute of Mining, Metallurgical, and Petroleum Engineers.
38. World Health Organization, International Agency for Research on Cancer. 1986. IARC Monographs on the Evaluation of the Carcinogenic Risk of Chemicals to Humans: Tobacco Smoking, Vol. 38. Lyon, France: International Agency for Research on Cancer.

ANNEX 2A
The Committee's Analyses of Four Cohorts of Miners

PART 1: Data Analysis

INTRODUCTION

The statistical methodology for epidemiological cohort analysis is evolving rapidly. Breslow[1] recently has prepared an excellent review of these developments. The new methodology is, to some extent, a generalization of the traditional standardized mortality ratio (SMR) methods which improves and formalizes examination of the dependence of the relative risk on exposure level, time since exposure, age at risk, age at exposure, gender, and other relevant risk factors. It also provides a unified approach for testing the validity of models and for estimating the value of parameters. For the purposes of interest here, a particular strength of the new methods is that they permit the analysis of cohort data with purely internal comparisons, rather than comparison with external population rates, as in traditional SMR methods. Heretofore, these new statistical methods have not been widely applied to the studies of cohorts of miners exposed to radon and its decay products. The committee applied these methods to the data for four cohorts of miners in order to analyze the data sets with a common approach and extend those analyses already published. Our analysis provides the basis for the committee's computed estimates of lung-cancer risk following exposure to radon and its progeny.

Lung cancer is caused by many different agents that may cause the disease on their own or through combined effects. The process of carcinogenesis is undoubtedly very complex, and any model to describe it will remain, at best, a rather rough approximation. Further, even with fairly simple models that incorporate some of the factors determining lung-cancer risk, the available data are not always strong enough to provide clear answers to some of the most important questions, such as the effect of age at exposure or time since exposure. We have carried out a combined analysis of data from four cohorts to estimate certain effects thought to be important, but not adequately estimated from any single cohort.

Analysis of data combined from several cohorts can lead to serious biases. The combined analysis described here was done carefully, with that possibility in mind, and only after fairly thorough analysis of each cohort separately. Allowance was made in the combined analysis for unexplained differences in the general level of apparent risk in the different studies, and effects were fitted in common to the studies only when they were strongly and similarly suggested in each. In general these effects were too poorly estimated in the individual studies to allow useful inferences. The final models selected are certain to be rough approximations that may ignore potentially important factors, which could not be analyzed with the available data.

Our analysis is done in terms of models that describe lung-cancer risk relative to age-specific background rates. The rationale for this choice is explained in some detail later; it does not amount to a priori acceptance of what is commonly meant by a relative-risk model in the literature on this subject, i.e., that the relative risk is constant in age, and perhaps in other factors such as gender, smoking habits, and locality. To the extent possible, we have avoided assumptions about the form of the relative risk, and determined from the cohort data those factors upon which it depends. We believe that viable models for the relative risk may be simpler than those for the absolute excess risk, because the latter has been found in other studies to increase substantially with age and/or time since exposure in a way that requires complex modeling. The relative risk, even if not constant in these factors, often depends on them in a less substantial way, and thus can be modeled more simply. Moreover, statistical methods for analysis of the relative risk are more fully developed than those for analysis of the absolute excess risk. When the relative risk and the absolute excess risk are allowed to depend in rather arbitrary ways on other appropriate factors, then there is no issue of which model is correct, because the two models are simply alternative expressions of the excess risk.

Our analysis places substantial emphasis on comparisons purely internal to the cohorts, in contrast to comparisons with external population rates. Because the latter method, based on calculation of SMRs, is conventional and indeed important for cohort analyses, special efforts were made to compare the two methods. Even though the cohorts were not designed to include control groups, it will be seen that remarkably useful inferences can be drawn from the internal comparisons. Reliance on internal comparisons has the advantage of avoiding potential biases due to differences, other than the exposure of interest, between the cohort and the comparison population. A major effort has been made in this report to lay out a proper statistical methodology for such internal comparisons. This is done through a unified approach that includes both types of analyses.

The following two sections discuss some general issues in the modeling of excess cancer risks and scientific issues of statistical methodology appropriate for such models. The results of such analyses for the cohort data available to us are given in a subsequent section.

MODELS FOR EXCESS CANCER RISK

It is not the committee's intention to present a single proper way for modeling excess cancer risk. The issues involved in the modeling of cancer risk are very difficult. Given the limited human data, models cannot be derived that can be shown to be unequivocally the most appropriate for describing all the variables that are clearly relevant. Thus, definition of a true model is not the aim here. Progress has been continually made, however, in obtaining increasingly clear descriptions of cohort experience. The discussion below lays out the essentials of our particular approach, so that the reader can more easily understand the basis of the results given in Chapter 2 of the committee's analyses of the radon-exposed cohorts.

The presentation here describes only statistical models, as opposed to more mechanistic ones such as multistage models. This distinction is not clear-cut; a multistage model can, in principle, be used for statistical fitting. What we mean by a *statistical model* is the minimal regression-type model necessary for proper data analysis.

BASIC MODELS FOR RISK

First, consider models that incorporate only effects of *exposure level* (dose),* *age at exposure, time since exposure,* and *age at risk,* ignoring other factors such as gender, smoking, and calendar time trends. Consider also, for the present, only exposures of fairly short duration, for example, less than 2 or 3 yr, so that dose and age at exposure might adequately characterize the exposure. In the models below we will use the symbols d, a, t, and e to denote the variables dose, age at risk, time since exposure, and age at exposure, respectively. Let $r(d,a,t,e)$ be the rate of lung-cancer mortality per person-year at risk as a function of these variables. In statistical terminology, r is essentially what is called a hazard function, but we will refer to it, rather imprecisely, simply as risk. For an individual, r is thus the age-specific risk, that is, the chance of dying of lung cancer in 1 yr, at age a, given that he is at risk (is still alive) at that age.

A useful model for the risk can be written as:

$$r(d, a, t, e) = r_0(a)[1 + f(d)\rho(a, t, e)], \qquad (2A\text{-}1)$$

where $r_0(a)$ denotes the age-specific background risk; $f(d)$ represents the effect of dose, where $f(0) = 0$; and $\rho(a,t,e)$ is the excess relative risk for unit increase of $f(d)$. The only assumption involved in such a model, assuming that these four explanatory variables are the only relevant ones, is that the excess relative risk factors into the product of the effect of d and that of $\rho(a,t,e)$. That is, if one had expressed the risk in the additive form:

$$r(d, a, t, e) = r_0(a) + f(d)\tau(a, t, e), \qquad (2A\text{-}2)$$

then it could be reexpressed by a model in the form of Equation 2A-1 by taking $\rho(a,t,e) = \tau(a,t,e)/r_0(a)$. Thus, at this level of generality,

---

*Our analysis is conducted using working-level month (WLM) as the unit of exposure. It is technically not a dose in either the strict radiological sense nor in the more general toxicological sense. To simplify presentation, we often use the term *dose* to represent the cumulative WLM over a defined period of exposure.

a model for the relative risk and a model for the additive excess risk are just two different ways of expressing the same relationship. The distinction between the two types of models only begins to arise when more restrictive forms are taken for the functions ρ and τ given above.

That the additive and relative excess risks factor into functions of d and of (a,t,e) is indeed an assumption. Although this assumption can, in principle, be checked by analysis of cohort data, modeling of excess risk becomes a morass of complexity without a tentative assumption of this nature. In particular, without it there would be no concept of a dose effect independent of the values of the other variables. Departures from this type of assumption can be useful, however, for example, by allowing the latent period of the excess risk to depend on dose. Factorization can be tentatively accepted at the outset, and specific alternatives to it can be considered as needed.

Models in which $f(d)$ is linear in d are of particular interest, and for clarity in this report, we focus on that case. While the linear approximation is useful in the analysis of cohort data, biological considerations suggest it should not be expected to hold for extremely high doses, and whether it holds for very low doses cannot really be determined from epidemiological data. Linearity does appear to be a reasonable approximation in the intermediate range of doses for the radon data considered by this committee. In this linear case, we will express Equation 2A-1 as:

$$r(d,a,t,e) = r_0(a)[1 + \beta(a,t,e)d], \tag{2A-3}$$

so that $\beta(a,t,e)$ is a dose-response slope, possibly depending on the values of a, t, and e.

In these terms, what is ordinarily referred to as a relative risk model in the radiation effects literature usually means Equation 2A-3, with $\beta(a,t,e)$ depending at most upon e, the age at exposure. The critical issue is that ordinarily β would not be taken to depend on age at risk (a) or time since exposure (t) so that a distinction between relative-risk models and additive (or absolute)-risk models becomes relevant. A model of the form given in Equation 2A-2 depending on how $\tau(a,t,e)$ was formulated, would ordinarily be different from that in Equation 2A-3, where β depended only upon e.

Absolute-risk models of the form given in Equation 2A-2, where τ does not depend on a or t, will not be discussed at length here, because it is now evident that the absolute excess risk increases with age at risk, at least fairly generally. This behavior is very clear, for example, in the Japanese atomic-bomb survivors for cancers other than leukemia.[17,18]

Although our models for the cohorts of radon-daughter-exposed miners are not taken directly from studies of the atomic-bomb survivors or other cohorts exposed to external low linear energy transfer (LET) radiation, for perspective, we review those findings briefly. Ignoring effects of gender and calendar time, the Radiation Effects Research Foundation (RERF) Life Span Study[17] shows that models roughly of the form:

$$r(d, a, t, e) = r_0(a)[1 + \beta(e)d] \qquad \text{(2A-4)}$$

fit the data remarkably well, with $\beta(e)$ substantially larger for those people exposed as children than as adults. Such models seem adequate for epithelial cancer quite generally, but not for leukemia, for the case of acute, low-LET external exposures. Darby et al.[5] have shown that the patterns of excess risk in the ankylosing spondilitics are consistent with those in the atomic-bomb survivors, and Land and Tokunaga[10] have shown that this type of model is appropriate for breast cancer. In the radioepidemiologic tables[14] this type of model was chosen, after very careful consideration, for describing the temporal patterns of radiation-induced cancers other than leukemia. Because the background rates of epithelial cancers generally increase roughly as a power of age, with an exponent of perhaps 5, the absolute excess risk increases very sharply with age as well.

On the other hand, Pierce and Preston[15] have shown that epithelial cancer in the atomic-bomb survivors also fits a model for the absolute excess risk of the form:

$$(d, a, t, e) = r_0(a) + \beta(e)\tau(t)d, \qquad \text{(2A-5)}$$

where $\tau(t)$ is a power (roughly 3) of time since exposure t, and $\beta(e)$ is a slope (different than that in Equation 2A-4), which is much greater for those exposed as adults than as children. In the atomic-bomb survivors, the relative validity of Equation 2A-4 and 2A-5 cannot be established purely on the grounds of adequacy of fit. The two models differ greatly, of course, in interpretation, with Equation 2A-4 meaning that age at risk is a primary determinant of excess risk, and Equation 2A-5 meaning that time since exposure is a primary determinant. Note that $\beta(e)$ in Equation 2A-4 decreases with age at exposure e and in Equation 2A-5 increases with e. That is, at a given time since exposure, the absolute excess risk increases with age at exposure, but not as fast as in a constant-relative-risk model that does not account for age at exposure. Although both models provide about the same fit to the atomic-bomb survivor data, the relative risk model in Equation 2A-4 is preferable for many purposes because it is simpler and does not require extensive analysis to estimate proper form for $\tau(t)$.

Models for Prolonged Exposures

Modeling is greatly complicated in moving from brief to prolonged exposures. Doses over a prolonged period must be considered in the above models by a summarization of exposure experience up to the age at risk considered. The simplest and most commonly used approach is to take the dose variable to be cumulative exposure up to the age at risk, except for a few years' lag. The lag allows for a minimal period between the exposure and the expression of the risk, as manifested by diagnosis or death. Such a summary exposure variable is almost certainly a very rough approximation, because very different temporal exposure histories can lead to the same cumulative dose.

For example, both the rate and duration of exposure may be important. These effects can be studied with a model of form given in Equation 2A-3, with d representing cumulative (lagged) dose up to age a and with β allowed to depend on dose rate or duration as well as the other variables indicated.

A more difficult issue to examine is the effect of time since exposure. The variable t in the models described thus far simply has no meaning for the case of a prolonged exposure. It must be replaced by some measure of the time that has elapsed since each part of the exposure. That is, a more complex quantity needs to be estimated, the effect of each part of the exposure in terms of how long ago it occurred.

Various workers have considered this issue: Lundin et al.,[11] Harley and Pasternack,[6] and in their Appendix M, Thomas and McNeill.[22] These researchers have suggested using either some form of probability distribution for the latent period associated with each increment of dose or a specified form of the decay with time of the effect of each increment. These methods involve assumptions about the mathematical form of the functions involved, and the assignment of numerical values to parameters in them. Although these approaches are useful in explaining the general features of the data, we used a more formal method of estimation of the functions and the parameters involved in our analyses of the cohort data.

The essence of the following approach, which fits in well with the modeling and data analysis used here, was suggested to us by Jan Muller and Robert Kusiak of the Ontario Ministry of Labour, Toronto, Ontario, Canada. Let $d_1, d_2, \ldots, d_k$ be the parts of the dose that are incurred in k fixed windows of time prior to age a, so that $d = d_1 + d_2 + \ldots + d_k$. Consider models of the form:

$$r(a, d_1, \ldots, d_k) = r_0(a)[1 + \gamma(a)(\beta_1 d_1 + \beta_2 d_2 + \ldots + \beta_k d_k)]. \quad (2A\text{-}6)$$

If $\beta_1 = \beta_2 = \ldots = \beta_k$, then this reduces to a model of the form in Equation 2A-3, using only cumulative dose to age a (lagged). The term $\gamma(a)$ above is an expression of the possible dependence of the risk on age, and plays a similar role to the dependence of β on a, t, and e in Equation 2A-3. If the β_i are unequal, then these coefficients describe the variation in effect for exposures of differing times prior to age a. It will be helpful at times to express this model as:

$$r(a, d_1, \ldots, d_k) = r_0(a)[1 + \gamma(a)\beta(d_1 + \theta_2 d_2 + \ldots + \theta d_k)] \quad (2A\text{-}7)$$

if d_1 is the dose in the most recent window, then the θ_i represents the weight that is given to doses in earlier windows, in relation to recent doses. The entire term $d_1 + \theta_2 d_2 + \ldots + \theta_k d_k$ can be thought of as an effective cumulative dose at age a.

Models Incorporating Other Risk Factors

Factors such as gender or smoking that are generally independent of dose history but that affect the risk or modify the effect of dose in some way must also be considered. In models for the relative risk, two methods are primarily used to introduce an effect of another factor; they are usually referred to as multiplicative and additive. These terms are not sufficiently descriptive, and the concepts can best be understood in terms of simple examples. Ignoring effects of age at exposure and time since exposure, suppose that one were considering a very simple model of the form: $r(d, a) = r_0(a)[1 + \beta d]$ and wanted to incorporate an effect of smoking, s. Although there are alternative expressions, the multiplicative type of model can take the form:

$$r(d,a,s) = r_0(a)[1+\theta(s)](1+\beta d), \tag{2A-8}$$

and the additive type of model would take the form

$$r(d,a,s) = r_0(a)[1+\theta(s)+\beta d], \tag{2A-9}$$

where θ is set to 0 for nonsmokers. In these models, $r_0(a)$ corresponds to the age-specific lung-cancer rates among nonsmokers who are not exposed to radon. For those not exposed, the ratio of disease rates for smokers to nonsmokers, $r(0,a,s)/r(0,a,0)$, is equal to $1 + \theta(s)$. Simultaneously, the relative risk at dose d among nonsmokers is $1 + \beta d$ for either of these models. Among smokers, however, the relative risk for dose is now dependent on whether Equation 2A-8 or 2A-9 is more appropriate.

The relationship between smoking and radon exposure is very complex, and the best description of the association could vary from supramultiplicative to subadditive, although analyses presented in Appendix VII clearly suggest that the relationship is not additive or subadditive. As discussed in Appendix VII, one way to consider such relationships is to form the following richer model which defines a smooth transformation between and beyond the additive and multiplicative models,[20] namely:

$$r(d,a,s) = r_0(a)\{[1+\theta(s)][1+\beta d]\}^{\lambda}[1+\theta(s)+\beta d]^{1-\lambda}, \tag{2A-10}$$

where $\lambda = 1$ corresponds to the multiplicative model and $\lambda = 0$ to the additive model. One can then obtain maximum likelihood estimates for all parameters, including λ (see Appendix VII).

The effects of gender can be considered in a similar way by replacing $\theta(s)$ by $\theta(g)$, with $r_0(a)$ corresponding to a given sex.

An example from the atomic-bomb survivor data illustrates the effects of gender and the distinction between the two models.[17,18] If models of the form given in Equation 2A-8 are fit to the data, then it is found that the parameter β depends significantly on gender. However, until recently it was not pointed out that the β values differ in essentially a ratio to cancel $1 + \theta(g)$; that is, the model represented by Equation 2A-9 fits very adequately without the dependence of β on gender. Thus, the additive excess due to radiation, which for a given age at exposure is reasonably proportional to $r_0(a)$, does not depend on gender.

In contrast to smoking, there are no data to assess the effects of radon exposure among females. As discussed in the section "Projecting Risks Associated with Radon-Daughter Exposure" in Chapter 2, simply applying the basic time-since-exposure model separately to male and female background mortality rates to obtain lifetime risk estimates is equivalent to assuming a multiplicative model for gender effects. As illustrated by the example given in that section, use of the model represented by Equation 2A-9 for estimating sex-specific lifetime risks can result in substantially different estimates of the lifetime risk for females.

METHODS OF STATISTICAL ANALYSIS FOR COHORT DATA

We have used statistical methods appropriate for analysis of the relative risk. Our approach was to fit and compare models to identify the factors that determine the relative risk, rather than simply to fit a predetermined type of constant-relative-risk model. These methods allow comparisons with known population rates external to the cohort as well as background rates estimated from within the cohort itself. The standard SMR methods are special cases of the former, so their use is covered as well. In fact, our methods for comparison to external rates amount to an appropriate regression analysis of stratified SMRs, calculated separately within ranges of the explanatory variables of interest. We give a moderately detailed description of the statistical calculations because it seems important for scientists other than statisticians to grasp what was done with the data.

BASIC APPROACH TO FITTING MODELS

Our approach is a version of what statisticians call a relative-risk regression, which seems to us the best currently available method for analysis of the data on radon exposures of miners. Previously, this methodology appears to have been used for the cohorts of subjects exposed to radon daughters only for the Colorado Plateau data, by Whittemore and McMillan[24,25] and Hornung and Meinhardt.[7] We believe that a thorough analysis of these cohort data required methods of essentially the same statistical nature as those described here. There are many variations on these approaches, for example, use of ungrouped data or more emphasis on analysis of the absolute excess risk. Other useful methods could place more explicit emphasis on certain random variables such as latent period; for an example, see Thomas.[21]

Our internal analysis is essentially a grouped-data version of the Cox regression method.[2] For large-scale cohorts, the analysis of grouped data seems preferable, as it enables easier exploratory analysis. Further, the rationale for our methods is more transparent for grouped data, in terms familiar from ordinary regression analysis. We believe that in view of the approximate nature of models and the errors inherent in estimation of exposure levels, approximation due to grouping is not an issue warranting any concern.

For simplicity, we consider first the case of the risk modeled only in terms of age, calendar period, and dose, where dose is the cumulative exposure as a function of age, with possibly some lag time. A simple example of the model can be written, following the notation of the previous section with the addition of the variable p for calendar period and the omission of t for time since exposure, as:

$$r(a,p,d) = r_0(a,p)[1 + \rho(a,d)]. \tag{2A-11}$$

The extension to situations in which the relative risk depends on additional factors will be given later. First consider the special case in which $\rho(a,d) = \beta(a)$d, where $\beta(a)$ denotes a slope that may depend upon age. This model is particularly important because it allows for departures from a model with relative risk constant in age. The dependence of β on age might be formulated simply by choosing a few intervals of age and letting $\beta(a)$ take on free values for each of these; alternatively, one might specify $\beta(a)$ as a smooth function of age, with parameters in the function to be estimated. In either case the model to be fitted is:

$$r(a,p,d) = r_0(a,p)[1 + \beta(a)d]. \tag{2A-12}$$

The data summary to which this model is fitted consists of a cross-classification, in cells defined by specified intervals of a, p, and d, of the numbers of deaths due to lung cancer, the numbers of person-years at risk, and the mean dose for the cell. The intervals of age and calendar period should be fairly narrow, such as 5 yr. The grouping on dose is less critical, for reasons discussed below, and about six to eight intervals will ordinarily be adequate.

This type of tabulation is standard in the classical SMR computations, but it is a rather sophisticated construct that needs to be well-understood. Over the follow-up period, each individual effectively traces a line of slope 1 in what is called the Lexis diagram, a two-dimensional diagram with the axes of calendar time and age. With increase in age, an individual passes through rectangular cells determined by the categories of time and age. As the individual moves along this line, he may also be accumulating exposure, which can be considered a third dimension of the table. Ordinarily, the individual's exact cumulative dose at some suitable point in the age-time cell should be calculated, rather than using grouped doses at the outset. This dose level then places the individual in a series of cells in the three-dimensional cross-classification. For each of these cells, then, we record the length of time the individual spent in the cell, whether the individual became a "case" there, and the actual dose for the chosen point in the cell. Finally, the data are accumulated over all individuals by adding up for each cell the times spent and the number of cases and by computing the mean dose for that cell. In computing that mean, each individual's contribution is weighted by the time he spends in that cell. If a person dies from another cause or becomes lost to follow-up for any reason, he simply accumulates no more time at risk. This type of calculation is standard in epidemiology, and computer programs are available for doing it.

Let $t(a,p,d)$ and $c(a,p,d)$ be the total time at risk and number of cases for each cell, and let $r'(a,p,d) = c(a,p,d)/t(a,p,d)$ be the raw estimate of risk for each cell for which the denominator is not 0. If the number of cells is large, these quotients will be poor estimators of risk because the observed time at risk will not be long enough to provide stable estimates; indeed, the numerators of these ratios will usually be small integers, often 0. The method basically smooths the raw risks by fitting a model to them, much as in ordinary regression analysis. As long as the model involves fitting only a relatively small number of parameters, the instability of the raw rates does not cause problems.

Estimation Using External Population Rates

The parameters $\beta(a)$ in Equation 2A-12 above are estimated by fitting the regression model

$$r'(a,p,d) = r_0(a,p)[1 + \beta(a)d] + \text{error} \qquad \text{(2A-13)}$$

by weighted least squares, using weights inversely proportional to the variance of the error term. Variance is given by $r(a,p,d)/t(a,p,d)$, where the numerator is the fitted value of the right-hand side. These calculations are quite standard, and can be done in any of several widely available statistical computing packages, such as SAS, BMDP, or GLIM.

The general strategy is to first fit a model without allowing $\beta(a)$ to vary with a and then allow β to depend upon a in some specified manner. Essentially the same approach is used when the risk is modeled in terms of more variables and with more complicated models for the relative risk. Other factors are added to the set (a,p,d) in carrying out the cross-tabulation of cases and person-years at risk. For example, in some of the analyses done for this report, the cross-tabulation was by age, calendar period, dose category, age at first exposure, average exposure rate (up to age a), and time since cessation of exposure. When several such factors are included, the number of cells in the complete cross-classification becomes large, but most of the cells will have 0 person-yr at risk and can be omitted.

The primary advantage of the relative-risk regression methods is that they can be used to examine systematically the dependence of the relative risk on factors of interest. In the regression analysis, the excess relative risk $\beta(a)d$ in Equation 2A-13 can be replaced by various expressions that depend on these factors, so that their effect on the relative risk can be tested. For example, it is straightforward to examine whether significant nonlinearity in dose occurs by adding terms such as dose squared to the models. In modeling effects of factors unrelated to exposure, such as gender and smoking, the technical aspects of the statistical methods are the same. These factors are just added into the cross-classification of cases and person-years. Any given model for the risk can be fitted by the methods described above, but the choice of models is substantially more difficult as variables are added.

This method is based on a natural probability model for the data: the number of cases in a given cell, $c(a,p,d)$, is a Poisson random variable with a mean given by $t(a,p,d)\ r(a,p,d)$. The weights for the regression follow from this Poisson model. This method leads to maximum likelihood estimates of the parameters in the model. Commonly held concerns in a regression analysis of enumerative data, such as the treatment of cells with zero frequencies or those with very small expected values, are not relevant with this method.

The Poisson probability model is important in significance tests to compare models. Maximum likelihood estimation entails finding the values of unknown parameters in a model that maximize the probability of having obtained the data observed. This model and the maximum likelihood calculations provide a proper means of judging whether one model fits the data significantly better than another. This is carried out by likelihood ratio tests.[26] As an example in one of our analyses of the Eldorado data, described below, we used the model defined by Equation 2A-12. If β is not allowed to depend on a, the estimate of β is 0.026; when β is allowed to take on different values for a in intervals <55, 55–64, and 65+, the estimates are $\beta = 0.025$, 0.036, and 0.017 for these intervals, respectively (see Table 2A-3). The likelihood (that is, the maximum probability of observing the given data) is, of course, greater in this last case because there are more free parameters in the model. In this case the likelihood increased by a factor of 2.1. The likelihood ratio test for determining whether this is a significantly better fit is given by a statistic, distributed as chi-squared, which is twice the natural logarithm of the ratio of the likelihoods, i.e., $2 \times \ln(2.1) = 1.5$, and the degrees of freedom (d.f.) for this statistic is the difference in the number of parameters estimated by the two models, i.e., $3 - 1 = 2$. The improvement in fit is significantly better only if this statistic is large; so in this example there is no evidence for a dependence of β on a, since the P value for a chance occurrence being as large as that observed, 1.5, is about 50%.

The procedure can be used for comparing the fit of any two models, provided that the simpler one merely restricts the parameters of the richer one, for example, restricting the three β values to be equal. The analysis based on the Poisson model also provides approximate standard errors for the parameter estimates. These are useful for certain purposes, but the likelihood ratio method of significance testing should be used to select between models because it is less sensitive to the approximations involved.

The statistical approach described above is closely related to standard SMR calculations. A commonly used manner of calculating the dose response is to compute an SMR for each dose category, and then to regress the SMRs in a suitable way on the mean dose levels for each category. The dose-specific SMRs are computed from the data summary described above as $c(d)/b(d)$, the ratio of observed to expected cases for each dose category, where $c(d)$ and $b(d)$ are the sums over a and p of $c(a,p,d)$ and the product $r_0(a,p)\ t(a,p,d)$, respectively. The slope β obtained by suitably regressing these SMRs on (weighted) mean doses is the same as would be obtained by our relative-risk regression method for the case that $\beta(a)$ in Equation 2A-13 is taken independent of a. Thus, the SMR method is formally equivalent to a maximum likelihood fitting for the constant-relative-risk model. Furthermore, the type of significance testing described above can be used to decide whether the constant-relative-risk-model is adequate. We note that the SMR calculations model can be thought of as estimating a weighted average of $\beta(a)$ with respect to a, even when β is not constant.

It is often unwise to assume uncritically that the cohort under study differs from the external population chosen for comparison only with regard to radiation exposure. For example, the well-known healthy worker effect is often evident in occupational cohorts; that is, disease incidence in a working cohort is lower than that in the general population simply because cohort members are healthy enough to be working. On the other hand, a cohort of miners may experience higher risks than a comparison population because of factors not analyzed. Differences due to locality and socioeconomic factors are also possible, especially when the external comparison is to national rates. A simple and effective adjustment to allow for such problems is to estimate from the cohort an additional parameter that multiplies all the $r_0(a,p)$ values. This has precisely the same effect as the commonly used device of estimating an intercept term in the regression of dose-specific SMRs on dose, or equivalently, replacing the 1 in the model represented by Equation 2A-12 by an intercept parameter to be estimated.

Estimation Without Use of External Rates

We can go further and carry out analyses with purely internal comparisons within the cohort, as opposed to comparison to external population rates. This approach simply treats each of the quantities $r_0(a,p)$ as free parameters to be estimated in the fitting process. The calculations can, at least in principle, be done by the same iterative weighted least-squares algorithm appropriate for the regression model represented in Equation 2A-13, except that many more parameters need to be estimated. One may be concerned that fitting so many parameters will prevent adequate estimation of the factors involved in the relative-risk function, which are of greater interest. In fact, it is often the case that this issue does not cause serious problems. The estimates of parameters in the relative-risk term are nearly as precise, in many situations, for the internal analysis as for the external analysis, provided that a one-parameter adjustment is made to the external rates, as suggested above. This conclusion is supported by theoretical investigations summarized by Kalbfleisch and Prentice[9] and Cox and Oakes,[3] and is demonstrated practically by the analyses described later in this annex. This very favorable comparability of the internally based estimates relative to the externally based ones may degrade somewhat as the model for the relative risk becomes more complex.

Estimating the $r_0(a,p)$ parameter as described here is equivalent to the more sophisticated rationale that underlies the Cox regression methodology, in which it is not necessary to think of the internal analysis as actually estimating the background risk. More precisely, the inferences regarding the parameters in the model for the relative risk are exactly the same whether one formally estimates the $r_0(a,p)$ parameters or carries out an analysis conditional on the total number of cases in each of these strata. This latter rationale provides the more sound theoretical basis for the method.

Some rather special computational methods have been used for the internal analyses. Numerical methods quite suitable for the external analysis, for example, those which would be used in a program such as SAS, are not appropriate when a very large number of parameters $r_0(a,p)$ need to be estimated as well. We used a method in which the parameters in the relative risk part of the model were held fixed at some trial value and the parameters $r_0(a,p)$ were estimated simply, with no numerical search. One can alternate in an iterative algorithm between (1) updating the parameters in the relative risk while treating the $r_0(a,p)$ parameter as fixed and using standard iterative methods, and (2) updating the $r_0(a,p)$ parameter while treating the parameters in the relative risk part as fixed. A convenient interactive program called AMFIT, developed by Pierce and Preston,[15] has been used for the calculations in this report. This program is also useful for the analysis based on external comparisons.

THE COMMITTEE'S ANALYSIS OF MINER DATA

The approach outlined above was used to carry out separate but parallel analyses of four cohorts of exposed miners. The aim in this was to examine the dependence of the relative risk on several factors: age, time since exposure, age at first exposure, and exposure rate (expressed as duration of exposure for a given cumulative exposure). In particular, analyses were carried out to examine the adequacy of models where the relative risk is taken to be constant for a given cumulative exposure. Comparisons to external population rates and comparisons purely within the cohorts were done in parallel. After determining from these analyses which variables seem to affect the relative risk, a more formal analysis of the combined cohort data was carried out. The first aim here was to obtain better estimates of the effects which seemed important in the separate analyses, including better assessment of whether these effects were common over the cohorts. In these investigations, separate general levels of risk in the four cohorts were allowed, in order to avoid confounding of these differences with the effects under investigation.

Finally, after arriving at a model to describe the patterns of relative risk, attention was turned to combining over cohorts the estimation of the general levels of risk. A single model, more complex than ordinarily used in the past, but as simple as we felt possible, was arrived at to describe the level and patterns of risk in these cohort studies.

RESULTS

GENERAL ISSUES AND SUMMARY OF DATA

This section contains a moderately detailed summary of the analysis of the four miner cohorts as well as a justification for the choice of the final model given in Chapter 2. The protocols for defining data from these studies were taken essentially as described in the reports by Howe et al.,[8] Muller et al.,[12,13] Radford and Renard,[19] and Hornung and Meinhardt.[7] Data on individual miners were made available to us for the first three of these cohorts. For the Colorado cohort they were not, but special detailed cross-tabulations were made available to us by the National Institute of Occupational Safety and Health. Descriptions of the cohorts are given in Appendix IV. For convenience, the term *dose* is sometimes used here for WLM.

Person-years at risk are used differently here than in the SMR calculations by some of the original investigators, and the values used are correspondingly different. This is largely because in SMR calculations, which are implicitly based on constant-relative-risk models, time at risk is often excluded during periods when the excess risk is known a priori to be small, such as during the first few years after exposure begins. The methods used here do not assume constant relative risk and, in particular, model the excess risk as 0 during such periods. Further, exclusions of time at risk to deal with the healthy worker effect are avoided here by relying on internal cohort comparisons and estimation of a background SMR for each cohort. Thus, all time at risk is included in the values reported below.

The person-years at risk were computed from the beginning of the follow-up period of each miner. For the Ontario and Colorado miners, we have taken this to be the date of the first annual physical during the study period. Cumulative exposures were computed with a 5-yr lag of current age, to allow for a minimal period when the exposures would not be expected to result in a substantial risk of lung-cancer mortality. It is not possible to estimate precisely the duration of the latent period, and this value is somewhat arbitrary. However, if the actual latent period is longer, its effect will be accounted for in the final model.

For the Malmberget cohort, Radford and Renard[19] used special methods for computing final cumulative exposures for the lung-cancer cases. Since the analysis here required computing exposures at each age, it was necessary to use the same method for cases as for noncases. For the Ontario cohort, Muller et al.[12,13] have computed two sets of exposure estimates, called standard and special, which are discussed in Appendix IV. Because the committee believes that the method used to obtain the special WLM provided only an upper bound on exposure, the analyses here are in terms of the standard WLM. Our analyses using the special WLM exhibited essentially the same patterns of risk but at an overall level that was about half as large. For the Ontario data, we have included those workers referred to by Muller et al.[13] as having no prior gold mine experience. It is known that some of the Ontario miners included in the zero-exposure group had some exposures in aboveground work, but ignoring these exposures has no substantial effect on the results here. For the external analyses, the comparison populations for the cohorts other than the Colorado cohort were taken as given by the original investigators: all Canadian males (Eldorado), all Ontario males (Ontario), all Swedish males (Malmberget). For Colorado, these were taken as all U.S. white males. In each case some adjustment, as described later, was made to allow for possibly different background rates between the study cohorts and the comparison population.

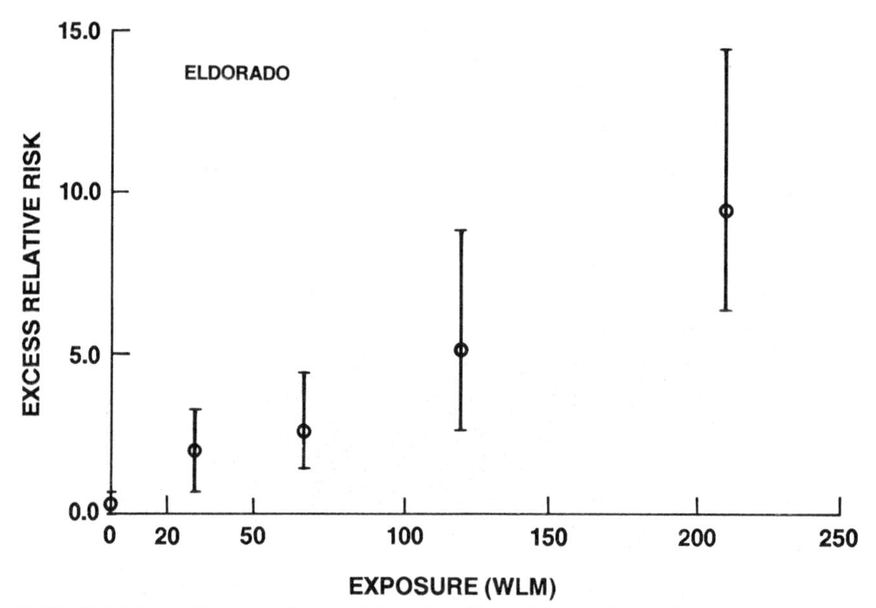

FIGURE 2A-1 Excess relative risk and 67% confidence limits by exposure category as observed in the Eldorado uranium miners at Beaverlodge, Saskatchewan, Canada. The comparison is to external population rates.

Figures 2A-1 through 2A-4 show the excess relative risk in exposure categories for each cohort. These results were computed by ordinary SMR methods, comparing the cohort experience to the external population rates without the adjustment of the external rates just mentioned. Such an adjustment, in the context of these figures, would amount only to estimation of a possibly nonzero intercept at zero exposure in fitting some smooth model to these points. Table 2A-1 gives the numerical data corresponding to Figures 2A-1 through 2A-4. With the exception of the Colorado cohort, where the range of exposures is very much greater, there is no suggestion of nonlinearity in dose, either in the figure or in formal tests in any of the analysis to follow. As will be discussed later, for most of the analysis here it was decided to use only the Colorado cohort experience for cumulative exposures of 2,000 WLM or less.

For each of the cohorts, the follow-up periods, total number of lung-cancer deaths, and person-years at risk are as follows:

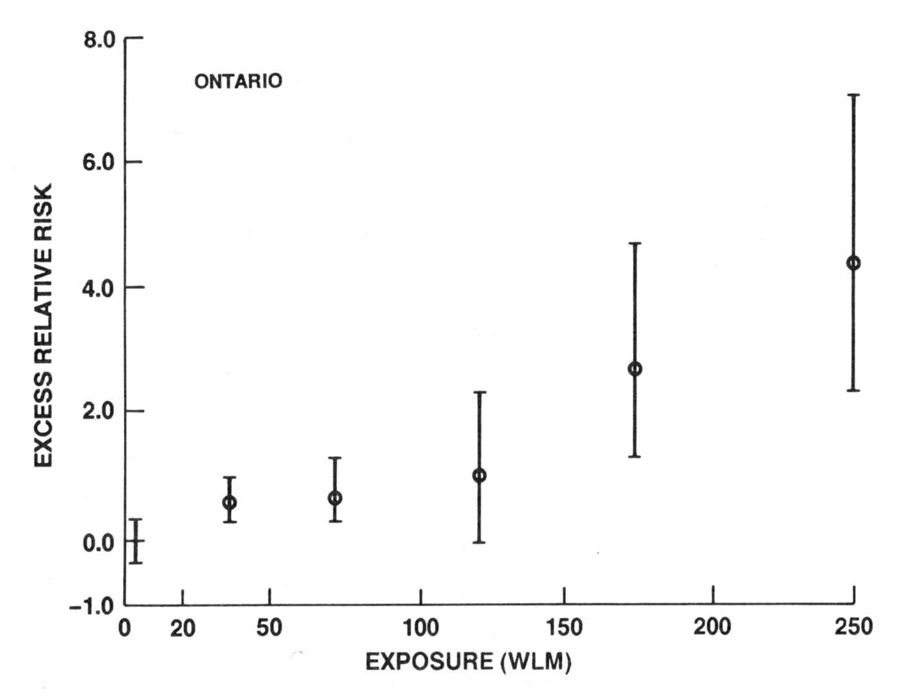

FIGURE 2A-2 Excess relative risk and 67% confidence limits by exposure category as observed in the uranium miners in Ontario, Canada. The comparison is to external population rates.

Eldorado (January 1, 1950, to December 31, 1980)
  Lung-cancer deaths: 65
  Person-years at risk: 114,170
Ontario (January 1, 1955, to December 31, 1981)
  Lung-cancer deaths: 87
  Person-years at risk: 217,810
Malmberget (January 1, 1951, to December 31, 1976)
  Lung-cancer deaths: 51
  Person-years at risk: 27,397
Colorado (January 1, 1951, to December 31, 1982)
  Total
    Lung-cancer deaths: 256
    Person-years at risk: 73,642
  Under 2000 WLM
    Lung-cancer deaths: 157
    Person-years at risk: 66,237

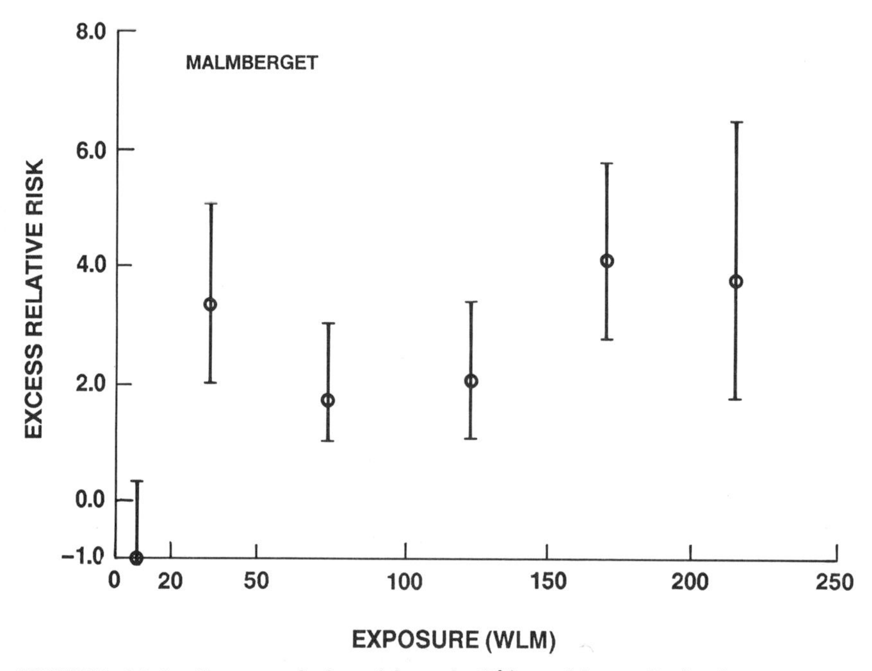

FIGURE 2A-3 Excess relative risk and 67% confidence limits by exposure category as observed in the iron ore miners in Malmberget, Sweden. The comparison is to external population rates.

Tables 2A-6 through 2A-9 at the end of Part 1 of this annex provide a rather detailed description of the data for each of these cohorts, with several cross-classifications of exposure and other factors of interest. Although this amount of detail will ordinarily be of interest only after seeing the results of our analysis, it will be convenient to refer to these tables occasionally in what follows.

As described in the statistical methods section, the analysis was done in terms of grouped data that define the cells for which dose, cases, person-years, mean doses, for example, are tabulated. The age groups were as follow: under 30 yr, 5-yr intervals from 30 to 79 yr, and over 80 yr. The calendar periods were approximately 5-yr intervals. The dose groups were those shown in Tables 2A-6 through 2A-9. Because means of ungrouped doses for each cell were used, we think this grouping has had very little effect. Age at first exposure and duration of exposure were categorized as indicated in Tables 2A-2 and 2A-3.

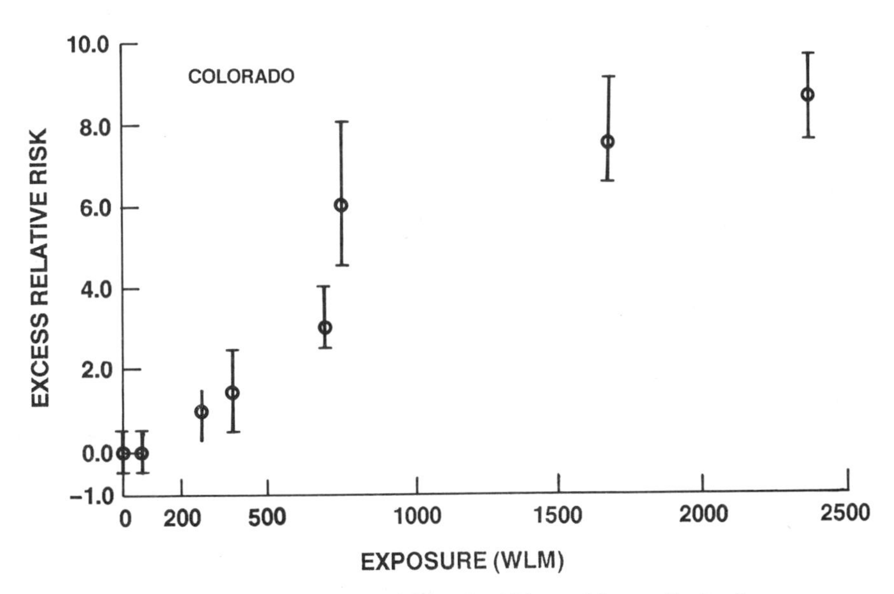

FIGURE 2A-4 Excess relative risk and 67% confidence limits by exposure category as observed in the uranium miners on the Colorado Plateau. The comparison is to external population rates.

SEPARATE ANALYSES OF THE COHORTS

Initially, separate analyses were done for each cohort. The aim was to determine how the relative risk depends on dose history and other factors, by methods explained in the previous sections. It was found that the data from individual cohorts were only strong enough to investigate the effect of factors, in addition to dose, one at a time. After finding in this way those factors that seem to have consistent and substantial effects, we will turn to investigating their joint effects by analyzing the combined data from the cohorts.

Part of this analysis was done in terms of models of the form:

$$r(\text{age, period, dose, other factors}) = r_0(\text{age, period})(1+\beta d), \quad \text{(2A-14)}$$

where β is in turn taken as constant, and then is allowed to depend on intervals of age and other factors. Here d means cumulative exposure in WLM at a time 5 yr prior to the age at which risk is being estimated. The other factors considered, in addition to age, were (1) age at first exposure; (2) duration of exposure (essentially adjusted for cumulative exposure because of the d in the model); and

TABLE 2A-1 Summary Data for the Four Miner Cohorts

Exposure Category	Mean Exposure	No. of Cases	Expected No. of Cases	Person-Years at Risk	Excess Relative Risk	67% Confidence Limits[a]	
Eldorado							
0–20	1	33	21.36	97,959	0.54	0.23	0.88
20–50	32	9	3.31	9,243	1.72	0.84	2.96
50–100	66	8	2.27	4,127	2.53	1.31	4.26
100–150	120	5	0.85	1,479	4.87	2.35	8.83
>150	208	10	0.94	1,351	9.69	6.38	14.22
Ontario							
0–20	5	34	31.49	149,387	0.08	−0.10	0.30
20–50	35	21	12.59	38,574	0.67	0.31	1.11
50–100	71	13	7.14	17,571	0.82	0.32	1.48
100–150	123	5	2.47	6,068	1.03	0.15	2.39
150–200	174	6	1.67	3,842	2.60	1.18	4.74
>200	250	8	1.50	2,364	4.32	2.49	6.94

Malmberget							
0–20	8	0	1.26	5,559	−1.00	—	0.46
20–50	34	11	2.61	6,132	3.22	1.97	4.90
50–100	75	11	3.98	7,105	1.76	0.95	2.86
100–150	123	8	2.77	4,594	1.89	0.90	3.31
150–200	171	15	2.94	3,067	4.10	2.81	5.78
>200	217	6	1.31	937	3.57	1.77	6.29
Colorado							
0–60	14	9	9.00	18,048	0.00	−0.33	0.46
60–120	86	5	4.85	6,342	0.03	−0.41	0.73
120–360	254	32	14.31	18,711	1.24	0.84	1.70
360–480	374	9	3.78	3,545	1.38	0.61	2.46
480–720	674	34	8.26	8,450	3.12	2.42	3.95
720–960	756	23	3.28	4,027	6.02	4.58	7.80
960–2,000	1,682	45	5.17	7,111	7.70	6.41	9.19
>2,000	2,411	99	10.46	7,405	8.46	7.52	9.51

[a]The 67% confidence limits for excess relative risk are an analog of ±1 standard error, but are computed by more exact methods using the Poisson distribution.

(3) in preliminary analyses not reported here, time since cessation of exposure. Cumulative dose up to the age at which risk is being evaluated is a very crude summary of dose history for the case of prolonged exposures. The other model used for this part of the analysis, in order to more adequately assess the effects of prolonged exposures, was of the form:

$$r(\text{age}, \text{period}, d_1, d_2, d_3) = r_0(\text{age}, \text{period})[1 + (\beta_1 d_1 + \beta_2 d_2 + \beta_3 d_3)] \qquad (2A\text{-}15)$$

where d_1, d_2, and d_3 are the WLM exposures incurred in respective periods of 5–10, 10–15, and 15 or more* years prior to the age at which risk is being estimated. As discussed previously, this was an attempt to investigate an effect of time since exposure which was suggested in preliminary analyses by an apparent decrease in risk with time since cessation of exposure. Note that in a model of this form, the assumption of a minimal latent period of 5 yr is not critical since a possibly longer latent period will be reflected by a relatively smaller value of β_1.

Tables 2A-2 and 2A-3 summarize the main results from analysis with these types of models. For the Colorado cohort only experience where exposures are less than 2,000 WLM have been used; the rationale for this is fairly apparent from the lack of linearity seen in Figure 2A-4, and will be discussed in more detail below. Table 2A-2 is based on comparisons internal to the cohorts, and Table 2A-3 is based on comparisons to external population rates. That is, in Table 2A-2 the functions r_0(age, period) are estimated from the data. In Table 2A-3, r_0(age, period) is taken as the external population rates multiplied by a single factor for each cohort, which was estimated as a parameter in fitting the model. These factors are called background SMRs and are given in Table 2A-3 for the fit to the model where β is not allowed to depend on other factors. The background SMRs change very little in fitting the other models given in Table 2A-3. The external SMR for the Malmberget cohort has been fixed, however, at a value somewhat different than estimated from the data. This does not have a substantial effect on the general conclusions under discussion; the reason for doing this will be presented after discussion of the results in these tables.

---

*5–10 means from the beginning of the 5th year until the end of the 9th year, etc.

TABLE 2A-2 Internal Analysis of Relative Risk as a Function of Age at Risk, Age at Start of Exposure, Duration, and Time Since Exposure

β[a] Depending on:	Increase in Relative Risk/100 WLM[a]			
	Eldorado	Ontario	Malmberget	Colorado[b]
Constant	2.6 (1.5)[c]	1.4 (1.6)	1.4 (2.6)	0.6 (1.5)
Age (yr)				
<55	2.0 (2.0)	2.1 (1.8)	3.0 (13.0)	0.8 (2.1)
55-64	3.9 (1.7)	0.5 (2.5)	1.3 (3.5)	1.2 (2.4)
65+	1.6 (2.4)	11.6 (12.0)	1.0 (5.8)	0.1 (2.7)
Chi-squared[d]	1.0; $P > 0.5$	2.4; $P = 0.30$	0.2; $P > 0.5$	6.0; $P = 0.05$
Age (yr) at start				
<30	0.3 (7.4)	1.5 (2.1)	1.4 (2.4)	0.4 (1.7)
30+	3.7 (1.5)	1.2 (1.6)	0.2 (75)	0.9 (1.5)
Chi-squared	4.0; $P = 0.05$	0.1; $P > 0.5$	2.1; $P = 0.15$	4.9; $P = 0.03$
Duration (yr)[e]				
<2.5	0.9 (2.1)	1.5 (2.1)	1.4 (2.4)	0.1 (2.5)
2.5-5	0.4 (30)	1.7 (1.9)	1.7 (2.4)	0.3 (1.5)
>5	3.2 (1.4)	0.6 (3.1)	1.3 (2.5)	0.7 (1.4)
Chi-squared	5.0; $P = 0.08$	0.2; $P > 0.5$	2.9; $P = 0.23$	20; $P < 0.001$
Time (yr) since exposure[f]				
5-10	17.7 (1.5)	1.1 (5.2)	10.0 (2.4)	0.5 (1.6)
10-15	1.4 (6.0)	3.6 (1.6)	0.0 (—)	0.5 (1.6)
>15	1.1 (2.0)	0.3 (6.6)	1.0 (4.0)	0.6 (1.4)
Chi-squared	11; $P = 0.004$	5.3; $P = 0.07$	2.1; $P = 0.35$	0.1; $P > 0.5$

[a] β in Equation 2A-14.
[b] Cumulative exposures >2,000 WLM are excluded.
[c] Numbers in parentheses are multiplicative standard errors.
[d] Significance of the dependence of the relative risk on the factor in comparison with a constant relative risk model.
[e] Durations for Malmberget are <20, 20-29, and >30 yr.
[f] Coefficients to multiply the three components of exposure in Equation 2A-15.

Except for the rows labeled "time since exposure," the numbers given in Tables 2A-2 and 2A-3 are estimates of β in the model represented by Equation 2A-14, that is, age-specific excess relative risk per 100 WLM (lagged by 5 yr) for the specified levels of the variables indicated. For the case where the relative risk is not taken to depend on other factors, labeled "constant" in the tables, the estimates in Table 2A-3 for the external comparisons are essentially those that arise from conventional SMR calculations. That is, if SMRs were computed by dose category and appropriately regressed on the mean dose for each category, then the estimated slope is the

TABLE 2A-3 External Analysis of Relative Risk as a Function of Age at Risk, Age at Start of Exposure, Duration, and Time Since Exposure

β[a] Depending on:	Increase in Relative Risk/100 WLM[a]			
	Eldorado	Ontario	Malmberget	Colorado[b]
Constant	2.6 (1.4)[c]	1.2 (1.5)	1.6 (2.5)	0.6 (1.5)
Age (yr)				
<55	2.5 (1.7)	1.6 (1.6)	2.5 (3.2)	0.7 (1.4)
55-65	3.6 (1.5)	1.2 (1.7)	2.4 (2.2)	0.6 (1.5)
<65	1.7 (2.0)	0.0 (—)	1.1 (3.0)	0.3 (1.6)
Chi-squared[d]	1.5; $P = 0,47$	2.8; $P = 0.25$	3.2; $P = 0.20$	6.6; $P = 0.04$
Age (yr) at start				
<30	0.7 (4.0)	1.4 (2.9)	1.7 (2.4)	0.6 (1.5)
>30	3.2 (1.4)	1.2 (1.6)	1.3 (6.0)	0.6 (1.5)
Chi-squared	2.9; $P = 0.09$	0.0; $P > 0.5$	0.2; $P > 0.5$	0.0; $P > 0.5$
Duration (yr)[e]				
<2.5	1.0 (2.0)	1.4 (5.0)	1.2 (—)	0.1 (2.5)
2.5-5	0.5 (12.0)	1.6 (1.9)	2.3 (2.6)	0.3 (1.5)
>5	3.2 (1.4)	1.2 (1.6)	1.3 (3.4)	0.7 (1.4)
Chi-squared	4.3; $P = 0.12$	0.2; $P > 0.5$	1.7; $P = 0.43$	21.0; $P < 0.001$
Time (yr) since exposure				
5–10	11.2 (1.5)	0.7 (5.9)	8.6 (2.3)	0.7 (1.5)
10–15	1.2 (3.8)	3.5 (1.5)	0.7 (—)	0.8 (1.5)
>15	1.7 (1.8)	0.2 (6.0)	1.0 (3.7)	0.5 (1.4)
Chi-squared	7.1; $P = 0.03$	8.2; $P = 0.02$	3.8; $P = 0.15$	1.6; $P = 0.45$
Background SMR	1.4 (1.2)	1.1 (1.2)	1.25 (fixed)	0.8 (1.3)

[a] β in Equation 2A-14.
[b] Cumulative exposures >2,000 WLM are excluded.
[c] Numbers in parentheses are multiplicative standard errors.
[d] Significance of the dependence of the relative risk on the factor in comparison with a constant relative risk model.
[e] Durations for Malmberget are <20, 20–29, and >30 yr.

β for a constant relative excess risk in Table 2A-3. The estimates of the time-since-exposure effect at the bottom of Tables 2A-2 and 2A-3 are the parameters β_1, β_2, β_3 of the model represented by Equation 2A-15.

The numbers in parentheses are multiplicative standard errors, which arose in the following way. These parameter estimates are more nearly normally distributed on a logarithmic scale. A standard error of the ordinary plus or minus type for estimates on a log scale becomes a (multiplicative) factor given by its antilogarithm upon changing back to the original scale. For example, the first entry in Table 2A-2 was calculated on a log scale as 0.956 ± 0.400, corresponding to a risk of exp(0.956 ± 0.400) = 2.6 (×/÷ 1.5). This is close in meaning to saying that the coefficient of variation of this estimate is 50%, but with the special interpretation that, for example, a 95% confidence interval corresponding to ±2 standard errors on the log scale becomes [2.6 ×/÷ $(1.5)^2$]. These standard errors represent only the uncertainty of the estimates due to sampling variation, not that, for example, due to any errors in dosimetry ascertainment of deaths.

Another primary feature of the tables are chi-squared statistics and *P* values for the statistical significance of the dependence of relative risk on each of the factors investigated. As outined above, these statistics are based on likelihood ratio tests, and are preferable on several grounds to assessing the significance in terms of the variation in risk estimates and their standard errors. For example, in Table 2A-2 for Ontario and variation in relative risk with age, there would be approximately a 30% chance of obtaining this much apparent variation in risk if in fact it were truly constant in age, due simply to inherent random variation in estimation. The chi-squared value from which this *P* value is computed is a measure of the improvement of fit in moving from the constant-relative-risk model at the top of Table 2A-2 to the one where the relative risk is allowed to depend upon age as indicated. The risks specified to levels of factors studied are clearly not very well-estimated, and it is important to be attentive to these significance levels.

Turning to the general results indicated in Tables 2A-2 and 2A-3, the internal analyses over cohorts yield no clear evidence for dependence of the relative risk on age. In contrast, the external analyses show a general pattern of a decrease in excess relative risk with age, which would appear to be significant. This is followed up in the subsequent analysis of pooled data described below.

In none of the analyses does there appear to be a consistent or significant effect of age at first exposure. Consideration of this is important since there does appear to be an effect of age at exposure among the Japanese atomic-bomb survivors, particularly for those exposed before age 20,[17,18] and it is rather widely held that such an effect should be expected more generally.[14] It should be realized, though, that in prolonged exposures such at these, age at first exposure is not a critical variable. Also, the range of ages at first exposure that we examined may not be great enough to give sufficient statistical power to detect a real effect. The results in Tables 2A-2 and 2A-3 certainly do not suggest following up on this aspect of risk variation for these particular cohort data.

Turning to the effect of duration of exposure, it is emphasized that since the relationship examined here is, in effect, adjusted for cumulative exposure, this could equally be called an effect of exposure rate. The only significant or consistent such effect found here is in the Colorado cohort. The direction of this effect is that short high-level exposures are associated with lower risks per unit cumulative exposure, a result also noticed by Hornung and Meinhardt.[7] Because this effect is not seen in the other cohorts, and because we found it difficult to incorporate this rough assessment of it into further modeling, it is not followed up here. It would be important to do so in future work.

Except for the Colorado cohort, the parameter estimates at the bottom of Tables 2A-2 and 2A-3 show a quite consistent and possibly significant decrease in relative risk with time since exposure. The committee considered, through various more detailed investigations, whether this apparent effect might be an artifact of some other time-dependent factor in the cohort experience. Although this may be possible, no alternative explanation could be found. A possible concern is that when an apparent decrease of relative risk with age and time since exposure factor are investigated separately, as in Tables 2A-2 and 2A-3, they might reflect the same aspect of the data. This is because at a time when a miner's substantial exposure was many years ago he will also tend to be older. The separation of these effects is considered in the analysis of their joint effect in the combined data (see below).

Several aspects of the Colorado data require further discussion. If the full dose range is used, a quadratic departure from linearity in the model given in Equation 2A-14 is marginally significant ($P = 0.06$). The linear term in this quadratic fit is essentially the same as the value of 0.60 given in Tables 2A-2 and 2A-3 for the linear fit to the restricted dose range. If a linear model is fitted to the entire dose range, the slope decreases to about 0.40. These considerations led us to conclude that for risk estimation in the moderate dose range, it would be better to restrict the range of exposures used. The decision to use exposures under 2,000 WLM was based on the observations that this is roughly the widest range where these nonlinearity problems are not present, and that the slope is essentially the same for this range as for restricting the exposure to under 1,000 WLM. Moreover, current interest centers on risks at relatively low levels of exposure. The analyses of the Colorado data in Table 2A-2 were also carried out with a linear model both by restricting the exposures to less than 1,000 WLM and by using all exposures. The results were quite similar to those in Table 2A-2 with three exceptions. As noted, the constant risk estimate is substantially smaller when the entire dose range is used, and also in that case the decrease of relative risk with age and with time since exposure become larger.

The reason for fixing the Malmberget background SMR at 1.25 was as follows. For this cohort the background SMR is very poorly estimated at a questionably high value of about 2.0, with a multiplicative standard error of 2.5. The reason for this can be seen by inspection of Figure 2A-3, where estimation of this background SMR corresponds to estimation of a zero-dose intercept. (In Figure 2A-3, the intecept would be 1.0 less than the SMR.) The large relative risk of 3.2 in the 20–50-WLM category is very poorly estimated but highly influential in estimating the value of the intercept. When the SMR is estimated as 2.0, the slope is about 0.7. However, fixing the SMR at even 1.0 does not cause a significantly poorer fit (chi-squared = 1.0 on 1 d.f.), and the slope estimate increases to 1.8. This means that estimation of the background SMR for that cohort results in a very imprecise estimate of the slope, which is highly influenced by the large but poorly estimated relative risk in the 20–50-WLM category. It was our judgment, considering the pattern of dose response shown in Figure 2A-3, that the slope estimate of 0.7 was unreasonably small; and thus, it was best in this case to fix the background SMR at some arbitrary but reasonable value. In line with the other two cohorts where the value was greater than 1, it was fixed at a value of 1.25. This element of arbitrariness does not have a substantial effect on the overall conclusions presented here. The inferences from Tables 2A-2 and 2A-3 regarding effects of factors other than cumulative dose are affected very little by fixing this SMR. In the analyses of the combined data sets to obtain our final risk estimates, the Malmberget estimate is given relatively small weight because these data are less extensive than are those for the other cohorts.

For those interested in more careful scrutiny of the above results, Tables 2A-6 through 2A-21 at the end of Part 1 of this annex provide a fairly detailed description of the data for each of the four cohorts. These tables give observed and expected lung-cancer deaths, in several cross-classifications of dose and other factors. No adjustment was made to the background rates in computing the expected numbers in Tables 2A-6 through 2A-21 that were obtained by conventional SMR methods. The background SMRs from Table 2A-3 can be applied to these if desired.

Summarizing the results of the analyses of individual cohorts, there is substantial evidence that the relative risk depends on age at risk and/or time since exposure, and there is little clear evidence that it depends on the other factors considered. The Colorado cohort does not exhibit the time since exposure effect at under 2,000 WLM; but, as with all the effects examined here, this aspect is not well-estimated from that cohort alone. Moreover, this effect does emerge when the wider range of exposures there are considered. The duration (or dose-rate) effect seen in the Colorado data is of interest, but will not be pursued further here.

ANALYSIS OF COMBINED COHORTS

In this section the effects of age and time since exposure are investigated further by formally estimating them in common over the four cohorts. It is well-known that serious biases can occur in estimation of effects by combining cohort studies with very different levels of apparent risk or by involving very different ranges of dose and other variables whose effects are under consideration. We feel that we have taken reasonable care to avoid this by paying close attention to the comparison of analyses by cohort and the combined analysis. Further, in the analysis to follow a different level of overall risk for each cohort is allowed, largely to minimize the possibility of such biases.

The basic model used for the combined analysis is of the form:

$$r(\text{age, period, cohort}, d_1, d_2, d_3) = r_0(\text{age, period, cohort}) \cdot [1 + \beta(\text{cohort})\gamma(\text{age})(d_1 + \theta_2 d_2 + \theta_3 d_3)], \quad (2A\text{-}16)$$

where, d_1, d_2, and d_3 are the WLM exposures incurred in respective periods of 5–10, 10–15, and 15 yr or more prior to the age at which risk is being estimated. The parameter $\beta(\text{cohort})$ represents the cohort-specific level of risk, which is defined more precisely below. The parameter $\gamma(\text{age})$ represents an effect of age on the relative risk, taken to be the same over cohorts. This parameter is allowed to take on distinct values for age in each of the three intervals <55, 55–64, and 65+ yr. Due to the parameterization here, one age interval can be arbitrarily defined as a baseline, and it is taken to be 1.0 for 55–64 yr of age. The parameters θ_2 and θ_3 represent an effect of time since exposure, also taken as the same over cohorts. Factoring out β(cohort) and reexpressing the time-since-exposure effect in this way amounts to requiring that the ratios of the three β's for each cohort in Equation 2A-16 are constant over cohorts, while their general level can depend on cohort. In view of these definitions, it follows that β(cohort) is specifically the excess relative risk per 100 WLM incurred during the period 5–10 yr before the current age for one of age 55–64 yr.

Both internal and external analyses follow, with r_0(age, period) for each case taken as explained in the previous section. The Malmberget background SMR is fixed at 1.25, as before, and only the Colorado data at less than 2,000 WLM of cumulative (lagged) exposure are used. For each of the analyses, four models related to that in Equation 2A-16 are fitted to the combined cohort data. The results are reported in Tables 2A-4 and 2A-5. First, all three γ's and both θ's are fixed at 1.0, resulting in a constant-relative-risk model as in Tables 2A-2 and 2A-3. Next, the models were fitted by estimating the γ's to represent an age effect but holding θ's at 1, that is, no time-since-exposure effect. Then, similarly, models were fitted with a time-since-exposure effect but no age effect, and finally, models were fitted with both effects estimated. The primary purpose of fitting four models in this way is to investigate whether these are to a large extent separate effects, or just different ways of explaining the same aspect of the data.

A number of things can be learned from studying Tables 2A-4 and 2A-5 and comparing them with Tables 2A-2 and 2A-3. These will first be discussed without going into technical statistical issues, which will be explained later. The fit to the combined data with common age effects is not significantly poorer than that when each cohort is allowed its own estimated age effect; that is, the age effects do not differ significantly between cohorts. In this same sense, the time-since-exposure effects are marginally significantly different between cohorts, primarily because of the rather sharp and significant effect in the Eldorado cohort and the lack of an apparent effect (for exposures under 2,000 WLM) in the Colorado cohort. It is nevertheless our judgment that it is reasonable to pool the cohorts for estimation of this effect, and the way of doing this will be explained in the following section. Discussion of the variation of

TABLE 2A-4 Internal Analysis of Combined Data for Four Cohorts: Analysis of Joint Effects of Age and Time Since Exposure

	Excess Relative Risk/100 WLM[a]			
	Constant Relative Risk	Age Effect[b]	Effect of TSE[c]	Both Age and TSE Effect
Eldorado	2.6 (1.5)[d]	3.3 (1.6)	6.2 (1.6)	7.0 (1.7)
Ontario	1.4 (1.6)	1.3 (1.7)	2.7 (1.6)	2.3 (1.8)
Malmberget	1.4 (2.6)	2.2 (2.6)	3.5 (2.4)	5.1 (2.7)
Colorado	0.6 (1.5)	0.7 (1.7)	1.0 (1.6)	1.1 (1.8)
γ parameters at:				
<55 yr	—	1.1 (1.7)	—	1.1 (1.6)
65+ yr	—	0.3 (2.0)	—	0.3 (2.0)
Theta parameters (TSE):				
Exp 10-15	—	—	0.7 (1.4)	0.8 (1.4)
Exp 15+	—	—	0.3 (1.4)	0.4 (1.4)
Comparison of fits:				
Chi-squared comparison to column 1	—	4.1 (2 df[e]); $P = 0.13$	6.0 (2 df); $P = 0.05$	9.6 (4 df); $P = 0.05$

[a]Parameter estimates for Equation 2A-15 fitted with neither age nor time effects, with each one alone, and with both.
[b]Age effect = 1 for age 55-64.
[c]TSE = Time since exposure.
[d]Numbers in parentheses are multiplicative standard errors.
[e]df = Degrees of freedom.

general levels of risk between the cohorts is also postponed until the following section.

The age and time effects do appear to be substantially independent effects, rather than just capturing a single effect in the data that is correlated with both variables. Simply stated, the evidence for this is that the chi-squared statistics for the joint effect are substantially larger than those for either of the effects taken separately, and also that the standard errors in the model with both effects are not substantially larger than those in the models with a single effect.

The significance of the age and time effects is somewhat greater in the external analysis than in the internal one. This is to be expected on general statistical grounds. Even the combined cohort data are rather weak for estimating such detailed aspects of the relative risk. Use of known external rates is of substantial help in estimating such detail, even though the estimates for simpler models

TABLE 2A-5 External Analysis[a] of Combined Data for Four Cohorts: Analysis of Joint Common Effects of Age and Time Since Exposure in All Four Cohorts Together

	Excess Relative Risk/100 WLM[b]			
	Constant Relative Risk	Age Effect[c]	Effects of TSE[d]	Both Age and TSE
Eldorado	2.6 (1.4)[e]	3.2 (1.5)	4.1 (1.5)	3.8 (1.6)
Ontario	1.3 (1.5)	1.4 (1.5)	1.9 (1.6)	1.6 (1.6)
Malmberget	1.6 (2.5)	2.2 (2.5)	2.5 (2.4)	2.7 (2.4)
Colorado	0.6 (1.5)	0.6 (1.5)	0.8 (1.5)	0.7 (1.6)
γ parameters at:				
<55 yr	—	1.1 (1.2)	—	1.2 (1.2)
65+ yr	—	0.5 (1.3)	—	0.6 (1.3)
Theta parameters (TSE):				
Exp 10–15	—	—	1.1 (1.4)	1.3 (1.4)
Exp 15+	—		0.5 (1.3)	0.6 (1.3)
Comparison of fits:				
Chi-squared comparison to column 1	—	12.5 (2 df[f]); $P = 0.002$	9.9 (2 df[f]); $P = 0.007$	19.2 (4 df[f]); $P < 0.001$

[a]The background SMRs are essentially the same as in Table 2A-3.
[b]Parameter estimates for Equation 2A-15 fitted with neither age nor time effects, with each one alone, and with both.
[c]Age effect = 1 for age 55–64.
[d]TSE = Time since exposure.
[e]Numbers in parentheses are multiplicative standard errors.
[f]df = Degrees of freedom.

such as the constant-relative-risk model are almost equally precise by either method. Counter-balancing this gain from the use of external rates is the concern that they may not be entirely appropriate for these cohorts.

The decline in relative risk with age is quite consistent and significant. It is estimated as essentially the same in both the internal and external comparisons and whether or not the time-since-exposure effect is simultaneously estimated. Our interpretation of this effect is that excess cancer begins to appear within a few years of exposure, regardless of age, and can lead to a rather substantial relative risk in middle age because the background risk is not as great there as at older ages (65+ yr).

Point estimates of the parameters for the time-since-exposure effect differ somewhat under the two types of analyses. In particular, it is of concern that, in the external analyses, the parameter for exposures 10–15 yr back is greater than that for the 5- to 10-yr interval. This may be due, in part, to the fact that the 5–10 yr category includes latency years, which reduces the apparent risk. However, it should be understood that these individual parameters are not very precisely estimated from these data. Eliminating the Colorado cohort from the analysis shows that most of the difference between the external and internal analyses comes from this cohort. In the external analysis of the Colorado cohort alone, the risk due to exposures 10–15 yr back is estimated as only 0.8/0.7 of that due to exposures 5–10 yr back (Table 2A-3), a ratio smaller than 1.3 from the combined analysis. Thus it seems likely that some bias due to combining cohorts may lead to the estimate of 1.3 in the external analysis. Recall also that analysis of the full dose range for the Colorado cohort shows an effect of time since exposure similar to that in the other cohorts. We conclude that there is a trend in relative risk with time since exposure, as well as with age at risk, and merge the evidence from all aspects of the analyses.

Finally, we turn to some of the more technical statistical aspects of analyses to support the above conclusions. Formal statistical assessment of the significance of the time-since-exposure effect adjusted for the age effect can be made as follows. In the internal analysis (Table 2A-4), the difference in chi-square for both effects compared to chi-square for the age effect is $9.6 - 4.1 = 5.5$ on 2 d.f. ($P = 0.07$), about the same level of significance as for the time-since-exposure effect alone. Similar calculations apply to the external analysis and for assessing the effect of age adjusted for time since exposure. These considerations show that the two effects are substantially independent of one another.

Information given in Tables 2A-2 through 2A-5 also provides for assessment of the commonality of the age and time effects over cohorts. This forms the basis for this analysis and underlies any generalization from these studies. For example, consider the question of equality of the time-since-exposure effects over the cohorts from the internal analysis. The sum $11 + 5.3 + 2.1 + 0.1 = 18.5$ of the chi-squared statistics for time-since-exposure effects in Table 2A-2, less the value 6.0 for testing this effect from Table 2A-4, is a chi-squared statistic of 12.5 on 6 d.f. for testing the equality over cohorts of the time-since-exposure effect. (The 6 d.f. is the result of the fact that there are eight parameters for this effect when they are not constrained to be equal and two parameters for it when they are.) The *P* value for this test is about 0.05, indicating some evidence for a difference. However, from the external analysis, the *P* value is about 0.10, and the corresponding tests for equality of age effects are not at all significant. Our interpretation of these results is that it is reasonable to rely on estimates of these effects from the combined analysis, taking them to be an estimation of a quantity that is roughly constant over cohorts.

ESTIMATION OF A FINAL MODEL

Our conclusions from the above analysis are that it is best to describe these data in terms of a model that incorporates effects on the relative risk of both age and time since exposure, and that effects of age at first exposure and duration of exposure should not be included. Further, we conclude that the dose-modification effects that we have included can reasonably be estimated as the same in all four cohorts.

How to handle the apparent variation of the general levels of risk over cohorts, as shown by the cohort-specific parameters of Tables 2A-4 and 2A-5, is a somewhat different matter. For the constant-relative-risk models, the likelihood ratio test of equal relative risk over the four cohorts yields chi-squared statistics of approximately 6.6 and 6.5, on 3 d.f., for the internal and external analyses, respectively. These correspond to a *P* value of about 9%; that is, there would be 9 chances in 100 of obtaining estimates this different if, in fact, the risks were the same for all four cohorts and the dosimetry and ascertainment of mortality were exact. For the other models incorporating age and time effects taken to be the same over cohorts, testing for the equality of cohorts shows about the same level of statistical significance. This quite marginal statistical significance of the cohort differences might seem surprising, in view of what are, from a risk assessment perspective, quite disparate estimates; it is worth looking at this in another way. It is a routine and general statistical calculation that if four independent estimates of the same quantity are obtained, estimates whose logarithms are normally distributed and whose multiplicative standard errors are 1.5, in line with the estimates here, then there is a 10% chance that the ratio of the largest to the smallest will be at least 4.6, which is about the same as the variation found here.

Thus, particularly in view of the inevitable systematic differences in dosimetry and other substantial inadequacies in studies such as this, the variation seen here is as small as could reasonably be expected. This is not to say that the differences are of no interest and may not reflect inadequacies in the data, which would be important to eliminate. There is certainly no reason, though, to discount the results from any of these studies purely on the grounds that the estimate from it is too extreme in relation to the others.

Returning to the age and time effects, the decline in relative risk with time since exposure seems to us to be indeed quite different from the experience of the Japanese atomic-bomb survivors[17,18] and some other cohorts exposed to external low-LET radiation.[14] Darby et al.[4] have, however, recently reported a similar effect in the ankylosing spondylitis data. Other investigators have found this general sort of effect in the radon data on miners: Lundin et al.,[11] Harley and Pasternack,[6] Thomas and McNeill,[22] Thomas et al., [23] and Hornung and Meinhardt.[7] We feel that the differences between our results and those of these investigators are much less important than the possible differences between the results here and those found in the acute, low-LET external exposures.

A decline of relative risk with age is not easily distinguished from a decline with time since exposure, and these effects are likely to become confused in many cohort analyses. Evidence presented here does suggest that both effects act somewhat independently. The analyses of the Japanese atomic-bomb survivor data are commonly held to support no such effect. Those analyses, however, include a substantial decrease in relative risk with age at exposure, and this is difficult to distinguish from a decline with age. If models are fitted to the Japanese atomic-bomb survivor data with no (or substantially less) effect of age at exposure, then there is an apparent decrease in relative risk with age.[17,18] Moreover, it has recently been pointed out that there is some indication of a decrease in relative risk with age for those exposed at an early age.[17,18] Thus, the evidence of a difference between the data analyzed here and those on external, acute, low-LET exposures with regard to age at risk is not totally clear.

In view of the results shown in Tables 2A-4 and 2A-5, it seems reasonably clear that the parameter θ_2 in Equation 2A-16 for the ratio of effects of exposures 10–15 yr back to that of 5–10 yr back should be taken as fairly close to unity. Our judgment is that it is best for present purposes, with the lack of clearer information, to take this as 1.0. If the combined data are refitted with this specification, the remaining estimates for age/time effects are:

	Internal Analysis	External Analysis
Time since exposure ≥15 yr	0.4	0.6
Age, <55 yr	1.2	1.2
Age, 65+ yr	0.3	0.6

The committee decided to take the TSE parameter as 0.5, and the age parameters for less than 55 yr and 65 or more yr as 1.2 and 0.4, respectively. Refitting the data with the three parameters fixed at these values yields for β, the excess relative risk per 100 WLM cohort-specific risk for the baseline categories of age and time as:

Cohort	Internal Analysis	External Analysis
Eldorado	5.1	5.0
Ontario	1.8	2.0
Malmberget	3.6	3.7
Colorado	0.9	0.7

When the combined data are fitted by replacing these cohort-specific parameters by a common parameter for all cohorts, the estimate of β for the internal analysis is 2.2 and for the external analysis is 2.6. The difference between these values is small relative to the precision of estimation. The committee selected 2.5 as the values of this parameter. Thus, the final model chosen to describe these data is

$$r(\text{age, period, dose history}) = r_0(\text{age}) \cdot [1 + 0.025\gamma(\text{age})(W_1 + 0.5W_2)], \quad \text{(2A-17)}$$

where γ(age) is 1.2 for age <55 yr, 1.0 for age 55–64 yr, 0.4 for age 65 yr or more; W_1 is the cumulative WLM incurred between 5 and 15 yr before age a; and W_2 is the WLM incurred 15 yr or more before this age.

We do not think that standard errors of parameter estimates in a model of this complexity are very useful. Some discussion of this is given in Chapter 2. It may be of interest, though, in consideration of this, that if the age and time effects are taken as fixed at the above values, then formal multiplicative standard errors of the two estimates 2.2 and 2.6 per 100 WLM given above for the internal and external analyses are 1.3 and 1.2, respectively. Omission of the Colorado data, keeping the same effects of age and time since exposure in Equation 2A-17, would increase the estimated risk by about 33%. The formal multiplicative standard is essentially unchanged. Again, this indicates that the Colorado data are not exceptional in terms of statistical significance.

Finally, it may of interest on a number of grounds to be able to compare the results of using this model with those obtained by assuming a constant relative risk, not depending on age or time since exposure. The cohort-specific relative risks are given in Tables 2A-4 and 2A-5. If the combined data are fitted to a constant-relative-risk model without this cohort dependence, the corresponding estimates are 1.34 and 1.50 per 100 WLM from the internal and external analyses, respectively. The multiplicative standard errors for these estimates are 1.3 and 1.2.

TABLE 2A-6a Observed and Expected Lung Cancer Deaths and Excess Relative Risk[a] Cross-Classified by Age and Cumulative Dose for the Eldorado Cohort

Dose (WLM)	Age (yr)								
	<55			55-64			65+		
0	4	4.5815	0.00[b]	4	3.5484	0.13	3	2.8984	0.04
0-20	10	5.6468	0.77	9	4.1125	1.19	3	2.1597	0.39
20-50	5	1.4687	2.40	4	1.2265	2.26	0	0.6107	0.00
50-100	1	0.7423	0.35	4	0.8130	3.92	3	0.7134	3.21
100-150	2	0.2704	6.40	3	0.2818	9.64	0	0.2991	0.00
150-200	2	0.1322	14.13	1	0.1460	5.85	0	0.0606	0.00
>200	1	0.1155	7.66	3	0.1941	14.45	3	0.2875	9.44

[a]The data in each cell are arranged as follows:
Observed (O)
Expected (E)
Excess relative risk [(O/E) − 1]

[b]Excess relative risks computed to be less than zero are reported as zero.

TABLE 2A-6b Observed and Expected Lung Cancer Deaths and Excess Relative Risk[a] Cross-Classified by Age at First Exposure and Cumulative Dose for the Eldorado Cohort

Dose (WLM)	Age at First Exposure (yr)					
	<30			30+		
0	6	8.8084	0.00[b]	5	2.2199	1.25
0-20	1	2.4188	0.00	21	9.5002	1.21
20-50	1	0.7544	0.33	8	2.5515	2.14
50-100	0	0.4183	0.00	8	1.8505	3.32
100-150	1	0.1612	5.20	4	0.6902	4.80
150-200	1	0.0844	10.84	2	0.2543	6.87
200-250	0	0.0870	0.00	7	0.5100	12.72

[a]The data in each cell are arranged as follows:
Observed (O)
Expected (E)
Excess relative risk [(O/E) − 1]

[b]Excess relative risks computed to be less than zero are reported as zero.

TABLE 2A-6c Observed and Expected Lung Cancer Deaths and Excess Relative Risk[a] Cross-Classified by Time since Cessation of Exposure and Cumulative Dose for the Eldorado Cohort

	Time since Cessation (yr)					
Dose (WLM)	<5		5-15		15+	
0	11		0		0	
	11.0280		0.0000		0.0000	
		0.00[b]		0.00		0.00
0-20	5		8		9	
	1.4089		4.6136		5.8965	
		2.55		0.73		0.53
20-50	3		3		3	
	0.5411		1.1169		1.6479	
		4.54		1.69		0.82
50-100	5		1		2	
	0.4467		0.8863		0.9358	
		10.19		0.13		1.14
100-150	4		1		0	
	0.2334		0.3575		0.2606	
		16.14		1.80		0.00
150-200	2		1		0	
	0.0992		0.0985		0.1411	
		19.16		9.16		0.00
200-250	2		4		1	
	0.1053		0.2975		0.1943	
		18.00		12.45		4.15

[a]The data in each cell are arranged as follows:
Observed (O)
Expected (E)
Excess relative risk [(O/E) − 1]

[b]Excess relative risks computed to be less than zero are reported as zero.

TABLE 2A-6d Observed and Expected Lung Cancer Deaths and Excess Relative Risk[a] Cross-Classified by Duration of Exposure and Cumulative Dose for the Eldorado Cohort

	Duration of Exposure (yr)					
Dose (WLM)	<2.5		2.5-4		5+	
0	11		0		0	
	11.0280		0.0000		0.0000	
		0.00[b]		0.00		0.00
0-20	18		1		3	
	7.9415		2.9194		1.0582	
		1.27		0.00		1.84
20-50	5		3		1	
	1.1935		1.3431		0.7693	
		3.19		1.23		0.30
50-100	1		1		6	
	.3635		0.7939		1.1114	
		1.75		0.26		4.40
100-150	0		0		5	
	.0439		0.2464		0.5611	
		.00		0.00		7.91
150-200	0		1		2	
	.0017		0.1527		0.1843	
		.00		5.55		9.85
200-250	0		0		7	
	.0002		0.0469		0.5500	
		.00		0.00		11.73

[a]The data in each cell are arranged as follows:
Observed (O)
Expected (E)
Excess relative risk [(O/E) − 1]

[b]Excess relative risks computed to be less than zero are reported as zero.

TABLE 2A-7a Observed and Expected Lung Cancer Deaths and Excess Relative Risk[a] Cross-Classified by Age and Cumulative Dose for Ontario Uranium Miners

Dose (WLM)	Age (yr) <55			55–64			≥65		
	O	E	Excess relative risk	O	E	Excess relative risk	O	E	Excess relative risk
0	6	2.7659	1.17	4	1.7359	1.30	0	0.8634	0.00[b]
0–20	13	13.2130	0.00	11	9.1468	0.20	0	3.7619	0.00
20–50	10	6.5829	0.52	10	4.6142	1.17	1	1.3922	0.00
50–100	6	3.3926	0.77	5	2.7187	0.84	2	1.0245	0.95
100–150	3	1.2711	1.36	1	0.9486	0.05	1	0.2485	3.02
150–200	4	0.8159	3.90	2	0.6682	1.99	0	0.1830	0.00
>200	5	0.5569	7.98	3	0.6862	3.37	0	0.2596	0.00

[a]The data in each cell are arranged as follows:
Observed (O)
Expected (E)
Excess relative risk [(O/E) − 1]

[b]Excess relative risks computed to be less than zero are reported as zero.

TABLE 2A-7b Observed and Expected Lung Cancer Deaths and Excess Relative Risk[a] Cross-Classified by Age at First Exposure and Cumulative Dose for Ontario Uranium Miners

Dose (WLM)	Age at First Exposure (yr) <30			30+		
	O	E	Excess relative risk	O	E	Excess relative risk
0	5	3.4983	0.43	5	1.8669	1.68
0–20	7	5.9525	0.00[b]	17	20.1690	0.00
20–50	4	3.2594	0.23	17	9.3300	0.82
50–100	3	1.7091	0.76	10	5.4379	0.84
100–150	1	0.6584	0.52	4	1.8098	1.21
150–200	2	0.4028	3.97	4	1.2643	2.16
>200	2	0.2539	6.88	6	1.2488	3.80

[a]The data in each cell are arranged as follows:
Observed (O)
Expected (E)
Excess relative risk [(O/E) − 1]

[b]Excess relative risks computed to be less than zero are reported as zero.

TABLE 2A-7c Observed and Expected Lung Cancer Deaths and Excess Relative Risk[a] Cross-Classified by Time since Cessation of Exposure and Cumulative Dose for Ontario Uranium Miners

	Time since Cessation of Exposure								
Dose[b]	<5			5-15			>15		
	O	E	ERR	O	E	ERR	O	E	ERR
0	10	5.3651	0.86	0[c]	0.0000	0.00	0	0.0000	0.00
0-20	3	1.4673	1.03	9	7.9903	0.13	12	16.6640	0.00
20-50	2	0.9276	1.16	11	4.6285	1.38	8	6.9882	0.14
50-100	2	1.1368	0.76	5	3.0798	0.62	6	2.9231	1.05
100-150	0	0.5268	0.00	3	1.1654	1.57	2	0.7761	1.58
150-200	4	0.3121	8.61	2	0.9429	1.12	0	0.4219	0.00
>200	3	0.7477	3.01	5	0.5827	7.58	0	0.1673	0.00

[a]The data in each cell are arranged as follows:

Observed (O)
Expected (E)
Excess relative risk [(O/E) − 1]

[b]Dose is average of standard and special.

TABLE 2A-7d Observed and Expected Lung Cancer Deaths and Excess Relative Risk[a] Cross-Classified by Duration of Exposure and Cumulative Dose for Ontario Uranium Miners

	Duration of Exposure (yr)								
Dose (WLM)	<2.5			2.5-5			>5		
	O	E	ERR	O	E	ERR	O	E	ERR
0	10	5.3651	0.86	5	0.0000	(infinity)	0	0.0000	0.00[b]
0-20	19	16.9260	0.12	15	8.9149	0.68	0	0.2708	0.00
20-50	4	1.9200	1.08	7	8.4669	0.00	2	2.2075	0.00
50-100	0	0.3205	0.00	1	2.7238	0.00	6	4.0955	0.47
100-150	0	0.0008	0.00	0	0.4043	0.00	4	2.0632	0.94
150-200	0	0.0001	0.00	1	0.0808	11.38	6	1.5863	2.78
>200	0	0.0000	0.00	0	0.0144	0.00	8	1.4883	4.38

[a]The data in each cell are arranged as follows:

Observed (O)
Expected (E)
Excess relative risk [(O/E) − 1]

[b]Excess relative risks computed to be less than zero are reported as zero.

TABLE 2A-8a Observed and Expected Lung Cancer Deaths and Excess Relative Risk[a] Cross-Classified by Age and Cumulative Dose for the Malmberget Cohort

	Age (yr)		
Dose (WLM)	<55	55–64	≥65
0	0 0.0811 0.00[b]	0 0.0285 0.00	0 0.0000 0.00
0–20	0 0.2545 0.00	0 0.5035 0.00	0 0.3886 0.00
20–50	2 0.4840 3.13	5 1.1737 3.26	4 0.9520 3.20
50–100	0 0.4530 0.00	6 1.5886 2.78	5 1.9495 1.55
100–150	3 0.3503 7.56	3 1.0392 1.89	2 1.3760 0.45
150–200	0 0.0671 0.00	9 1.1658 6.72	6 1.7075 2.51
200–250	0 0.0000 0.00	1 0.2372 3.22	5 1.0754 3.65

[a] The data in each cell are arranged as follows:

Observed (O)
Expected (E)
Excess relative risk [(O/E) − 1]

[b] Excess relative risks computed to be less than 0 are reported as 0.

TABLE 2A-8b Observed and Expected Lung Cancer Deaths and Excess Relative Risk[a] Cross-Classified by Age at First Exposure and Cumulative Dose for the Malmberget Cohort

	Age at First Exposure (yr)	
Dose (WLM)	<30	30+
0	0 0.0002 0.00[b]	0 0.1094 0.00
0–20	0 0.5912 0.00	0 0.5551 0.00
20–50	8 1.5958 4.01	3 1.0139 1.96
50–100	8 2.6185 2.06	3 1.3646 2.10
100–150	6 2.1286 1.82	2 0.6369 2.14
150–200	13 2.4394 4.33	2 0.5011 2.99
200–250	6 1.3126 3.57	0 0.0000 0.00

[a] The data in each cell are arranged as follows:

Observed (O)
Expected (E)
Excess relative risk [(O/E) − 1]

[b] Excess relative risks computed to be less than 0 are reported as 0.

TABLE 2A-8c Observed and Expected Lung Cancer Deaths and Excess Relative Risk[a] Cross-Classified by Time since Cessation of Exposure and Cumulative Dose for the Malmberget Cohort

	Time since Cessation of Exposure (yr)		
Dose (WLM)	<5	5-15	≥15
0	0	0	0
	0.1096	0.0000	0.0000
	0.00[b]	0.00	0.00
0-20	0	0	0
	0.2015	0.1385	0.8063
	0.00	0.00	0.00
20-50	4	2	5
	0.6920	0.5237	1.3941
	4.78	2.82	2.59
50-100	5	2	4
	1.0014	1.0783	1.9033
	3.99	0.85	1.10
100-150	5	2	1
	0.9468	0.8051	1.0136
	4.28	1.48	0.00
150-200	10	4	1
	1.1033	1.4989	0.3383
	8.06	1.67	1.96
200-250	1	5	0
	0.3325	0.8941	0.0860
	2.01	4.50	0.00

[a]The data in each cell are arranged as follows:
Observed (O)
Expected (E)
Excess relative risk [(O/E) − 1]

[b]Excess relative risks computed to be less than 0 are reported as 0.

TABLE 2A-8d Observed and Expected Lung Cancer Deaths and Excess Relative Risk[a] Cross-Classified by Duration of Exposure and Cumulative Dose for the Malmberget Cohort

	Duration of Exposure (yr)		
Dose (WLM)	<20	20-30	>30
0	0	0	0
	0.1096	0.0000	0.0000
	0.00[b]	0.00	0.00
0-20	0	0	0
	1.1463	0.0000	0.0000
	0.00	0.00	0.00
20-50	6	5	0
	2.1973	0.4124	0.0000
	1.73	11.37	0.00
50-100	7	1	3
	3.0249	0.3404	0.6177
	1.31	1.94	3.86
100-150	2	6	0
	0.8479	1.8332	0.0844
	1.36	2.27	0.00
150-200	0	9	6
	0.0000	1.4479	1.4926
	0.00	5.22	3.02
200-250	0	0	6
	0.0000	0.0000	1.3126
	0.00	0.00	3.57

[a]The data in each cell are arranged as follows:
Observed (O)
Expected (E)
Excess relative risk [(O/E) − 1]

[b]Excess relative risks computed to be less than 0 are reported as 0.

TABLE 2A-9a Observed and Expected Lung Cancer Deaths and Excess Relative Risk[a] Cross-Classified by Age and Cumulative Dose for the Colorado Cohort

	Age (yr)		
Dose (WLM)	<55	55-64	≥65
0-60	2 2.85 0.00[b]	2 3.48 0.00	5 2.68 0.87
60-120	2 1.25 0.60	2 1.76 0.14	1 1.84 0.00
120-240	4 2.00 1.02	4 2.68 0.49	6 2.41 1.49
240-480	10 2.75 2.64	10 4.13 1.42	7 4.14 0.69
480-960	13 2.71 3.79	23 4.54 4.07	21 4.28 3.90
>960	54 3.35 15.10	62 5.94 9.44	28 6.35 3.41

[a] The data in each cell are arranged as follows:
Observed (O)
Expected (E)
Excess relative risk [(O/E) − 1]

[b] Excess relative risks computed to be less than 0 are reported as 0.

TABLE 2A-9b Observed and Expected Lung Cancer Deaths and Excess Relative Risk[a] Cross-Classified by Age at First Exposure and Cumulative Dose for the Colorado Cohort

	Age at First Exposure (yr)	
Dose (WLM)	<30	30+
0-60	0 1.02 0.00[b]	9 8.00 0.13
60-120	0 .85 0.00	5 4.00 0.25
120-240	2 1.33 0.51	12 5.74 1.10
240-480	6 2.32 1.59	21 8.70 1.41
480-960	11 2.87 2.84	46 8.68 4.31
>960	70 6.87 9.19	74 8.76 7.44

[a] The data in each cell are arranged as follows:
Observed (O)
Expected (E)
Excess relative risk [(O/E) − 1]

[b] Excess relative risks computed to be less than 0 are reported as 0.

TABLE 2A-9c Observed and Expected Lung Cancer Deaths and Excess Relative Risk[a] Cross-Classified by Time since Cessation of Exposure and Cumulative Dose for the Colorado Cohort

Dose (WLM)	Time since Cessation (yr) 5	5-15	>15
0-60	2 2.53 0.00[b]	3 2.30 0.30	4 4.16 0.00
60-120	1 0.67 0.49	2 1.74 0.15	2 2.44 0.00
120-240	3 1.00 2.00	7 2.83 1.48	4 3.25 0.23
240-480	6 1.82 2.29	14 4.89 1.87	7 4.31 0.62
480-960	11 1.97 4.58	30 5.27 4.69	16 4.30 2.72
>960	47 3.33 13.10	76 7.34 9.35	21 4.97 3.23

[a]The data in each cell are arranged as follows:
Observed (O)
Expected (E)
Excess relative risk [(O/E) − 1]

[b]Excess relative risks computed to be less than 0 are reported as 0.

TABLE 2A-9d Observed and Expected Lung Cancer Deaths and Excess Relative Risk[a] Cross-Classified by Duration of Exposure and Cumulative Dose for the Colorado Cohort

Dose (WLM)	Duration of Exposure (yr) <2.5	2.5-5	>5
0-60	9 8.75 0.03	0 0.21 0.00[b]	0 0.03 0.00
60-120	3 3.94 0.00	2 0.81 1.46	0 0.09 0.00
120-240	10 3.89 1.57	3 2.50 0.20	1 0.68 0.47
240-480	4 4.14 0.00	12 4.40 1.73	11 2.49 3.43
480-960	5 1.73 1.90	17 5.31 2.20	35 4.50 6.78
>960	1 0.54 0.85	18 3.33 4.41	125 11.77 9.62

[a]The data in each cell are arranged as follows:
Observed (O)
Expected (E)
Excess relative risk [(O/E) − 1]

[b]Excess relative risks computed to be less than 0 are reported as 0.

PART 2: Measures of Lifetime Risk

The committee uses the model developed in Part 1 of this annex to estimate the risk due to exposure to radon progeny. In this part of Annex 2A, we present the mathematical basis for the risk estimates given in the section "Projecting Risks Associated with Radon-Daughter Exposure" in Chapter 2.

Two measures that can be used to assess an individual's risk of death from lung cancer are: (1) lifetime probability of lung cancer given that a person is alive at a given age t_0 and (2) expected years of life beyond t_0. If $t_0 = 0$, then items (1) and (2) above are, respectively, the lifetime probability of lung cancer and expected years of life at birth.

To calculate the probability of dying of lung cancer, suppose q_i is the probability of surviving year i when all causes are acting, conditional on surviving through year $i - 1$; h_i^* is the mortality rate due to all causes; and h_i is the lung-cancer mortality rate at year i. Then, $q_i = \exp(-h_i^*)$ and the probability of death in i is $1 - q_i$. The probability of surviving up to year i is the product of surviving each prior year:

$$q_1 \times q_2 \times \ldots \times q_i - 1 = \prod_{k=1}^{i-1} q_k = S(1,i), \qquad \text{(2A-18)}$$

with $S(1,1) = 1.0$. The probability of surviving through year $i - 1$ and dying in year i is:

$$\prod_{k=1}^{i-1} q_k(1 - q_i) = S(1,i)(1 - q_i). \qquad \text{(2A-19)}$$

Multiplying by the proportion of the cause of interest, that is, lung cancer among all causes, gives the probability of surviving $i - 1$ years and dying of lung cancer at i:

$$h_i/h_i^* S(1,i)(1 - q_i). \qquad \text{(2A-20)}$$

The lifetime probability of lung-cancer mortality is then the summation over all years, $t = 1$ to $t = 110$:

$$R_0 = \sum_{i=1}^{110} h_i/h_i^* \prod_{k=1}^{i-1} q_k(1 - q_i), \qquad \text{(2A-21)}$$

where maximum life is assumed to be 110 yr. In this formulation there is a small nonzero probability of living more than 110 yr, $S(1, 110)$. Therefore, to be precise Equation 2A-21 should be divided by the probability of a death event prior to 110, $1 - S(1,110)$. For the mortality rates for U.S. males given in Table 2A-10, this quantity is $1 - 0.0042\exp[-5(0.18794)] = 0.9984$.

The additional risk of lung cancer due to exposure to radon progeny is incorporated into these risk calculations through the age-specific lung-cancer mortality rates. Assuming a proportional hazards model, the lung-cancer mortality rate for an exposed individual is $h_i(1 + e_i)$, and the overall mortality rate is $h_i^* + h_i e_i$, where e_i is the excess relative risk for year i. The e_i term can be complicated expressions of, for example, age, dose rate, time since exposure, and tobacco use patterns. In the presence of exposure, the probability of surviving year i becomes $q_i \exp(h_i e_i)$. Corresponding to Equation 2A-21, the lifetime chance of lung cancer for an individual with excess risk profile $e_1, \ldots, e_{110}$ is:

$$R_e = \sum_{i=1}^{110} \frac{h_i\ (1 + e_i)}{h_i^* + h_i\ e_i}\ S(1,i)\ [1 - q_i\ \exp(-h_i e_i)] \cdot \exp(-\sum_{k=1}^{i-1} h_k e_k). \qquad (2A\text{-}22)$$

The ratio of lifetime risks for a specified exposure history compared to the case for no exposure is R_e/R_0.

In some instances it is important to condition all events on survival after a given age t_0, for example, lifetime lung-cancer risk for an occupationally exposed worker who is known to be alive at age 60. Then, conditional on this knowledge, the probability of lung cancer at $i \geq t_0$ is:

$$\sum_{i=t_o}^{110} h_i/h_i^*\ S(t_0,i)(1 - q_i). \qquad (2A\text{-}23)$$

Table 2A-10 demonstrates the calculation of lifetime lung-cancer risk using 1980–1984 data on lung cancer and overall mortality for U.S. males in columns 2 and 3, respectively. Using columns 2 through 5 and applying Equation 2A-21, the probability of lung cancer is the sum of the age-specific probabilities in column 6, $R_0 = 0.06734$. (Note that in this example q_i is the 5-yr survival

TABLE 2A-10 Probability of Lung-Cancer Mortality in the Presence of Other Causes of Death[a]

Age Interval	Mortality Rates		Probability of Surviving		Probability of Lung Cancer[b]
i	Lung-Cancer (h, $\times 10^4$ yr^{-1})	All Causes (h^*, $\times 10^4$ yr^{-1})	Interval i (q)	Up to Interval i (S)	P
0–4	0.006	36.552	0.9819	1.0000	0.000003
5–9	0.001	3.500	0.9983	0.9819	0.000001
10–14	0.003	3.833	0.9981	0.9802	0.000002
15–19	0.003	14.139	0.9930	0.9784	0.000001
20–24	0.014	20.347	0.9899	0.9715	0.000007
25–29	0.039	19.616	0.9902	0.9617	0.000019
30–34	0.124	19.595	0.9902	0.9523	0.000059
35–39	0.579	24.513	0.9878	0.9430	0.000272
40–44	1.925	36.426	0.9820	0.9314	0.000886
45–49	4.956	58.409	0.9712	0.9147	0.002235
50–54	10.133	94.292	0.9539	0.8883	0.004401
55–59	17.076	146.070	0.9296	0.8474	0.006974
60–64	26.387	223.096	0.8944	0.7877	0.009839
65–69	36.452	339.381	0.8439	0.7045	0.011812
70–74	45.769	507.750	0.7758	0.5946	0.012017
75–79	49.701	747.904	0.6880	0.4613	0.009564
80–84	47.391	1,124.224	0.5700	0.3173	0.005751
85–89	36.801	1,879.436	0.3907	0.1809	0.002158
90–94	36.801	1,879.436	0.3907	0.0707	0.000844
95–99	36.801	1,879.436	0.3907	0.0276	0.000329
100–104	36.801	1,879.436	0.3907	0.0108	0.000129
105+	36.801	1,879.436	0.3907	0.0042	0.000050

[a] Based on 1980–1984 U.S. male lung-cancer and all causes of mortality rates.
Risks computed in 5-yr intervals, $q_i = \exp(-5h_i)$.
[b] Probability of lung-cancer in interval i given survival to that interval.

probability, $\exp[-5h_i^*]$. In computing the results given in this report, yearly increments of risk were used.)

Conditioning on survival can result in changes in risk estimates. Suppose a person is known to have survived $t_0 = 60$ yr. Then the probability of disease must be divided by the probability that the individual is alive, $S(1, 60) = 0.7877$. The conditional lifetime risk of lung cancer (between 60 and 110 yr) increases to 0.06734/0.7877 = 0.08551.

Years of life lost is the difference in expected years of life between those exposed and not exposed. For each i, the probability of an event is assumed to follow an exponential distribution with rate h_i^*. The expected years of life are:

$$
\begin{aligned}
L_0 &= \int uh^*(u)P(\text{survival to } u)du \\
&= \sum_{i=1}^{110} S(1,i) \int_{i-1}^{i} u\, h_i^* \exp\,[-uh_i^* + (i-1)h_i^*]du \\
&= \sum_{i=1}^{110} S(1,i)[(1/h_i^* + i - 1)(1 - q_i) - q_i]. \qquad \text{(2A-24)}
\end{aligned}
$$

As above, expected years of life for an exposed individual L_e is obtained by replacing h_i^* by $h_i^* + h_i e_i$ in Equation 2A-24. The decrease in life span due to exposure is $L_0 - L_e$.

Expected years of life can also be conditioned to t_0, $L_0(t_0)$, and is similar in form to Equation 2A-24:

$$
\begin{aligned}
L_0 t_0 &= \int uh^*(u)P(\text{survival to } u|\text{alive at } t_0)du \\
&= \sum_{i=t_0}^{110} S(t_0,i) \int_{i-1}^{i} uh_i^* \exp[-uh_i^* + (i-1)h_i^*]du \\
&= \sum_{i=t_0}^{110} S(t_0,i)[1/h_i^* + i - 1)(1 - q_i) - q_i]. \qquad \text{(2A-25)}
\end{aligned}
$$

Using Equation 2A-25 and the data in Table 2A-22, the expected years of life for U.S. males is 69.72. Conditional on survival for 60 yr, the expected years of life increases to 77.30 yr.

REFERENCES

1. Breslow, N. E. 1985. Cohort analysis in epidemiology. Chapter 6 in A Celebration of Statistics: The ISI-Centenary Volume, A. C. Atkinson and S. E. Fienberg, eds. New York: Springer-Verlag.
2. Cox, D. R. 1972. Regression models and life-tables (with discussion). J. R. Stat. Soc. B 34:187–220.
3. Cox, D. R., and D. Oakes. 1984. Analysis of survival data. London: Chapman and Hall.
4. Darby, S. C., R. Doll, S. K. Gill, and P. G. Smith. 1987. Long term mortality after a single treatment course with x-rays in patients treated for ankylosing spondylitis. Br. J. Cancer 55:179–190.
5. Darby, S. C., E. Nakashima, and H. Kato. 1985. A parallel analysis of cancer mortality among atomic bomb survivors and patients with ankylosing spondylitis given x-ray therapy. J. Natl. Cancer Inst. 75:1–21.
6. Harley, N. H., and B. S. Pasternack. 1981. A model for predicting lung cancer risks induced by environmental levels of radon daughters. Health Phys. 40:307–316.
7. Hornung, R. W., and T. J. Meinhardt. 1987. Quantitative risk assessment of lung cancer in U.S. uranium miners. National Institute for Occupational Safety and Health, Centers for Disease Control, Cincinnati, Ohio. June 16, 1986.
8. Howe, G. R., R. C. Nair, H. B. Newcombe, A. B. Miller, andJ. D. Abbatt. 1986. Lung cancer mortality (1950–1980) in relation to radon daughter exposure in a cohort of workers at the Eldorado Beaverlodge uranium mine. J. Natl. Cancer Inst. 77(2):357–362.
9. Kalbfleisch, J. D., and R. L. Prentice. 1980. The Statistical Analysis of Failure Time Data. New York: John Wiley & Sons.
10. Land, C. E., and M. Tokunaga. 1984. Induction period. In Radiation Carcinogenesis. Epidemiology and Biological Significance, J. D. Boice, Jr., and J. E. Fraumeni, Jr., eds. New York: Raven.
11. Lundin, F. E., V. E. Archer, and J. K. Wagoner. 1979. An exposure-time-response model for lung cancer mortality in uranium miners—effects of radiation exposure, age, and cigarette smoking. Pp. 243–264 in Proceedings of the Work Group at the Second Conference of the Society for Industrial and Applied Mathematics, N. E. Breslow and A. S. Whittemore, eds. Philadelphia: Society for Industrial and Applied Mathematics.
12. Muller, J. 1984. Study of mortality of Ontario miners 1955–1977, Part I. Pp. 335–343 in Proceedings of the International Conference on Occupational Radiation Safety in Mining, Vol. 1., H. Stocker, ed. October 14-18, 1984. Toronto: Canadian Nuclear Association.
13. Muller, J., W. C. Wheeler, J. F. Gentleman, G. Suranyi, and R. A. Kusiak. 1983. Study of Mortality of Ontario Miners, 1955–1977, Part I. Toronto: Ontario Ministry of Labour, Ontario Workers Compensation Board.
14. National Institutes of Health. 1985. Report of the NIH Ad Hoc Working Group to Develop Radioepidemiological Tables. Publication No. 85-2748. Bethesda, Md.: National Institutes of Health.
15. Pierce, D. A., and D. L. Preston. 1985. Analysis of cancer mortality in the A-bomb survivor cohort. Pp. 1-13 in Proceedings of the 45th Session of the International Statistical Institute, Vol.3, No. 3. Amsterdam: International Statistical Institute.

16. Preston, D. L, K. J. Kopecky, and H. Kato. 1984. Analyses of mortality and disease incidence among atomic bomb survivors. In Statistical Methods in Cancer Epidemiology, W. J. Blot, T. Hirayama, and D. G. Hoel, eds. Hiroshima, Japan: Radiation Effects Research Foundation.
17. Preston, D. L., H. Kato, K. J. Kopecky, and S. Fujita. 1986. Life span study, Report 10, Part 1. Cancer mortality among A-bomb survivors in Hiroshima and Nagasaki, 1950–1982. Techincal Report 1-86. Hiroshima, Japan: Radiation Effects Research Foundation.
18. Preston, D. L., H. Kato, K. J. Kopecky, and S. Fujita. In press. Studies of the mortality of A-bomb survivors. Radiat. Res.
19. Radford, E. P., and K. G. St. Clair Renard. 1984. Lung cancer in Swedish iron miners exposed to low doses of radon daughters. N. Engl. J. Med. 310(23):1485–1494.
20. Thomas, D. C. 1981. General relative risk models for survival time in matched case control analysis. Biometrics 37:673–686.
21. Thomas, D. C. 1983. Statistical methods for analyzing effects of temporal patterns of exposure on cancer risks. Scand. J. Work Environ. Health 9 353–366.
22. Thomas, D. C., and K. G. McNeill. 1982. Risk Estimates for the Health Effects of Alpha Radiation. Info-0081. Ottawa: Atomic Energy Control Board.
23. Thomas, D. C., K. G. McNeill, and C. Dougherty. 1985. Estimates of lifetime lung cancer risks resulting from Rn progeny exposures. Health Phys. 5:825–846.
24. Whittemore, A. S., and A. McMillan. 1983. Lung cancer mortality among U.S. uranium miners: A reappraisal. J. Natl. Cancer Inst. 1(3):489–499.
25. Whittemore, A. S., and A. McMillan. 1983. Lung cancer mortality among U.S. uranium miners: A reappraisal. Technical Report 68. Stanford, Calif.: Stanford University, Department of Statistics.
26. Wilks, S. S. 1962. Mathematical Statistics. New York: John Wiley & Sons.

ANNEX 2B

Radon Dosimetry

This annex gives background information about the radon-222 decay chain and its dosimetry, the entry and deposition of radon daughters in human lungs, and the factors influencing dose per unit exposure in underground mines and homes.

DECAY OF RADON AND ITS DAUGHTERS

The decay scheme of radon-222, starting with its parent isotope radium-226 and ending with stable lead-206, is shown in Figure 2-1 of Chapter 2. Although this decay scheme appears complicated, the frequency of decay along some of the branches is so low that ordinarily they can be neglected. When these branches are omitted, the result is a simple, linear chain. The first four progenies of radon, ^{218}Po, ^{214}Pb, ^{214}Bi, and ^{214}Po (or RaA, RaB, RaC, and RaC′, respectively) have half-lives that are short (all less than 30 min) compared to the 22 yr of the fifth progeny ^{210}Pb (RaD). As a result, under most circumstances only these short-lived alpha-emitting daughters are of consequence in the respiratory dosimetry of the radon chain.

Table 2B-1 gives more information about the parent and these important daughters. The second column of Table 2B-1 gives the symbols that were assigned to the short-lived progeny before the present standard symbols were adopted. Notice the very short half-life of ^{214}Po. It is so short that for all practical purposes it is always in equilibrium with its parent, ^{214}Bi, and decay chain calculations only need to be made for the chain ^{218}Po, ^{214}Pb, and ^{214}Bi.

TABLE 2B-1 Some Physical Properties of ^{222}Rn and Its Short-Lived Decay Products

Element	Historical Symbol	Principal Radiation(s)	Decay Energies (MeV)	Half-Life	No. of Atoms	
					Per μCi	Per Bq
^{226}Ra	Ra	α	4.8	1,620 yr	2.7×10^{15}	7.4×10^{10}
^{222}Rn	Rn	α	5.5	3.82 day	1.8×10^{10}	4.8×10^{5}
^{218}Po	RaA	α	6.0	3.05 min	9.77×10^{6}	2.6×10^{2}
^{214}Pb	RaB	β,γ	1.0 max	26.8 min	8.58×10^{7}	2.3×10^{3}
^{214}Bi	RaC	β,γ	3.3 max	19.7 min	6.31×10^{7}	1.7×10^{3}
^{214}Po	RaC	α	7.7	164 μs	8.8	2.4×10^{-4}

Uranium-238, the head of the chain of which radon is a part, has a half-life of 4.5 × 10^9 yr. Since this time is comparable to the age of the earth, the ^{238}U now present was in existence when the earth was formed. A typical concentration of uranium in ordinary soil is 20 Bq/kg (0.6 pCi/g), with a range of roughly 7 to 40 Bq/kg (0.2–1 pCi/g).[32]

The fifth progeny of ^{238}U and the immediate parent of radon is radium, ^{226}Ra. Radium's half-life is 1,620 yr, long enough that the activity of a deposit of radium does not change appreciably during our lifetimes except by erosion and transport. Typically, the concentrations of radium in ordinary soil are about the same as those of ^{238}U, because the two are approximately in radioactive equilibrium. Only rarely do natural processes tend to separate radium from the uranium in the soil. Wastes from uranium and phosphate mining and processing contain significantly higher concentrations of radium; uranium tailings piles may have 100 to 500 times the concentrations of normal soils; in some cases this may be up to 14,000 times. Phosphate tailings may have an order of magnitude higher concentration of radium than do soils.[33]

A typical value for the radon concentration in air over average soil is 4 Bq/m^3 (100 pCi/m^3), but diurnal and seasonal variations are a factor of 2 or more. The radon escapes from considerable depths in soil, but less efficiently with greater depth. Kraner et al.[29] estimated that half of the escaping radon comes from the first meter and 75% comes from the top 2 m. The rate of escape of radon from soil differs widely between different locations and between different times at the same location. It depends on the porosity of the containment material, the radon concentrations of the interstices in that material, the wetness of the material, the barometric pressure, and the wind velocity at the surface of the material. The radon concentration over material with a of high radium and radon content, such as in and near mines, is roughly proportional to its concentration in the soil—as high as 2 orders of magnitude more than typical values.

Radon decays into elements that are solids at ordinary temperatures. There are electrical charges left on these atoms as a result of the decay processes. If the radon progenies are produced in air, most of the charged atoms rapidly become attached to aerosol particles. Because the proportion of ions that do not become so attached is particularly important in the dosimetry of radon progeny, it has been given a special designation, that is, the unattached fraction *f*.

The working level (WL) was introduced by Holaday et al.[16] as a convenient one-parameter measure of the concentration of radon progeny in uranium mine air that can be employed as a measure of exposure. They defined 1 WL to be any combination of ^{218}Po, ^{214}Pb, ^{214}Bi, and ^{214}Po (the short-lived progeny of radon) in 1 liter of air, under ambient temperature and pressure, that results in the ultimate emission of 1.3×10^5 MeV of alpha-particle energy. This is about the total amount of energy released over a long period of time by the short-lived daughters in equilibrium with 100 pCi of radon.

Only the short-lived daughters are included in the definition of the WL because they give most of the dose to the lung. The dose due to beta particles is small, and alpha particles due to radon itself are unlikely to be emitted within the body because almost all inhaled radon is exhaled. Most of the ^{210}Pb (22-yr half-life) and subsequent progeny are probably eliminated from the body before they decay (but ^{210}Pb has been measured in people with many years of exposure). Notice that the WL definition allows any combination of radon daughters that results in a certain total decay energy. Thus, the daughters need not be in equilibrium. Table 2B-2 shows the distribution of alpha energy for the different radon progeny. From the last column in this table, the number of WL for an atmosphere containing A pCi/liter of ^{218}Po, B pCi/liter of ^{214}Pb, and C pCi/liter of ^{214}Bi is

$$\mathrm{WL} = 0.001A + 0.0052B + 0.0038C \qquad \text{(2B-1)}$$

whether or not the daughters are in equilibrium. Note also that the working level is independent of the unattached fraction. These independent factors are important when dose to the lung is considered.

The working-level month (WLM) was introduced so that both the duration and level of exposure could be taken into account. The WLM is defined as the sum of the products of the WL times the duration of exposure during some specified total period. The unit WLM is equal to 170 WL hours, which corresponds to an exposure of 1 WL for 170 h (approximately 1 working month).

LUNG MODELS FOR RADON EXPOSURE

A *dosimetry model* is a collection of mathematical functions used to calculate absorbed doses. The dose delivered by radon daughters to the lung depends on both the aerosol involved and the physiology of the lung. The aerosol factors include the size distribution

TABLE 2B-2 Definition of the Working Level[a]

Nuclide	No. of Atoms/ 100 pCi of ^{222}Rn	Ultimate α Energy/atom (MeV)	Ultimate α Energy/ 100 pCi (MeV $\times$ 10^5)	Fraction of Total α Energy
^{218}Po	977	6.0 + 7.7 = 13.7	0.134	0.10
^{214}Pb	8,500	7.7	0.660	0.52
^{214}Bi	6,310	7.7	0.485	0.38
^{214}Po	0	7.7	0	0
Totals			1.279	1.00

[a] Explanation of the calculation: From Table 2B-1, 977 atoms of ^{218}Po are in equilibrium with 100 pCi of Rn. Each atom gives 6.0 MeV when it decays plus another 7.7 MeV when its daughter ^{214}Pb decays. There will also be 8,500 atoms of ^{214}Pb; no alpha particle energy will be released until these atoms turn into RaC which releases 7.7 MeV when it decays, etc.

of the particles, the fraction unattached, and the equilibrium factor. The physiological factors which determine the dose include the lung morphometry, the depth of the target cells, the amount of air moved through the lung per minute, the particle deposition fraction, the mucus thickness, and the mucus transport rate. This section describes some of the recent dosimetric models for radon daughters.

In 1964, Altshuler et al.[1] and Jacobi[21] introduced the first modern lung models which described the inhalation, deposition, and retention of the radon daughters. Since then, many other investigators authors have developed improvements, reflecting the growing interest in exposure to radon.[6,14,23,25,34,40–42] The Nuclear Energy Agency (NEA)[35] gives a good review of the problems of modeling radon daughter dose, and its report is particularly valuable in reporting results of numerical comparisons of two commonly used models, the models described by Jacobi and Eisfeld[23] and James et al.[28] using similar values for the parameters in these models.

Radon is an inert gas. Essentially, it does not react with anything. It does dissolve in body fluids, but the resulting concentration is so low that the dose to tissues from radon itself is negligible. Figure 2B-1 shows roughly how the regions of the body relevant to radon-daughter dosimetry are considered for mathematical modeling. Each block in the model, in principle, gives rise to a differential equation; in practice, however, only limiting forms of the equations, representing equilibrium conditions, are used.

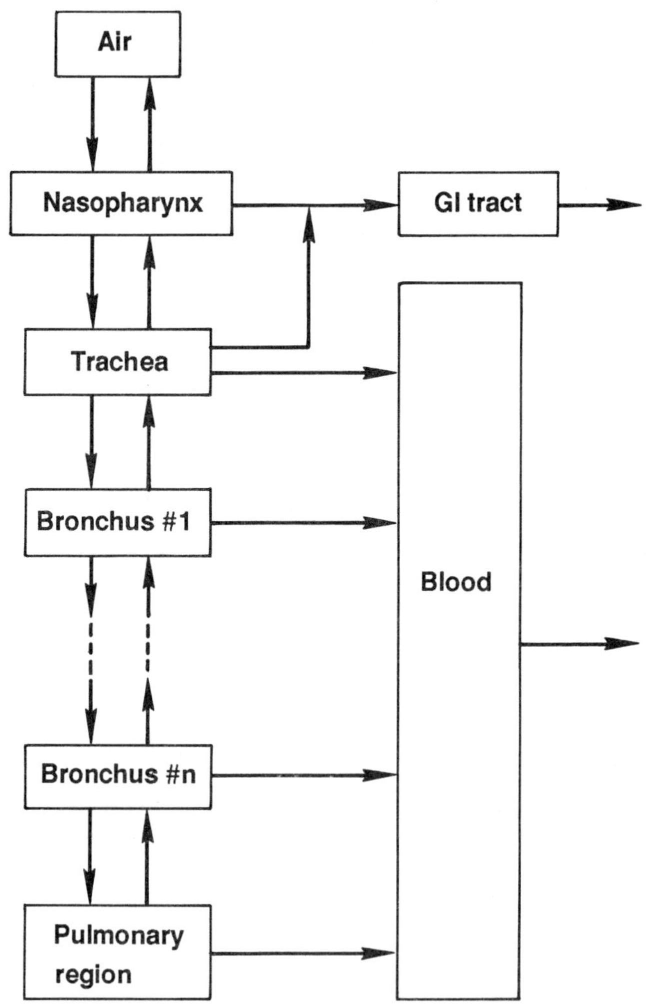

FIGURE 2B-1 A simple compartmental model for radon-daughter dosimetry.

The differential equations are of the same form as those used for radioactive decay chains because the two theories are based on similar assumptions. Present dosimetry models assume that every radionuclide of a given type has the same probability per unit time for removal from a compartment (i.e., first-order kinetics) such as those shown in Figure 2B-1. Similar models used in the study of radioactive tracers have been successful in many cases, probably because the tracers undergo very rapid mixing in the body and thus give each nuclide a similar chance to be removed. But alpha-emitters deposited in an organ, for example, the lung, may have different removal probabilities in different parts of the organ—in different parts of each compartment in Figure 2B-1. One can divide the organ into smaller parts in the hope that the spread of probabilities will thus be reduced; but such division requires that the probabilities be supplied for each part. Often the probabilities are poorly known even for the whole organ, and objections are raised to the speculations involved in assigning probabilities to smaller parts. Many think it best to keep the physiological part of the model fairly simple and to accept the inaccuracies that result from oversimplification—the uncertainties may be no greater than those resulting from inaccuracies in the physiological data.

Figure 2B-1 was simplified for the present discussion. One simplification is that only one set of arrows (the arrows represent processes of deposition and removal) is shown; actually, each arrow must be replaced by several, for example, for individual radionuclides, for free atoms of the radionuclide, for atoms bound to aerosols. Each arrow is associated with a probability per unit time for transfer of material from one compartment to another. These probabilities are determined from detailed studies of the processes.

The first step for the mathematical modeling of the respiratory tract is compartmentalization, as shown in Figure 2B-1. The dimensions and relative orientations of the airways these compartments represent are based on measurements of human lungs.[17,43,44] To obtain a computer-manageable structure for the respiratory tract, the measurements are averaged to give a symmetrical branching structure; that is, each division of an airway was assumed to give two new airways of equal size at the same angle to the original airway. The change in lung structure during the breathing cycle is not included in the dosimetry models; rather, the structure at mid-breath is used. The parts of the tract are numbered, and the parameters describing each part are stored in the computer. For purposes of discussion, the airways are also divided into three functional compartments: the nasopharyngeal, the tracheobronchial, and the pulmonary regions. The first 16 generations are included in the tracheobronchial region, and the pulmonary region extends beyond the 16th generation.

Yeh and Schum[44] used a computer program to trace all available pathways from the trachea to the terminal bronchioles in their measured data for human lungs. They then averaged the parameters describing the pathways to develop a typical path model. Their data were for a large man with lungs at full inflation, so the dimensions for smaller persons had to be scaled down.[35] James and colleagues[28,35] included use of the Yeh and Schum model[44] as an option in their dosimetry model.

There are three principal models that have been applied to study the dosimetry of radon daughters in the lung. These models are those Harley-Pasternak (H-P),[14] Jacobi-Eisfeld (J-E),[23] and James and colleagues (J-B).[28] While these models differ in several important respects, including clearance mechanisms, cells at risk, and lung morphology, the effects of these different assumptions on the calculated dose per unit exposure for any specific conditions to be found in a mine or a home are reported to differ by only a factor of 3 or less.[26] Although there is not yet a scientific basis for choosing a best model, it is shown below that the predictions made from these three models for the ratio of the dose per unit exposure in a home as compared to that in a mine are similar.

In the H-P model, clearance is by mucus only and uptake in the blood is ignored. It incorporates the recent model of the lung by Yeh and Schum[44] and assumes that the only cells at risk are the basal cells of the tracheobronchial epithelium. Dose is calculated at a fixed depth of 22 μm below the surface for the first 10 generations and at 10 μm beyond the 10th generation.

The J-E model assumes the symmetrical lung model of Weibel[43] and considers the effect of variable depth of target cells. Clearance is by both mucus and by uptake in the blood.

The J-B model makes use of the Yeh and Schumm[44] lung morphometry and a uniform probability distribution of depths for target cells; that is, the dose is averaged over the range of the alpha particle (mean dose). It includes the dose to epithelial tissue as dissolved activity is transported through the bronchial membranes into blood. This assumption results in a higher mean dose per unit of exposure of the unattached fraction than that given by the other two models.

Each of these models can be applied to the determination of the dose to the tracheobronchial region of the lung, where most lung cancers in humans are found. The differences between the models for different particle sizes, unattached fractions, and minute volumes will be seen to be only a factor of 2 or 3 within the normal range of these factors. A comparison of the models gives a measure of one source of uncertainty in the determination of dose from radon daughters.

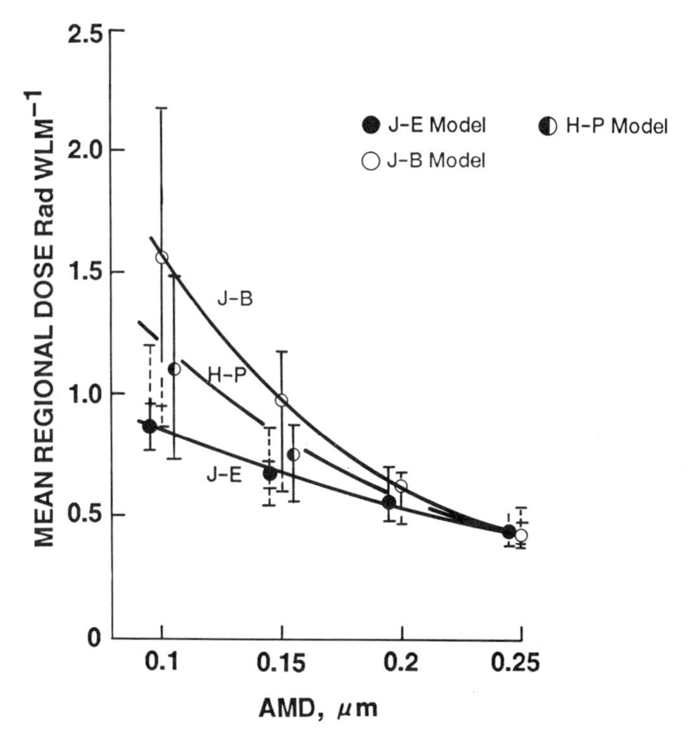

FIGURE 2B-2 Mean dose to the tracheobronchial region as a function of particle size (minute volume, 1.2 m^3/h).[26]

Figure 2B-2 shows how the mean dose (averaged over the range of the alpha particles) to the bronchi varies as a function of the particle size for each of these models. The J-E and the J-B models differ for small particles by about a factor of 2 in the tracheobronchial region, while the mean dose for the H-P model is intermediate. For the dose to basal cells at 22 μm, the H-P model gives a higher value for 0.1 activity median aerodynamic diameter (AMAD) particles than do the other two models. For small particles the variation between the models is about a factor of 3. For the pulmonary region, both the J-E and J-B models give much smaller doses that do not vary appreciably with the size of the particles.

In most circumstances, inhaled aerosols have a fairly wide distribution of sizes. For aerosols with attached radon daughters in mines, George et al.[9] measured activity median diameters of 0.1 to 0.3 μm with geometric standard deviations σ_g of about 2.7. Jackson et al.[20] found four components of mine atmospheres: (1) free atoms;

(2) condensation nuclei with activity mean aerodynamic diameters (AMAD) of less than about 0.1 μm; (3) diesel soot with AMAD between 0.2 and 0.4 μm, with a σ_g of about 2; and (4) ore dust with AMAD of 0.6 μm and a σ_g of 2. Davies[5] also concluded that the broad spectrum of aerosol sizes could be represented by components with σ_g's of about 2.

A second factor affecting the bronchial dose, the fraction of radon progeny that is not attached to dust particles (f_p), also varies between different kinds of occupational and domestic environments. The behavior of unattached atoms is not well understood. It is often modeled by assuming the atoms diffuse with a diffusion coefficient of 0.05 cm^2/s (from the work of Chamberlain and Dyson[3]). Jacobi and Eisfeld[23] and James and colleagues[26,35] investigated the use of diffusion coefficients as low as 0.02 and as high as 0.08 in their models. More recently, Harley and Pasternack[14] have suggested a value of 0.005 cm^2/s for the diffusion coefficient (corresponding to the formation of clusters with diameters of approximately 0.004 μm). The variation in the calculated doses between the models is greater for larger diffusion coefficients. In the range 0.02–0.08 cm^2/s the deposition changes by as much as a factor of 2 in the regions of highest deposition and by 1 order of magnitude in the lower airways. In no case did any appreciable deposition occur in the pulmonary lung.

Each of the three models treats the unattached fraction differently. Figure 2B-3 shows how the mean dose per unit exposure varies as a function of percent unattached fraction in each model. The means of dose per WLM given by the various models differ by about a factor of 2 for an unattached fraction of 0.1 but agree better for lower values (Figure 2B-3). The mean doses shown in Figure 2B-3 are again the average doses throughout the full depth of tissue penetrated by the alpha particles. The dose is larger at the surface of the bronchi. Consequently, calculations using the H-P model for doses to the basal cells at 22 μm in the segmented bronchi are larger than those shown in Figure 2B-3. For example, at $f_p = 0.1$, this model yields doses that are about twice the basal dose calculated with the J-B and J-E models.[26] At smaller values of the unattached fraction, differences between models are more modest.

Mouth breathing increases the dose due to the unattached fraction. Measurements by George and Breslin[7] indicate that for nasal breathing about 60% of a charged aerosol is deposited in the nose.

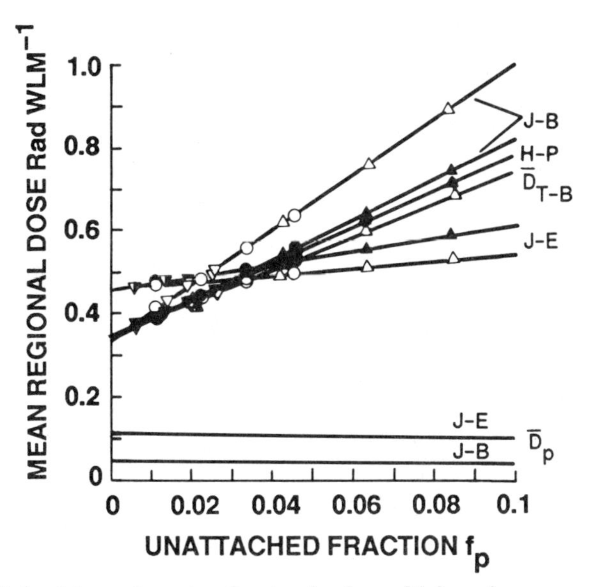

FIGURE 2B-3 Mean dose to the tracheobronchial region as a function of percent unattached fraction.[26] Symbols: △, diffusion coefficient = 0.005 cm^2/s; ▲, diffusion coefficient = 0.05 cm^2/s

In contrast to charged particulates in the unattached fraction, little deposition of uncharged particulates occurs in nasal passages.

A third factor affecting the dose is the amount of radioactivity that is inhaled and deposited in the bronchi. The entry of radioactivity into the lung depends on the amount of air inhaled per unit time (the *minute volume*), which is the product of two quantities: the *tidal volume*, the volume of air inhaled per breath, and the *breathing rate*, the number of breaths per unit time. The International Commission on Radiological Protection (ICRP) report on reference man[18] gives a considerable range of minute volumes that depend chiefly on the level of activity of a person. For reference values it gives 7.5 liters/min for men at rest and 20 liters/min for light activity. The ICRP assumes that the value for light activity is a reasonable average over the day's activities for most occupations, although it is recognized that there is a wide variation. The 20 liters/min corresponds to a tidal volume of 1.25 liters and a breathing rate of 16 breaths per minute.

The NEA study[35] compared minute volumes of 12.5 and 20 liter/min (0.75 and 1.2 m^3/h) in two models. Increased breathing rates can decrease the fractional deposition in the earlier bronchial

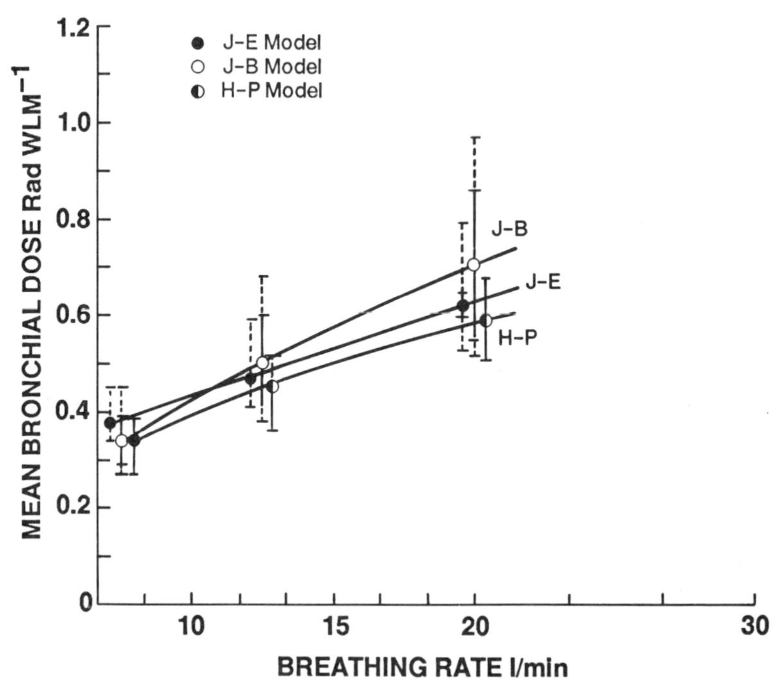

FIGURE 2B-4 Mean bronchial dose as a function of minute volume.[26]

generations by as much as 50%, so that the total deposition increases less rapidly than the increased alpha energy inhaled. Recently, James[27] has reported that the total deposition is proportional to the square root of the minute volume over the range of particle sizes in which deposition is due to particle diffusion.

The variation in the mean bronchial dose as a function of minute volume for the three models is shown in Figure 2B-4 for particles with an AMAD of 0.15 μm. Note that the curves are concave downward because of the square root variation of dose with minute volume.

In summary, the results of efforts to model the dose to the lung show significant differences in the estimates of the dose per unit of exposure in the tracheobronchial region of the lung: factors of 2 to 3 are typical. Unfortunately, there are few experimental data with which these estimates can be compared, so it is not possible to make a fully informed choice among the models on the basis of measurements. Recently, Cohen[4] has reported on measurements leading to correction factors of about 2 in an updated H-P five-lobe model.[13]

Furthermore, some recent experiments indicate that the dose at bifurcations is not uniform and may be considerably larger than those calculated with current models.[31] These observations have not, however, been confirmed by Cohen's studies.[4] Although the deposition pattern predicted by any particular model cannot be validated in vivo, the average deposition patterns (and hence, the average dose) for all the models are sufficiently similar so that any of them can be used with about equal confidence.

Even though these three lung models give different values for any particular case, the dose estimates are converging as more information becomes available. A survey of dose calculations for miners made in the period 1951–1981 by the National Council on Radiation Protection and Measurements (NCRP)[34] included 28 calculations of the dose per WLM to the bronchial tissue. The results ranged from 0.7 to 140 mGy/WLM (0.07 to 14 rad/WLM). In addition to differences in the loci of the bronchi for which the dose was calculated, differences in input values of the parameters describing the radon daughters in the air and of the parameters characterizing the dosimetry models were responsible for much of the spread. Jacobi[22] estimated that input of the same aerosol characteristics into these models would reduce the spread to about 3 to 10 mGy/WLM (0.3 to 1 rad/WLM). Agreement is even better when the average dose to the trachealobronchial tree is calculated. For the most advanced calculational models, the variation in mean doses can be as little as 20%.[35] However, the degree of agreement is better for some aerosol distributions then others, being best for air in mines and poorer for the cleaner air in homes where the unattached fraction may be higher.

EXTRAPOLATION OF DOSES FROM MINES TO HOMES

The committee's estimates of lung-cancer risks due to the inhalation of radon and its progeny are based on data derived from mining populations. These estimates can be used to calculate the risk per WLM in nonoccupational situations, but the attendant assumptions and uncertainties associated with this change in exposure conditions must be considered. There are potentially important differences between the environmental conditions in which exposure is sustained in a mine and a home and between the physical characteristics of miners and members of the general population. Just as there is no single lung model that has been shown to be best in all cases of radon exposure, the committee finds that there is no single characterization of the environment in mines or homes that can be called typical. For the same working-level exposure, there can be a wide variation of particle size, equilibrium factors, and percent unattached fraction. In addition, a miner working underground and a person living in a home have different levels of activity and, thus, different breathing patterns.

Rather than postulating some average or typical values for the various factors that affect the dosimetry of radon progeny in mines and homes and calculating appropriate dose per WLM conversion factors, the committee proposes a methodology that can be applied when any radon environment is sufficiently characterized. By using the ratio of the dose per WLM in mines to that in homes obtained by means of the lung models described above, it is possible to extrapolate the risk values obtained for underground miners to people living in homes, at a reasonable level of approximation.

Let K be a dimensionless factor that, when multiplied by the risk to miners per WLM, will give the risk to an individual in a home per WLM.

$$K \frac{(\text{Risk})_m}{(\text{WLM})_m} = \frac{(\text{Risk})_h}{(\text{WLM})_h}. \qquad (2B\text{-}2)$$

While miners exposed to 1 WL receive 1 WLM during a 170-h working month, a person living in a home at 1 WL would normally be exposed to more than 1 WLM in a month. If an occupancy factor (O.F.) is defined as the fraction of a 720-h month that is spent in a house, then 1 WL will result in 720/170 O.F. WLM per month of occupancy. Since risk is proportional to dose, K varies as the ratio of the dose per WLM in houses and mines.

$$K = \frac{\text{Risk}_h/\text{WLM}_h}{\text{Risk}_m/\text{WLM}_m} \propto \frac{\text{Dose}_h/\text{WLM}_h}{\text{Dose}_m/\text{WLM}_m}. \qquad (2B\text{-}3)$$

As described above, the dose to the lung depends upon both aerosol and physiological factors. Therefore, to a first approximation, the dimensionless proportionality constant K can be expressed in terms of the partial dose conversion factors obtained from a specified lung model.

$$K \propto \frac{P_h}{P_m} \frac{f_h}{f_m} \frac{F_h}{F_m} \frac{(\text{MV})_h}{(\text{MV})_m} \frac{\ldots/\text{WLM}_h}{\ldots/\text{WLM}_m}, \qquad (2B\text{-}4)$$

where P is dose conversion factor for particle size, f is dose conversion factor for the unattached fraction, F is dose conversion factor for the equilibrium factor, MV is dose conversion factor for the minute volume; all these parameters have units of rad per WLM. Physiological factors such as mucus thickness, mucus transport rate, bronchial morphometry, depth of the target cells, and the particle deposition fraction can be assumed to be approximately equal in miners and persons living in homes. These factors undoubtedly vary from person to person; but except for the possibility of job-related damage to the lung, they should be about equal in miners and others, and therefore would largely cancel out in Equation 2B-4. Differences between smokers and nonsmokers, however, may be important (see Part 3 of Appendix VII).

The committee considered the possibility of systematic differences between miners and others, for example, prevalence of inhalation via the mouth, but found that data were not available for evaluation. However, sex and age are factors to be considered. Although most calculations of aerosol deposition are based on the anatomy of the male, the female airway geometry is similar, and scaling factors can be used for estimating the dimensions of individual airways.[34] In the case of the growing child, the alveolar area is not fully developed at birth. However, the ciliated airways are complete. Calculated deposition patterns[15,36] indicate higher relative deposition in the first few generations in the lungs of children for aerosols of 0.2-μm diameter or less.

For adults of working age and the same smoking status, it is likely that mucociliary clearance rates are similar in miners and in the general population. However, there is little information concerning clearance in other groups in the population and none for children. It is known that clearance in the elderly is delayed.[10] Most lung models show an increase of bronchial dose per unit exposure until about age 6, after which it falls off and becomes nearly constant after age 10. The NEA Organisation for Economic Co-operation and Development has recommended that age dependency of dose equivalent per unit exposure be neglected.[35]

It is instructive to examine the variations of the proportionality factor K, given by Equation 2B-4, under the various dosimetry models, as illustrated by the results shown in Figures 2B-2, 2B-3, and 2B-4. In doing so, it will be shown that for a range of conditions typical in mines and homes, under a given model, K is reasonably stable, even though the variation between models in the calculated dose per WLM varies by a factor of 2 or more.

The first three terms in Equation 2B-4, the conversion factors for particle size, unattached fraction, and equilibrium factor, are representative of the physical environment in the mine or home and are, to some extent, interconnected. For example, with increasing aerosol concentration the dose per WLM conversion factor for equilibrium (F) generally increases somewhat, while that for the unattached fraction (f) decreases. However, these factors are independent to a first order and K can be approximated by the product of the ratios given in Equation 2B-4.

George and colleagues.[8,9] made particle size measurements in homes in New York and New Jersey and in mines in Canada and Colorado. They found that particle size was log normally distributed with an activity mean diameter of 0.12 μm in homes and 0.17 μm in the mines. From Figure 2B-3 it is seen that all three lung models give about 1 rad/WLM for 0.12-μm particles compared to approximately 0.7 rad/WLM for 0.17-μm particles. For these particle distributions P_h/P_m is about 1.4.

The amount of activity that is unattached depends on the ambient environment. Because of low ventilation, the mass concentration of airborne dust was probably high in early mines and the unattached fraction low. More recently, the ventilation in mines has been greatly increased. However, the number concentration of airborne dust particles may have become higher due to the widespread use of underground diesel equipment. In a careful study of several uranium mines, George et al.[9] found that the unattached fraction ranged from 0.004 to 0.16, with a mean value of 0.04.

Recent investigations have provided some preliminary data on the fractions of unattached RaA and the size distribution of the carrier aerosol for the attached fraction in the home environment.[8] The number and size of particles in the air of a home vary considerably throughout the day, depending, among other factors, on the presence of cigarette smoke and the cooking of food. This, in turn, has a large effect on how many ions become attached to particulates. Particulate concentrations are generally lower at night but rarely fall below 10^4 particles/ml. In the absence of specific aerosol sources in the home, Reineking et al.[38] calculated, on the basis of measured concentrations of particulates, that the unattached fraction ranged between 0.06 and 0.15 (mean value, 0.1). With additional aerosol sources, the unattached fraction fell below 0.05. This value for the unattached fraction (0.1) is somewhat higher than those previously assumed,[19,23] and is higher than the mean value (0.04) measured in uranium mines.[7,9] Although the unattached fraction of RaA may be perhaps twice as high in homes compared to mines, extensive data on homes and mines have yet to be published.

Assuming that the mines used to obtain the risk numbers for radon had an unattached fraction (f_m) of about 4%, the three models shown in Figure 2B-4 give a conversion factor to the bronchial region of about 0.5–0.6 rad/WLM. For a diffusion coefficient of 0.005 cm^2/s and an unattached fraction of 7% in homes, the same curves give conversion factors between 0.5 and 0.7 rad/WLM. The ratio f_h/f_m would then be about 1.2 (1.0–1.4), depending on which lung model is chosen.

The equilibrium factor F is dependent on the ventilation rate. High ventilation rates result in low values of F. Jacobi and Eisfeld[24] have shown that variation of the equilibrium factor has very little effect on the dose per unit exposure to the bronchial cells. Therefore, even though F_h and F_m may be different, the ratio of the conversion factors is about 1.

Physiological factors must be considered also. Underground miners, usually adult males, spend about 40 h weekly engaged in moderate to heavy activity in the course of their work. Nonoccupational exposures of adults, both male and female, and children generally occur in the home during light activity or while asleep. So, in addition to differing aerosols in homes and mines, the different breathing patterns of miners and the population in homes must be considered.

The increased minute volume required for the metabolic cost of exercise is achieved by an increase in both the tidal volume and the frequency of breathing. The increased frequency of breathing decreases the mean residence time of aerosols in the lung and, by so doing, reduces the time available for diffusion to deposit particles on the bronchial airways.

Exercise also has a role in how people inhale. With increasing ventilation there is a shift from nasal to oral breathing. For example, with exercise requiring ventilation of 35 liters/min, there is a shift from a pattern of 80% nasal breathing in the resting subject[2] to about 50% nasal breathing.[37] Moreover, many normal people breath oronasally,[39] and those with any form of nasal obstruction have a mainly oral form of breathing. The proportion of oral and nasal breathing influences the bronchial dose since, as noted above, about half of the unattached fraction is assumed to be deposited in the nose. Increased oral breathing, therefore, increases bronchial doses, whereas increased nasal breathing reduces this dose. It can be argued that the larger fraction of unattached RaA in the home compared to that in a mine increases the dose to the bronchus for a given concentration of radon progeny in air. However, the increased minute ventilation of the miner is associated with more mouth breathing, which would tend to increase his dose. The committee could not find data on the pattern of nose or mouth breathing in miners and therefore could not examine this factor analytically. It is an obvious problem for further field investigation.

ICRP,[19] NEA,[35] and NCRP[33] assign a breathing rate of 20 liters/min to underground miners and base their dose estimates on this minute volume. Ventilation rates in working coal miners have been measured;[11] values obtained ranged from about 30 to 40 liter/min when averaged throughout the working shift. This is about 1.5 to 2 times the 20-liter/min value assigned to miners and 3 to 4 times the 12.5 liter/min value assigned to the general population in the NEA[35] model. Moreover, individuals undertaking the same task have differing minute ventilation.[11] Heavier miners breathe more, as do older miners. The very high ventilation rates found in coal miners might not be representative of values for uranium miners.

In the case of the home environment, there are considerable uncertainties in attempting to estimate overall deposition patterns throughout the 24-h cycle. Not only is the distribution of oral/nasal breathing a relevant factor, but the tidal volume and frequency of breathing change continuously. The minute volume of a sleeping person is low, and the convective diffusion component of bronchial deposition may be insignificant.[30]

For a miner who averages a minute volume of 30 liters/min, the conversion factors from the three models shown in Figure 2B-4 would be on the order of 0.8 rad/WLM. The average person in a home breathing, 12.5 liters/min would receive a dose of 0.45 rad/WLM. Thus, the conversion factor ratio due to these different minute volumes (MV_h/MV_m) would be 0.56. If the miners do a significantly larger amount of mouth breathing than occurs in homes, this ratio would be smaller.

For the mean bronchial dose, differences between the models are small. The ratio of the conversion factors due to the different minute volumes would remain nearly the same for any of the three models because the shapes of the curves in Figure 2B-4 are very similar. However, for basal cells at 22-μm depths as the target cells, the H-P model gives somewhat higher doses per unit exposure than do the other two models, and the variation in the ratio of the factor, MV_h/MV_m, in Equation 2B-2 due to the choice of the model becomes larger.

As outlined in Equation 2B-4, the four conversion factors making up K have ratios in homes to mines of 1.4, 1.2, 1.0, and 0.56. Therefore, for this specific case, the product of these factors is 0.94 and the mean dose received by a person exposed in a home to 1 WL of radon for 170 h is very nearly the same as the dose to a miner exposed to 1 WLM. On the other hand, if the minute volume for miners is taken as 20 liters/min, the ratio K is increased to about 1.3. In homes where the unattached fraction is low (4%), this ratio is less, about 0.8.

Other investigators have reached similar conclusions.[12] After reviewing the parameters relevant to a comparison of the WLM in a mine and in the home, the NCRP concluded that the alpha dose to the surface of the bronchial tree (rad per WLM) may be somewhat higher for environmental exposure than for underground exposure, largely due to a higher fraction of unattached RaA in the home.[34] In contrast, the NEA has estimated smaller doses per WLM for adult members of the general population than for miners.[35] The NEA assumed a smaller unattached fraction (3%) than that assumed by the NCRP (7%) and a slightly lower breathing rate than that assumed by the NCRP. Both the NCRP and the NEA analyses assume that miners inhale 20 liters/min, that is, light activity. Upon their review of the possible range of the input parameters for the dosimetric models, the NCRP concluded that the dose per WLM in homes, as compared to that in mines, differs by less than a factor of 2.

In summary, the committee believes that with the present state of scientific knowledge, it can neither choose between the three major radon lung models nor specify the best values of the factors which characterize the dose in a home or in a mine. However, all of the models are in agreement within about a factor of 3 or so. While the method of ratios used here does not directly calculate a dose to the lung tissues, it does allow the extrapolation of known risks in a mine to a home where dosimetry factors have been established.

Future work on this methodology should be concentrated in areas that will improve the quantification of the constant K. This will include improving our knowledge of the environment in the mines from which the risk estimates were obtained, characterizing the indoor environment of homes more completely, and determining breathing characteristics of uranium miners by on-site measurements. Finally, high priority should be given to creating and validating a lung model which retains the best features of the models now in use.

REFERENCES

1. Altshuler, N., N. Nelson, and M. Kuschner. 1964. Estimation of lung tissue dose from the inhalation of radon daughters. Health Phys. 10:1137–1161.
2. Camner, P. 1981. Influence of nose and mouth breathing on particle deposition in the lung. Health Phys. 40:99–100.
3. Chamberlain, A. C., and E. D. Dyson. 1956. The dose to the trachea and bronchi from the decay products of radon and thoron. Br. J. Radio. 29:317–325.
4. Cohen, B. S. In press. Deposition of ultrafine particles in human tracheobronchial tree. In Radon and Its Decay Products: Its Occurrence Properties and Health Effects, P. K. Hopke, ed. Symposium Series 331. Washington, D.C.: American Chemical Society.
5. Davies, C. N. 1974. Size distribution of atmospheric particles. Aerol. Sci. 5:293–300.
6. Fry, R. M. 1970. Radon and its hazards. Pp. 13–32 in Personal Dosimetry and Area Monitoring Suitable for Radon and Daughter Products. Paris: Nuclear Energy Agency, Organisation for Economic Co-operation and Development.
7. George, A. C., and A. J. Breslin. 1969. Deposition of radon daughters in humans exposed to uranium mine atmospheres. Health Phys. 17:115–124.
8. George, A. C., and A. J. Breslin. 1975. The distributions of ambient radon and radon daughters in residential buildings in the New Jersey-New York area. In The Natural Radiation Environment II, T. F. Gesell and W. M. Lowden, eds. Washington, D.C.: Technical Information Center, U.S. Department of Energy.
9. George, A. C., L. Hinchcliffe, and R. Sladowski. 1975. Size distribution of radon daughter particles in uranium mine atmospheres. Am. Ind. Hyg. Assoc. J. 34:484–490.
10. Goodman, R. M., B. M. Yergin, J. K. Landa, M. H. Golinuaux, and M. A. Sacker. 1978. Relationship of smoking history and pulmonary function tests to tracheal mucous velocity in nonsmokers, young smokers, ex-smokers and patients with chronic bronchitis. Am. Rev. Respir. Dis. 117:205–214.
11. Hadden G. G., C. O. Jones, and D. C. Morgan. 1967. A study of volumes of refined air in relation to dust exposures of coal miners. Pp. 37–48 in Inhaled Particles and Vapours II, C. N. Davies, ed. Elmsford, N.Y.: Pergamon.
12. Harley, N. H. 1984. Comparing radon daughter dose: Environmental versus underground exposure. Rad. Protect. Dosimetry 7:371–375.
13. Harley, N. H., and B. S. Cohen. In press. Updating radon daughter bronchial dosimetry. In Radon and Its Decay Products: Its Occurrence Properties and Health Effects, P. K. Hopke, ed. Symposium Series 331. Washington, D.C.: American Chemical Society.
14. Harley N. H., and B. S. Pasternack. 1982. Environmental radon daughter alpha factors in five-lobed human lung. Health Phys 42:789–799.

15. Hislop, A., D. C. F. Muir, M. Jacobsen, G. Simon, and L. Reid. 1972. Postnatal growth and function of the pre-acinar airways. Thorax 27:265–274.
16. Holaday, D. A., D. E. Rushing, R. D. Coleman, P. F. Woolrich, H. L. Kusnetz, and W. F. Bale. 1957. Control of Radon and Daughters in Uranium Mines and Calculations of Biological Effects. U.S. Pulic Health Service Publication No. 494. Washington, D.C.: U.S. Government Printing Office.
17. Horsfield, K., F. G. Relea, and G. Cumming. 1976. Diameter, length and branching ratios in the bronchial tree. Respir. Physiol. 26:351–356.
18. International Commission on Radiological Protection (ICRP). 1975. Report of the Task Group on Reference Man. ICRP Publication 23. Oxford: Pergamon.
19. International Commission on Radiological Protection (ICRP). 1981. Limits for the Inhalation of Radon Daughters by Workers. ICRP Publication 32. Oxford: Pergamon.
20. Jackson, P. O., J. A. Cooper, J. C. Langford, and M. R. Peterson. 1982. Characteristics of attached radon-222 daughters under both laboratory and underground uranium mine environments. Pp. 1031–1042 in Radiation Hazards in Mining: Control, Measurement, and Medical Aspects, M. Gomez, ed. New York: Society for Mining Engineers of the American Institute of Mining, Metallurgical, and Petroleum Engineers, Inc.
21. Jacobi, W. 1964. The dose to the human respiratory tract by inhalation of short-lived ^{222}Rn- and ^{220}Rn-decay products. Health Phys. 10:1163–1174.
22. Jacobi, W. 1977. Interpretation of measurements in uranium mines: Dose evaluation and biomedical aspects. Pp. 33–48. in Personal Dosimetry and Area Monitoring Suitable for Radon and Daughter Products. Paris: Nuclear Energy Agency, Organisation for Economic Co-operation and Development.
23. Jacobi, W., and K. Eisfeld. 1980. Dose to Tissue and Effective Dose Equivalent by Inhalations of Radon-222 and Their Short-Lived Daughters. GSF Report S-626. Neurherberg, Federal Republic of Germany: Geschelscheft für Strahlen and Umwelfforschung.
24. Jacobi, W., and K. Eisfeld. Internal Dosimetry of Radon-222, Radon-220 and their Short-Lived Daughters. Pp. 131-143 in Proc. Second Special Symposium on Natural Radiations Environment, K. G. Vohra et al., eds., January 1981, Bhabha Atomic Research Centre, Bombay. New York: John Wiley & Sons.
25. James, A. C. 1977. Bronchial deposition of free ions and submicron particle studies in excised lung. Pp. 203–218 in Inhaled Particles IV, W. H. Walton, ed. New York: Pergamon.
26. James, A. C. 1986. Dosimetric approaches to risk assessment for indoor exposure to radon daughters. Radiat. Prot. Dosimetry 7:353–366.
27. James, A. C. In press. A reconsideration of cells at risk and other key factors in radon daughter dosimetry. In Radon and Its Decay Products: Its Occurrence Properties and Health Effects, P. K. Hopke, ed. Symposium Series 331. Washington, D.C.: American Chemical Society.
28. James, A. C., J. R. Greenhalgh, and A. Birchall. 1980. A domestic model for tissues of the human respiratory tract at risk from inhaled radon and thoron daughters. Pp. 1045–1048 in Radiation Protection. A Systematic Approach to Safety, Proceedings of the 5th Congress of International

Radiation Protection Association (IRPA), Vol. 2, Jerusalem. New York: Pergamon.

29. Kraner, H. W., G. L. Schroder, and R. D. Evans. 1964. Measurements of the effects of atmospheric variables on radon-222 flux and soil-gas concentrations. Pp. 191–215 in The Natural Radiation Environment, J. A. S. Adams and W. M. Lowder, eds. Chicago: University of Chicago Press.
30. Martin, D., and W. Jacobi. 1972. Diffusion deposition of small-sized particles in the bronchial tree. Health Phys. 23:23–29.
31. Martonen, T. B., W. Hofman, and J. E. Lowe. 1987. Cigarette smoke and lung cancer. Health Phys. 52:213–217.
32. National Council on Radiation Protection and Measurements (NCRP). 1975. Natural Background Radiation in the United States. NCRP Report 45. Washington, D.C.: National Council on Radiation Protection and Measurements.
33. National Council on Radiation Protection and Measurements. (NCRP). 1984. Exposures from the Uranium Series with Emphasis on Radon and Its Daughters. NCRP Report 77. Washington, D.C.: National Council on Radiation Protection and Measurements.
34. National Council on Radiation Protection and Measurements (NCRP). 1984. Evaluation of Occupational and Environmental Exposures to Radon and Radon Daughters in the United States. NCRP Report 78. Washington, D.C.: National Council on Radiation Protection and Measurements.
35. Nuclear Energy Agency (NEA), Group of Experts of the Organisation for Economic Co-operation and Development. 1983. Dosimetry Aspects of Exposure to Radon and Thoron Daughter Products. Paris: Nuclear Energy Agency, Organisation for Economic Co-operation and Development.
36. Phalen, R. F., M. J. Oldham, M. T. Kleinaman, and T. T. Crocker. In press. Tracheobronchial deposition predictions for infants, children and adolescents. Anat. Respir.
37. Proctor, D. F. 1981. Oronasal breathing and studies of effects of air pollutants on the lung. Am. Rev. Respir. Dis. 123:242.
38. Reineking, A., K. H. Becker, and J. Porstendorfer. 1985. Measurements of unattached fractions of radon daughters in houses. Sci. Total Environ. 45:261–270.
39. Sheppard, D., J. A. Nadel, and H. A. Boushey. 1981. Oronasal breathing and studies of effects of air pollutants on the lung (correspondence). Am. Rev. Respir. Dis. 123:242–243.
40. United Nations Scientific Committee on the Effects of Atomic Radiation (UNSLEAR). 1977. Sources and Effects of Ionizing Radiation. Report E.77.IX.1. New York: United Nations, 1977.
41. Walsh, P. J. 1970. Radiation dose to the respiratory tract of uranium miners—a review of the literature. Environ. Res. 3:14–36.
42. Walsh, P. J., and P. E. Hamrick. 1977. Radioactive materials—determinants of dose to the respiratory tract. Pp. 233–242 in Handbook of Physiology, D. H. K. Lee, ed. Bethesda, Md.: American Physiological Society.
43. Weibel, E. R. 1963. Morphometry of the Human Lung. New York: Academic Press.
44. Yeh, H. C., and G. M. Schum. 1980. Models of human lung airways and their application to inhaled particle deposition. Bull. Math. Biol. 42:462–480.

3
Polonium

INTRODUCTION

Polonium was the first radioactive element that Marie and Pierre Curie separated from the uranium ore pitchblende. It has over 25 isotopes with mass numbers of 192–218. All are radioactive, most are predominantly alpha-emitters, and many have very short half-lives. The important isotopes and their nuclear properties are listed in Table 3-1. Only ^{208}Po, ^{209}Po, and ^{210}Po have half-lives long enough to permit useful biomedical research; of these, ^{210}Po has been used most. ^{214}Po and ^{218}Po are also important, because they are daughters of radon and contribute a substantial portion of the radiation dose from inhaled radon. They have such short half-lives that no experimental biomedical work can be done with them, and their effects must be inferred from the effects of isotopes with longer half-lives. ^{212}Po and ^{216}Po are in the thorium-thoron decay chain.

One reason for interest in the alpha particles from polonium is their existence as radon daughters; indeed, with respect to important radiation dose, the radon problem is due largely to polonium. There are other reasons for interest in polonium; of primary interest is primarily the isotope ^{210}Po, which has a half-life of about 138 days and decays to a stable lead isotope (^{206}Pb) by almost pure alpha-particle emission. These properties led to its use in much experimental work that required an alpha-particle source with useful energy and a convenient half-life.

TABLE 3-1 Selected Polonium Isotopes and Their Alpha Emissions

Isotope	Half-Life	Emission	α Energy (MeV)
^{208}Po	2.93 yr	α (99%)	5.114
^{209}Po	103 yr	α (99%)	4.882
^{210}Po	138.4 days	α	5.305
^{211}Po	0.52 s	α (99%)	7.448
^{212m}Po	45 s	α (97%)	11.65
^{213}Po	4×10^{-6} s	α (99%)	8.38
^{214}Po	1.6×10^{-4} s	α (99%)	7.69
^{215}Po	1.8×10^{-3} s	α	7.38
^{216}Po	0.15 s	α	6.77
^{217}Po	<10 s	α (80%)	6.55
^{218}Po	3.05 min	α	6.11

Polonium has been used extensively as an alpha-particle source for the production of neutrons by interaction with beryllium. Many large alpha-particle sources were prepared before plutonium was available in sufficient quantities to supplant polonium for this application. In fact, ^{210}Po was the alpha-particle source in the neutron-producing initiators of at least the first generation of atomic weapons. During the Manhattan Project days of World War II, large quantities of ^{210}Po were produced at a plant in Dayton, Ohio, and protection of workers and the environment was needed.

^{210}Po has found wide application in static-eliminator devices, for example, in paper and textile plants. The high specific ionization around such devices is effective in reducing static electricity buildup. However, polonium is difficult to contain, and there have been instances of contamination, not only from static-eliminator bars but from solutions left in the open, because of polonium's marked tendency to "creep."

Polonium occurs naturally in the environment. Airborne radon decays into polonium isotopes that can be deposited on terrain and vegetation, for example, on tobacco leaves. Some have postulated that the alpha-particle radiation from polonium, volatilized from smoking tobacco, plays an important role in the genesis of lung cancer in smokers.[37]

PROPERTIES

The chemical behavior of polonium was described many years ago by Haissinsky[18,19] and others, later by Moyer,[33] and in the Russian literature by Moroz and Parfenov.[30] It has a complex solution and electrochemistry. Properties of prime importance for understanding its behavior in living systems are discussed in a volume prepared by Stannard and Casarett[52,53] and are discussed specifically by Morrow et al.,[31,32] Thomas and Stannard,[49] Thomas,[51] and Feldman and Saunor.[12] Subcellular distribution was investigated by Lanzola et al.[26]

Polonium is chemically different from most of the alpha-emitting elements discussed in this report. It has many of the characteristics of the rare-earth elements, is amphoteric, and tends to form hydroxides and radiocolloids both in vitro and in vivo. As a result of the latter, polonium is phagocytized readily by cells of the reticuloendothelial system and deposits substantially in the spleen, lymph nodes, bone marrow, and liver (in that order) after parenteral administration. Major deposition also occurs in the kidneys. Tissue distribution is influenced considerably by the route of administration.

Autoradiographic studies, begun in the 1920s by Lacassagne and Lattes,[25] have characteristically demonstrated the presence of much aggregated polonium both in solutions at or near neutral pH and in vivo. These aggregates demonstrate the presence of radiocolloids. They are not seen in vivo after oral administration of polonium,[5] and they become disorganized and gradually disappear. In contrast, nonaggregated (semi-ionic or ionic) polonium is more uniformly distributed to tissues and less influenced by the route of administration. The nonaggregated form, although less striking autoradiographically, can account for a substantial fraction of the radiation dose.

Polonium has less tendency to form specific complexes with biomolecules than do radium, plutonium, americium, or other transuranic elements, although relatively loose combinations with numerous moieties are common; for example, polonium combines with the globin portion of hemoglobin and other blood constituents and binds nonspecifically to proteins. It does not exchange for calcium in bone, as does radium, nor does it combine with osteoid, as does plutonium. Polonium is relatively volatile and is easily vaporized from a solid source.

The only other major element in the alpha-emitter series with physical and chemical properties somewhat resembling those of polonium is thorium. However, the situation with thorium isotopes is much more complex, partly because of the ingrowth of daughter products and the decay of the parent isotopes.

DOSIMETRY

In contrast with many of the alpha-emitters discussed in this report, ^{210}Po is at the end of a decay chain and disintegrates to a stable isotope of lead. The short-lived isotopes in the radon chain decay to a long-lived intermediate, ^{210}Pb.

Substantial radiation doses from polonium can be expected in many tissues of the body. Indeed, it supplies a more nearly whole-body dose than any other alpha-emitter except radon gas, but only by contrast to the highly localized doses imparted by bone seekers, such as plutonium and radium. In general, the spleen and kidneys concentrate polonium more than other tissues except for temporary deposition in the lung after inhalation of an insoluble form. Effects are more common in the kidney than in the spleen, despite a nominally higher dose in the spleen. The lymph nodes and the liver are also affected.

A high concentration of polonium is found in blood cells after oral administration and long after parenteral administration.[13,43,45,47] Polonium can enter red blood cells in nonaggregate form, but apparently not in aggregate, that is, colloidal, form. Because the polonium that enters the body by intestinal absorption does so largely in nonaggregate form, a much larger fraction of the dose is found associated with red blood cells combined with the globin of hemoglobin.[4,51] A considerably smaller fraction appears in the cells of the reticuloendothelial system, and colloidal aggregates are notably absent. Thus, dose distribution after oral administration involves a larger contribution to radiation dose from the circulating blood and from nonaggregated polonium. The distribution of polonium after inhalation is intermediate between those after parenteral and oral administration. The relative concentration increases over time after intravenous administration; by 250 days, a substantial fraction of the body burden of polonium is associated with red blood cells.[46] However, this shift occurs long after the major biological elimination and does not alter the cumulative dose very much. It could be pertinent to calculations of dose rate over long-term conditions.

These characteristics of dose do not seem to be sufficient to have a large effect on the early toxicity of polonium. Indeed, for periods of up to 200 days, a gross plot of toxicity as measured by lethality in rats shows that it is nearly the same for all routes of administration.[10] This simplifies dosage calculations, but applies largely to doses considerably higher than those of primary interest.

Calculations of long-term low to intermediate radiation doses should consider the differences influenced by route of administration.[6,7]

Because retention half-times in tissues range from as little as 11 days in liver to 153 days in testes, the relative contribution to dose rate in different tissues depends markedly on time after a single administration. The largest part of the alpha dose is delivered over the first 100 days after a single administration, so these differences have less effect later on total dose than on the dose rate.

Many situations, even continuous intake, involve a series of doses. In a large experiment with rats,[48] it was shown that the metabolism of polonium was not the same after a multiple-dose regimen as after a single-dose regimen. Excretion was slower, effective half-time in the body rose from about 30 days to nearly 40 days, and other evidence indicated that a more tenacious retention of polonium was received in multiple doses. Thus, a somewhat higher radiation dose per microcurie disintegrating within the body is expected for the important organs (spleen, liver, and kidney) than after a single dose. Anthony et al.[1] discussed the influence of these variations on establishing maximal permissible exposures to polonium. Long-term pathologic developments are affected by differences in dosage regime.[8]

The presence of aggregates of polonium in tissues after parenteral administration raises the question of whether special account should be taken of the nominally much larger potential dose surrounding a large aggregate in a given tissue compared with the dose from a comparable amount of diffusely distributed polonium. Results of an investigation of the comparability of the "hot spot" problem in bone and the "hot-particle" problem for plutonium in the lungs seem to indicate that the diffuse pattern can be as effective or even more effective in carcinogenesis. This is because of the cell-killing component of the aggregates and because a substantial portion of the dose is derived from monomeric or weakly polymeric forms. Therefore, special account need not be taken of the dose from aggregates in calculating polonium doses; the average dose to tissue is considered the pertinent quantity.

ANIMAL STUDIES

TISSUE DISTRIBUTION AND EXCRETION

Much effort has been devoted over many decades to animal studies of the distribution and excretion of polonium.[13,25,30,33,47,49] After intravenous administration, most polonium is excreted in the feces. More appeared in urine in rabbits and consistently less in the urine of dogs. Except for the contribution of polonium not absorbed from the gut after oral administration, the effect of route of administration is not large. There is a tendency for greater urinary excretion of inhaled polonium and especially of orally administered polonium. Nevertheless, clearances of polonium from the blood through the kidney are generally low, for example, 0.005 ml/min after intravenous administration in the cat and 0.01 ml/min on absorption from the cat's stomach.[32] (Clearance rates for man are 0.01–0.08 ml/min.) In contrast, clearance rates in the cat for radium are 1 and 2 ml of plasma per min, and those for strontium and calcium about 0.91 and 0.17 ml/min, respectively. The amounts of polonium bound to protein and in colloidal form are considered sufficient to account for its low urinary clearance rate.

The role of the liver in removing polonium to the feces by the bile was confirmed in bile duct ligation experiments in animals by Fink[13] and in the early work of Lacassagne summarized by Fink. It has also been postulated that the intestinal wall has a role.[11]

Polonium is secreted in the milk of lactating animals.[40,54] Equations for its concentration in milk as a function of time after intake have been developed. The effective half-life for excretion in cows' milk over a 20-day period was about 3.7 days, increasing to 33 days at longer times after uptake. The transfer coefficients to milk (in cows or goats) depend on the species and the nature of the compound ingested, but are always well below 1.0 (maximum, 0.18; minimum, 0.0089).

In the context of the important relative toxicity experiments with alpha-emitters, ^{210}Po has acute toxicity far greater, on a per microcurie basis, than either plutonium or radium. However, if lethality is measured at 300 days in rats, polonium is only about twice as toxic.

Blair[2] compared the long-term life-span-shortening effects of single doses of ^{210}Po, ^{239}Pu, and ^{226}Ra in rats. ^{210}Po administered at 1 μCi/kg of body weight shortened the life span of the animals by 4.3 weeks. Comparable life-span shortening was brought about by ^{239}Pu at 0.9 μCi/kg and ^{226}Ra at 5 μCi/kg. Thus, the long-term life-span-shortening effects of polonium and plutonium appear to be comparable and about 5 times as great as that of radium. Life-span shortening with a multiple-dose regimen was almost identical, on the basis of effective dose.[50] That was taken as evidence that most of the injury (perhaps about 80%) produced by polonium alpha particles was irreversible. In contrast, life shortening in the mouse by strontium-89 is much reduced if the dose is divided.[17] Indeed, the effect of beta particles from ^{89}Sr is comparable with that of x rays and has been interpreted as being due to the presence of a much higher fraction of reversible injury caused by low linear energy transfer radiation.

The increase in effectiveness caused by protraction of the dose from alpha particles, described elsewhere in this report, has not been seen with polonium. The female rat shows a tendency toward more life-span shortening on a multiple-dose regimen, but the male does not. These findings might arise from the design of the experiments with polonium and perhaps do not contradict the general observation, especially inasmuch as these experiments predated those which stimulated the idea that the alpha-emitter effect was increased by protraction.

HISTOPATHOLOGY AND CARCINOGENESIS

Specific short- and long-term pathologic effects of polonium have been described by investigators at Argonne National Laboratory, Argonne, Ill.; the Mound Laboratory, Maimisburg, Ohio; the University of Rochester, N.Y.; and the USSR. Most follow the sequences of acute or chronic radiation injury. Most of the reports have been in reasonable agreement, except for a description of extensive liver damage in rats in a multiple-dose experiment at the Mound Laboratory. The liver damage, not seen in other studies, might have been attributable to a strain difference.

Increases in the incidences of cancer in experimental animals attributable to polonium were not reported until the early 1950s. Finkel and Hirsch[16] reported the presence of lymphomas in mice by 250 days after injection of polonium at about 1.5 and 0.9 μCi/kg. They also reported a significant increase in bone tumors (presumably arising from bone marrow deposition, inasmuch as polonium is not a bone-seeking element) at 8 and 0.46 μCi/kg.[14,15,47] Although tumors appeared in the animals at the Mound Laboratory, the incidence did not differ significantly between control and experimental animals.

Casarett[6–8] has described neoplastic changes in rats that received single or multiple oral doses of polonium. Over 40 soft-tissue tumors appeared in 175 rats (23%) on a single-dose regimen and only three tumors appeared in 34 controls (9%). Some of the tumors were malignant. Many were primary and there was a considerable variety. Tumor incidence was maximal at the middle doses (5 and 10 μCi/kg). At the highest dose (20 μCi/kg), the life span was too short for much expression of tumor growth. At the lowest dose (1 μCi/kg), there were somewhat fewer tumors. Tumors were both increased in incidence and advanced temporally by polonium. Neoplastic effects were present, but in different incidences and distributions. After oral administration, less hyperplastic change occurred in the hematopoietic organs and testes; this finding was compatible with the lower concentrations of polonium in these organs after oral administration.

Among the nonneoplastic degenerative effects of polonium is the development of sclerotic changes in blood vessels, which might be due to bloodborne polonium. This effect was seen particularly in the testes and the kidneys. In the kidneys, the process could be followed through proliferation of the arteriolar endothelium to the blocking of blood vessels and ischemia of portions of the kidney.[6] This lesion, like many others in this and other experiments, depended heavily on dose and was never as marked in the multiple-dose experiment.

General nonneoplastic changes from polonium administration included atrophy of the seminiferous epithelium and hyperplasia of interstitial (Leydig) cells in the testes. In addition, hypoplasia and atrophy of lymph nodes, thymus and spleen, and bone marrow occurred. There was involution of growing cartilage, arteriolar nephrosclerosis, vacuolization of adrenal cortical cells, atrophy of the pancreas, hypoplastic and hyperplastic changes in pulmonary lymphoid tissue, obstructive pulmonary emphysema with partial obstruction of bronchioles, and general arteriosclerosis.[6] Some of these changes were direct radiation effects; others were indirect degenerative sequelae.

Nearly all these detailed histopathological findings have occurred at relatively high doses, well above those of primary interest in this report. Few of the doses are relevant to possible population or environmental exposure.

Observations in animals in the USSR[30] placed considerable emphasis on dogs and on changes in the nervous, the endocrine, and the immune systems and the general stress syndrome, including changes in the sympathoadrenal system. Extrapolation of the phenomena to lower than the experimental doses is important, but cannot readily be quantitated. The increased incidence of lymphomas, lung cancers, kidney tumors, and neoplasms of the mammary and sex glands has been described in the report from the USSR.[30]

FUNCTIONAL CHANGES IN ANIMALS

With the evidence of polonium-associated arteriosclerosis and hypertension, Sproul et al.[44] used an indirect noninvasive technique to measure blood pressure in animals bearing polonium under conditions similar to those in the 10-μCi/kg histopathology experiment. Increases in blood pressure were clearly evident and were a function of time after injection of the polonium. The investigators found cataracts with a slit-lamp microscope. The incidence was high—nearly 100% by 1 yr at some doses. Cataracts might be produced even at lower doses, but the data are inconclusive.

Other functional studies have appeared in the Russian literature,[30] including changes related to the blood and cardiovascular systems, such as changes in coagulation times, changes in capillary strength, increased vascular permeability, and changes in cardiac function and blood-pressure stability. Disturbances of protein and nucleic acid metabolisms and impairment of activity of the nervous system and the immune system were observed.

HUMAN STUDIES

PATIENTS AND WORKERS

In the work of the Manhattan Project during World War II and as part of the broad study of distribution and excretion of polonium, five patients hospitalized with lymphatic cancer or leukemia received tracer doses of polonium.[13] Excretion rates and partition between urine and feces in all five patients were comparable with those findings in animal experiments. The tissue distribution of the isotope measured in one patient who died on the sixth day was also comparable, except for a suggestion that more polonium was deposited in the liver than was usually seen in animals.

Workers at the Dayton Project in World War II were tested weekly for polonium excretion. The effective half-life in the human body was calculated as about 30 days (an average for 18 employees) and was comparable with findings in rats and dogs.

Three chemists who had increased urinary excretion of polonium were studied in some detail.[34,35] They were estimated to have received maximal doses of 10 μCi in the body (0.13 μCi/kg). These doses were well above the maximal allowable body concentrations at the time and the limits imposed after World War II (0.04 μCi in the body). No evidence of kidney damage was evident, but subclinical depression of the hematopoietic system was suspected in association with the two higher doses. There was no long-term follow-up.

Hematologic changes; impairment of the liver, of the kidney, and in reproductive organs; and changes in protein, carbohydrate, and pigment metabolism have been reported in Russian workers who had incorporated 1–5 μCi of ^{210}Po.[24] This result is consistent with findings in animal experiments and could be the only documented instance of effects of polonium in man.

A different source of polonium in workers is its gradual ingrowth from the decay of ^{226}Ra deposited in the skeleton of luminous-dial painters and radium chemists, or as an end product of radon-222 exposure and the deposition of ^{210}Pb in bone. In both instances, the polonium is formed in situ and may or may not be transferred away from the site of deposition of the precursor. As a result, more polonium is found in bone in these cases, and the potential for effects in bone is greater than that after direct uptake. Hill[21] analyzed both ^{210}Po and ^{210}Pb concentrations in tissues of several former dial painters. The ratio of bone concentrations of polonium to soft-tissue concentration was much higher than was ever found when it entered the body directly. In one case, the bone of a former radium-dial painter contained ^{210}Po at 1,500 pCi/kg and ^{226}Ra at about 4,000 pCi/kg. In the absence of exposure to specific precursors, the ^{210}Pb concentrations in soft tissues have been found to be quite low. Holtzman[22] postulated that normal people acquire only a small fraction of their ^{210}Po burden in soft tissues from the decay of skeletal ^{226}Ra or ^{210}Pb. The highest concentrations of polonium in any tissue in the radium-dial painters were found in hair.[21] One sample contained 25 μCi/kg. High concentrations in the pelts of animals have also been reported sporadically.

OTHER EXPERIENCES IN HUMANS

Shantyr and coworkers[41] reported on clinical and laboratory investigations in 10 children who were contaminated accidentally with ^{210}Po from a damaged polonium-beryllium neutron source. Analysis of excreta indicated body burdens of 0.2–7 μCi, which is far above the existing maximal permissible burden of 0.04 μCi. Yet no noticeable changes in general health, blood, or kidney function were observed throughout a 46-month observation period. There was some impairment of protein formation in the liver beginning at about 21 months.

Experience in man has amply confirmed the metabolic behavior of polonium as observed in investigations with animals. However, experience with effects, although consistent with the findings in animals, is far too sparse to support any direct estimates of health risks in man at the present time.

POLONIUM IN THE ENVIRONMENT

After decay of radon or its daughters in air, polonium deposits on plants. Grazing animals take up appreciable amounts of ^{210}Po from the environment. Concentrations above 1,000 pCi/kg have been found in some animal tissues in the Arctic food chain and in areas of high rainfall. Hill[21] presented a summary of the ^{210}Po content of various human and animal foods, largely from the United Kingdom. The amounts ranged from 1 pCi/kg in carrots and potatoes in the United Kingdom to 10,000 pCi/kg in a sample of dry lichen and 16,000 pCi/kg in a sample of dried grass, both from the United Kingdom. These figures are subject to wide variation, but Hill concluded that appreciable amounts of ^{210}Po are available to humans in their diet. Adding the contribution of ^{210}Pb, which has a half-life of 21 yr and which is a precursor of ^{210}Po, Hill estimated that the average Western diet would probably include from 1 to 10 pCi of polonium per day. A check on this amount has been made by analyses of the fecal excretion of these nuclides. Holtzman[22] calculated a dietary intake of 1.8 pCi/day for a few otherwise unexposed subjects on an average American diet, and Hill[21] calculated 3.2 pCi/day for six people in the United Kingdom.

A more elaborate study of metabolic balances involved 12 unexposed men maintained in a metabolic ward for a month or more.[23] Both urine and feces were collected. Mean dietary concentrations over 5 months were 1.63 ± 0.05 pCi/day of ^{210}Po and 1.25 ± 0.04 pCi/day of ^{210}Pb. Urinary excretion of ^{210}Po was 0.269 ± 0.033 pCi/day and ^{210}Pb was 0.275 ± 0.026 pCi/day. Fecal excretion accounted for 1.89 ± 0.10 and 1.333 ± 0.062 pCi/day, respectively. Thus, the mean overall balances showed that larger amounts were excreted than were taken in through the diet. Ordinary atmospheric intake could not account for the differences, but intake from tobacco smoke, especially from cigarettes, was sufficient to make up the differences.

The usual dose contribution to humans from polonium in the natural environment has been estimated by the National Council on Radiation Protection and Measurements[36] to be 4.8–60 mrem/yr.

As evidence accumulated in laboratory research that alpha-emitters were potent pulmonary carcinogens when inhaled as aerosols of plutonium or polonium and on the health effects in uranium miners, attention turned to a possible role of naturally occurring polonium in the production of lung cancer in smokers. Radford and Hunt[37] calculated a minimal dose of 36 rem to bronchial epithelium from ^{210}Po inhaled on particles in cigarette smoke as a result of smoking two packs of cigarettes per day for 25 yr. Skrable et al.[42] used the mathematical model of the International Commission on Radiological Protection and calculated a much lower figure. Rajewsky and Stahlhoffen[39] considered the doses far too low to cause lung cancer.

Work done in several laboratories demonstrated that the polonium content of tobacco varies considerably in different types of tobacco and in different locations. The polonium was largely foliar; that implied deposition from the air rather than uptake from the soil. Ratios of ^{210}Pb to ^{210}Po were used to determine how long the insoluble particles from cigarette smoke remained on bronchial epithelium.[38] Polonium contents of many plants were measured, and no large differences in initial contents were found. However, the ^{210}Pb activity of tobacco was fixed on insoluble particles by the curing process. Also, the small trichomes on the surface of the tobacco leaf entered the smoke stream and were deposited in the lungs,[28] where they produced high local alpha-radiation sources by the ingrowth of ^{210}Po. Furthermore, the same trichomes are covered with a sticky hydrophobic substance that makes foliar deposits of ^{210}Pb or ^{210}Po stick to tobacco leaves, whereas they might wash off other plant leaves.

Moroz and Parfenov[30] observed that urinary polonium output was sometimes higher in tobacco smokers. Little et al.[27] showed little significant difference in general tissue contents of polonium between smokers and nonsmokers. Blanchard[3] measured average concentrations of polonium in several tissues and found them to be slightly higher in smokers than in nonsmokers. Only in lung concentrations were the differences statistically significant. Hill,[21] in contrast, reported a two- to threefold difference in lung-tissue concentration between smokers and nonsmokers, but this was not found at bronchial bifurcations.

Because tumors arise in the bronchi, the research focused on polonium concentrations in the bronchial epithelium. Not only did different investigators come up with different findings, but their models for dose calculation were different. Moroz and Parfenov[30] concluded that the smoking of two packs of cigarettes a day leads to supplementary alpha irradiation of the lungs of the smoker, equal to a 0.1 to 100-fold absorbed dose rate in comparison with that of the natural background. The highest dose of this wide range of possible doses approached potential biological significance, whereas the lowest dose was far too low to influence the statistics on lung-cancer incidence in smokers. Cohen et al.[9] and Harley et al.[20] reported that alpha activity in the bronchi from ^{210}Po in cigarette-smoke tar might be about 10 fCi per cigarette. From analyses of human lung tissue, they concluded that there could be areas of high concentration, but that the usual concentration was about 1 fCi/m^2. That translated to an average radiation dose of about 1 mrad/yr, with possible hot spots of up to about 1 rad/yr. Again, the doses cover the range from probably insignificant to possibly significant, depending upon the importance of hot spots.

Martell[29] explained the postulated effectiveness of a few picocuries of ^{210}Po and ^{210}Pb in the lung—on the basis of the insolubility of the compounds inhaled in cigarette smoke, in contrast with the relative solubility of the deposits from other natural sources—and adopted the hot-particle theory. Martell also proposed that chemical carcinogens in smoke could potentiate the effects of the low radiation levels. Martell extended his arguments to other carcinogenic agents, such as asbestos, and even to a role for polonium in the development of atherosclerosis.

RISK ESTIMATES

We have no direct measure of risk for most polonium isotopes based on experience in humans, but its health effects exemplify those of alpha-particle radiation in soft tissues. Risk estimates for humans must therefore be based on other alpha-particle emitters with appreciable components of dose to soft tissue. Studies on radon and its daughters and some thorium compounds may be useful in estimating polonium risks. A risk evaluation based on radon daughter exposure is probably that of a risk based on the short-half-life polonium isotopes. However, it would apply only to lung cancer, and it may or may not apply to the longer-lived ^{210}Po.

Risk evaluation based on thorium would introduce much uncertainty because of the complexities of the thorium decay chain. Some thorium isotopes are distributed in the body very differently from polonium. The best candidate is probably a colloidal form of thorium dioxide called Thorotrast, because it seeks the reticuloendothelial system and delivers most of its alpha dose to the tissue of deposition, usually the liver. Risk estimates based on Thorotrast would be largely those for the development of liver tumors. If these could be expanded to the reticuloendothelial system in general, there would be some potential for correlation.

An entirely different approach could be based on the empirical toxicity ratios among plutonium, polonium, and radium developed in the large experiments with mice by Finkel.[14,15] In these experiments the long-term toxicity of polonium was roughly comparable to that of plutonium and about 5 times that of radium. However, the time span for these experiments was still short, relative to the life span of humans and other long-lived animals. In view of the short radiologic and effective half-time of polonium, it might be overconservative to assume that the ratio reported in the animal experiments for plutonium could apply to long-term risk from polonium in humans.

REFERENCES

1. Anthony, D. S., R. K. Davis, R. N. Cowden, and W. D. Jolley. 1956. Experimental data useful in establishing maximum permissible single and multiple exposures to polonium. Pp. 215–218 in Proceedings of the International Conference on Peaceful Uses of Atomic Energy, Vol. 13. New York: United Nations.
2. Blair, H. A. 1964. The shortening of life span by a single injection of radium, plutonium or polonium. Radiat. Res. Suppl. 5:216–277.
3. Blanchard, R. L. 1967. Concentrations of ^{210}Pb and ^{210}Po in human soft tissues. Health Phys. 13:625–632.
4. Campbell, J. E., and L. H. Talley. 1954. Association of polonium-210 with blood. Proc. Soc. Exp. Biol. Med. 87:221–244.
5. Casarett, L. J. 1964. Distribution and excretion of polonium-210. V. Autoradiographic study of effects of route of administration on distribution of polonium-210. Radiat. Res. Suppl. 5:93–105.
6. Casarett, G. W. 1964. Pathology of single intravenous doses of polonium. Radiat. Res. Suppl. 5:246–321.
7. Casarett, G. W. 1964. Pathology of orally administered polonium. Radiat. Res. Suppl. 5:361–372.
8. Casarett, G. W. 1964. Pathology of multiple intravenous doses of polonium. Radiat. Res. Suppl. 5:347–360.
9. Cohen, B. S., M. Eisenbud, and N. H. Harley. 1980. Measurement of the alpha radioactivity on the mucosal surface of the human bronchial tree. Health Phys. 39:619.

10. Della Rosa, R. J., and J. N. Stannard. 1964. Acute toxicity as a function of route of administration. Radiat. Res. Suppl. 5:205–215.
11. Erleksova, E. K. 1972. The distribution of Some Radioactive Elements in Animals (^{210}Po, ^{222}Ra, Th, ^{239}Pu and Sr). Moscow: Atlas Medgiz. (in Russian) (Quoted by Moroz and Parfenov.[39])
12. Feldman, I., and P. Saunor. 1964. Some in vitro studies of polonium-210 binding by blood constituents. Radiat. Res. Suppl. 5:40–48.
13. Fink, R. M., ed. 1950. Biological Studies with Polonium, Radium and Plutonium. New York: McGraw-Hill.
14. Finkel, M. P. 1956. Relative biological effectiveness of internal emitters. Radiology 67:665–672.
15. Finkel, M. P. 1959. Late effects of internally deposited radioisotopes in laboratory animals. Radiat. Res. Suppl. 1:265–279.
16. Finkel, M. P., and G. M. Hirsch. 1954. Pp. 80–92 in Progress of the Polonium Mouse Experiment. II. Analysis at 500 Days. Argonne National Laboratory Report ANL-4531. Argonne, Ill.: Argonne National Laboratory.
17. Finkel, M. P., A. M. Brues, and H. Lisco. 1952. The Toxicity of Sr^{89} in Mice. Quarterly Report ANL 5247 of the Biological and Medical Division, Argonne National Laboratory. Also Progress Report, Design of Experiment and Survival Following Single Injection. Quarterly report ANL 4840, July 1952. Argonne, Ill.: Argonne National Laboratory.
18. Haissinsky, M. 1932. Electrochemical research on polonium. J. Chim. Res. 29:453–473.
19. Haissinsky, M. Electrochemistry of polonium. Trans. Electrochem. Soc. 70:343–371.
20. Harley, N. H., B. S. Cohen, and T. C. Tso. 1980. Polonium-210: A Questionable Risk Factor in Smoking Related Carcinogenesis. Banbury Report 3: A Safe Cigarette?
21. Hill, C. R. 1965. Polonium-210 in man. Nature 208:423–428.
22. Holtzman, R. B. 1963. Measurement of the natural contents of radium D (Pb^{210}) in human bone—estimates of whole-body burdens. Health Phys. 9:385–400.
23. Holtzman, R. B., H. Spencer, F. H. Ilcewicz, and L. Kramer. 1974. Metabolic balances of ^{210}Pb and ^{210}Po in unexposed men. Pp. 1406–1411 in Third International Congress of the International Radiation Protection Association. AEC CONF-730907-P2. Washington, D.C.: Atomic Energy Commission.
24. Kauranen, P., and J. K. Miettinen. 1967. ^{210}Po and ^{210}Pb in environmental samples in Finland. Pp. 275-280 in Radioecological Concentration Processes. Proceedings of a Symposium. New York: Pergamon.
25. Lacassagne, A., and J. Lattes. 1924. Methode auto-histo-radiographique pour la detection dans les organes du polonium injecte. Compt. Rend. Soc. Biol. 178:488–490.
26. Lanzola, E., M. Allegnini, and D. M. Taylor. 1973. The binding of polonium-210 to rat tissues. Radiat. Res. 56:370–384.
27. Little, J. B., E. P. Redford, Jr., H. L. Mccombs, V. R. Hun, and C. Nelson. 1964. Polonium-210 in lungs and soft tissues of cigarette smokers. Radiat. Res. 22:209 (abstract).
28. Martell, E. A. 1974. Radioactivity of tobacco trichomes and insoluble cigarette smoke particles. Nature 249:214–217.

29. Martell, E. A. 1975. Tobacco radioactivity and cancer in smokers. Am. Sci. 63:404–412.
30. Moroz, B., and Y. Parfenov. 1972. Metabolism and biological effects of polonium-210. Atomic Energy Rev. 10:175–232.
31. Morrow, P. E., R. J. Della Rosa, L. J. Casarett, and G. J. Miller. 1964. Investigations of the collidal properties of polonium-210 solutions using molecular filters. Radiat. Res. Suppl. 5:1–15.
32. Morrow, P. E., F. A. Smith, R. J. Della Rosa, L. J. Casarett, and J. N. Stannard. 1964. Distribution and excretion of polonium-210. II. The early fate in cats. Radiat. Res. Suppl. 5:60–66.
33. Moyer, H. 1956. Chemical Properties of Polonium. In Polonium, H. Moyer, ed. Report TID 5221. Washington, D.C.: U.S. Atomic Energy Commission.
34. Naimark, D. H. 1948. Acute Exposure to Polonium (Medical Study on Three Human Cases). Mound Laboratory Report MLM-67. (Declassified per TID 1182.) Dayton, Ohio: Monsanto Chemical Co.
35. Naimark, D. H. 1949. Effective Half-Life of Polonium in the Human. Mound Laboratory. Report MLM-272. (Declassified per ACR/TID-1153.) Dayton, Ohio: Monsanto Chemical Co.
36. National Council on Radiation Protection and Measurements (NCRP). 1975. Natural Background Radiation in the United States. NCRP Report 45. Washington, D.C.: National Council on Radiation Protection and Measurements.
37. Radford, E. P., Jr., and V. R. Hunt. 1964. Polonium-210: A volatile radioelement in cigarettes. Science 143:247–249.
38. Radford, E. P., and E. A. Martell. 1976. Polonium-210 lead-210 ratios as an index of residence times of insoluble particles from cigarette smoke in bronchial epithelium. Pp. 567–581 in Inhaled Particles IV, Part 2. New York: Pergamon.
39. Rajewsky, B., and W. Stahlhoffen. 1966. Polonium-210 activity in the lungs of smokers. Nature 209:1312–1313.
40. Schreckhise, R. G., and R. L. Watters. 1969. The internal distribution and miln secretion of ^{210}Po after oral administration to a lactating goat. J. Diary Sci. 52.
41. Shantyr, V. I. et al. 1969. Med. Radiologija 10:57. (Cited by Moroz and Parfenov.[39])
42. Skrable, K. W., F. J. Haughey, and E. L. Alexander. 1964. Polonium-210 in cigarette smokers. Science 146:86.
43. Spoerl, E., and D. S. Anthony. 1956. Biological research related to polonium. Chapter 5 in Polonium, H. V. Moyer, ed. Report TID-5221. Oak Ridge, Tenn: Technical Information Division, U.S. Atomic Energy Commission.
44. Sproul, J. A., R. C. Baxter, and L. W. Tuttle. 1964. Some late physiological changes in rats after polonium-210 alpha-particle irradiation. Radiat. Res. Suppl. 5:373–388.
45. Stannard, J. N. 1964. Distribution and excretion of polonium-210. I. Comparison of oral and intravenous routes in the rat. Radiat. Res. Suppl. 5:49–59.
46. Stannard, J. N. 1964. Distribution and excretion of polonium-210. III. Long-term retention and distribution in the rat. Radiat. Res. Suppl. 5:67–79.

47. Stannard, J. N. In press. Polonium and thorium. Chapter 4 in Radioactivity and Health—A History. Washington, D.C.: National Technical Information Services, U.S. Department of Energy.
48. Stannard, J. N., and R. C. Baxter. 1964. Distribution and excretion of polonium-210. IV. On a multiple dose regimen. Radiat. Res. Suppl. 5:80–92.
49. Stannard, J. N., and G. W. Casarett. 1964. Metabolism and biological effects of an alpha-particle emitter, polonium-210. Radiat. Res. Suppl. (The Polonium Supplement), Vol. 5. 442 pp.
50. Stannard, J. N., H. A. Blair, and R. C. Baxter. 1964. Mortality, life span and growth of rats with a maintained body burden of polonium. Radiat. Res. Suppl. 5:228–245.
51. Thomas, R. G. 1964. The binding of polonium by red cells and plasma proteins. Radiat. Res. Suppl. 5:29–39.
52. Thomas, R. G., and J. N. Stannard. 1964. Influence of physicochemical state of intravenously administered polonium-210 on uptake and distribution. Radiat. Res. Suppl. 5:16–22.
53. Thomas, R. G., and J. N. Stannard. 1964. Some characteristics of polonium solutions of importance in biological experiments. Radiat. Res. Suppl. 5:23–28.
54. Watters, R. L., and J. F. McInroy. 1969. The transfer of ^{210}Po from cattle to milk. Health Phys. 16:221.

4
Radium

INTRODUCTION

Four isotopes of radium occur naturally and several more are man-made or are decay products of man-made isotopes. Radium is present in soil, minerals, foodstuffs, groundwater, and many common materials, including many used in construction. In communities where wells are used, drinking water can be an important source of ingested radium. Radium has been used commercially in luminous paints for watch and instrument dials and for other luminized objects. It has also been used for internal radiation therapy.

The primary sources of information on the health effects and dosimetry of radium isotopes come from extensive studies of ^{224}Ra, ^{226}Ra, and ^{228}Ra in humans and experimental animals. These studies were motivated by the discovery of cancer and other debilitating effects associated with internal exposure to ^{226}Ra and ^{228}Ra. Later, similar effects were also found to be associated with internal exposure to ^{224}Ra. The purpose of this chapter is to review the information on cancer induced by these three isotopes in humans and estimate the risks associated with their internal deposition.

All members of the world's population are presumably at risk, because each absorbs radium from food and water; as a working hypothesis, radiation is assumed to be carcinogenic even at the lowest dose levels, although there is no unequivocal evidence to support this hypothesis. Before concern developed over environmental exposure, attention was devoted primarily to exposure in the workplace, where the potential exists for the accidental uptake of radium at levels known to be harmful to a significant fraction of exposed individuals. As the practical concerns of radiation protection have shifted and knowledge has accumulated, there has been an evolution in the design and objectives of experimental animal studies and in the methods of collection, analysis, and presentation of human health effects data.

The first widespread effort to control accidental radium exposure was the abandonment of the technique of using the mouth to tip the paint-laden brushes used for application of luminous material containing ^{226}Ra and sometimes ^{228}Ra to the often small numerals on watch dials. This change occurred in 1925–1926 following reports and intensive discussion of short-term health effects such as "radium jaw" in some dial painters. Shortly thereafter, experimental animal studies and the analysis of case reports on human effects focused on the determination of tolerance doses and radiation protection guides for the control of workplace exposure. These limits on radium intake or body content were designed to reduce the incidence of the then-known health effects to a level of insignificance. The question remained open, however, whether the health effects were threshold phenomena that would not occur below certain exposure or dose levels, or whether the risk would continue at some nonzero level until the exposure was removed altogether. The issue remains unresolved, but as a matter of philosophy, it is now commonly assumed that the so-called stochastic effects, cancer and genetic effects, are nonthreshold phenomena and that the so-called nonstochastic effects are threshold phenomena. Practical limitations imposed by statistical variation in the outcome of experiments make the threshold-nonthreshold issue for cancer essentially unresolvable by scientific study. For nonstochastic effects, apparent threshold doses vary with health endpoint. Low-level endpoints have not been examined with the same thoroughness as cancer. There is evidence that ^{226,228}Ra effects on bone occur at the histological level for doses near the limit of detectability. Whether these effects magnify other skeletal problems is unknown, but issues such as these leave the threshold-nonthreshold question open to further investigation.

Current efforts focus on the determination of risk, as a function of time and exposure, with emphasis on the low exposure levels where there is the greatest quantitative uncertainty. The presentation and analysis of quantitative data vary from study to study, making precise intercomparisons difficult. Occasionally, data from several studies have been analyzed by the same method, and this has helped to illuminate similarities and differences in response among ^{224}Ra, ^{226}Ra, and ^{228}Ra.

Human health studies have grown from a case report phase into epidemiological studies devoted to the discovery of all significant health endpoints, with an emphasis on cancer but always with the recognition that other endpoints might also be significant. This chapter focuses on bone cancer and cancer of the paranasal sinuses and mastoid air cells because these effects are known to be associated with ^{224}Ra or ^{226,228}Ra and are thought to be nonthreshold phenomena.

Several general sources of information exist on radium and its health effects, including portions of the reports from the United Nations Scientific Committee on the Effects of Atomic Radiation; *The Effects of Irradiation on the Skeleton* by Janet Vaughan; *The Radiobiology of Radium and Thorotrast*, edited by W. Gössner; *The Delayed Effects of Bone Seeking Radionuclides*, edited by C. W. Mays et al.; Volume 35, Issue 1, of *Health Physics*; the Supplement to Volume 44 of *Health Physics*; and publications of the Center for Human Radiobiology at Argonne National Laboratory, the Radioactivity Center at the Massachusetts Institute of Technology, the New Jersey Radium Research Project, the Radiobiology Laboratory at the University of California, Davis, and the Radiobiology Division at the University of Utah.

CHEMISTRY AND PHYSICS OF RADIUM

When injected into humans for therapeutic purposes or into experimental animals, radium is normally in the form of a solution of radium chloride or some other readily soluble ionic compound. Little research on the chemical form of radium in body fluids appears to have been conducted. The radium might exist in ionic form, although it is known to form complexes with some compounds of biological interest under appropriate physiological conditions; it apparently does not form complexes with amino acids.

Each isotope of radium gives rise to a series of radioactive daughter products that leads to a stable isotope of lead (Figure 4-1a and 4-1b). In addition to the primary radiation—alpha, beta, or both—indicated in the figures, most isotopes emit other radiation such as x rays, gamma rays, internal conversion electrons, and Auger electrons. In the analysis of radiation-effects data, the alpha particles emitted are considered to be the root cause of damage. This is because of the high linear energy transfer (LET) associated with alpha particles, compared with beta particles or other radiation, and the greater effectiveness of high-LET radiations in inducing cancer and various other endpoints, including killing, transformation, and mutation of cells.

The decay products of radium, except radon, are atoms of solid materials. Radon is gaseous at room temperature and is not chemically reactive to any important degree. Unless physically trapped in a matrix, radon diffuses rapidly from its site of production. For ^{222}Rn (whose half-life is very long compared with the time required for untrapped atoms within the body to diffuse into the blood supply), this rapid diffusion results in a major reduction of the radiation dose to tissues.

RETENTION AND DISTRIBUTION

Following entry into the circulatory system from the gut or lungs, radium is quickly distributed to body tissues, and a rapid decrease in its content in blood occurs. It later appears in the urine and feces, with the majority of excretion occurring by the fecal route. Retention in tissues decreases with time following attainment of maximal uptake not long after intake to blood. The loss is more rapid from soft than hard tissues, so there is a gradual shift in the distribution of body radium toward hard tissue, and ultimately, bone becomes the principal repository for radium in the body. The fundamental reason for this is the chemical similarity between calcium and radium. Because of its preference for bone, radium is commonly referred to as a bone seeker.

Various radiation effects have been attributed to radium, but the only noncontroversial ones are those associated with the deposition of radium in hard tissues. Two compartments are usually identified in the skeleton, a bone surface compartment in which the radium is retained for short periods and a bone volume compartment in which it is retained for long periods. A third compartment, which is not a repository for radium itself but which is relevant to the induction of health effects, consists of the pneumatized portions of the skull bones, that is, the paranasal sinuses and the air cells of the temporal bone (primarily the mastoid air cells), where radon and its progeny, the gaseous decay products of radium, accumulate.

Direct observation in vivo of retention in these three compartments is not possible, and what has been learned about them has

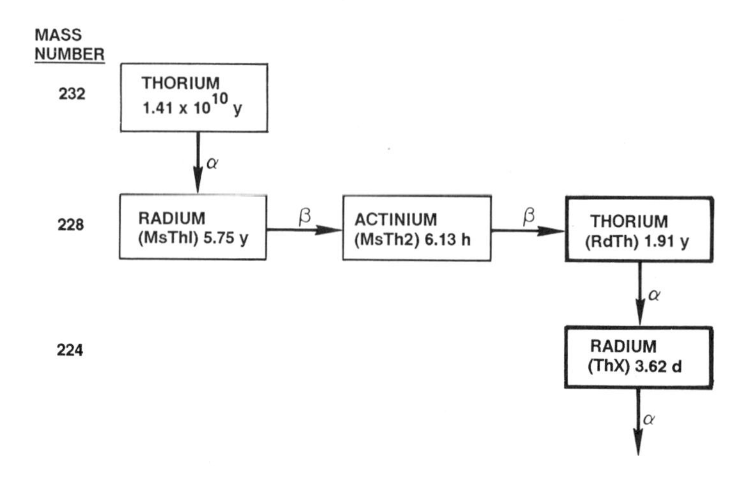

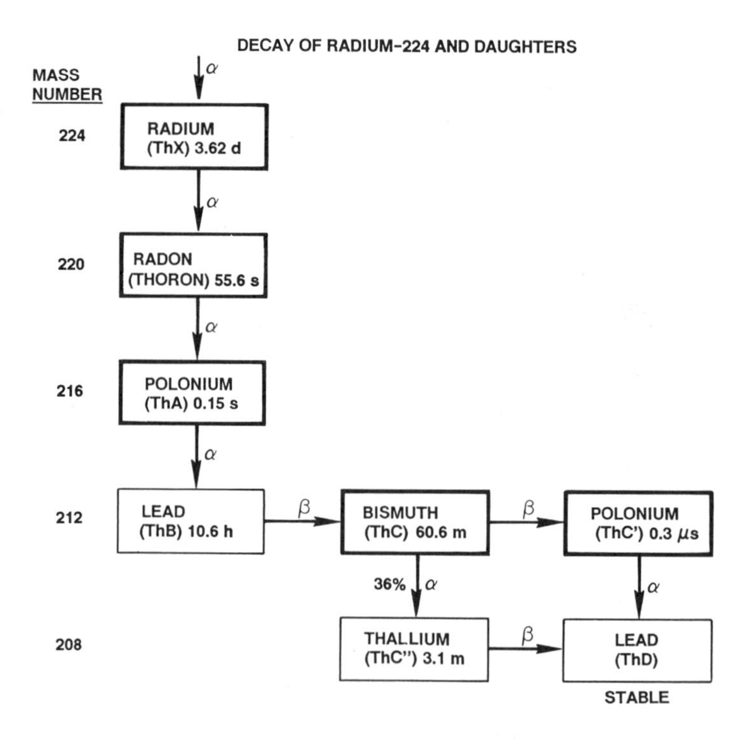

FIGURE 4-1 a. Decay series for radium-228, a beta-particle emitter, and radium-224, an alpha-particle emitter, showing the principal isotopes present, the primary radiations emitted (α, β, or both), and the half-lives (s = second, m = minute, h = hour, d = day, y = year). b. Decay series for radium-226 showing the primary radiations emitted and the half-lives.

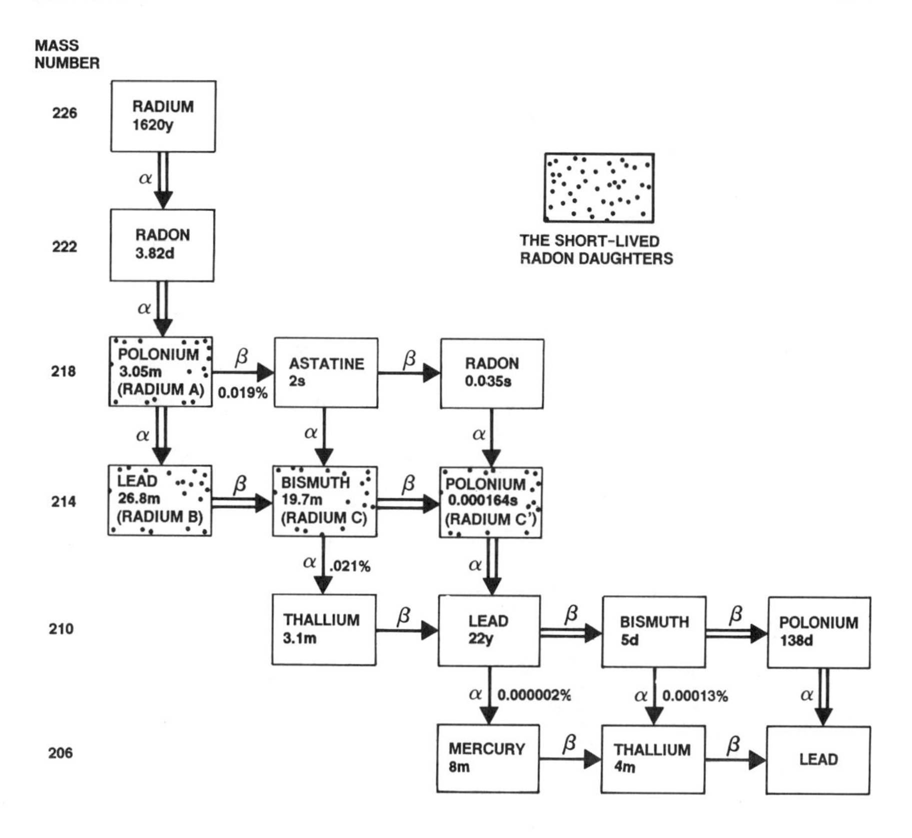

been inferred from postmortem observations and modeling studies. During life, four quantities that can be monitored include whole-body content of radium, blood concentration, urinary excretion rate, and fecal excretion rate. These are supplemented by postmortem measurements of skeletal and soft-tissue content, observations of radium distribution within bone on a microscale, and measurements of radon gas content in the mastoid air cells.

For humans and some species of animals, an abundance of data is available on some of the observable quantities, but in no case have all the necessary data been collected. In general, the data from humans suffice to establish radium retention in the bone volume compartment. Animal data supplemented by models are required to estimate retention in the human bone surface, and human data combined with models of gas accumulation are applied to the pneumatized space compartment.

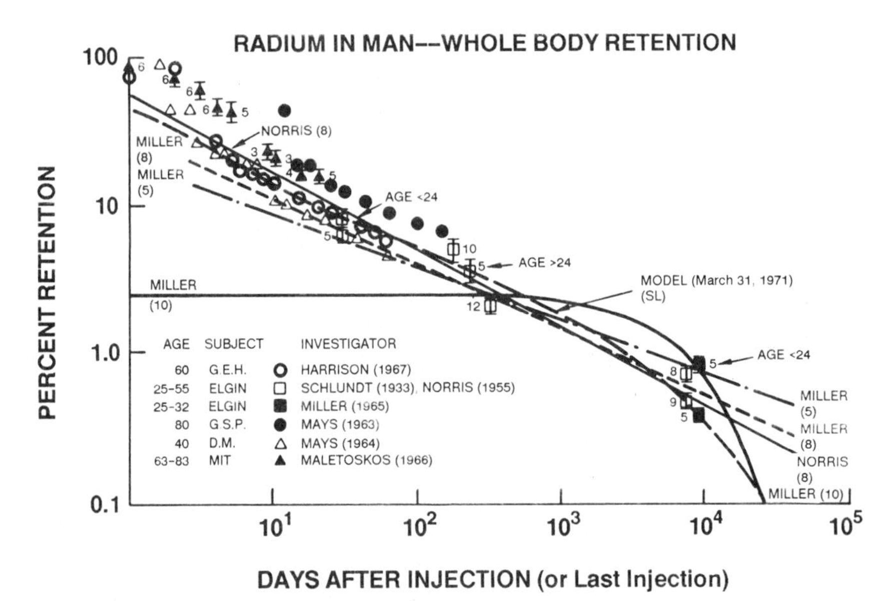

FIGURE 4-2 Whole-body radium retention in humans. Summary of virtually all available data for adult man. The heavy curve represents the new model. Most of the points lie above the model curve for the first 1–2 days because no correction for fecal delay has been made. SOURCE: International Commission on Radiological Protection (ICRP).[29]

Figure 4-2 is a summary of data on the whole-body retention of radium in humans.[29] Whole-body retention diminishes as a power function of time. This observation has also been made for the retention of radium and other alkaline earths in animals. Marshall and Onkelix[39] explained this retention in terms of the diffusion characteristics of alkaline earths in the skeleton.

The excretion rate of radium can be determined by direct measurement in urine and feces or by determining the rate of change in whole-body retention with time. When radium levels in urine and feces are measured, by far the largest amount is found in the feces. In people with radium burdens of many years' duration, only 2% of the excreted radium exits through the kidneys. The other 98% passes out through the bowel.

At high radiation doses, whole-body retention is dose dependent. This observation was originally made on animals given high doses where retention, at a given time after injection, was found to increase with injection level. The most likely explanation is that tissue damage to the skeleton, at high doses, alters the retention pattern, primarily through the reduction in skeletal blood flow that results from the death of capillaries and other small vessels and through the inhibition of bone remodeling, a process known to be important for the release of radium from bone. A recent examination of data on whole-body radium retention in humans revealed that the excretion rate diminished with increasing body burden.[70] Absolute retention could not be studied, because the initial intake was unknown, but the data imply the existence of a dose-dependent retention similar to that observed in animals. Subnormal excretion rate can be linked with the apparent subnormal remodeling rates in high-dose radium cases.[77]

Radium has an affinity for hard tissue because of its chemical similarity to calcium. It does, however, deposit in soft tissue and there is a potential for radiation effects in these tissues. The data on human soft-tissue retention were recently reviewed.[74] The rate of release from soft tissue exceeds that for the body as a whole, which is another way of stating that the proportion of total body radium that eventually resides in the skeleton increases with time.

Postmortem skeletal retention has been studied in animals and in the remains of a few humans with known injection levels. Otherwise, the retention in bone is estimated by models.

Autoradiographic studies[37] of alkaline earth uptake by bone soon after the alkaline earth was injected into animals revealed the existence of two distinct compartments in bone (see Figure 4-3), a short-term compartment associated with surface deposition, and a long-term compartment associated with volume deposition. The uptake and release of activity into and out of the surface compartment was studied quantitatively in animals and was found to be closely related to the time dependence of activity in the blood.[65] Mathematical analysis of the relationship showed that bone surfaces behaved as a single compartment in constant exchange with the blood.[37] This model for the kinetics of bone surface retention in animals was adopted for man and integrated into the ICRP model for alkaline earth metabolism, in which it became the basis for distinguishing between retention in bone volume and at bone surfaces. This is an instance in which an extrapolation of animal data to humans has played an important role.

A mechanistic model for alkaline earth metabolism[29] was developed by the ICRP to describe the retention of calcium, strontium, barium, and radium in the human body and in human soft tissue, bone volume, bone surfaces, and blood. Separate retention functions

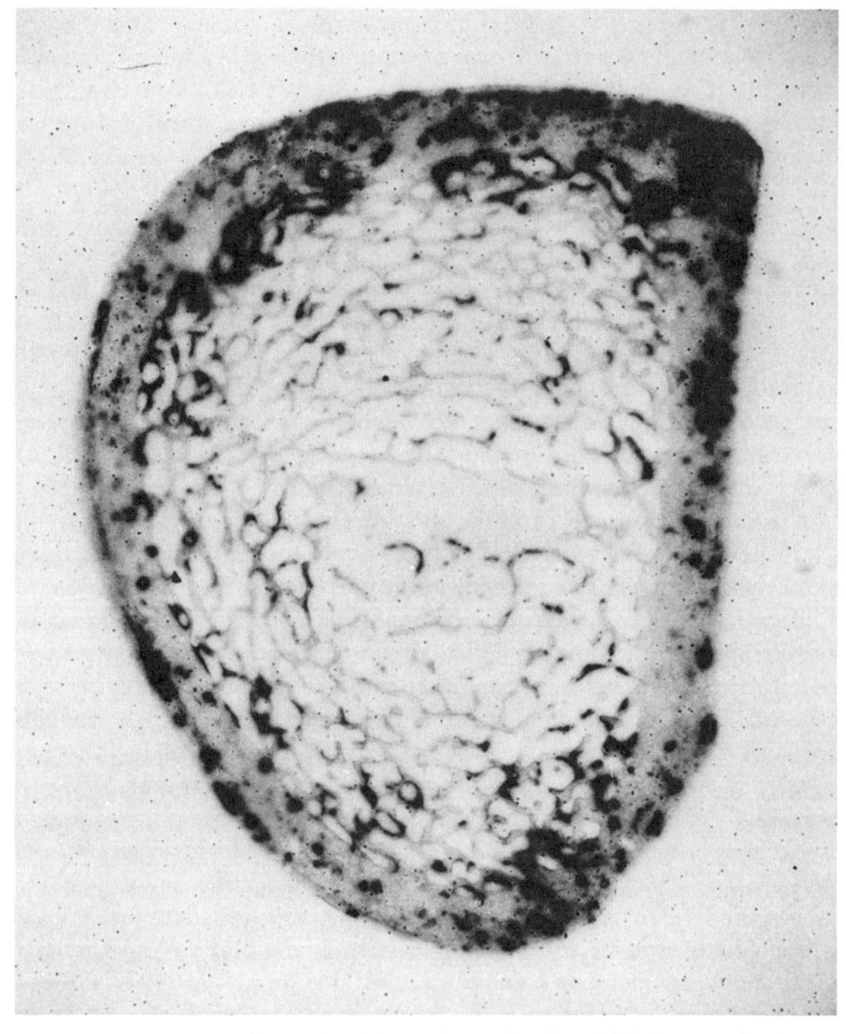

FIGURE 4-3 Autoradiograph of bone from the distal left femur of a former radium-dial painter showing hotspots (black areas) and diffuse radioactivity (gray areas).

are given for each of these compartments. When the model is used for radium, careful attention should be paid to the constraints placed on the model by data on radium retention in human soft tissues.[74] Because of the mathematical complexity of the retention functions, some investigators have fitted simpler functions to the ICRP model.

These simpler functions have no mechanistic interpretation, but they do make some calculations easier.

The kinetics of radon accumulation in the pneumatized air spaces are determined by the kinetics of radium in the surrounding bone, the rate of diffusion from bone through the intervening tissue to the air cavity, and the rate of clearance through the ventilatory ducts and the circulatory system. Diffusion models for the sinuses have not been proposed, but work has been done on the movement of ^{220}Rn through tissue adjacent to bone surfaces. Clearance through the ventilatory ducts is rapid when they are open. The eustachian tube provides ventilation for the middle ear and pneumatized portions of the temporal bone. This duct is normally closed, and clearance by this pathway is negligible. The sinus ducts are normally open but can be plugged by mucus or the swelling of mucosal tissues during illness. When these ducts are open, clearance is almost exclusively through them. Clearance half-times for the frontal and maxillary sinuses are a few minutes when the ducts are open. Otherwise, clearance half-times are about 100 min and are determined by the blood flow through mucosal tissues.[73] The radioactive half-lives of the radon isotopes—55 s for ^{220}Rn and 3.8 days for ^{222}Rn—are quite different from their clearance half-times. In effect, essentially all the ^{220}Rn that diffuses into the pneumatized air space decays there before it can be cleared, but essentially all the ^{222}Rn that reaches the pneumatized air space is cleared before it can decay. These relationships have important dosimetric implications.

BONE CANCER

FREQUENCY AND CELL TYPE

Radium deposited in bone irradiates the cells of that tissue, eventually causing sarcomas in a large fraction of subjects exposed to high doses. The first case of bone sarcoma associated with ^{226,228}Ra exposure was a tumor of the scapula reported in 1929, 2 yr after diagnosis in a woman who had earlier worked as a radium-dial painter.[42] Bone tumors among children injected with ^{224}Ra for therapeutic purposes were reported in 1962 among persons treated between 1946 and 1951.[87]

Spontaneously occurring bone tumors are rare. Sarcomas of the bones and joints comprise only 0.24% of microscopically confirmed malignancies reported by the National Cancer Institute's

TABLE 4-1 Locations of Bone Sarcomas among Persons Exposed to ^{224}Ra and $^{226,228}Ra$ for Whom Skeletal Dose Estimates Are Available

Location	^{224}Ra[a]	$^{226,228}Ra$
Axial skeleton	8.5	8
Appendicular skeleton	35.5	58
Unspecified or widespread	5	—

[a] One tumor located in the left sacroiliac joint has been assigned half to the appendicular skeleton and half to the axial skeleton.

Surveillance, Epidemiology, and End Results (SEER) program.[52] The chance of contracting bone sarcoma during a lifetime is less than 0.1%.

Some 87 bone sarcomas have occurred in 85 persons exposed to $^{226,228}Ra$ among the 4,775 persons for whom there has been at least one determination of vital status. Multiple sarcomas not confirmed as either primary or secondary are suspected or known to have occurred in several other subjects. A total of 66 sarcomas have occurred in 64 subjects among 2,403 subjects for whom there is an estimate of skeletal dose; fewer than 2 sarcomas would be expected. Many of the 2,403 subjects are still alive. Tumor frequencies for axial and appendicular skeleton are shown in Table 4-1. The frequencies for different bone groups are axial skeleton-skull (3), mandible (1), ribs (2), sternebrae (1), vertebrae (1), appendicular skeleton-scapulae (2), humeri (6), radii (2), ulnae (1), pelvis (10), femora (22), tibiae (7), fibulae (1), legs (2; bones unspecified), feet and hands (5; bones unspecified).

Some 55 sarcomas of bone have occurred in 53 of 898 ^{224}Ra-exposed patients whose health status is evaluated triennially.[46] Two primary sarcomas occurred in 2 subjects. Locations are shown in Table 4-1 for 49 tumors among 47 subjects for whom there is an estimate of skeletal dose.

In Table 4-1 note the low tumor yield of the axial compared with the appendicular skeleton. In an earlier summary for 24 ^{224}Ra-induced osteosarcomas,[90] 21% occurred in the axial skeleton. These percentages contrast sharply with the results for beagles injected with ^{226}Ra, in which osteosarcomas were about equally divided between the axial and appendicular skeletons and one-quarter of the tumors appeared in the vertebrae.[90]

TABLE 4-2 Relative Frequencies for Radium-Induced and Naturally Occurring Tumors by Age Group

Age and Tumor Type	Natural	^{224}Ra	$^{226,228}Ra$
Under 30			
Osteosarcoma	1.0	1.0	1.0
Chondrosarcoma	0.08	0.10	0.0
Fibrosarcoma	0.06	0.0	0.16
Over 30			
Osteosarcoma	1.0	1.0	1.0
Chondrosarcoma	1.1	0.10	0.03
Fibrosarcoma	0.42	0.10	0.44

Histologic type has been confirmed by microscopic examination of 45 tumors from 44 persons exposed to $^{226,228}Ra$ for whom dose estimates are available; there were 27 osteosarcomas, 16 fibrosarcomas, 1 spindle cell sarcoma, and 1 pleomorphic sarcoma. The distributions of histologic types for the 47 subjects exposed to ^{224}Ra with bone sarcoma and a skeletal dose estimate are 39 osteosarcomas, 1 fibrosarcoma, 1 pleomorphic sarcoma, 4 chondrosarcomas, 1 osteolytic sarcoma, and 3 bone sarcomas of unspecified type. The distribution of tumor types is not likely to undergo major changes in the future; the group of $^{226,228}Ra$-exposed patients at high risk is dwindling due to the natural mortality of old age and the rate of tumor appearance among ^{224}Ra-exposed patients has dropped to zero in recent years.[46]

The distribution of histologic types for radium-induced tumors is compared in Table 4-2 with that reported for naturally occurring bone tumors.[11] The data have been divided into two groups according to age of record for the tumor. In some cases, this is the age at death and in others this is the age at which the presence of the tumor can be definitely established from the information available. The data have been normalized to the frequency for osteosarcoma and limited to the three principal radiogenic types: osteosarcoma, chondrosarcoma, and fibrosarcoma.

Under age 30, the relative frequencies for radiogenic tumors are about the same as those for naturally occurring tumors. The total numbers of tumors available are too small to assign significance to the small differences in relative frequencies for a given histologic type. Over age 30, the situation is different. Distinctly lower relative frequencies occur for chondrosarcoma and fibrosarcoma induced by ^{224}Ra compared with these same types that occur spontaneously. The relative frequencies for fibrosarcomas induced by ^{224}Ra and $^{226,228}Ra$

are also different, as are the relative frequencies for chondrosarcomas induced by $^{226,228}Ra$ and naturally occurring chondrosarcomas. Thus, the spectrum of tumor types appears to be shifted from the naturally occurring spectrum when the tumors are induced by radium.

DOSIMETRY

The weight of available evidence suggests that bone sarcomas arise from cells that accumulate their dose while within an alpha-particle range. These cells are within 30–80 μm of endosteal bone surfaces, defined here as the surfaces bordering the bone-bone marrow interface and the surfaces of the forming and resting haversian canals. The identities of these cells are uncertain, and their movements and life cycles are only partly understood. Since it is not yet possible to realistically estimate a target cell dose, it has become common practice to estimate the dose to a 10-μm-thick layer of tissue bordering the endosteal surface as an index of cellular dose. This discussion will be devoted to matters that have a quantitative effect on the estimation of endosteal tissue dose.

During the first few days after intake, radium concentrates heavily on bone surfaces and then gradually shifts its primary deposition site to bone volume. Because of its short radioactive half-life, about 90% of the ^{224}Ra atoms that decay in bone decay while on the surfaces.[40]

The extreme thinness of the surface deposit has been verified in dog bone, but the degree of daughter product retention at bone surfaces is in question.[76] Schlenker and Smith[80] have reported that only 5–25% of ^{220}Rn generated at bone surfaces by the decay of ^{224}Ra is retained there 24 h after injection into beagles. The rest diffuses into surrounding tissue. Most of the ^{220}Rn (half-life, 55 s) that escapes bone surfaces decay nearby, as will ^{216}Po (half-life 0.2 sec). This will extend the zone of irradiation out into the marrow, beyond the region that is within alpha particle range from bone surfaces.

Schlenker and Smith[80] also reported incomplete retention for ^{212}Pb and concluded that the actual endosteal dose rate 24 h after injection varied between about one-third and one-half of the equilibrium dose rate for their experimental animals. If this reduction factor applied to the entire period when ^{224}Ra was resident on bone surfaces and was applicable to humans, it would imply that estimates of the risk per unit endosteal dose, such as those presented in the Biological Effects of Ionizing Radiation (BEIR) III report,[54] were low by a factor of 2–3.

^{226}Ra and ^{228}Ra are also heavily concentrated on bone surfaces at short times after intake. Roughly 20% of the total lifetime endosteal dose deposited by ^{226}Ra and its daughters is contributed by the initial surface deposit. These estimates are based on retention integrals[74] and relative distribution factors[40] that originate from retention and dosimetry models.

There is more information available on the dosimetry of the long-term volume deposit. The principal factors that have been considered are the nonuniformity of deposition within bone and its implications for cancer induction and the implications for fibrotic tissue adjacent to bone surfaces.

The nonuniform deposition in bones and the skeleton is mirrored by a nonuniformity at the microscopic level first illustrated with high-resolution nuclear track methods by Hoecker and Roofe for rat[27] and human[28] bone. The intense deposition in haversian systems and other units of bone formation (Figure 4-3) that were undergoing mineralization at times of high radium specific activity in blood are called hot spots and have been studied quantitatively by several authors.[25–28,65,77]

Hoecker and Roofe[28] determined the dose rate produced by the highest concentrations of radium in microscopic volumes of bone from two former radium-dial painters, one who died in 1927 with an estimated terminal radium burden of 50 μg 7 yr after leaving the dial-painting industry, and one who died in 1931 with an estimated terminal burden of 8 μg 10 yr after last employment as a dial painter. These body burden estimates presumably include contributions from both ^{226}Ra and ^{228}Ra. They reported that about 50% of the Haversian systems in the os pubis were hot spots, while hot spots constituted only about 2% of the Haversian systems in the femur shaft. They conclude from their microscopic measurements that the average density of radium in the portions of the pubic bone studied was about 35 times as great as that in the femur shaft; this subject developed a sarcoma in the ascending and descending rami of the os pubis.

In a study of microscopic volumes of bone from a radium-dial painter, Hindmarsh et al.[26] found the ratio of radium concentrations in hot spots to the average concentration that would have occurred if the entire body burden had been uniformly distributed throughout the skeleton to range between 1.5 and 14.0, with 3.5 being the most frequent value. In a similar study on bone from a man who had been exposed to radium for 34 yr, they found concentration ratios in the range of 1–16.[25] Rowland and Marshall [65] reported the maximum hot-spot and average concentrations for 12 subjects. The ratios of maximum to average lay in the range 8–37. The higher values of the ratios were associated with shorter exposure times, usually the order of a year or less.

Marshall[37] summarized results of limited studies on the rate of diminution of ^{226}Ra specific activity in the hot-spot and diffuse components of beagle vertebral bodies that suggest that the rates of change with time are similar for the maximum hot-spot concentration, the average hot-spot concentration, and the average diffuse concentration. If this is true for all dose levels and all bones, this would ensure that the ratio of lifetime doses for these different components of the radium distribution was about the same as the ratio of terminal dose rates determined from microdistribution studies. This is not a trivial point since rate of loss could be greatly affected by the high radiation doses associated with hot spots.

According to Hindmarsh et al.[26] the most frequent ratio of hot-spot to average concentration in bone from a radium-dial painter was 3.5. When combined with the mean value for diffuse to average concentration of about 0.5,[65,77] this indicates that the hot-spot concentration is typically about 7 times the diffuse concentration and that typical hot-spot doses would be roughly an order of magnitude greater than typical diffuse doses. This large difference has prompted theoretical investigations of the time dependence of hot-spot dose rate and speculations on the relative importance of hot-spot and diffuse components of the radioactivity distribution for tumor induction. Marshall[36] showed that bone apposition during the period of hot-spot formation, following a single intake of radium, would gradually reduce the dose rate to adjacent bone surface tissues far below the maximum for the hot spot and concluded that the accumulated dose from a hot spot would be no more than a few times the dose from the diffuse component.[37] Later, Marshall and Groer[38] stated that most hot spots are buried by continuing appositional bone growth and do not deliver much of their dose to endosteal cells that may lie within the alpha-particle range. This, plus the high level of cell death that would occur in the vicinity of forming hot spots relative to that of cell death in the vicinity of diffuse radioactivity and the increase of diffuse concentration relative to hot-spot concentration that occurs during periods of prolonged exposure led them to postulate that it is the endosteal dose from the diffuse radioactivity that is the predominant cause of osteosarcoma induction.

A different hypothesis for the initiation of radiogenic bone cancer has been proposed by Pool et al.[59] They suggest that the cells at risk are the primitive mesenchymal cells in osteons that are being formed. Because of internal remodeling and continual formation of haversian systems, these cells can be exposed to buried radioactive sites.

It should be borne in mind that hot-spot burial only occurs to a significant degree following a single intake or in association with a series of fractions delivered at intervals longer than the time of formation of appositional growth sites, about 100 days in humans. When the average exposure period is several hundred days, as it was for humans exposed to $^{226,228}Ra$, there will be only a minor reduction of hot-spot dose rate because the blood level is maintained at a high average level for the whole period of formation of most hot spots.[67] Autoradiographs from radium cases with extended exposures such as those published by Rowland and Marshall[65] bear this out and form a sharp contrast to autoradiographs of animal bone following single injection[36] on which the model of hot-spot burial was based.

The increase of diffuse activity relative to hot-spot activity, which is suggested by Marshall and Groer[38] to occur during prolonged intake, has a strong theoretical justification. As stated earlier, average hot-spot concentrations are about an order of magnitude higher than average diffuse concentrations, leading to the conclusion that the doses to bone surface tissues from hot spots over the course of a lifetime would also be about an order of magnitude higher than the doses from diffuse radioactivity.

If cell survival is an exponential function of alpha-particle dose in vivo as it is in vitro, then the survival adjacent to the typical hot spot, assuming the hot-spot-to-diffuse ratio of 7 derived above, would be the 7th power of the survival adjacent to the typical diffuse concentration. If the survival adjacent to the diffuse component were 37%, as might occur for endosteal doses of 50 to 150 rad, the hot-spot survival would be 0.09%. When one considers that endosteal doses from the diffuse component among persons exposed to $^{226,228}Ra$ who developed bone cancer ranged between about 250 and 25,000 rad, it becomes clear that the chance for cell survival in the vicinity of the typical hot spot was infinitesimal. For this reason, diffuse radioactivity may have been the primary cause of tumor induction among those subjects in whom bone cancer is known to have developed.

As dose diminishes below the levels that have been observed to induce bone cancer, cell survival in the vicinity of hot spots increases, thus increasing the importance of hot spots to the possible induction of bone cancer at lower doses. The picture that emerges from considerations of cell survival is that hot spots may not have played a role in the induction of bone cancers among the ^{226,228}Ra-exposed subjects, but they would probably play a role in the induction of any bone cancer that might occur at significantly lower doses, for example, following an accidental occupational exposure.

With life-long continuous intake of dietary radium, the distinction between hot spot and diffuse activity concentrations is diminished; if dietary intake maintains a constant radium specific activity in the blood, the distinction should disappear altogether because blood and bone will always be in equilibrium with one another, yielding a uniform radium specific activity throughout the entire mineralized skeleton.

In summary, hot spots may not have played a role in the induction of bone cancer among members of the radium population under study at Argonne National Laboratory because of excessive cell killing in tissues which they irradiate, and the carcinogenic portion of the average endosteal dose may have been about one-half of the total average endosteal dose. With the occasional accidental exposures that occur with occupational use of radium, both hot-spot and diffuse radioactivity are probably important to cancer induction, and the total average endosteal dose may be the most appropriate measure of carcinogenic dose. For animals given a single injection, hot spots probably played a role similar to that played by diffuse radioactivity. This was because the dose rate from most hot spots is rapidly reduced by the overgrowth of bone with a lower and lower specific activity during the period of appositional bone growth that accompanies hot spot formation. For this reason, the total average endosteal dose is probably the best measure of carcinogenic dose. For exposure at environmental levels, the distinction between hot spots and diffuse radioactivity is reduced or removed altogether. Therefore, the total average endosteal dose should be taken into account when the potential for tumor induction is considered.

A common reaction to intense radiation is the development of fibrotic tissue. Lloyd and Henning[33] described a fibrotic layer adjacent to the endosteal surface and the types and locations of cells within it in a radium-dial painter who had died with fibrosarcoma 58 yr after the cessation of work and who had developed an average skeletal dose of 6,590 rad, roughly the median value among persons who developed radium-induced bone cancer. The layer was 8- to 50-μm thick, was sometimes acellular, and sometimes contained cells or cell remnants within it. Cells with a fibroblastic appearance similar to that of the cells lining normal bone were an average distance of 14.9 μm from the bone surface compared with an average distance of 1.98 μm for normal bone. The probability of survival for cells adjacent to the endosteal surface and subjected to the estimated average endosteal dose for this former radium-dial painter was extremely small. The authors concluded that bone tumors most likely arise from cells that are separated from the bone surface by fibrotic tissue and that have invaded the area at long times after the radium was acquired. Such cells could accumulate average doses in the range of 100–300 rad, which is known to induce transformation in cell systems in vitro. If Lloyd and Henning[33] are correct, current estimates of endosteal dose for ^{226}Ra and ^{228}Ra obtained by calculating the dose to a 10-μm-thick layer over the entire time between first exposure and death may bear little relationship to the tumor-induction process.

The time course for development of fibrosis and whether it is a threshold phenomenon that occurs only at higher doses are unknown. Therefore, no judgment can be made as to whether such a layer would develop in response to a single injection of ^{224}Ra or whether the layer could develop fast enough to modify the endosteal cell dosimetry for multiple ^{224}Ra fractions delivered over an extended period of time.

TIME TO TUMOR APPEARANCE AND TUMOR RATE

The times to tumor appearance for bone sarcomas induced by ^{224}Ra and ^{226,228}Ra differ markedly. For ^{224}Ra tumors have been observed between 3.5 and 25 yr after first exposure, with peak occurrence being at 8 yr. The mean and standard deviation in appearance times for persons first injected at ages less than 21 are 10.4 $\pm$ 5.1 yr and for persons exposed at age 21 and above, the mean and standard deviation are 11.6 $\pm$ 5.2 yr.[46] In contrast, tumors induced by ^{226,228}Ra have appeared as long as 63 yr after first exposure.[1] The average and standard deviation of tumor appearance times for female radium-dial workers for whom there had been a measurement of radium content in the body, was reported as 27 $\pm$ 14 yr; and for persons who received radium as a therapeutic agent, the average and standard deviation in appearance times were 29 ± 8 yr.[69]

Spiess and Mays[85,86] have shown that the distributions of appearance times for leukemias among Japanese atomic-bomb survivors and bone sarcomas induced by ^{224}Ra lie approximately parallel with one another when plotted on comparable scales. For the atomic-bomb survivors and the ^{224}Ra-exposed patients, the exposure periods were relatively brief. Leukemias induced by prolonged irradiation from Thorotrast (see Chapter 5) have appeared from 5 to more than 40 yr after injection, similar to the broad distribution of appearance times associated with the prolonged irradiation with ^{226,228}Ra. It is not known whether the similarity in appearance time distribution for the two tumor types under similar conditions of irradiation of bone marrow is due to a common origin.

Groer and Marshall[20] estimated the minimum time for osteosarcoma appearance in persons exposed to high doses of ^{226}Ra and ^{228}Ra. Among these individuals the minimum observed time to osteosarcoma appearance was 7 yr from first exposure. They used the method of hazard plotting, which corrects for competing risks, and concluded that the minimum time to tumor appearance was 5.4 yr with a 95% confidence interval of 1.3–7.0 yr. In addition, they reported a tumor rate of 1.8%/yr for these subjects exposed to high doses and suggested that the sample of tumor appearance times investigated had been drawn from an exponential distribution.

DOSE-RESPONSE RELATIONSHIPS

Cancer induction by radiation is a multifactorial process that involves biological and physical variables whose importance can vary with time and with age of the subject. For the presentation of empirical data, two-dimensional representations are the most convenient and easiest to visualize. Thus, most data analyses have presented cancer-risk information in terms of dose-response graphs or functions in which the dependent variable represents some measure of risk and the independent variable represents some measure of insult. There is no common agreement on which measure is the most appropriate for either variable, making quantitative comparisons between different studies difficult.

Three-dimensional representation of health effects data, although less common, is more realistic and takes account simultaneously of incidence, exposure, and time. This is sometimes in the form of a three-dimensional dose-time-response surface, but more often it is in the form of two-dimensional representations that would result from cutting a three-dimensional surface with planes and plotting the curves where intersections occur.

Dose is used here as a generic term for the variety of dosimetric variables that have been used in the presentation of cancer incidence data. Among these are the injected activity, injected activity normalized to body weight, estimated systemic intake, body burden, estimated maximal body burden, absorbed dose to the skeleton, time-weighted absorbed dose, and pure radium equivalent (a quantity similar to body burden used to describe mixtures of ^{226}Ra and ^{228}Ra). The type of dose used is stated for each set of data discussed.

^{224}Ra, ^{226}Ra, and ^{228}Ra all produce bone cancer in humans and animals. Because of differences in the radioactive properties of these isotopes and the properties of their daughter products, the quantity and spatial distribution of absorbed dose delivered to target cells for bone-cancer induction located at or near the endosteal bone surfaces and surfaces where bone formation is under way are different when normalized to a common reference value, the mean absorbed dose to bone tissue, or the skeleton. Since it is the bombardment of target tissues and not the absorption of energy by mineral bone that confers risk, the apparent carcinogenic potency of these three isotopes differs markedly when expressed as a function of mean skeletal absorbed dose, which is a common way of presenting the data.

The dosimetric differences among the three isotopes result from interplay between radioactive decay and the site of radionuclide deposition at the time of decay. As revealed by animal experiments and clearly detailed by metabolic models, alkaline earth elements deposit first on bone surfaces and then within the volume of bone. The radioactive half-life of ^{224}Ra is short enough that most of the absorbed dose to target tissues is delivered while it is resident on bone surfaces, a location from which absorbed dose delivery is especially efficient. In contrast, ^{226}Ra delivers most of its dose while residing in bone volume, from which dose delivery is much less efficient. With ^{228}Ra, dose delivery is practically all from bone volume, but the ranges of the alpha particles from this decay series exceed those from the ^{226}Ra decay series, allowing ^{228}Ra to go deeper into the bone marrow and, possibly, to irradiate a larger number of target cells.

RADIUM-226 AND RADIUM-228 BONE CANCER

The original cases of radium poisoning were discovered by symptom, not by random selection from a defined population. This method of selection, therefore, made such cases of questionable suitability for inclusion in data analyses designed to determine the probability of tumor induction in an unbiased fashion. To circumvent this problem, two strategies have been developed: (1) classification of the cases according to their epidemiological suitability, on a scale of 1 to 5, with 5 representing the least suitable and therefore the most likely to cause bias and 1 representing the most suitable and therefore the least likely to cause bias; and (2) definition of subgroups of the whole population according to objective criteria presumably unrelated to tumor risk, for example, by year of first exposure and type of exposure. The latter method does not, in effect, correct for selection bias because there is no way to select against such cases. For radium-dial painters, however, the number of persons estimated to have worked in the industry is not too much greater than the number of subjects that have been located and identified by name.[67] This fact implies that coverage of the radium-dial painter segment of the population is reasonably good, thus reducing concerns over selection bias.

The first comprehensive graphical presentations of the dose-response data were made by Evans.[15] In that study both tumor types (bone sarcoma and head carcinoma) were lumped together, and the incidence data were expressed as the number of persons with tumor divided by the total number known to have received the same range band of skeletal radiation dose. These were plotted against a variety of dose variables, including absorbed dose to the skeleton from ^{226}Ra and ^{228}Ra, pure radium equivalent, and time-weighted absorbed dose, referred to as cumulative rad years. This type of analysis was used by Evans[15] in several publications, some of which employed epidemiological suitability classifications to control for case selection bias. Regardless of the dose variable used, the scatter diagram indicated a nonlinear dose-response relationship, a qualitative judgment that was substantiated by chi-squared tests of the linear functional form against the data.

Concern over the shape of the dose-response relationship has been a dominant theme in the analyses and discussions of the data related to human exposure to radium. In simple terms, the main issue has been linear or nonlinear, threshold or nonthreshold. Evans et al. provided an interesting and informative commentary on the background and misapplications of the linear nonthreshold hypothesis.[17]

Concurrently, Mays and Lloyd[44] analyzed the data on bone tumor induction by using Evans' measures of tumor incidence and dosage without correction for selection bias and presented the results in a graphic form that leaves a strong visual impression of linearity, but which, when subjected to statistical analysis, is shown to be nonlinear with high probability. Although the conclusions to be drawn from Evans' and Mays' analyses are the same—that a linear nonthreshold analysis of the data significantly overpredicts the observed tumor incidence at low doses—there is a striking difference in the appearance of the data plots, as shown in Figure 4-4, in which the results of studies by the two authors are presented side by side.

Following consolidation of U.S. radium research at a single center in October 1969, the data from both studies were combined and analyzed in a series of papers by Rowland and colleagues.[66–69] Bone tumors and carcinomas of the paranasal sinuses and mastoid air cells were dealt with separately, epidemiological suitability classifications were dropped, incidence was redefined to account for years at risk, and dose was usually quantified in terms of a weighted sum of the total systemic intakes of ^{226}Ra and ^{228}Ra, although there were analyses in which mean skeletal dose was used. The use of intake as the dose parameter rested on the fact that it is a time-independent quantity whose value for each individual subject remains constant as a population ages. In contrast, mean skeletal dose changes with time, causing a gradual shift of cases between dose bands and confusing the intercomparison of data analyses carried out over a period of years. The outcome of the analyses of Rowland and colleagues was the same whether intake or average skeletal dose was employed, and for comparison with the work of Evans and Mays and their coworkers, analyses based on average skeletal dose will be used for illustration. Another difference between the analyses done by Rowland et al. and those done earlier was division of the radium-exposed subjects into subpopulations defined by type of exposure, that is, radium-dial workers (mostly dial painters), those medically exposed, and others. In this way, some problems of selection bias could be avoided, because most radium-dial workers were identified by search, and coverage of the radium-dial worker groups was considered to be high. Coverage of other groups, especially those with medical exposure, was considered low, and many subjects were selected by symptom. Dose-response data were fitted by a linear-quadratic-exponential expression:

$$(C + \alpha D + \beta D^2)\exp(-\gamma D), \tag{4-1}$$

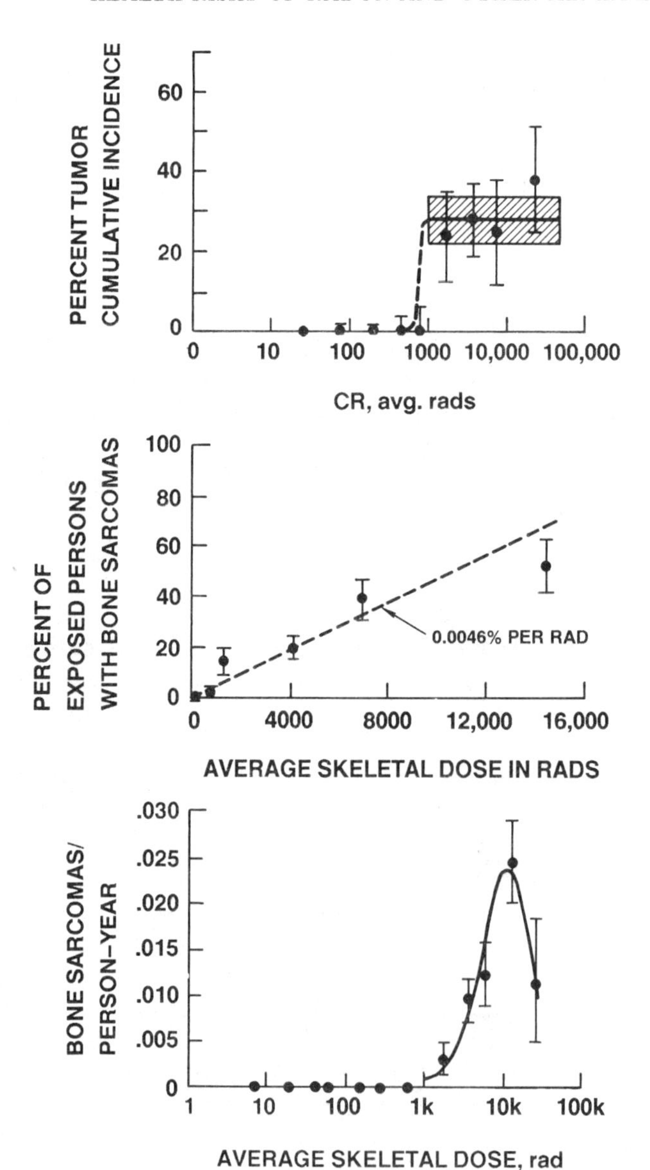

FIGURE 4-4 Dose-response relationships of Evans et al.[17] (a), Mays and Lloyd[44] (b), and Rowland et al.[68] (c).

where D is estimated systemic intake. Equation 4-1 was modified from the general form adopted in the BEIR III report:[54]

$$(\alpha_0 + \alpha_1 D + \alpha_2 D^2)\exp(-\beta_1 D - \beta_2 D^2). \quad (4\text{-}2)$$

As with Evans et al.'s work,[17] the data were plotted against the logarithm of dose so that the low-dose region was not obscured.

A plot of the bone sarcoma data for a population subgroup defined as female radium-dial workers first exposed before 1930 is shown in Figure 4-4. The dissimilarities, primarily between the plots of Evans et al. and Rowland et al., are from the use of person-years at risk in the definition of tumor incidence, from the inclusion of both groups of radium-induced tumor, and the use of different weighting factors in the summation of ^{226}Ra and ^{228}Ra dose. In the Evans et al. analysis, ^{226}Ra and ^{228}Ra dose contributions were weighted equally; in Rowland et al.'s analysis, the ^{228}Ra dose was given a weight 1.5 times that of ^{226}Ra.

The functional form in the analysis of Rowland et al. that provided the best fit to the data as judged by the chi-squared test, was $(C + \beta D^2)\exp(-\gamma D)$, although three other forms provided acceptable fits: $C + \alpha D + \beta D^2$, $(C + \alpha D)\exp(-\gamma D)$, and $(C + \alpha D + \beta D^2)\exp(-\gamma D)$. If forms with negative coefficients are eliminated, as postulated by the model, then only $(C + \alpha D)\exp(-\gamma D)$ from this latter group provided an acceptable fit, but it had a chi-squared probability (0.06) close to the rejection level (0.05). Forms with positive coefficients, which were rejected on the basis of goodness of fit, were $C + \alpha D$ and $C + \beta D^2$. Rowland et al. concluded that a linear dose-response function was incapable of describing the data over the full range of doses.

Three other analyses of the data relevant to the shape of the dose-response curve are noteworthy. The first is that of Rowland et al.[67] in which estimated systemic intake (D) rather than average skeletal absorbed dose was used as the dose parameter and functions of the form $(C + \alpha D + \beta D^2)\exp(-\gamma D)$ were fitted to the data. The findings were similar to those described above. For female radium-dial workers first employed before 1930, the only acceptable fit to the data on bone sarcomas per person-year at risk was provided by the functional form $(C + \beta D^2)\exp(-\gamma D)$, which was obtained from the more general expression by setting $\alpha = 0$. When the population was later broadened to include all female radium-dial workers first employed before 1950[69] for whom there was an estimate of radium exposure based on measurement of body radioactivity, a much larger group than female radium-dial workers first employed before 1930 (1,468 versus 759), the only acceptable fit was again provided by the functional form $(C + \beta D^2) \exp(-\gamma D)$. When the size of the study group was reduced by changing the criterion for acceptance into the group from year of first entry into the industry to year of first measurement of body radioactivity while living, the observed number of bone tumors dropped from 42 to 13, because radioactivity in many persons was first measured after death. Under these circumstances, the forms $C + \alpha D$ and $(C + \beta D^2) \exp(-\gamma D)$ gave acceptable fits.

The second analysis is that of Marshall and Groer,[38] in which a carefully constructed theoretical model was fitted to bone-cancer incidence data. The model was based on a series of three differential equations that described the dynamics of cell survival, replacement, and transformation when bone is irradiated by alpha particles. The outcome of the fitting procedure was presented in graphic form, with total unweighted estimated systemic intake of ^{226}Ra and ^{228}Ra normalized to body weight as the dose parameter. Cumulative incidence, which is the total number of tumors per intake group divided by the numbers of persons alive in that group at the start of observation, was the response parameter. An acceptable fit, as judged by a chi-squared criterion, was obtained. At low doses, the model predicts a tumor rate (probability of observing a tumor per unit time) that is proportional to the square of endosteal bone tissue absorbed dose. In the model, this dose is directly proportional to the average skeletal dose, and tumor rate is an analog of the response parameter, which is bone sarcomas per person-year at risk. Thus, the model and the Rowland et al. analysis are closely parallel and, as might be expected, lead to the same general conclusion that the response at low doses [where $\exp(-\gamma D) \simeq 1$] is best described by a function that varies with the square of the absorbed dose. The analysis of Marshall and Groer[38] is noteworthy, not only because it provides a good fit to the data but also because it links dose and events at the cellular level to epidemiological data, an essential step if the results of experimental research at the cellular level are to play a serious role in the estimation of tumor risk at low doses.

The third analysis was carried out by Raabe et al.,[61,62] with time to death by bone cancer and average skeletal dose rate as the response and dose parameters, respectively. The analysis is most relevant to the question of practical threshold and will be discussed again in that context. Raabe et al. employed a log-normal dose-rate, time-response model that was fitted to the data and that could be used to determine bone-cancer incidence, measured as a percentage of those at risk, versus absorbed skeletal radiation dose. When plotted, the model shows a nonlinear dose-response relationship for any given time after exposure.

Evans, Mays, and Rowland and their colleagues presented explicit numerical values or functions based on their fits to the radium tumor data. For Evans' analysis, the percent tumor cumulative incidence for bone sarcomas plus head carcinomas is constant at 28 ± 6% for mean skeletal doses between 1,000 and 50,000 rad. No fitted value is given for doses below 1,000 rad, but all data points in this range are at zero incidence. Error bars on the points vary in size, and are all less than about 6% cumulative incidence (Figure 4-4). It is clear, therefore, that a nonzero function could be fitted to these data but would have numerical values substantially less than 28%.

For the percent of exposed persons with bone sarcomas, Mays and Lloyd[44] give 0.0046% D_s, where D_s is the sum of the average skeletal doses for ^{226}Ra and ^{228}Ra, in rad.

In the analysis by Rowland et al. [67,68] based on dose, equations that give an acceptable fit are:

$$I = [10^{-5} + (2.0 \pm 0.4) \times 10^{-6} D_s] \exp[-(3.1 \pm 1.8) \times 10^{-5} D_s], \text{ and} \quad (4\text{-}3)$$

$$I = [10^{-5} + (9.8 \pm 1.4) \times 10^{-10} D_s^2] \exp[-(1.5 \pm 0.1) \times 10^{-4} D_s], \quad (4\text{-}4)$$

where the risk coefficient I equals the number of bone sarcomas per person-year at risk that begin to appear after a 5 yr latent period, and D_s is the average skeletal dose from ^{226}Ra plus 1.5 times the average skeletal dose from ^{228}Ra, expressed in rad. For the analyses based on intake, the equation that gives an acceptable fit is:

$$I = [10^{-5} + (6.8 \pm 0.6) \times 10^{-8} D_i^2] \exp[-(1.1 \pm 0.1) \times 10^{-3} D_i], \quad (4\text{-}5)$$

where I is bone sarcomas per person-year at risk, and D_i is the total systemic intake of ^{226}Ra plus 2.5 times the total systemic intake of ^{228}Ra, expressed in microcuries. In the latter analysis,[69] the only acceptable fit based on year of entry into the study is:

$$I = [0.7 \times 10^{-5} + (7.0 \pm 0.6) \times 10^{-8} D_i^2] \exp[-(1.1 \pm 0.1) \times 10^{-3} D_i], \quad (4\text{-}6)$$

where I and D_i are as defined above. The equations based on year of first measurement of body radioactivity are:

$$I = 1.75 \times 10^{-5} + (2.0 \pm 0.6) \times 10^{-5} D_i, \text{ and} \quad (4\text{-}7)$$

$$I = [1.75 \times 10^{-5} + (1.8 \pm 0.4) \times 10^{-7} D_i^2] \exp[-(1.5 \pm 0.2) \times 10^{-3} D_i]. \quad (4\text{-}8)$$

With attention now focused on exposure levels well below those at which tumors have been observed, it is natural to exploit functions such as those presented above for radiogenic risk estimation. The radiogenic risk equals the total risk given by one of the preceding expressions minus the natural tumor risk. This may lead to negative values at low exposures. For example, if D_s = 0.5 rad, which is approximately equal to the lifetime skeletal dose associated with the intake of 2 liters/day of water containing the Environmental Protection Agency's maximum concentration limit of 5 pCi/liter, the expression of Mays and Lloyd[44] would predict a total risk of 0.0023%. With a lifetime natural tumor risk of 0.1%, the radiogenic risk would be −0.0977%. Comparable examples can be given for each expression of Rowland et al. that contains an exponential factor. Such negative values follow logically from the mathematical models used to fit the data and underscore the inaccuracy and uncertainty associated with evaluating the risk far below the range of exposures at which tumors have been observed.

Negative values have been avoided in practical applications by redefining the dose-response functions at low exposure levels. For the Mays and Lloyd[44] function, this consists of setting the radiogenic risk equal to the total risk rather than to the total risk minus the natural risk. For the functions of Rowland et al. that contain an exponential factor, the natural tumor rate is set equal to zero, and the resulting expression is then defined as the radiogenic risk. For example, when the risk coefficient is:

$$I = [0.7 \times 10^{-5} + (7.0 \pm 0.6) \times 10^{-8} D_i^2] \exp[-(1.1 \pm 0.1) \times 10^{-3} D_i], \quad (4\text{-}9)$$

the radiogenic risk would be:

$$R = [(7.0 \pm 0.6) \times 10^{-8} D_i^2] \exp[-(1.1 \pm 0.1) \times 10^{-3} D_i]. \quad (4\text{-}10)$$

For functions that lack an exponential factor, such as $I = 1.75 \times 10^{-5} + (2.0 \pm 0.6) \times 10^{-5} D_i$, redefinition is not required to avoid negative expected values, and radiogenic risk is set equal to the difference between total risk and natural risk.

There have been two systematic investigations of the $^{226,228}Ra$ data related to the uncertainty in risk at low doses. Rowland et al.[69] examined the class of functions $I = (C + \alpha D_i + \beta D_i^2) \exp(-\gamma D_i)$ with positive coefficients, not all of which were determined by least-square fitting to the data, based on year of entry and found that:

$$I_u = (0.7 \times 10^{-5} + 1.3 \times 10^{-5} D_i + 4.3 \times 10^{-8} D_i^2) \exp(-0.9 \times 10^{-3} D_i), \text{ and} \quad (4\text{-}11)$$

$$I_l = (0.7 \times 10^{-5} + 7.0 \times 10^{-8} D_i^2) \exp(-1.1 \times 10^{-3} D_i) \quad (4\text{-}12)$$

determined the upper and lower boundaries (I_u and I_l, respectively) of an envelope of curves that provided acceptable fits to the data, as judged by a chi-squared criterion. When the radiogenic risk functions $(I_u - 0.7 \times 10^{-5})$ and $(I_l - 0.7 \times 10^{-5})$ are used to determine a range of values based on the envelope boundaries, a measure of the uncertainty in estimated bone sarcoma risk at low doses can be formed as:

$$(I_u - I_l)/(I - 0.7 \times 10^{-5}), \quad (4\text{-}13)$$

where I is the best-fit function $[0.7 \times 10^{-5} + 7.0 \times 10^{-8} D_i^2] \exp(-1.1 \times 10^{-3} D_i)$, based on year of entry. This ratio increases monotonically with decreasing intake, from a value of 1.5 at $D_i = 100\ \mu Ci$ to a value of 480 at $D_i = 0.5\ \mu Ci$.

Schlenker[74] presented a series of analyses of the $^{226,228}Ra$ tumor data in the low range of intakes at which no tumors were observed but to which substantial numbers of subjects were exposed. For each of the seven intake groupings in this range (e.g., 0.5–1, 1–2.5, 2.5–5), there was about a 5% chance that the true tumor rate exceeded 10^{-3} bone sarcomas per person-year when no tumors were observed, and there was a 48% chance that the true tumor rate, summed over all seven intake groups exceeded the rate predicted by the best-fit function $I = (10^{-5} + 6.8 \times 10^{-8} D_i^2) \exp(-1.1 \times 10^{-3} D_i)$. With smooth curves, this analysis defined envelopes for which there was a 9, 68, or 95% chance that the true tumor rate summed over the seven intake groups fell between the envelope boundaries when no tumors were observed. The 9% envelope was obtained by allowing the parameters

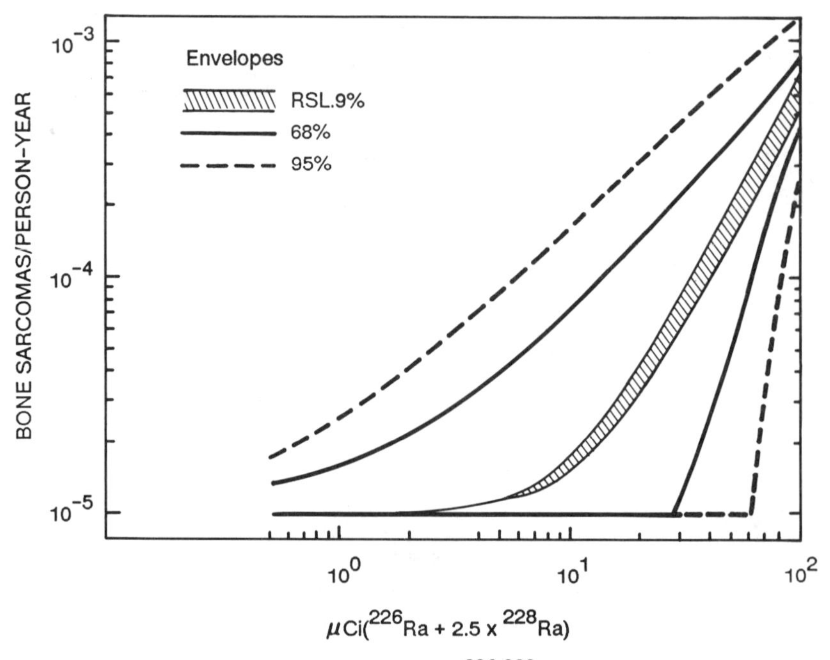

FIGURE 4-5 Dose-response envelopes for ^{226,228}Ra.

in the function to vary by 2 standard errors on either side of the mean and emphasizes that the standard errors obtained by least-square fitting underestimate the uncertainty at low doses. Figure 4-5 shows the results of this analysis, and Table 4-3 gives the equations for the envelope boundaries.

TABLE 4-3 Equations for the Functions I_u and I_l That Define the Dose-Response Envelopes in Figure 4-5

Envelope	Function
9%	$I_u = (10^{-5} + 8.0 \times 10^{-8} D_i^2)\exp(-1.1 \times 10^{-3} D_i)$
	$I_l = (10^{-5} + 5.6 \times 10^{-8} D_i^2)\exp(-1.1 \times 10^{-3} D_i)$
68%	$I_u = 10^{-6} + 6.1 \times 10^{-6} D_i + 2.1 \times 10^{-8} D_i^2$
	$D_i < 26.8$ μCi: $I_l + 10^{-5}$
	$D_i \geq 26.8$ μCi: $I_l = 10^{-5} + 7.9 \times 10^{-8} (D_i - 26.8)^2$
95%	$P = 10^{-5} + 1.6 \times 10^{-5} D_i - 3.6 \times 10^{-8} D_i^2$
	$D_i < 57.5$ μCi: $I_l = 10^{-5}$
	$D_i \geq 57.5$ μCi: $I_l = 10^{-5} + 1.3 \times 10^{-7} (D_i - 57.5)^2$

This work allows one to specify a central value for the risk, based on the best-fit function and a confidence range based on the envelopes. For example, the central value of total risk, including that from natural causes, is $I = (10^{-5} + 6.8 \times 10^{-8} D_i^2)\exp(-1.1 \times 10^{-3} D_i)$ with 95% confidence that total risk lies between $I_l = 10^{-5}$ and $I_u = 10^{-5} + 1.6 \times 10^{-5}\ D_i - 3.6 \times 10^{-8}\ D_i^2$ for D_i between 0.5 and 100 μCi. The ratio of the 95% confidence interval range, for radiogenic risk, to the central value,

$$\frac{I_u - I_l}{I - 10^{-5}}, \tag{4-14}$$

increases with decreasing intake from 1.7 at $D_i = 100\ \mu$Ci to 700 at $D_i = 0.5\ \mu$Ci, the lower boundary of the lowest intake cohort used when fitting functions to the data.

When radiogenic risk is determined by setting the natural tumor rate equal to 0 in the expressions for total risk and by eliminating the natural tumor rate (10^{-5}/yr) from the denominator in Equation 4-14, the value of the ratio increases more slowly, reaching 470 at $D_i = 0.5\ \mu$Ci. At $D_i = 0.05\ \mu$Ci, the total systemic intake in 70 yr for a person drinking 2 liters of water per day at the Environmental Protection Agency's maximum contaminant level of 5 pCi/liter, the ratio is 4,700. These high ratios emphasize, in quantitative terms, our ignorance of risk at low exposure levels.

The risk envelopes defined by these analyses are not unique. Other functions can be determined that meet this 95% probability criterion. This emphasizes that there is no unique way to specify the uncertainty in risk at low exposures when the shape of the dose-response curve is unknown. Regardless of the functions selected as envelope boundaries, however, the percent uncertainty in the risk cannot be materially reduced.

^{224}Ra Bone Cancer

Internal radiation therapy has been used in Europe for more than 40 yr for the treatment of various diseases. Between 1944 and 1951 it was injected in the form of Peteosthor, a preparation containing ^{224}Ra, eosin, and colloidal platinum, primarily for the treatment of tuberculosis and ankylosing spondylitis. Its use with children came to an end in 1951, following the realization that growth retardation could result and that it was ineffective in the treatment of tuberculosis. Since then it has been used with adults as a clinically successful treatment for the debilitating pain of ankylosing spondylitis. Platinum and eosin, once thought to focus the uptake of ^{224}Ra at sites of disease development, have been proven ineffective and are no longer used.

Two extensive studies of the adverse health effects of ^{224}Ra are under way in Germany. Roughly 900 persons who were treated with Peteosthor as children or adults during the period 1946–1951 have been followed by Spiess and colleagues[84–86] for more than 30 yr and have shown a variety of effects, the best known of which is bone cancer. To supplement these investigations of high-level exposure, a second study was initiated in 1971 and now includes more than 1,400 individuals treated with small doses of ^{224}Ra for ankylosing spondylitis and more than 1,500 additional patients with ankylosing spondylitis treated with other forms of therapy who serve as controls.

As with other studies, the shape of the dose-response curve is an important issue. Based on their treatment of the data, Mays et al.[49] made the following observation: "We have fit a variety of dose-response relationships through our follow-up data, including linear ($y = ax$), linear multiplied by a protraction factor, dose-squared exponential ($y = ax^2e^{-kx}$), and a threshold function. None can be rejected because of the scatter in our human data." Rowland[64] published linear and dose-squared exponential relationships that provided good visual fits to the data. Recent analyses with a proportional hazards model led to a modification of the statement about the adequacy of the linear curve, as will be discussed later. However, the change was not so great as to alter the basic conclusion that the data have too little statistical strength to distinguish between various mathematical expressions for the dose-response curve. As a convenient working hypothesis, in several papers it has been assumed that the linear form is the correct one, leading to analyses that are illuminating and easily understood.

In the first dose-response analyses, average skeletal dose was adopted as the dose parameter, and details of the dose calculations were presented. With only two exceptions, average skeletal dose computed in the manner described at that time has been used as the dose parameter in all subsequent analyses. As a response parameter, the number of bone sarcomas that have appeared divided by the number of persons known to have been exposed within a dose group was used. The data for persons exposed as juveniles (less than 21 yr of age) were analyzed separately from the data for persons exposed as adults, and different linear dose-response functions that fit the data adequately over the full range of doses were obtained.[85] The linear slope for juveniles, 1.4%/100 rad, was twice that for adults, 0.7%/100 rad. The analysis took into account tumors appearing between 14 and 21 yr after the start of exposure in 43 subjects that received a known dose. These constitute about 85% of the subjects with bone sarcoma on which the most recent analyses have been based. The importance of this work lies in the fact that it shows the maximum difference in radiosensitivity between juvenile and adult exposures for this study. In later work, juvenile-adult differences have not been reported.

The removal of the difference came in two steps associated with analyses of the influence of dose protraction on tumor induction. Based on a suggestion by Muller drawn from his observations of mice, Speiss and Mays[86] reanalyzed their ^{224}Ra data in an effort to determine whether there was an association between dose protraction and tumor yield. In the analyses, a linear dose-response relationship was postulated, and the data were sorted according to the time period over which ^{224}Ra was administered. The found that the slope of the linear dose-response curve increased with increasing time period, suggesting that bone-cancer incidence increased with decreasing average skeletal dose rate, in accordance with results in mice. Although the change of tumor incidence with exposure duration was not statistically significant, an increase did occur both for juveniles and adults. In a subsequent analysis,[46] the data on juveniles and adults were merged, and an additional tumor was included for adults, bringing the number of subjects with tumors and known dose to 48. A single function was fitted to these data to describe the change of the dose-response curve slope with the length of time over which injections were given:

$$y = 40 + 160(1 - e^{-0.09x}) \tag{4-15}$$

where y is the number of bone sarcomas per millon person-rad and x is the length of the injection span, in months. The asymptotic value of this function is 200 bone sarcomas/million person-rad, which is considered applicable both to childhood and adult exposure. For comparison with the values given previously for juveniles and adults separately, this is 2.0% incidence per 100 rad, which is somewhat higher than either of the previous values.

The case for a dose rate or dose-protraction effect rests on the observation of an association of the linear dose-response slope with dose rate in humans and the unequivocal appearance of a dose-protraction effect in mice and rats. Though one might wish to dispute its existence in humans on statistical grounds in order to defend a claim for greater childhood radiosensitivity, it would seem uneconomical to do so until there is clear evidence of greater radiosensitivity to alpha radiation for the induction of bone cancer in the young of another species.

The first analysis to take account of competing risks and loss to follow-up[74] was based on a life-table analysis of data collected[88] for persons 16 yr of age and older. Cumulative incidence, computed as the product of survival probabilities in the life table,[10] was used as the measure of response with errors based on approximations by Stehney. As the dose parameter, absorbed dose in endosteal tissue was used, computed from the injection levels, in micrograms per kilogram, using conversion factors based on body weight and relative distribution factors similar to those of Marshall et al.[40] but altered to take into account the dependence of stopping power on energy. The functional form found to provide a best fit to the data was:

$$\nu/N = 1 - \exp(-0.00003D_e), \qquad (4\text{-}16)$$

where ν/N is the cumulative incidence, and D_e is the endosteal dose.

Not long afterward, Mays and Spiess[45] published a life-table analysis in which cumulative incidence was computed annually from the date of first injection by summing annual tumor occurrence probabilities. For each year, the cumulative incidence so obtained was divided by the average value of the mean skeletal dose for subjects within the group, in effect yielding the slope of a linear dose-response curve for the data. Adults and juveniles were treated separately. The resultant graph of dose-response curve slopes versus years of follow-up is shown in Figure 4-6. Although the points for adults always lie below those for juveniles, there is always substantial statistical overlap. In a subsequent life-table analysis, in which the same methods were used but 38 cases for whom there were not dose estimates were excluded, the points for juveniles and adults lie somewhat further apart. This type of analysis updates the one originally conducted for this group of subjects in which juvenile radiosensitivity was reported to be a factor of 2 higher than adult radiosensitivity. According to the latest life-table analysis, the risk to juveniles (188 $\pm$ 32 bone sarcomas/10^6 person-rad) is 1.4 times the risk to adults (133 $\pm$ 36 bone sarcomas/10^6 person-rad). Presumably, if dose protraction were taken into account by the life-table analysis, the difference between juveniles and adults would vanish.

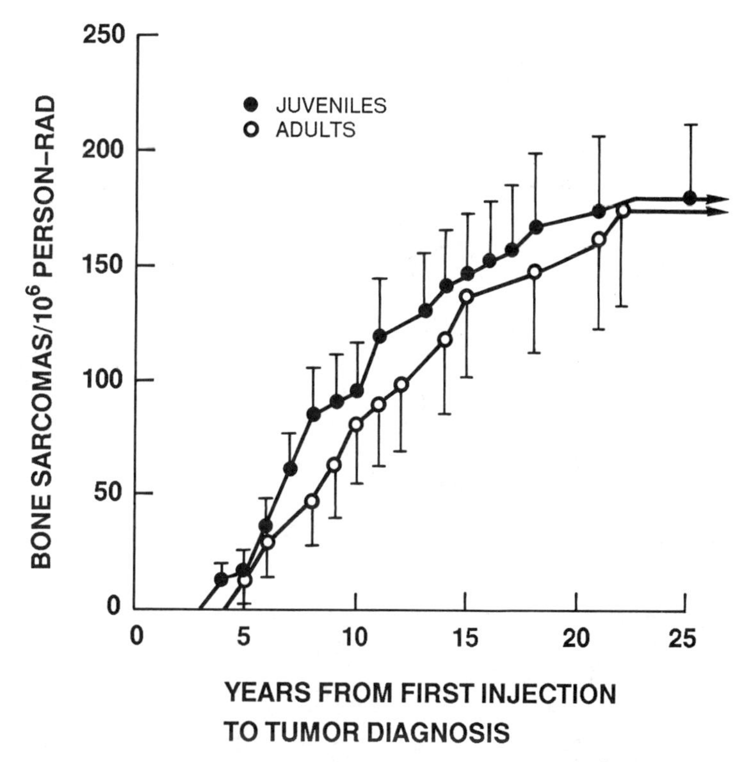

FIGURE 4-6 The cumulative tumor risk (bone sarcomas/10^6 person-rad) was similar in the juvenile and adult patients under the dosimetric assumptions used. The standard deviation for each point is shown. SOURCE: Mays and Spiess.[45]

The third analysis that corrects for competing risks was performed by Chemelevsky et al.[9] using a proportional hazards model. The data for juveniles and adults was separated into different dose groups, a step not taken with the life-table analysis of Mays and Spiess.[45] This, in effect, frees the analysis from the assumption of a linear dose-response relationship, implicit in the Mays and Spiess analysis. Estimates of the cumulative tumor rate (incidence) versus time after first injection were obtained, and when those for juveniles and adults in comparable dose groups were compared, no difference in either the magnitude or the growth of cumulative tumor rate with time was found between the two age groups. The cumulative tumor rate for juveniles and adults at 25 yr after injection, a time after which, it is now thought, no more tumors will occur, were merged into a single data set and fitted with a linear-quadratic exponential relationship:

$$R = (0.0085D_s + 0.0017D_s^2)\exp(-0.025D_s), \qquad (4\text{-}17)$$

where R is the probability that a tumor will occur per person-gray and D_s is the average skeletal dose in gray (1 Gy is 100 rad). This curve and the data points are shown in Figure 4-7. The error bars on each point are a greater fraction of the value for the point here than in Figure 4-6, because the subdivision into dose groups has substantially reduced the number of subjects that contributes to each datum point. A linear function was fitted to the data over the full range of doses, but the fit was rejected by a statistical test for goodness of fit that yielded a P value of 0.02. This is the first report of an explicit test of linearity that has resulted in rejection. The best-fit function, however, does contain a linear term, in contrast to the best-fit functions for the data on $^{226,228}Ra$. This means that when doses are low enough, the risk varies linearly with dose. The same observation can be made for the function $1 - \exp(-0.00003D)$ for the probability of tumor induction developed from the life-table analysis of Schlenker.[74]

The data points in Figure 4-7 for juveniles and adults are not separable from one another, and the difference between juvenile and adult radiosensitivity has completely disappeared in this analysis. From this, we can conclude that much, and perhaps all, of the difference in radiosensitivity between juveniles and adults originally reported was due to the failure to take into account competing risks and loss to follow-up. If a dose-protraction effect were included in the analysis, there might be a reversal of the original situation, with adults having the greater radiosensitivity. The results of this series of studies of bone sarcoma incidence among ^{224}Ra-exposed subjects extending over a period of 15 yr underscore the importance of repeated scrutiny of unique sets of data. Whether due to competing risks, dose protraction, or a combination, it is clear that differential radiosensitivity for this group of subjects is a hypothesis that cannot be supported.

As of December 1982, the average followup time was 16 yr for patients injected after 1951 with lower doses of ^{224}Ra for the treatment of ankylosing spondylitis.[93] Of 1,426 patients who had been traced, the vital status for 1,095 of them was known. Of these, 363 died and three bone cancers, one fibrosarcoma, one reticulum cell sarcoma, and one multiple myeloma were recorded. The average

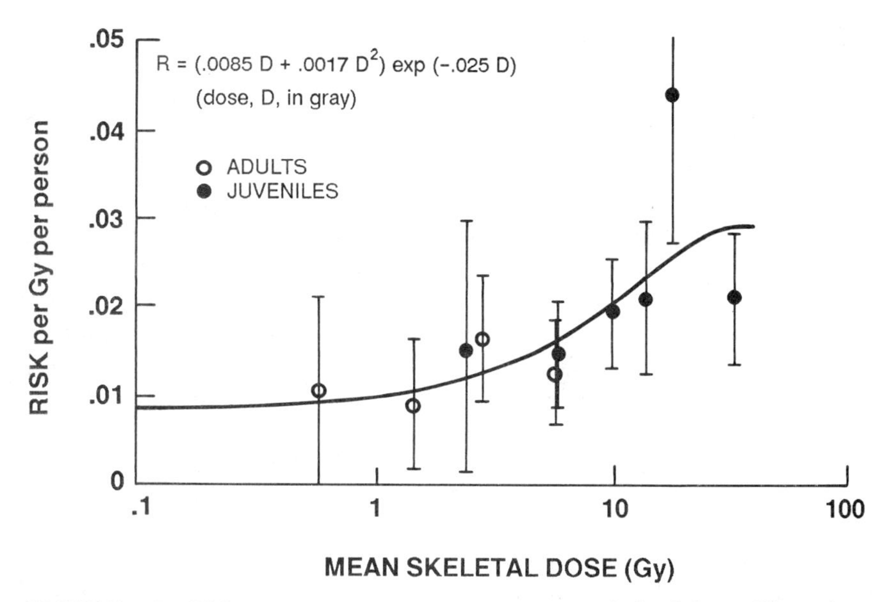

FIGURE 4-7 Risk per person per gray versus mean skeletal dose. The points with their standard errors result from the proportional hazards analysis of Chemelevsky et al.[9]

dose for the exposed group, based on patients for whom there were extant records of treatment level, was 65 rad. In discussing these cases, Wick and Gössner[93] noted that three cases of bone cancer were within the range expected for naturally occurring tumors and also within the range expected from a linear extrapolation downward to lower doses from the Spiess et al.[88] series. However, 80% of the bone tumors in the this series, for which histologic type is known, are osteosarcomas, while fibrosarcomas and reticulum cell sarcomas each represent only about 2% of the total, and multiple myeloma was not observed at all. Based on this, the chance of randomly selecting three tumors from the this distribution and coming up with no osteosarcomas is about $(0.2)^3 = 0.008$, throwing the weight of evidence in favor of a nonradiogenic origin for the three bone cancers found in this study.[93,94] However, this could occur if there were a dramatic change in the distribution of histologic types for tumors induced by ^{224}Ra at doses below about 90 rad, which is approximately the lower limit for tumor induction in the Spiess et al.[88] series. If the tumors are nonradiogenic, then the linear extrapolation gives a substantial over prediction of the risk at low doses, just as a linear extrapolation of the ^{226,228}Ra data overpredict the risk from these isotopes at low doses.[17,44]

Schlenker[74] has provided a confidence interval analysis of the Spiess et al.[88] data in the region of zero observed tumor incidence to parallel that for ^{226,228}Ra. The results are shown in Figure 4-8. There is a 14% probability that the expected number of tumors lies within the shaded region, defined by allowing the parameter value in Equation 4-16 to vary by 2 standard errors about the mean, and a 68% probability that it lies between the solid line that is nearly coincident with the upper boundary of the shaded region and the lower solid curve. The shaded region emphasizes that standard errors obtained by least-square fitting underestimate the uncertainty in risk at low doses. There is a 95% probability that the expected number lies between the dashed boundaries. As with ^{226,228}Ra, the curves in Figure 4-8 can be used to establish confidence limits for risk estimates at low doses, although it is to be understood that these limits are not unique, because the shape of the dose-response curve is unknown. As an example, the upper boundaries of the 95% confidence envelope for total cumulative incidence corrected for competing risks are:

$$(\nu/N)_u = 0.0218, D_e \geq 235 \text{ rad, and}$$
$$(\nu/N)_u = 3 \times 10^{-5} + 1.8 \times 10^{-4} D_e - 4.0 \times 10^{-7} D_e{}^2,$$
$$D_e < 235 \text{ rad;} \tag{4-18}$$

Those for the lower boundary are:

$$(\nu/N)_l = 3 \times 10^{-5} + 1.2 \times 10^{-7}(D_e - 314)^2, D_e \geq 314 \text{ rad, and}$$
$$(\nu/N)_l = 3 \times 10^{-5}, D_e < 314 \text{ rad.} \tag{4-19}$$

The ratio of the 95% confidence interval range for radiogenic risk to the radiogenic risk defined by the central value function

$$(\nu/N) = 3 \times 10^{-5} - 2.1 \times 10^{-7} D_e + 4.0 \times 10^{-8} D_e{}^2 \tag{4-20}$$

is:

$$[(\nu/N)_u - (\nu/N)_l]/[(\nu/N) - 3 \times 10^{-5}],$$

where 3×10^{-5} is the natural risk adapted here. This ratio increases monotonically with decreasing endosteal dose, from 1.8 at 500 rad to

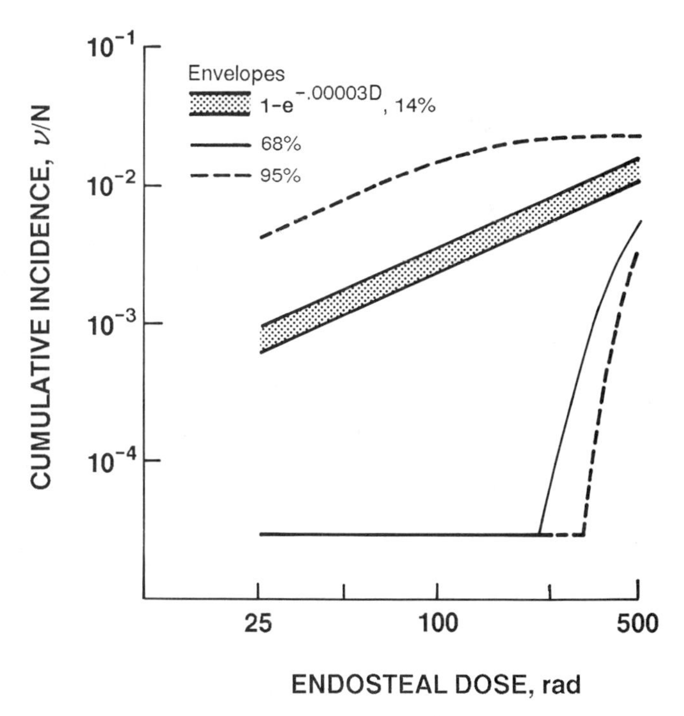

FIGURE 4-8 Dose-response envelopes for ^{224}Ra from equation 4-16. The upper curve of the 68% envelope is nearly coincident with the upper boundary of the shaded envelope.

220 at 25 rad, which is the lower boundary of the lowest dose cohort used in Schlenker's[74] analysis.

PRACTICAL THRESHOLD

The term practical threshold was introduced into the radium literature by Evans,[15] who perceived an increase of the minimum tumor appearance time with decreasing residual radium body burden and later with decreasing average skeletal dose.[16] A plot showing tumor appearance time versus average skeletal dose conveys the impression that the minimum tumor appearance time increases with decreasing dose. The practical threshold would be the dose at which the minimum appearance time exceeded the maximum human life span, about 50 rad. Below this dose level, the chance of developing a radium-induced tumor would be very small, or zero, as the word threshold implies. Evans et al.[17] suggested an increase of median tumor appearance time with decreasing dose based on observations of tumors in a group of radium-dial painters, radium chemists, and persons who had received or used radium for medicinal purposes. This trend was subsequently verified by Polednak[57] for bone tumors in a larger, all female group of radium-dial workers. Polednak cautioned that the shorter median appearance time at high doses might simply reflect the shorter overall median survival time. Mays et al.[47] showed that mean survival time increased with decreasing dose in beagles that had contracted osteosarcoma following radionuclide injection.

Raabe et al. demonstrated an increase of median tumor appearance time with decreasing average skeletal dose rate for a subset of radium-induced bone tumors in humans[61] and for bone tumors induced in experimental animals by a variety of radionuclides.[60] The validity of the analysis of mouse data has been challenged,[62] but not the analysis of human and dog data. As suggested by Polednak's analysis,[57] the reduction of median appearance time at high dose rates in the work by Raabe et al.[61,62] may be caused by early deaths from competing risks. It is striking, however, that the graph for radium in humans[61,62] lies parallel to the graphs for all long-lived nuclides in dogs,[60] where death from bone tumor tends to occur earlier than death from other causes. This is evidenced by the fact that bone tumor incidence rises to 100% with increasing dose. This suggests that competing risks exert no major influence on the analysis by Raabe et al.[61,62].

The work by Raabe et al.[61,62] permits the determination of a practical threshold dose and dose rate. They based their selection on the point of intersection between the line representing the human lifetime and "a cancer risk that occurs three geometric standard deviations earlier than the median." This yielded a dose rate of 0.0039 rad/day for humans and a cumulative dose of 80 rads to the skeleton.[61]

The increase of median tumor appearance time with decreasing dose rate strengthens the case for a practical threshold. Whether the practical threshold represents a dose below which the tumor risk is zero, or merely tiny, depends on whether the minimum tumor appearance time is an absolute boundary below which no tumors can occur or merely an apparent boundary below which no tumors have been observed to occur in the population of about 2,500 people for whom radium doses are known. The data provide no answer.

The theory of bone-cancer induction by alpha particles[38] offers some insights. The theory postulates that two radiation-induced initiation steps are required per cell followed by a promotion step not dependent on radiation. The chance that two independent initiations will occur close enough together to permit a short tumor appearance time increases with increasing dose rate, in agreement with the observations of Raabe et al.[61,62] When the total dose is delivered over a period of time much shorter than the human life span, both initiations must occur within the period of dose delivery, and there is a high probability of short tumor appearance times, regardless of dose level, as confirmed by the human ^{224}Ra data.[46] Reasoning from the theory, there is always a nonzero chance for both initiations to occur close together, regardless of dose rate or total dose. Therefore, the minimum observed tumor appearance time is not an absolute lower bound, and there is a small nonzero chance for tumors to occur at doses less than the practical threshold.

CARCINOMA OF THE PARANASAL SINUSES AND MASTOID AIR CELLS

The paranasal sinuses are cavities in the cranial bones that exchange air and mucus with the nasal cavity through a small ostium. The sinuses are present as bilateral pairs and, in adulthood, have irregular shapes that may differ substantially in volume betwen the left and right sides. The ethmoid sinuses form several groups of interconnecting air cells, on either side of the midline, that vary in number and size between individuals.[92] The sinus surfaces are lined with a mucous membrane that is contiguous with the nasal mucosa and consists of a connective tissue layer attached to bone along its lower margin and to a layer of epithelium along its upper margin. The cilia transport mucus in a more or less continuous sheet across the epithelial surface toward the ostium.[13]

The mastoid air cells, like the ethmoid sinuses, are groups of interconnecting air cavities located bilaterally in the left and right temporal bones. The mastoid air cells communicate with the nasopharynx through the middle ear and the eustachian tube. Radium-induced carcinomas in the temporal bone are always assigned to the mastoid air cells, but the petrous air cells cannot be logically excluded as a site of origin.

The mucosal lining of the mastoid air cells is thinner than the lining of the sinuses. The epithelium is of squamous or cuboidal type with scattered ciliated cells but no goblet cells. It shows no signs of significant secretory activity but is always moist. The typical adult maxillary cavity has a volume of about 13 cm^3; one frontal sinus has a volume of about 4.0 cm^3, and one sphenoid sinus has a volume of about 3.5 cm^3. The collective volume of one set of ethmoid air cells is about 3.5 cm^3; there are nine cells on the average,[92] for an average volume per cell of 0.4 cm^3. The pneumatized portion of one mastoid process has a volume of about 9.2 cm^3. The individual cells range from 0.1 to more than 1 cm across and are too numerous to be counted. The distance across a typical air cell is 0.2 cm,[73] equivalent to a volume of about 0.004 cm^3 if the cell were spherical.

The total thickness of the mucosa, based on the results of various investigators, ranges from 0.05 to 1.0 mm for the maxillary sinuses, 0.07 to 0.7 mm for the frontal sinuses, 0.08 to 0.8 mm for the ethmoid sinuses, and 0.07 to 0.7 for the sphenoid sinuses. The thickness of the simple columnar epithelium, including the cilia, is between 30 and 45 μm.

The difference between mucosal and epithelial thickness gives the thickness of the lamina propria a quantity of importance for dosimetry. In the simple columnar epithelium, the thicknesses for the lamina propria implied by the preceding information range from about 10 μm upward to nearly 1 mm. Direct observations of the lamina propria indicate that the thickness lies between 14 and 541 μm.[21]

Mucosal dimensions for the mastoid air cells have been less well studied. Littman et al.[31] report a single value of 17 μm for the lamina propria in a person who had contracted mastoid carcinoma. In a more complete series of measurements on normal persons and persons exposed to low $^{226,228}Ra$ doses, Harris and Schlenker[21] reported total mucosal thicknesses between 22 and 134 μm, with epithelial thicknesses in the range of 3 to 14 μm and lamina propria thicknesses in the range of 19 to 120 μm.

The normally functioning sinus is ventilated; that is, its ostium or ostia are open, permitting the free exchange of gases between the sinus and nasal cavities. When the sinus becomes unventilated due to ostial closure, the gas composition of the sinus cavity changes and slight overpressure or underpressure may occur.[13] When radioactive gases (radon) are present, as with persons exposed to $^{226,228}Ra$, there is the potential for a much higher concentration of those gases in the air of the sinus when unventilated than when ventilated. Ventilation of the mastoid air cells occurs through the eustachian tube which normally allows little air to move. Thus, there is a potential for the accumulation of large quantities of radon.

Cancer of the paranasal sinuses and mastoid air cells has been associated with $^{226,228}Ra$ exposure since the late 1930s[43] following the death of a radium-dial painter who had contracted epidermoid carcinoma of the epithelium lining of the ethmoid air cells.[3]

The natural tumor rate in these regions of the skull is very low, and this aids the identification of etiological agents. Malignancies of the auditory tube, middle ear, and mastoid air cells (ICD 160.1) make up only 0.0085% of all malignancies reported by the National Cancer Institute's SEER program.[52] Those of the ethmoid (ICD 160.3), frontal (ICD 160.4), and sphenoid (ICD 160.5) sinuses together make up 0.02% of all malignancies, or if the nonspecific classifications, other (ICD 160.8) and accessory sinus, unspecified (ICD 160.9), are added as though all tumors in these groups had occurred in the ethmoid, frontal, or sphenoid sinuses, the incidence would be increased only to 0.03% of all malignancies. In 1977 it was estimated that only 15 people died in the United States from cancers of the auditory tube, middle ear, and mastoid air cells.[53] Comparable statistics are lacking for cancers of the ethmoid, frontal, and sphenoid sinuses; but mortality, if scaled from the incidence data, would not be much greater than that caused by cancers of the auditory tube, middle ear, and mastoid air cells.

Carcinomas of the paranasal sinuses and mastoid air cells may invade the cranial nerves, causing problems with vision or hearing[3,23] prior to diagnosis. Littman et al.[31] have presented a list of symptoms in tabular form gleaned from a study of the medical records of 32 subjects who developed carcinoma of the paranasal sinuses or mastoid air cells following exposure to $^{226,228}Ra$. The most frequent clinical symptoms for paranasal sinus tumors were problems with vision, pain (not specified by location), nasal discharge, cranial nerve palsy, and hearing loss. The most frequent symptoms for mastoid air cell tumors were ear blockage or discharge and hearing loss.

Some 35 carcinomas of the paranasal sinuses and mastoid air cells have occurred among the 4,775 $^{226,228}Ra$-exposed patients for whom there has been at least one determination of vital status. For 31 of the tumors, estimates of skeletal dose can and have been made. Data on tumor locations and histologic type are presented in Table 4-4. No maxillary sinus carcinomas have occurred, but 69% of the tumors have occurred in the mastoids. For tumors of known histologic type, 56% are epidermoid, 34% are mucoepidermoid, and 10% are adenocarcinomas. For the sinuses alone, the distribution of types is 40% epidermoid, 40% mucoepidermoid, and 20% adenocarcinoma,

TABLE 4-4 Carcinomas of the Paranasal Sinuses and Mastoid Air Cells among Persons Exposed to $^{226,228}Ra$ and Currently Under Study at Argonne National Laboratory

		Histologic Type			
Site	No.	Epidermoid	Mucoepidermoid	Adenocarcinoma	Unknown
Mastoid	24	14	7	1[a]	2
Ethmoid	2	2	—	—	—
Ethmoid/sphenoid	2	—	1	—	1
Sphenoid	6	1	3	2	—
Frontal	1	1	—	—	—
Totals	35	18	11	3	3

[a]Mucin-producing.

compared with 37, 0, and 24%, respectively, of naturally occurring carcinomas in the ethmoid, frontal, and sphenoid sinuses.[4] Among all microscopically confirmed carcinomas with known specific cell type in the nasal cavities, sinuses and ear listed in the National Cancer Institute SEER report,[52] 75% were epidermoid, 1.6% were mucoepidermoid, and 7% were adenocarcinoma. The rarity of naturally occurring mucoepidermoid carcinoma, contrasted with its frequency among $^{226,228}Ra$-exposed subjects, suggests that alpha-particle radiation is capable of significantly altering the distribution of histologic types. The complete absence of other, less-frequent types of naturally occurring carcinoma that represent 16% of the carcinomas of specific cell type in the SEER[52] study and 39% of the carcinomas in the review by Batsakis and Sciubba[4] provides further evidence for perturbation of the distribution of carcinoma types by alpha radiation.

The cause of paranasal sinus and mastoid air cell carcinomas has been the subject of comment since the first published report,[43] when it was postulated that they arise ". . . in the mucosa . . . as result of the local effects of the radon . . . in the expiratory air. . . ." In 1952, Aub et al.[3] stated that the origin of these neoplasms in mucosal cells that were well beyond the range of the alpha particles emitted by radium, mesothorium, and their bone-fixed disintegration products is also interesting. Only the beta and gamma rays, which were of low intensity compared to the alpha rays, emitted by these radioactive materials in the adjacent bone could have reached these cells. However, the mucosa may have been irradiated by the alpha rays from the radiothorium that was fixed in the adjacent periosteum. Also, they were continuously subjected to alpha radiation from another source: the radon in expired breath. Restated in more modern terms, the residual range from bone volume seekers (^{226}Ra and ^{228}Ra) is too small for alpha particles to reach the mucosal epithelium, but the range may be great enough for bone surface seekers (^{228}Th), whose alpha particles suffer no significant energy loss in bone mineral;[78] long-range beta particles and most gamma rays emitted from adjacent bone can reach the mucosal cells, and free radon may play a role in the tumor-induction process. Hasterlik[22] and Hasterlik et al.[23] further elucidated the role of radon by postulating that it can diffuse from bone into the essentially closed airspaces of the mastoid air cells and paranasal sinuses and decay there with its daughters, adding an additional dose to the epithelial cells.

A significant role for free radon and the possibly insignificant role for bone volume seekers is not universally acknowledged; the ICRP lumps the sinus and mastoid mucosal tissues together with the endosteal bone tissues and considers that the dose to the first 10 μm of tissue from radionuclides deposited in or on bone is the carcinogenically significant dose, thus ignoring trapped radon altogether and taking no account of the epithelial cell locations which are known to be farther from bone than 10 μm.

The first attempts at quantitative dosimetry were those of Kolenkow[30] who presented a detailed discussion of frontal sinus dosimetry for two subjects, one with and one without frontal sinus carcinoma. In the subject with carcinoma, he observed a hot layer of bone beginning about 2 μm from the surface and extending inward a distance greater than the alpha-particle range. The radium concentration in this layer was 50 to 75 times the mean concentration for the whole skeleton. In the subject without carcinoma, the measured radium concentration in the layer adjacent to the bone surface was only about 3 times the skeletal average. Kolenkow[30] presented his results as depth-dose curves for the radiation delivered from bone but made no comment on epithelial cell location. The depth dose for radon and its daughters in the frontal sinus of the subject with carcinoma was based on a direct measurement of radon activity in the unaffected frontal sinus at the time surgery was performed on the diseased sinus. The calculated dose from this source was much less than the dose from bone. Based on Kolenkow's work,[30] Evans et al.[16] reported a cumulative dose of 82,000 rad to the mucous membrane at a depth of 10 μm for the subject with carcinoma. Kolenkow's work[30] illustrated many of the complexities of sinus dosimetry and emphasized the rapid decrease of dose with depth in the mucous membrane.

The first explicit description of the structure of the sinus and mastoid mucosa in the radium literature is probably that of Hasterlik,[22] who described it as "thin wisps of connective tissue," overlying which "is a single layer of epithelial cells. . . ." He placed the total thickness of connective tissue plus epithelium at between 5 and 20 μm. A clear implication of these data is that the connective tissue in the mastoid is thinner than the connective tissue in the paranasal sinuses. In a dosimetric study, Schlenker[73] confirmed this by determining the frequency with which the epithelium lay nearer to or farther from the bone surface than 75 μm, at which level more than 75% of the epithelial layer in the mastoids would be irradiated. Commenting on the mucosal thickness data of Ash and Raum,[2] Littman et al.[31] observed: "If the dimensions of the sinus walls are applicable to the radium cases, it would appear that only a relatively sparse population of epithelial cells in the submucosal glands of the paranasal sinuses would receive significant dose from alpha particles originating in bone."

Equations for the dose rate averaged over depth, based on a simplified model of alpha-particle energy loss in tissue, were presented by Littman et al.[31] for dose delivered by radium in bone and by radon and its daughters in an airspace with a rectangular cross section. They also presented an equation for depth dose from radon and its daughters in the airspace for the case of a well-ventilated sinus, in which the radon concentration was equal to the radon concentration in exhaled breath. For the 27 subjects for whom radium body burden information was available, they estimated that, for airspace thicknesses of 0.5 to 2 cm, the dose from radon and its daughters averaged over a 50-μm-thick mucous membrane would be 2 to 5% of the average dose from ^{226}Ra in bone. Clearly, under these assumptions, dose from radon and its daughters in the airspaces would be of little radiological significance.

The quantitative impact of cell location on dosimetry was emphasized by Schlenker[75] who focused attention on the relative importance of dose from radon and its daughters in the airspaces compared to dose from radium and its daughters in bone. He emphasized that current recommendations of the ICRP make no clear distinction between the locations of epithelial and endosteal cells and leave the impression that both cell types lie within 10 μm of the bone surface;

this leads to large overestimates of the dose to epithelial cells from bone.

In a more complete development, Schlenker[73] investigated the dosimetry of sinus and mastoid epithelia when ^{226}Ra or ^{228}Ra was present in the body. He used the same assumptions about linear energy transfer as Littman et al.;[31] adopted a spherical shape for the air cavities; and considered air cavity diameters from 0.2 mm, representing small mastoid air cells, up to 5 cm, representing large sinuses. He took into account the dose rate from ^{226}Ra or ^{228}Ra in bone, the dose rate from ^{222}Rn or ^{220}Rn in the airspaces, the impact of ventilation and blood flow on the residence times of these gases in the airspaces, measured values for the radioactivity concentrations in the bones of certain radium-exposed patients, and determined expected values for radon gas concentrations in the airspaces. For five subjects on whom he had autoradiographic data for the ^{226}Ra specific activity in bone adjacent to the mastoid air cells, the dose rate at death from ^{222}Rn and its daughters in the airspaces exceeded the dose rate from ^{226}Ra and its daughters in bone. On average, the dose rate from airspaces was about 4 times that from bone.

He also estimated dose rates for situations where there were no available autoradiographic data. The dose rate from the airspaces exceeded the dose rate from bone when ^{226}Ra or ^{228}Ra was present in the body except in one situation. For ^{228}Ra the dose rate from the airspace to the mastoid epithelium was about 45% of the dose rate from bone. These results are in marked contrast to those of Kolenkow[30] and Littman et al.[31] Under Schlenker's[73] assumptions, the airspace is the predominant source of dose, with the exception noted, whether or not the airspace is ventilated.

Working from various radium-exposed patient data bases, several authors have observed that carcinomas of the paranasal sinuses and mastoid air cells begin to occur later than bone tumors.[16,18,66,71] In the latest tabulation of tumor cases,[1] the first bone tumor appeared 5 yr after first exposure, and the first carcinoma of the paranasal sinuses or mastoid air cells appeared 19 yr after first exposure; among persons for whom there was an estimate of skeletal radiation dose, the first tumors appeared at 7 and 19 yr, respectively. The frequency distribution for appearance times shows a heavy concentration of paranasal sinus and mastoid carcinomas with appearance times of greater than 30 yr. For bone tumors there were approximately equal numbers with appearance times of less than or greater than 30 yr.[67] Based on the most recent summary of data, 32 bone tumors occurred with appearance times of less than 30 yr among persons with known radiation dose and 29 tumors had occurred with appearance times of 30 yr or greater. Within the same group, four carcinomas occurred with appearance times equal to or greater than 30 yr. Unless there is a bias in the reporting of carcinomas, it is clear that carcinomas are relatively late-appearing tumors.

Rowland et al.[66] plotted and tabulated the appearance times of carcinomas for five different dosage groups. On the basis of minimum and median appearance times, they concluded that the appearance times do not change with dose. Rowland et al.[67] performed a dose-response analysis of the carcinoma data in which the rate of tumor occurrence (carcinomas per person-year at risk) was determined as a function of radium intake. The linear relationship that provided the best fit to the data predicted a tumor rate lower than the rate that had been observed recently, and led the authors to suggest that the incidence at long times after first exposure may be greater than the average rate observed thus far.

An analysis of the tumor appearance time data for carcinomas based on hazard plotting has been as employed by Groer and Marshall[20] to analyze bone tumor rate in persons exposed to high doses from radium. The data are subdivided into three groups based on the ^{226}Ra intake. ^{228}Ra intake was excluded because it was assumed that ^{228}Ra is ineffective for the production of these carcinomas. Data points fall along a straight line when the tumor rate is constant. The intersection of the line with the appearance time axis provides an estimate of the minimum appearance time. The analysis shows that the minimum appearance time varies irregularly with intake (or dose) and that the rate of tumor occurrence increases sharply at about 38 yr after first exposure for intakes of greater than 470 μCi and may increase at about 48 yr after first exposure for intakes of less than 260 μCi.

As of the 1980 follow-up, no carcinomas of the paranasal sinuses and mastoid air cells had occurred in persons injected with ^{224}Ra, although Mays and Spiess[46] estimated that five carcinomas would have occurred if the distribution of tumor appearance times were the same for ^{224}Ra as for ^{226,228}Ra.

Carcinomas of the frontal sinus and the tympanic bulla, a portion of the skull comparable to the mastoid region in humans, have appeared in beagles injected with radium isotopes and actinides. Based on epizootiological studies of tumor incidence among pet dogs, Schlenker[73] estimated that 0.06 tumors were expected for 789 beagles from the University of Utah beagle colony injected with a variety of alpha emitters, while five tumors were observed. Three of the five tumors were induced by actinides that have no gaseous daughter products. Their induction, therefore, cannot be influenced by dose from the airspace as can the induction of carcinomas by ^{226}Ra in humans. The beagle data demonstrate that a gaseous daughter product is not essential for the induction of sinus and mastoid carcinomas, while Schlenker's[73] dosimetric analysis and the epidemiological data[16,67] indicate that it is an important factor in human carcinoma induction. The conclusion from this and information on tissue dimensions is that the sinuses, and especially the mastoids, are at risk from alpha emitters besides ^{226}Ra, but that the risk may be significantly lower than that from ^{226}Ra and its decay products.

Rowland et al.[67] have reported the only separate analyses of paranasal sinus and mastoid carcinoma incidence. As the response variable, they used carcinomas per person-year at risk and regressed it against a measure of systemic intake of ^{226}Ra and against average skeletal dose. They fit mathematical functions of the general form:

$$I = (C + \alpha D + \beta D^2)e^{-\gamma D}, \tag{4-21}$$

in which all three coefficients (α, β, γ) were allowed to vary or one or more of the coefficients were set equal to zero. In this expression, C is the natural carcinoma rate and D is the systemic intake or mean skeletal dose. The best fit of response against systemic intake was obtained for the functional form $I = C + \alpha D$, obtained from Equation 4-21 by setting $\beta = \gamma = 0$. The poorest fit, and one that is unacceptable according to a chi-squared criterion, was obtained for $I = C + \beta D^2$. All other functional forms gave acceptable fits. When persons that had entered the study after exhumation were excluded from the analysis, in an effort to control selection bias, all six forms of the general function gave acceptable fits to the data. The exclusion of exhumed subjects removed from analysis 23 of the 759 individuals in the population and 1 of the 21 carcinomas that had occurred among them. This change had no effect on the fitted value of α, the free parameter in the linear dose-response function.

The analysis of response as a function of ^{226}Ra dose was conducted with exhumed cases included. The best fit was obtained for the functional form $I = (C + \alpha D)\exp(-\gamma D)$, an unacceptable fit was obtained for $I = C + \beta D^2$, and all other forms provided acceptable fits.

The linear functions obtained by Rowland et al.[67] were:

$$I = 10^{-5} + (1.6 \pm 0.2) \times 10^{-5} D_i, \text{ and}$$
$$I = 10^{-5} + (1.6 \pm 0.3) \times 10^{-6} D_s, \qquad (4\text{-}22)$$

where D_i is ^{226}Ra intake, and D_s is ^{226}Ra skeletal dose. In the data analyses that lead to these equations, a 10-yr latent period is assumed for carcinoma induction. This latent period must be included when the equations are applied to risk estimation. For example, if a person is exposed to ^{226}Ra at time zero, the person is not considered to be at risk for 10 yr; the total number of carcinomas expected to occur among N people with identical systemic intakes D_i is IN (t − 10) for $t \geq 10$ yr and 0 for $t < 10$ yr. This is also true for N people, all of whom accumulate a skeletal dose D_s.

The analysis of Rowland et al.[67] assumes that tumor rate is constant with time for a given intake D_i, and when based on skeletal dose assumes that tumor rate is constant for a given dose D_s. The analysis also yields good fits to the data. It should be noted that if tumor rate were constant for a given dose, it could not be constant for a given intake because the dose produced by a given intake is itself a function of time; therefore, the tumor rate would be time dependent. The success achieved in fitting dose-response functions to the data, both as a function of intake and of dose, indicates that the outcome is not sensitive to assumptions about tumor rate. No firm conclusions about the constancy or nonconstancy of tumor rate should be drawn from this dose-response analysis. Recall that the preceding discussion of tumor appearance time and rate of tumor appearance indicated that tumor rate increases with time for some intake bands, verifying a suggestion by Rowland et al.[67] made in their analysis of the carcinoma data.

The subjects used in this analysis were all women employed in the radium-dial-painting industry at an average age of about 19 yr. There is no assurance that women exposed at a greater age or that men would have yielded the same results. This represents a nonquantifiable uncertainty in the application of the preceding equations to risk estimation.

The statistical uncertainty in the coefficient α is determined principally by the variance in the high-dose data, that is, at exposure levels for which the observed number of tumors is nonzero. In this analysis, there were one or more tumors in the six intake groups with intakes above 25 μCi and no tumors observed in groups with intakes below 25 μCi. Therefore, calculations of the uncertainty of risk estimates from the standard deviation will be accurate above 25 μCi but may be quite inaccurate and too small below 25 μCi. Schlenker[74] examined the uncertainties in risk estimates for bone tumor induction at low intakes and found it to be much greater than would be determined from the standard deviations in fitted risk coefficients. The analysis was not carried out for carcinoma risk, but the conclusions would be the same.

LEUKEMIA

Leukemia has not often been seen in the studies of persons who have acquired internally deposited radium. Nevertheless, the discussion of leukemia as a possible consequence of radium exposure has appeared in a number of published reports.

Martland,[42] summarizing his studies of radium-dial painters, mentioned the development of anemias. He also described the development of leukopenia and anemia, which appeared resistant to treatment.

Evans[15] listed possible consequences of radium acquisition, which included leukemia and anemia. However, no mention of such cases appear in his report.

In a report by Finkel et al.,[18] mention is made of seven cases of leukemia and aplastic anemia in a series of 293 persons, most of whom had acquired radium between 1918 and 1933. Five of these cases of leukemia were found in a group of approximately 250 workers from radium-dial painting plants in Illinois. The radium content in the bodies of 185 of these workers was measured. There were three cases of chronic myeloid leukemia (CML) and one of chronic lymphocytic leukemia (CLL). Because CLL is not considered to be induced by radiation, the latter case was assumed to be unrelated to the radium exposure. The remaining two cases were aplastic anemias; these latter two cases and one of the CML cases were not available for study, and hence no measurements of radium content in the workers' bodies were available.

In an additional group of 37 patients who were treated with radium by their personal physicians, two blood dyscrasias were found. One of these was panmyelosis, and the other was aplastic anemia; the radium measurements for these two cases showed body contents of 10.5 and 10.7 μCi, respectively. The average skeletal doses were later calculated to be 23,000 and 9,600 rad, respectively, which are rather substantial values.

Finkel et al.[18] concluded that the appearance of one case of CML in 250 dial workers, with about 40 yr of follow-up time, would have been above that which was expected. Two cases, by implication, might be considered significant.

In a review of the papers published in the United States on radium toxicity, and including three cases of radium exposure in Great Britan, Loutit[34] made a strong case "that malignant transformation in the lymphomyeloid complex should be added to the accepted malignancies of bone and cranial epithelium as limiting hazards from retention of radium." He pointed out that the reports of Martland[41–43] describe a regenerative leucopenic anemia, and he stated that "this syndrome has features of atypical (aleukemic) leukemia or myelosclerosis or both."

It should be noted, however, that the early cases of Martland were all characterized by very high radium burdens. The British patients that Loutit described[34] also may have experienced high radiation exposures; two were radiation chemists whose radium levels were reported to fall in the range of 0.3 to 0.5 μCi, both of whom probably had many years of occupational exposure to external radiation. The third patient was reported to contain 4–5 μg of radium.

Following the consolidation of the U.S. radium cases into a single study at the Argonne National Laboratory, Polednak[57] reviewed the mortality of women first employed before 1930 in the U.S. radium-dial-painting industry. This study examined a cohort of 634 women who had been identified by means of employment lists or equivalent documents. This cohort was derived from a total of about 1,400 pre-1930 radium-dial workers who had been identified as being part of the radium-dial industry of whom 1,260 had been located and were being followed up at Argonne. By 1954, when large-scale studies of the U.S. radium cases were initiated, 521 of the cohort of 634 women were still alive, and 360 of them had whole-body radium measurements made after that date while they were still living.

In the cohort of 634 women, death certificates indicated that there were three cases attributed to leukemia and aleukemia and four more to blood and blood-forming organs; both were above expectations. When the study was restricted to the 360 measured cases, one case of leukemia was found in a woman with a radium intake greater than 50 μCi. Similarly, only one death attributable to diseases of the blood, acquired hemolytic anemia, was found for a person with a very low radium intake.

TABLE 4-5 Incident Leukemia in Located Radium Workers

	Female			Male		
Year	No. Located	No. Observed	No. Expected	No. Located	No. Observed	No. Expected
Before 1930	1,285	4	5.44	133	1	0.83
1930–1949	1,185	4	2.27	59	0	0.34
1950–1969	226	1	0.26	52	0	0.10
Totals	2,696	9	7.97	244	1	1.27

Stebbings et al.[89] published results of a mortality study of the U.S. female radium-dial workers using a much larger data base. This study included 1,285 women who were employed before 1930. A comparison study included 1,185 women employed between 1930 and 1949, when radium contamination was considerably lower.

In this enlarged study, three cases of leukemia were recorded in the pre-1930 population, which yielded a standard mortality ratio of 73. All of these cases occurred among 293 women employed in Illinois; none were recorded among the employees from radium-dial plants in other states. An additional three cases were found in the 1930–1949 cohort, yielding a standard mortality ratio of 221. These authors concluded that there was no relationship between radium level and the occurrence of leukemia.

The most inclusive and definitive study of leukemia in the U.S. radium-dial workers was published by Spiers et al.[83] By including all the dial workers, male and female, who entered the industry before 1970, a total of 2,940 persons who could be located, they were able to document a total of 10 cases of leukemia. A total of 9.2 cases would be expected to occur naturally in such a population. Table 4-5, based on their report, illustrates their results.

Included in the above summary are four cases of chronic lymphocytic or chronic lymphatic leukemia. Spiers et al.[83] note that this number from a total of 10 is not dissimilar from the 3.6 expected in the general population. They conclude that the incidence of myeloid and other types of leukemia in this population is not different from the value expected naturally.

When the U.K. radium-luminizer study for the induction of myeloid leukemia is examined,[5] it is seen that among 1,110 women there are no cases to be found. The expected number, however, is only 1.31.

The above results, based on observations of several thousand individuals over periods now ranging well over 50 yr, make the recent report by Lyman et al.[35] on an association between radium in the groundwater of Florida and the occurrence of leukemia very difficult to evaluate. Florida has substantial deposits of phosphate, and this ore contains ^{238}U, which in turn produces ^{226}Ra and ^{222}Rn. The radium from this ore evidently finds its way into the groundwater supplies. By measuring the radium content of 50 private wells in 27 selected counties, the counties were divided into 10 low-exposure and 17 high-exposure groups. The high-exposure group was further divided into three graded groups. These divisions were made on the basis of the number of these private wells in each county that contained more than 5 pCi/liter of water.

Lyman et al.[35] show a significant association between leukemia incidence and the extent of groundwater contamination with radium. The majority of the leukemias were acute myeloid leukemias. Further, a dose-response relationship is suggested for total leukemia with increasing levels of radium contamination.

Lyman et al.[35] do not claim, however, to have shown a causal relationship between leukemia incidence and radium contamination. They point out that there is no information on individual exposure to radium from drinking water, nor to other confounding factors. Since leukemia rates are not elevated in the radium-dial worker studies, where the radium exposures ranged from near zero to many orders of magnitude greater than could be attributed to drinking water, it is difficult to understand how radium accounts for the observations in this Florida study.

In summary, the evidence indicates that acquisition of very high levels of radium, leading to long-term body contents of the order of 5 μCi or more, equivalent to systemic intakes of the order of several hundred microcuries, resulted in severe anemias and aleukemias. However, at lower radium intakes, such as those experienced by the British luminizers and the bulk of the U.S. radium-dial workers, incorporated ^{226}Ra does not appear to give rise to leukemia.

The 3.62-day half-life of ^{224}Ra results in a prompt, short-lived pulse of alpha radiation; in the case of the German citizens injected with this radium isotope, this pulse of radiation was extended by repeated injections. Nevertheless, the time that bone and adjacent tissues were irradiated was quite short in comparison to the irradiation following incorporation of ^{226}Ra and ^{228}Ra by radium-dial workers. In spite of these differences, ^{224}Ra has been found to be an efficient inducer of bone cancer.

In the case of leukemia, the issue is not as clear. Leukemia has been seen in the Germans exposed to ^{224}Ra, but only at incidence rates close to those expected in unexposed populations. When an excess has occurred, there exist confounding variables.

Mays et al.[50] reported on the follow-up of 899 children and adults who received weekly or twice-weekly intravenous injections of ^{224}Ra, mainly for the treatment of tuberculosis and ankylosing spondylitis. While five cases of leukemia were observed among 681 adults who received an average skeletal dose of 206 rad, none were observed among 218 1 - to 20-yr-olds at an average skeletal dose of 1,062 rad. The expected number of leukemias for the adult group was two, but the authors point out that the drugs often taken to suppress the pain associated with ankylosing spondylitis are suspected of inducing the acute forms of leukemia. Four of the five leukemias occurred in patients with ankylosing spondylitis; two were known to be acute; it is not known whether the other three were acute or chronic.

It is evident that leukemia was not induced among those receiving ^{224}Ra before adulthood, in spite of the high skeletal doses received and the postulated higher sensitivity at younger ages. There may be an excess of leukemia among the adults, but the evidence is weak.

Wick et al.[95] reported on another study of Germans exposed to ^{224}Ra. While the report of Mays et al.[50] dealt with persons injected with ^{224}Ra between 1946 and 1950, the study of Wick et al.[95] examined the consequences of lower doses as a treatment for ankylosing spondylitis and extended from 1948 to 1975. The average skeletal dose to a 70-kg male was stated to be 56 rad. There were 1,501 exposed cases and 1,556 ankylosing spondylitis controls. Each group consisted of about 90% males. There were 11 bone marrow failures in the exposed group, and only 4 in the control group. Similarly, there were six leukemias in the exposed group versus five in the control group. All five leukemias in the control group were acute forms, while three in the exposed group were chronic myeloid leukemia. The authors drew no conclusions as to whether the leukemias observed were due to ^{224}Ra, to other drugs used to treat the disease, or were unrelated to either.

This population has now been followed for 34 yr; the average follow-up for the exposed group is about 16 yr. A total of 433 members of the exposed group have died, leaving more than 1,000 still alive. It may be some time before this group yields a clear answer to the question of radium-induced leukemia. At this time, it is clear that it is not a primary consequence of radium deposited in human bones.

Thus, while leukemia and diseases of the blood-forming organs have been seen following treatment with ^{224}Ra, it is not clear that these are consequences of the radiation insult or of other treatments experienced by these patients. The extremely high radiation doses experienced by a few of the radium-dial workers were not repeated with ^{224}Ra, so clear-cut examples of anemias following massive doses to bone marrow are lacking.

RADIUM IN WATER

Since uranium is distributed widely throughout the earth's crust, its daughter products are also ubiquitous. As a consequence, many sources of water contain small quantities of radium or radon. In the United States there have been at least three attempts to determine whether the populations that drink water containing elevated levels of radium had different cancer experience than populations consuming water with lower radium levels.

A cooperative research project conducted by the U.S. Public Health Service and the Argonne National Laboratory made a retrospective study of residents of 111 communities in Iowa and Illinois who were supplied water containing at least 3 pCi/liter by their public water supplies. Control cities where the radium content of the public water supply contained less than 1 pCi/liter were matched for size with the study cities. A total of almost 908,000 residents constituted the exposed population; the mean level of radium in their water was 4.7 pCi/liter.

The final report of this study by Petersen et al.[56] reported on the number of deaths due in any way to malignant neoplasm involving bone. They found that, for the period 1950–1962, the age- and sex-adjusted rate for the radium-exposed group was 1.41/100,000/yr. The rate for the control group was 1.14; the probability of such a difference occurring by chance alone was reported as 8 in 100.

However, Petersen[55] wrote an interim report for a review board constituted to advise on a proposal for continued funding for this project. This report indicates that the age- and sex-adjusted osteosarcoma mortality rate for the total white population in the communities receiving elevated levels of radium for the period 1950–1962 was 6.2/million/yr; that of the control population was 5.5. The probability of such a difference occurring by chance was 51%. The authors concluded that "no significant difference could be detected between the osteosarcoma mortality rate in towns with water supplies having elevated levels of ^{226}Ra and matched control towns." The complexity of the problem is illustrated by their findings for Chicago. Although this city draws its water from Lake Michigan, where the radium concentration is reported as 0.03 pCi/liter, the age- and sex-adjusted osteosarcoma mortality rate was 6.3/million/yr, which is larger than that found for the towns with elevated radium levels in their water.

The mobility of populations in this country, the inability to document actual radium intakes, and the fact that water-softening devices remove radium from water all tend to make studies of this nature very difficult to evaluate.

A pair of studies relating cancer to source of drinking water in Iowa were reported by Bean and coworkers.[6,7] The first of these examined the source of water, the depth of the well, and the size of the community. This study was aimed at the role, if any, of trihalomethanes resulting from the disinfection of water by chlorination. The second, which used the deep-well data from the prior study, examined cancer incidence as a function of radium content of the water. Twenty-eight towns met the three criteria for the second study: a population between 1,000 and 10,000, water is obtained solely from wells greater than 500 ft (152 m) deep, and no water softening. These 28 towns had a total population of 63,689 people in 1970.

When the water supplies were divided into three groups levels of 0–2, 2–5, and > 5 pCi of ^{226}Ra per liter and the average annual age-adjusted incidence rates were examined for the period 1969–1978 (except for 1972), certain cancers were found to increase with increasing radium content. These were bladder and lung cancer for males and breast and lung cancer for females. Their data, plus the incidence rates for these cancers for all Iowa towns with populations 1,000 to 10,000 are shown in Table 4-6. When examined in this fashion, questions arise. There is no doubt that male and female lung cancers appear to increase with an increase in the radium content of the water, but in the case of female lung cancers the levels were never as great as observed for those who drank surface water. A similar situation exists for female breast cancer. For male bladder cancer only, the highest radium level produced a higher cancer rate than was observed for those consuming surface water. Were it not for

TABLE 4-6 Cancer Incidence Rate among Persons Exposed to Different Concentrations of Radium in Drinking Water

	Age-Adjusted Incidence Rate/1,000,000				
	pCi/liter			Water Supply from:	
Cancer Site	0–2	2–5	5	Surface[a]	Ground[b]
Male lung	64.9	85.6	108.8	82.8	78.2
Female lung	13.2	18.7	19.2	19.9	15.6
Male bladder	24.6	27.6	33.7	29.9	29.8
Female breast	75.0	89.4	101.5	104.5	95.6

[a]All towns, 1,000 to 10,000 population, with surface water supplies.
[b]All towns, 1,000 to 10,000 population, with groundwater supplies.

the fact that these cancers were not seen at radium intakes hundreds to thousands of times greater in the radium-dial painter studies, they might throw suspicion on radium. However, it is difficult to accept this hypothesis without an explanation of the lesser number of cancers found at higher radium intakes.

In summary, there are three studies of radium in drinking water, one of which found eleveted deaths due in any way to malignant neoplasm involving bone, the second found elevated incidences of bladder and lung cancer in males and lung and breast cancer in females, and the third found elevated rates of leukemia. None of these findings are in agreement with the long-term studies of higher levels of radium in the radium-dial workers.

RISK ESTIMATION

There is little evidence for an age or sex dependence of the cancer risk from radium isotopes, provided that the age dependence of dose that accompanies changes in body and tissue masses is taken into account. With the present state of knowledge, a single dose-response relationship for the whole population according to isotope provides as much accuracy as possible. For ^{224}Ra, ^{226}Ra, and ^{228}Ra the best-available relationships are based on different measures of exposure: absorbed skeletal dose for ^{224}Ra and systemic intake for ^{226}Ra and ^{228}Ra. Simple prescriptions for the skeletal dose from ^{224}Ra as a function of injection level have been given by Spiess and Mays[85] and can be used to estimate skeletal dose from estimated systemic intake. Because all of the data analysis for ^{224}Ra has been based on prescription of dose given by Spiess and Mays,[85] it is important that it be followed in applications of ^{224}Ra dose-response relationships for the estimation of cancer risk in the general population or in case of occupational or therapeutic exposure. Shifting to a different algorithm for dose calculation would, at a minimum, require demonstration that the new algorithm gives the same numerical values for dose as the Spiess and Mays[85] algorithm for subjects of the same age and sex. The alternative is to reanalyze all of the data on tumor induction for ^{224}Ra by using the new algorithm before it is applied it to dose calculations for risk estimation in a population group different from the subjects in the study by Spiess and Mays.[85]

For ingested or inhaled ^{224}Ra, a method for relating the amount taken in through the diet or with air to the equivalent amount injected in solution is required. The ICRP models for the gastrointestinal tract and for the lung provide the basis for establishing this relationship. A necessary first step for the estimation of risk from any route of intake other than injection is therefore to apply these models.

A similar issue exists for ^{226}Ra and ^{228}Ra. Here the available dose-response relationships are presented in terms of the number of microcuries that reach the blood. Intake by inhalation or ingestion must again account for transfer of radium across the intestinal or pulmonary membranes when the ICRP models are used.

For ^{224}Ra the dose-response relationship gives the lifetime risk of bone cancer following an exposure of up to a few years' duration. Because bone cancer is an early-appearing tumor, the risk, so far as is now known, disappears within 25 yr after exposure. It peaks about 5 yr after exposure following the passage of a minimum latent period. Thus, the absence of information on the tumor probability as a function of person-years at risk is not a major limitation on risk estimation, although a long-term objective for all internal-emitter analyses should be to reanalyze the data in terms of a consistent set of response variables and with the same dosimetry algorithm for both ^{224}Ra and for ^{226}Ra and ^{228}Ra. When the time dependence of bone tumor appearance following ^{224}Ra exposure is considered an essential component of the analysis, then an approximate modification of the dose-response relationship can be made by taking the product of the dose-response equation and an exponential function of time to represent the rate of tumor appearance:

$$M(D,t) = F(D)G(t) = F(D)(r)\exp[-r(t-5)], \qquad (4\text{-}23)$$

where $F(D)$ is the lifetime risk, as specified by the analyses of Spiess and Mays[85] and r is a coefficient based on the time of tumor appearance for juveniles and adults in the ^{224}Ra data analyses. The half-life for tumor appearance is roughly 4 yr in this data set, giving an approximate value for r of 0.18/yr. For t less than 5 yr, $M(D,t)$ is essentially 0 because of the minimum latent period. Thereafter, tumors appear at the rate $M(D,t)$.

The age structure of the population at risk and competing causes of death should be taken into account in risk estimation. An ideal circumstance would be to know the dose-response relationships in the absence of competing causes of death and to combine this with information on age structure and age-specific mortality for the population at large. With the analyses presently available, only part of this prescription can be achieved. An approximate approach would be to take the population as a function of age and exposure and apply the dose-response relationship to each age group, taking into account the projected survival for that age group in the coming years. At the low exposures that occur environmentally and occupationally, exposure to radium isotopes causes only a small contribution to overall mortality and would not be expected to perturb mortality sufficiently to distort the normal mortality statistics. Also, mortality statistics as they now exist include the effect of environmental exposures to radium isotopes.

Table 4-7 illustrates the effect, assuming that one million U.S. white males receive an excess skeletal dose of 1 rad from ^{224}Ra at age 40. The excess death rate due to bone cancer for $t > 5$ yr is computed from:

$$M(D,t) = (200 \times 10^{-6}/\text{rad}) \times (0.18/\text{yr})\exp[-0.18(t-5)]. \qquad (4\text{-}24)$$

This assumes the ^{224}Ra dose-response analyses described above and further assumes that tumors are fatal in the year of occurrence. After 25 yr, there would be 780,565 survivors in the absence of excess exposure to ^{224}Ra and 780,396 survivors with 1 rad of excess exposure at the start of the follow-up period, a difference of 169 excess deaths/person-rad, which is about 15% less than the lifetime expectation of 200×10^{-6}/person-rad calculated without regard to competing risks.

If there were a continous exposure of 1 rad/yr, the tumor rate would rise to an asymptotic value. If this were substituted for the tumor rate caused by ^{224}Ra exposure in Table 4-7 and the survival

TABLE 4-7 Effect of Single Skeletal Dose of 1 rad from ^{224}Ra Received by 1,000,000 U.S. White Males at Age 40[a]

Age (yr)	Natural Death Rate	Natural Survival Rate	No. of Survivors without ^{224}Ra Exposure	^{224}Ra Annual Tumor Rate	^{224}Ra Annual Survival Rate	No. of Survivors with ^{224}Ra Exposure
40	0.00240	0.998	998,000	0	1	998,000
41	0.00263	0.997	995,006	0	1	995,006
42	0.00289	0.997	992,021	0	1	992,021
43	0.00319	0.997	989,045	0	1	989,045
44	0.00353	0.996	985,089	0.000036	0.999964	985,054
45	0.00391	0.996	981,148	0.000030	0.999970	981,084
46	0.00434	0.996	977,224	0.000025	0.999975	977,136
47	0.00483	0.995	972,338	0.000021	0.999979	972,229
48	0.00538	0.995	967,476	0.000018	0.999982	967,351
49	0.00601	0.994	961,671	0.000015	0.999985	961,532
50	0.00669	0.993	954,939	0.000012	0.999988	954,790
51	0.00742	0.993	948,255	0.000010	0.999990	948,097
52	0.00820	0.992	940,669	0.0000085	0.9999915	940,504
53	0.00902	0.991	932,203	0.0000091	0.9999919	932,032
54	0.00989	0.990	922,881	0.0000060	0.9999930	922,705
55	0.01083	0.989	912,729	0.0000050	0.9999950	912,551
56	0.01184	0.988	901,776	0.0000041	0.9999959	901,597
57	0.01295	0.987	890,058	0.0000035	0.9999965	889,873
58	0.01416	0.986	877,592	0.0000029	0.9999971	877,412
59	0.01547	0.985	864,429	0.0000024	0.9999976	864,249
60	0.01685	0.983	849,733	0.0000020	0.9999980	849,555
61	0.01835	0.982	834,438	0.0000017	0.9999983	834,261
62	0.02004	0.980	817,749	0.0000014	0.9999986	817,575
63	0.02195	0.978	799,759	0.0000012	0.9999988	799,587
64	0.02407	0.976	780,565	0.0000010	0.9999990	780,396

[a]U.S. white male mortality rates for 1982 from *Statistical Abstract of the United States*, 106th ed., U.S. Department of Commerce, Washington, D.C., 1986.

rate of those exposed to ^{224}Ra were adjusted to the corresponding value (0.9998), survival in the presence of ^{224}Ra exposure after 25 yr would be 777,293, with 3,272 deaths attributable to the ^{224}Ra exposure.

Calculations for ^{226}Ra and ^{228}Ra are similar to the calculation with the asymptotic tumor rate for ^{224}Ra. For ^{226}Ra and ^{228}Ra the constant tumor rates given by Rowland et al.[68] as functions of systemic intake are computed for the intake of interest, and the results are worked out with a table such as Table 4-7. For continuous intake with the dose-squared exponential function for bone sarcoma induction, it is necessary to decide whether to add the cumulative dose and then take the square or to take the square for each annual increment of dose. Taking the former choice, it is implied that the doses given at different times interact; with the latter choice it is implied that the doses act independently of one another. On the microscale the chance of a single cell being hit more than once diminishes with dose; this would argue for the independent action of separate dose increments and the squaring of separate dose increments before the addition of risks. In the model of bone tumor induction proposed by Marshall and Groer,[38] however, two hits are required to cause transformation. This argues for the interaction of doses and in the extreme case for squaring the cumulative dose. Unless bone cancer induced by ^{226}Ra and ^{228}Ra is a pure, single-hit phenomenon, some interaction of dose increments is expected, although perhaps it is a less strong interaction than is consistent with squaring the total accumulated intake when intake is continuous.

The advantage of using a tabular form for the calculation of the effect of radiation is that it provides a general procedure that can be applied to more complex problems than the one illustrated above. With environmental radiation, in which large populations are exposed, a spectrum of ages from newborn to elderly is represented. Knowing the death rate as a function of time for each starting age then allows the impact of radiation exposure to be calculated for each age group and to be summed for the whole population. The use of a table for each starting age group provides a good accounting system for the calculation. The same goals can be achieved if normal mortality is represented by a continuous function and radiation-induced mortality is so represented, as for ^{224}Ra above, and the methods of calculus are used to compute the integrals obtained by the tabular method.

SUMMARY AND RECOMMENDATIONS

As documented above, research on radium and its effects has been extensive. With continued research the full fruits of these labors in terms of lifetime risk estimates for ^{226}Ra and other long-half-life alpha-emitters which are deposited in bone should be realized. In the case of ^{224}Ra, the relatively short half-life of the material permits an estimation of the dose to bone or one that is proportional to that received by the cells at risk. Correspondingly, relatively simple and complete dose-response functions have been developed that permit numerical estimates of the lifetime risk, that is, about 2 ×

10^{-2}/person-Gy for bone sarcoma following well-protracted exposure. In the case of the longer-half-life radium isotopes, the interpretation of the cancer response in terms of estimated dose is less clear. The dose is delivered continuously over the balance of a person's lifetime, with ample opportunity for the remodeling of bone tissues and the development of biological damage to modulate the dose to critical cells. Deposition (and redeposition) is not uniform and tissue reactions may alter the location of the cells and their number and radiosensitivity. Therefore, estimates of the cumulative average skeletal dose may not be adequate to quantitate the biological insult. Investigation of other dosimetric approaches is warranted.

Equally important is ensuring the availability of information on the rate at which tumors have occurred in the populations at risk. Hazard functions which consider the temporal appearance of tumors have shown some promise for delineating the kinetics of radium-induced bone cancers, and may provide insight into the temporal pattern of the effective dose. Combining this information with results observed with ^{224}Ra may lead to the development of a general model for bone cancer induction due to alpha-particle emitters.

Further efforts to refine dose estimates as a function of time in both man and animals will facilitate the interpretation of animal data in terms of the risks observed in humans. As indicated in Annex 7A, the radium-dial painter data can be a useful source of information for extrapolating to man the risks from transuranic elements that have been observed in animal studies. A more complete description of the radium-dial painter data and parallel studies with radium in laboratory animals, particularly the rat, would do much to further such efforts.

The committee believes a balanced program of radium research should include the following elements.

- The bone-cancer risk appears to have been completely expressed in the populations from the 1940s exposed to ^{224}Ra and nearly completely expressed in the populations exposed to ^{226}Ra and ^{228}Ra before 1930; the bone-cancer risk data from the two epidemiological studies should be integrated and analyzed with newer statistical methods to extend the usefulness of human data. The committee recommends that these studies continue to include dosimetric evaluation, especially at the tissue and cellular level, and evaluation of uncertainties from all sources.
- The committee recommends that the follow-up studies of the patients exposed to lower doses of ^{224}Ra since the 1940s now in

progress in Germany and of similar groups of patients exposed to ^{226}Ra and ^{228}Ra should continue. The detection of bone cancer or sinus and mastoid cancer at dose levels comparable to those encountered in occupational exposures would significantly reduce the uncertainties of bone-cancer risk estimation at low dose levels.

- Research should continue on the cells at risk for bone-cancer induction, on cell behavior over time, including where the cells are located in the radiation field at various stages of their life cycles, on tissue modifications which may reduce the radiation dose to the cells, and on the time behavior and distribution of radioactivity in bone. Meaningful estimates of tissue and cellular dose obtained by these efforts will provide a quantitative linkage between human and animal studies and cell transformation in vitro.
- The sinus and mastoid carcinomas in persons exposed to ^{226}Ra and ^{228}Ra are produced largely by the action of ^{222}Rn and its progeny; continued study may offer insights into the effects of occupational and environmental radon. The dosimetry of the mastoid air cell system is much simpler than the dosimetry of the bronchial tree; the mastoid mucosa may be the respiratory tissue for which the epithelial structure may permit accurate target cell dose estimates so that the risk to epithelial tissues per unit dose and the specific energy that has an impact on cells can be determined; this may improve our estimation of the carcinogenic risk in the epithelium of the respiratory tract.

REFERENCES

1. Argonne National Laboratory, Environmental Research Division. 1984. Annual Report No. ANL-84-103. Argonne, Ill.: Argonne National Laboratory.
2. Ash, J. E., and M. Raum. Undated. An Atlas of Otolaryngic Pathology, 4th ed. New York: Armed Forces Institute of Pathology.
3. Aub, J. C., R. D. Evans, L. H. Hempelmann, and H. S. Martland. 1952. The late effects of internally deposited radioactive materials in man. Medicine 31:221–329.
4. Batsakis, J. G., and J. J. Sciubba. 1985. Pathology. Pp. 74–113, in Surgery of the Paranasal Sinuses, A. Blitzer, W. Lawson, and W. H. Friedman, eds. Philadelpha: W. B. Saunders.
5. Baverstock, K. F., and D. G. Papworth. 1986. The U.K. radium luminiser survey: Significance of a lack of excess leukemia. Pp. 22–26 in The Radiobiology of Radium and Thorotrast, W. Gössner, G. B. Gerber, U. Hagen, and A. Luz, eds. Munich, West Germany: Urban and Schwarzenberg.
6. Bean, J. A., P. Isaacson, W. J. Hausler, and J. Kohler. 1982. Drinking water and cancer incidence in Iowa. I. Trends and incidence by source of drinking water and size of municipality. Am. J. Epidemiol. 116:912–923.

7. Bean, J. A., P. Isaacson, R. M. Hahne, and J. Kohler. 1982. Drinking water and cancer incidence in Iowa. II. Radioactivity in drinking water. Am. J. Epidemiol. 116:924–932.
8. Boege, K. 1902. Zur Anatomie der Stirnhohlen, Koniglichen Anatomischen Institut za Konigsberg Nr. 35. Konigsberg in Preussen: Hartangsche Buchdruckerei.
9. Chemelevsky, D., A. M. Kellerer, H. Spiess, and C. W. Mays. 1986. A proportional hazards analyis of bone sarcoma rates in German radium-224 patients. Pp. 32–37 in The Radiobiology of Radium and Thorotrast, W. Gössner, G. B. Gerber, U. Hagen, and A. Luz, eds. Munich, West Germany: Urban and Schwarzenberg.
10. Chiang, C. L. 1968. Introduction to Stochastic Processes in Biostatistics. New York: John Wiley & Sons.
11. Dahlin, D. C. 1978. Bone Tumors, 3rd ed. Springfield, Ill.: Charles C Thomas.
12. Davis, W. B. 1914. Development and Anatomy of the Nasal Accessory Sinuses in Man. Philadelphia: W. B. Saunders.
13. Drettner, B. 1982. The paranasal sinuses. Pp. 145–162 in The Nose: Upper Airway Physiology and the Atmospheric Environment, D. F. Proctor and I. Andersen, eds. Amsterdam: Elsevier Biomedical Press.
14. Evans, R. D. 1933. Radium poisoning; a review of present knowledge. Am. J. Public Health 23:1017–1023.
15. Evans, R. D. 1966. The effect of skeletally deposited alpha-ray emitters in man. Br. J. Radiol. 39:881–895.
16. Evans, R. D., A. T. Keane, R. J. Kolenkow, W. R. Neal, and M. M. Shanahan. 1969. Radiogenic tumors in the radium and mesothorium cases studied at M.I.T. Pp. 157–194 in Delayed Effects of Bone-Seeking Radionuclides, C. W. Mays, W. S. S. Jee, R. D. Lloyd, B. J. Stover, J. H. Dougherty, and G. N. Taylor, eds. Salt Lake City: University of Utah Press.
17. Evans, R. D., A. T. Keane, and M. M. Shanahan. 1972. Radiogenic effects in man of long-term skeletal alpha-irradiation. Pp. 431–468 in Radiobiology of Plutonium, B. J. Stover and W. S. S. Jee, eds. Salt Lake City: The J. W. Press.
18. Finkel, A. J., C. E. Miller, and R. J. Hasterlik. 1969. Radium-induced malignant tumors in man. Pp. 195–225 in Delayed Effects of Bone-Seeking Radionuclides, J. Mays, R. D. Lloyd, B. J. Stover, J. H. Dougherty, and G. N. Taylor, eds. Salt Lake City: University of Utah Press.
19. Frankel, B. 1906. Über die Beziehungen der Grossenvariationen der Highmorshohlen zum individuellen Schadelbau und deren praktische Bedeutung für die Therapie der Kieferhohleneiterungen. Arch. Laryngol. Rhinol. 18:229–257.
20. Groer, P. G., and J. H. Marshall. 1976. Hazard plotting and estimates for the tumor rate and the tumor growth time for radiogenic osteosarcomas in man. Pp. 17–21 in Radiological and Environmental Research Division Annual Report. Report No. ANL-76-88, Part II. Argonne, Ill.: Argonne National Laboratory.
21. Harris, M. J., and R. A. Schlenker. 1981. Quantitative histology of the mucous membrane of the accessory nasal sinuses and mastoid cavities. Ann. Otol. Rhinol. Laryngol. 90:33–27.

22. Hasterlik, R. J. 1960. Radiation neoplasia. Proceedings of the Institute of Medicine, Chicago: Vol. 23, No. 2.
23. Hasterlik, R. J., L. J. Lawson, and A. J. Finkel. 1968. Ophthalmologic aspects of carcinoma of the sphenoid sinus induced by radium poisoning. Am. J. Ophthalmol. 66:55–58.
24. Hentzer, E. 1970. Histologic studies of the normal mucosa in the middle ear, mastoid cavities and eustachian tube. Ann. Otol. Rhinol. Laryngol. 79:825–833.
25. Hindmarsh, M., M. Owen, J. Vaughan, L. F. Lamerton, and F. W. Spiers. 1958. The relative hazards of strontium 90 and radium-226. Br. J. Radiol. 31:518.
26. Hindmarsh, M., M. Owen, and J. Vaughan. 1959. A note on the distribution of radium and a calculation of the radiation dose non-uniformity factor for radium-226 and strontium-90 in the femur of a luminous dial painter. Br. J. Radiol. 32:183–187.
27. Hoecker, F. E., and P. G. Roofe. 1949. Structural differences in bone matrix associated with metabolized radium. Radiology 52:856–864.
28. Hoecker, F. E., and P. G. Roofe. 1951. Studies of radium in human bone. Radiology 56:89–98.
29. International Commission on Radiological Protection (ICRP). 1973. Alkaline Earth Metabolism in Adult Man. ICRP Publication 20. Oxford: Pergamon. (Also in Health Phys. 24:125–221, 1973.)
30. Kolenkow, R. J. 1967. Alpha-ray dosimetry of the bone-tissue interface with application to sinus dosimetry in the radium cases. Pp. 163–201 in Annual Progress Report No. MIT-952-4. Cambridge, Mass.: Radioactivity Center, Massachusetts Institute of Technology.
31. Littman, M. S., I. E. Kirsh, and A. T. Keane. 1978. Radium-induced malignant tumors of the mastoids and paranasal sinuses. Am. J. Roentgenol. 131:773-785, 1978.
32. Lloyd, E. The distribution of radium in human bone. Br. J. Radiol. 34:521–528.
33. Lloyd, E., and C. B. Henning. 1983. Cells at risk for the production of bone tumors in radium exposed individuals: An electron microscope study. Health Phys. 44 (Suppl. 1):135–148.
34. Loutit, J. F. 1970. Malignancy from radium. Br. J. Cancer 24:195–207.
35. Lyman, G. H., C. G. Lyman, and W. Johnson. 1985. Association of leukemia with radium groundwater contamination. J. Am. Med. Assoc. 254:621–626.
36. Marshall, J. H. 1962. Radioactive hotspots, bone growth and bone cancer: Self-burial of calcium-like hotspots. Pp. 35–50 in Radioisotopes and Bone, P. LaCroix and A. M. Bundy, eds. Oxford, England: Blackwell Scientific Publications.
37. Marshall, J. H. 1969. Measurements and models of skeletal metabolism. Pp. 1–122 in Mineral Metabolism, Vol. III, C. L. Comar and F. Bronner, eds. New York: Academic Press.
38. Marshall, J. H., and P. G. Groer. 1977. A theory of the induction of bone cancer by alpha radiation. Radiat. Res. 71:149–192.
39. Marshall, J. H., and C. Onkelix. 1968. Radial diffusion and the power function retention of alkaline earth radioisotopes in adult bone. Nature 217:742–744.

40. Marshall, J. H., P. G. Groer, and R. A. Schlenker. 1978. Dose to endosteal cells and relative distribution factors for radium-224 and plutonium-239 compared to radium-226. Health Phys. 35:91–101.
41. Martland, H. S. 1926. Microscopic changes of certain anemias due to radioactivity. Arch. Pathol. Lab. Med. 2:465-472.
42. Martland, H. S. 1931. The occurrence of malignancy in radioactive persons. Am. J. Cancer 15:2435–2516.
43. Martland, H. S. 1939. Occupational tumors, bones. In Encyclopedia of Health and Hygiene. Geneva: International Labor Organization.
44. Mays, C. W., and R. D. Lloyd. 1972. Bone sarcoma incidence vs. alpha particle dose. Pp. 409–430 in Radiobiology of Plutonium, B. J. Stover and W. S. S. Jee, eds. Salt Lake City: The J. W. Press.
45. Mays, C. W., and H. Spiess. 1983. Epidemiological studies of German patients injected with ^{224}Ra. Pp. 159–166 in Epidemiology Applied to Health Physics. Proceedings of the Sixteenth Mid-Year Topical Meeting of the Health Physics Society. CONF-830101. Springfield, Va.: National Technical Information Service Society.
46. Mays, C. W., and H. Spiess. 1984. Bone sarcomas in patients given radium-224. Pp. 241–252 in Radiation Carcinogenesis. Epidemiology and Biological Significance, J. B. Boice and J. F. Fraumeni, eds. New York: Raven.
47. Mays, C. W., T. F. Dougherty, G. N. Taylor, R. D. Lloyd, B. J. Stover, W. S. S. Jee, W. R. Christensen, J. H. Dougherty, and D. R. Atherton. 1969. Radiation-induced bone cancer in beagles. Pp. 387–408 in Delayed Effects of Bone-Seeking Radionuclides, C. W. Mays, W. S. S. Jee, R. D. Lloyd, B. J. Stover, J. H. Doughtery, and G. N. Taylor, eds. Salt Lake City: University of Utah Press.
48. Mays, C. W., H. Spiess, G. N. Taylor, R. D. Lloyd, W. S. S. Jee, S. S. McFarland, D. H. Taysum, T. W. Brammer, D. Brammer, and T. A. Pollard. 1976. Estimated risk to human bone from ^{239}Pu. Pp. 343–362 in The Health Effects of Plutonium and Radium, W. S. S. Jee, ed. Salt Lake City: The J. W. Press.
49. Mays, C. W., H. Spiess, and A. Gerspach. 1978. Skeletal effects following ^{224}Ra injections into humans. Health Phys. 35:83–90.
50. Mays, C. W., H. Spiess, D. Chmelevsky, and A. Kellerer. 1986. Bone sarcoma cumulative tumor rates in patients injected with ^{224}Ra. Pp. 27–31 in The Radiobiology of Radium and Thorotrast, W. Gössner, ed. Baltimore: Urban and Schwarzenberg.
51. Mygind, N., M. Pedersen, and M. H. Nielsen. 1982. Morphology of the upper airway epithelium. Pp. 71–97 in The Nose: Upper Airway Physiology and the Atmospheric Environment, D. F. Proctor and I. Andersen, eds. Amsterdam: Elsevier Biomedical Press.
52. National Cancer Institute. 1981. Surveillance, Epidemiology, and End Results: Incidence and Mortality Data, 1973–1977. Monograph No. 57. NIH Publication No. 81-2330. Bethesda, Md.: National Cancer Institute.
53. National Cancer Institute. 1982. Cancer Mortality in the United States: 1950–1977. Monograph No. 59. NIH Publication No. 82-2435. Bethesda, Md.: National Cancer Institute.
54. National Research Council, Committee on the Biological Effects of Ionizing Radiations (BEIR). 1980. The Effects on Populations of Exposure to Low

Levels of Ionizing Radiation. Washington, D.C.: National Academy Press. 524 pp.

55. Petersen, N. J. 1966. Midwest Environmental Health Study. Interim Report. Region V. Chicago: Environmental Protection Agency.
56. Petersen, N. J., L. D. Samuels, H. F. Lucas, and S. P. Abrahams. 1966. An epidemiologic approach to low-level ^{226}Ra exposure. Public Health Rep. 81:805–814.
57. Polednak, A. P. 1978. Bone cancer among female radium dial workers. Latency periods and incidence rates by time after exposre. J. Natl. Cancer Inst. 60:77–82 (brief communication).
58. Polednak, A. P., A. F. Stehney, and R. E. Rowland. 1978. Mortality among women first employed before 1930 in the U.S. radium dial-painting industry. Am. J. Epidemol. 107:179–195.
59. Pool, R. R., J. P. Morgan, N. J. Parks, J. Farnham, J. E. Littman, and M. S. Littman. Comparative pathogenesis of radium-induced intracortical bone lesions in humans and beagles. Health Phys. 44(Suppl. 1):155–177.
60. Raabe, O. G. 1984. Comparison of the carcinogenicity of radium and bone-seeking actinides. Health Phys. 46:1241–1258.
61. Raabe, O. G., S. A. Book, and N. J. Parks. 1980. Bone cancer from radium: Canine dose response explains data for mice and humans. Science 208:61–64.
62. Raabe, O. G., S. A. Book, and N. J. Parks. 1983. Lifetime bone cancer dose-response relationships in beagles and people from skeletal burdens of ^{226}Ra and ^{90}Sr. Health Phys. 44:33–48.
63. Rotblatt, J., and G. Ward. 1956. Analysis of the radioactive content of tissues by alpha-track autoradiography. Phys. Med. Biol. 1:57–70.
64. Rowland, R. E. 1975. The risk of malignancy from internally-deposited radioisotopes. Pp. 146–155 in Radiation Research, Biomedical, Chemical, and Physical Perspectives, O. F. Nygaard, H. I. Adler, and W. K. Sinclair, eds. New York: Academic Press.
65. Rowland, R. E., and J. H. Marshall. 1959. Radium in human bone: The dose in microscopic volumes of bone. Radiat. Res. 11:299–313.
66. Rowland, R. E., A. T. Keane, and P. M. Failla. 1971. The appearance times of radium-induced malignancies. Pp. 20–22 in Radiological Physics Division Annual Report, Report No. ANL-7860, Part II. Argonne, Ill.: Argonne National Laboratory.
67. Rowland, R. E., A. F. Stehney, A. M. Brues, M. S. Littman, A. T. Keane, B. C. Patten, and M. M. Shanahan. 1978. Current status of the study of ^{226}Ra and ^{228}Ra in humans at the Center for Human Radiobiology. Health Phys. 35:159–166.
68. Rowland, R. E., A. F. Stehney, and H. F. Lucas, Jr. 1978. Dose-response relationships for female radium dial workers. Radiat. Res. 76:368–383.
69. Rowland, R. E., A. F. Stehney, and H. F. Lucas. 1983. Dose-response relationships for radium-induced bone sarcomas. Health Phys. 44(Suppl. 1):15–31.
70. Rundo, J., A. T. Keane, and M. A. Essling. 1985. Long-term retention of radium in female former dial workers. Pp. 77–85 in Metals in Bone, N. D. Priest, ed. Lancaster, England: MTP Press.
71. Rundo, J., A. T. Keane, H. F. Lucas, R. A. Schlenker, J. H. Stebbings, and A. F. Stehney. 1986. Current (1984) status of the study of ^{226}Ra and ^{228}Ra in humans at the center for human radiobiology. Pp. 14–21 in Radiobiology

of Radium and Thorotrast, W. Gössner, G. B. Gerber, U. Hagen, and A. Luz, eds. Munich, West Germany: Urban and Schwarzenberg.

72. Schaeffer, J. P. 1920. The embryology, development and anatomy of the nose, paranasal sinuses, nasolacrimal passageways and olfactory organ in man. Philadelphia: P. Blakiston's Son and Co.
73. Schlenker, R. A. 1980. Dosimetry of paranasal sinus and mastoid epithelia in radium-exposed humans. Pp. 1–21 in Radiological and Environmental Research Division Annual Report, No. ANL-80-115, Part II. Argonne, Ill.: Argonne National Laboratory.
74. Schlenker, R. A. 1982. Risk estimates for bone. Pp. 153–163 in Critical Issues in Setting Radiation Dose Limits. Proceedings No. 3. Bethesda, Md.: National Council on Radiation Protection and Measurements.
75. Schlenker, R. A. 1983. Mucosal structure and radon in head carcinoma dosimetry. Health Phys. 44:556–562.
76. Schlenker, R. A. 1985. The distribution of radium and plutonium in human bone. Pp. 127–147 in Metals in Bone, N. D. Priest, ed. Lancaster, England: MTP Press.
77. Schlenker, R. A., and J. E. Farnham. 1976. Microscopic distribution of Ra-226 in the bones of radium cases: A comparison between diffuse and average Ra-226 concentrations. Pp. 437–449 in The Health Effect of Plutonium and Radium, W. S. S. Jee, ed. Salt Lake City: The J. W. Press.
78. Schlenker, R. A., and J. H. Marshall. 1975. Thicknesses of the deposits of plutonium at bone surfaces in the beagle. Health Phys. 29:649–654.
79. Schlenker, R. A., and B. G. Oltman. In press. High concentrations of Ra-226 and Am-241 at human bone surfaces: Implications for the ICRP 30 bone dosimetry model. Rad. Prot. Dosimetry.
80. Schlenker, R. A., and J. M. Smith. 1986. Argonne-Utah studies of Ra-224 endosteal surface dosimetry. Pp. 93–98 in The Radiobiology of Radium and Thorotrast, W. Gössner, G. B. Gerber, U. Hagen, and A. Luz, eds. Munich, West Germany: Urban and Schwarzenburg.
81. Schumacher, G. H., H. J. Heyne, and R. Fanghnel. 1972. Zur Anatomie der menschlichen Nasennebenhohlen. Anat. Anz. 130:113–157.
82. Silbiger, H. 1950–1951. Über das ausmass der Mastoidpneumatiation beim Menschen. Acta Anat. 11:215–245.
83. Spiers, F. W., H. F. Lucas, J. Rundo, and G. A. Anast. 1983. Leukemia incidence in the U.S. dial workers. Health Phys. 44(Suppl. 1):65–72.
84. Spiess, H. 1969. ^{224}Ra-induced tumors in children and adults. Pp. 227–247 in Delayed Effects of Bone-Seeking Radionuclides, C. W. Mays, W. S. S. Jee, R. D. Lloyd, B. J. Stover, J. H. Dougherty, and G. N. Taylor, eds. Salt Lake City: University of Utah Press.
85. Spiess, H., and C. W. Mays. 1970. Bone cancers induced by Ra-224 (ThX) in children and adults. Health Phys. 19:713–729.
86. Spiess, H., and C. W. Mays. 1973. Protraction effect on bone sarcoma induction of ^{224}Ra in children and adults. Pp. 437–450 in Radionuclides Carcinogenesis, CONF-720505, C. L. Sanders, R. H. Busch, J. E. Ballou, and D. D. Mahlum, eds. Springfield, Va.: National Technical Information Service.
87. Spiess, H., H. Poppe, and H. Schoen. 1962. Strahlenindizierte Knochentumoren nach Thorium X-Behandlung. Monatsschrift für Kinderheilkunde 110:198–201.

88. Spiess, H., A. Gerspach, and C. W. Mays. 1978. Soft-tissue effects following ^{224}Ra injections into humans. Health Phys. 35:61–81.
89. Stebbings, J. H., H. F. Lucas, and A. F. Stehney. 1984. Mortality from cancers of major sites in female radium dial workers. Am. J. Ind. Med. 5:435–459.
90. Thurman, G. B., C. W. Mays, G. N. Taylor, A. T. Keane, and H. A. Sissons. 1973. Skeletal location of radiation-induced and naturally occurring osteosarcomas in man and dog. Cancer Res. 33:1604–1607.
91. Tos, M. 1982. Goblet cells and glands in the nose and paranasal sinuses. Pp. 99–144 in The Nose: Upper Airway Physiology and the Atmospheric Environment, D. F. Proctor and I. Andersen, eds. Amsterdam: Elsevier Biomedical Press.
92. van Alyea, O. E. 1983. Ethmoid labyrinth. Arch. Otolaryngol. 29:881–902.
93. Wick, R. R., and W. Gössner. 1983. Followup study of late effects in ^{224}Ra-treated ankylosing spondylitis patients. Health Phys. 44(Suppl 1):187–195.
94. Wick, R. R., and W. Gössner. 1983. Incidence of tumours of the skeleton in ^{224}Ra-treated ankylosing spondylitis patients. Pp. 281–288 in Biological Effects of Low-Level Radiation. Vienna: International Atomic Energy Agency.
95. Wick, R. R., D. Chmelevsky, and W. Gössner. 1986. ^{224}Ra risk to bone and haematopoietic tissue in ankylosing spondylitis patients. Pp. 38–44 in The Radiobiology of Radium and Thorotrast, W. Gössner, G. B. Gerber, U. Hagen, and A. Luz, eds. Baltimore: Urban and Schwarzenberg.
96. Wolff, D., R. J. Bellucci, and A. A. Eggston. 1957. Mircoscopic anatomy of the temporal bone. Baltimore: The Williams & Wilkins Co.

5

Thorium

INTRODUCTION

Thorium-232 is a primordial element that is distributed throughout the environment. It has a very long physical half-life (1.41×10^{10} yr) and decays by emission of an alpha particle creating a series of radioactive daughters, many of which also emit alpha radiations. One of these daughters is an isotope of radon, ^{220}Rn, viz., thoron.

The high density and atomic number of thorium led to its use as a contrast agent in medical radiography, as commercially prepared Thorotrast, a 25% colloidal solution of thorium dioxide (ThO_2). Until after the end of World War II, Thorotrast was used extensively as an intravascular contrast agent for cerebral and limb angiography in Europe, the United States, and Japan. It was also injected directly into the spleen for hepatolienography and into abcess cavities in the brain and elsewhere. Direct instillation of Thorotrast into the nasal cavity and paranasal sinuses was also practiced in the past and resulted in a number of epithelial tumors.[13] Because of Thorotrast's colloidal characteristics, thorium and its decay products were deposited in body tissues and organs, most frequently in the reticuloendothelial tissues and in bone. Deposition resulted in continuous alpha-particle irradiation throughout life at a low dose rate.

Patients who received alpha-radiation exposure due to radiologically administered Thorotrast in the late 1920s through 1955 have been followed in epidemiological surveys in Germany,[51] Portugal,[5]

Denmark,[11] and Japan.[32] These studies, described below, demonstrate primarily an excess of liver cancer, including hemangiosarcomas and cholangiosarcomas, and acute myeloid leukemia. This is in contrast to the ^{224}Ra-exposed patients, discussed in Chapter 4, treated for tuberculosis and ankylosing spondylitis,[41] in whom no significant excess of liver cancer has occurred. The alpha-radiation dosimetry in the liver and bone marrow is complex, and precise quantification of risk in these patients is limited because of the nonuniform distribution of thorium dioxide in these tissues and the possible effects of the colloidal material on cancer risk. Moreover, the dose responsible for induction of neoplasia cannot be distinguished from the wasted radiation after initiation has occurred. Therefore, dose-response relationships are highly uncertain.

PROPERTIES AND DOSIMETRY

The long-lived isotope ^{232}Th is the parent of a naturally occurring radioactive decay series. The thorium decay series can be considered in two steps: (1) the formation of ^{224}Ra by the successive decays from ^{232}Th, and (2) the decay of ^{224}Ra and its daughters to stable lead (Figure 5-1). The isotope Ra (half-life, 3.62 days) is an important member of the thorium decay chain; its decay results in the ultimate emission of four alpha particles that release about 26.5 MeV. People with burdens of thorium administered for radiodiagnostic purposes are being irradiated by ^{224}Ra and its alpha-emitting progeny as a result of its continuous production in vivo from the ^{232}Th.[38] The radioisotopes in the thorium series and their physical characteristics are listed in Table 5-1.[39]

ENVIRONMENTAL PATHWAYS[45]

Thorium-232 is present in the soil at an average concentration of about 25 Bq/kg (1 Bq = 27 pCi). Because of its very low absorption through the gastrointestinal tract, natural thorium is mainly incorporated into the body by the inhalation of resuspended solid particles at a rate of about 0.1 Bq/yr. The average body content of thorium-232 is about 80 mBq, 60% of which can be found in the skeleton. Associated annual effective dose equivalent is estimated at about 3 μSv (1 Sv = 100 rem). The decay product of ^{232}Th, ^{228}Ra, is much more mobile environmentally, and unlike ^{232}Th, ingestion constitutes the major pathway for intake. The annual level is about 15

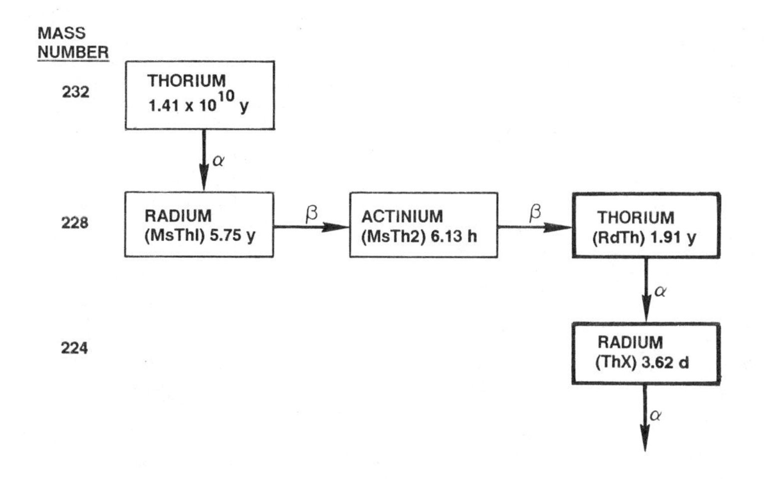

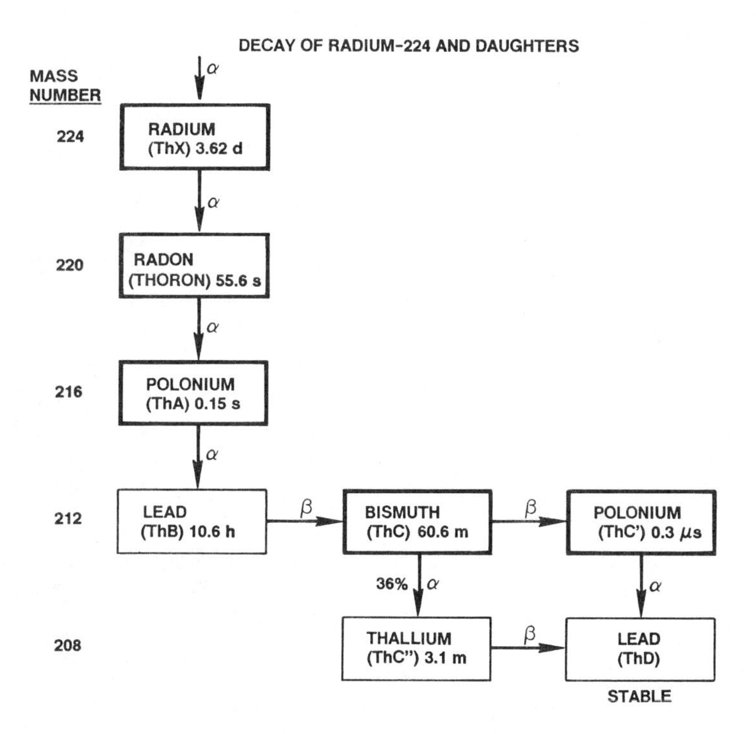

FIGURE 5-1 Formation of ^{224}Ra by successive decays from ^{232}Th. Heavily lined boxes have been drawn around the alpha-particle emitters important to dosimetry. SOURCE: Rundo.[38]

TABLE 5-1 Radioisotopes in the Thorium Series

Radioisotope (Historical Name)	Element	Half-Life	Particle Energy[a] (MeV)
Thorium	^{232}Th	1.4×10^{10} yr	α, 4.01 (76%) α, 3.95 (24%)
Mesothorium 1	^{228}Ra	5.7 yr	β−, 0.02
Mesothorium 2	^{228}Ac	6.13 h	β−, 0.45–2.18
Radiothorium	^{228}Th	1.91 yr	α, 5.42 (71%) α, 5.34 (28%)
Thorium *X*	^{224}Ra	3.64 days	α, 5.68 (95%) α, 5.44 (4.9%)
Thoron	^{220}Rn	55 s	α, 6.28 (99.7%) α, 5.75 (0.3%)
Thorium A	^{216}Po	0.16 s	α, 6.78
Thorium B	^{212}Pb	10.6 h	β−, 0.58 (12%) β−, 0.34 (84%)
Thorium C	^{212}Bi	60.5 min	β−, 0.08–2.27 (64%) α, 6.09, 6.05 (36%)
Thorium C′ (64%)	^{212}Po	0.30 μs	α, 8.78
Thorium C″ (36%)	^{208}Tl	3.1 min	β, 1.0–2.38
Thorium D	^{208}Pb	Stable	

[a]Where the β- or α-spectra contain many lines, only ranges of energy without abundances are given.

SOURCE: Spiers[45].

Bq by ingestion compared to approximately 0.01 Bq from inhalation of the suspended soil particles. Radium-228, on the average, concentrates in bone at a level of about 90 mBq/kg and in soft tissues at about 4 mBq/kg. The decay product of ^{228}Th, as is true of ^{228}Ra, is concentrated in bone, with about 80% of the body content of 300 mBq being found in the skeleton.

Radon-220 and its decay products (^{216}Po, ^{212}Pb, ^{212}Bi, ^{212}Po, and ^{208}Th) are responsible for an additional annual effective dose equivalent of about 0.22 mSv, 90% of which is a result from indoor exposure. Radon-220 and its decay products are generally present at levels about 10- to 20-fold lower than that of ^{222}Ra (from the decay of ^{226}Ra).

Biological Properties of the Thorium Series

Eight different chemical elements are represented in the thorium series. Three of them (Th, Ra, Po) are represented by two isotopes each (Figure 5-1 and Table 5-1). As the chemical identity of a given atom changes as a result of successive nuclear transformations, it may find itself situated at a metabolically inappropriate site, and there is the possibility that it may translocate. The recoil energy imparted to a nucleus on the emission of an alpha particle of several megaelectronvolts of energy is on the order of 100 keV, far greater than the strength of any chemical bond in the atom.[38]

Of the two isotopes of thorium in the precursors of ^{224}Ra, the parent of the series, ^{232}Th, is the only member of the chain that can exist in vivo in macroscopic quantities. The weight of 1 μCi of ^{232}Th is 9.13 g, while the weight of 1 μCi of ^{228}Th is 1.21×10^{-9} g. After intravenous injections of such quantities, the concentration of ^{228}Th in the blood would be about 6×10^{8} atoms/ml, but the concentration of ^{232}Th would be about 10^{10} times higher. Thorium appears to be held tenaciously at either its site of formation in vivo or its point of entry into the body (other than the bloodstream), regardless of the specific activity of the material. Following intravenous injection, thorium of high specific activity deposits mainly on bone surfaces, from which its release appears to be very slow. In the special case of Thorotrast, in which macroquantities of ^{232}Th in colloidal form are injected, it is the physical form that controls its deposition in the cells of the reticuloendothelial system rather than the chemical properties. The colloid aggregates in vivo into clumps as large as 100 μm across, and these aggregates are very stable. The ^{228}Th that is produced via ^{228}Ra and ^{228}Ac in these aggregates does have some mobility, and there is a small loss of activity from thorium deposits.[38,39]

ALPHA DOSIMETRY OF THORIUM IN HUMANS

Thorotrast was administered as a colloidal form of thorium dioxide; the colloidal particles agglomerate and pose a radiation risk to the reticuloendothelial system in which they are ultimately sequestered. The thorium is ultimately redistributed and produces a nonuniform irradiation. The range of the emitted alpha particles in unit density tissue is approximately 40 to 45 μm, that is, about four or five cell diameters. Microscopic and autoradiographic studies have shown that the colloidal aggregates can range to about 100 μm in diameter, producing a highly nonuniform dose-distribution pattern. Such a distribution is thought to be less biologically effective than a more uniform distribution of the same amount of alpha-particle energy[33] for two reasons. First, some of the alpha-particle energy is expended within the aggregate itself, and thus, that fraction of the radiation dose is unavailable to surrounding cells at risk. Second, the cells closest to the aggregates are subject to multiple alpha-particle traversals of critical targets within the cells. Such over irradiation of sensitive cells at risk increases the likelihood of cell killing or sterilization. To the extent that this occurs, it diminishes the opportunity for the same cells to be transformed later and adds to the overall oncogenic risk. This is illustrated in Figure 5-2, which shows a high-resolution autoradiograph of a Thorotrast aggregate surrounded by dense fibrotic tissue in the human liver.[13] Quantitative aspects of this exposure situation are discussed in Appendix I and Chapter 4.

Radioactivity and, therefore, dose increase with the size of the Thorotrast aggregation. However, this is offset to some extent by alpha-energy absorption within the aggregate. With increasing amounts of Thorotrast injected, an increase in the effective average aggregate diameter and a corresponding decrease in the fraction of alpha-energy emitted by the aggregate are found.[50] Table 5-2 shows the mean tissue doses in the liver and red bone marrow based on measurements from the German Thorotrast study[50] and indicates the magnitude of dose modification to tissue afforded by the self-absorption of the alpha particles in thorium dioxide aggregates. For example, in the case of the liver, an increase in the injected quantity of Thorotrast by a factor of 10 is associated with only a fourfold increase in annual radiation dose. The lower uptake of Thorotrast by the bone marrow and the consequent smaller mean aggregate size produced less of an effect.

Following Thorotrast injection and deposition within the body, a buildup of daughter products proceeds, but it never reaches equilibrium.[38] The lack of equilibrium is indicated by the relative excretion rates of ^{232}Th and its daughters; thorium is excreted at a slow rate relative to that of radium isotopes.[38]

Kaul and Noffz[22] calculated absorbed doses to the liver, spleen, red bone marrow, lungs, kidneys, and bone for long-term burdens of intravascularly injected Thorotrast. The estimates were performed for typical injection levels of 10, 30, 60, and 100 ml based on best estimates of ^{232}Th tissue distribution and steady-state activity ratios between subsequent daughters. The typical tissue distribution of ^{232}Th in patients was estimated in the German Thorotrast Study to be as follows: liver, 59%; spleen, 29%; red bone marrow, 9%; calcified bone, 2%; lungs, 0.7%; kidneys, 0.1%.[22] The thorium dioxide concentration in regional lymph nodes of the liver and spleen was high, but very low in other lymph nodes in the body.

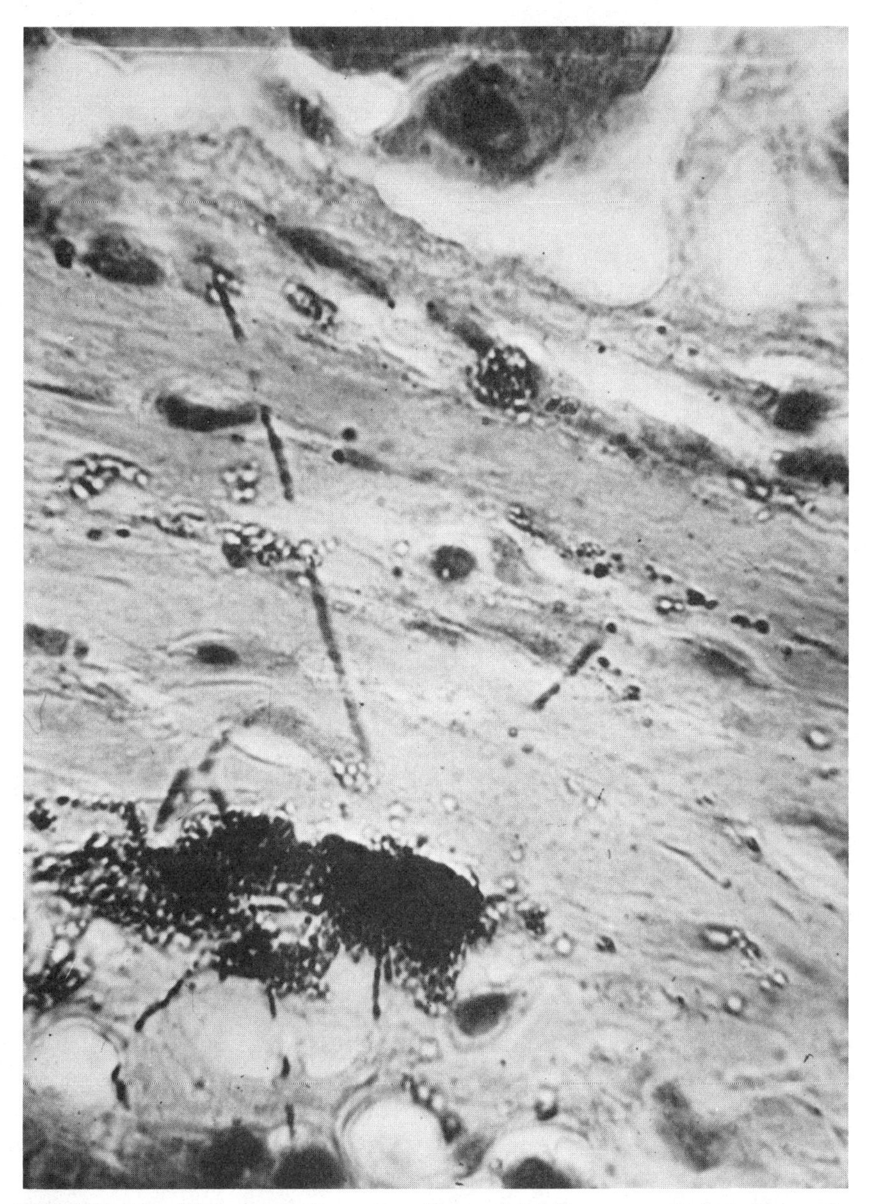

FIGURE 5-2 High-resolution autoradiograph of liver autopsy specimen of a 60-yr-old male who died of hemangioendothelioma in the liver 15 yr after a 75-ml Thorotrast injection for hepatolienography. Magnification, × 1,250; oil immersion.

TABLE 5-2 Mean Annual Alpha-Radiation Dose Following Thorotrast Injection

Organ	Volume of Thorotrast Injected (ml)	^{232}Th in Organ (Bq)	F^a	Mean Tissue Dose (Gy/yr)	Dose/ml (Gy/yr)
Liver	10	4,850	0.85	0.13	0.0130
	30	14,500	0.65	0.28	0.0093
	50	24,200	0.52	0.38	0.0076
	100	48,500	0.38	0.55	0.0055
Red bone marrow	10	740	0.97	0.04	0.0040
	30	2,220	0.92	0.11	0.0037
	50	3,700	0.87	0.17	0.0034
	100	7,400	0.77	0.30	0.0030

[a] *F* is the fraction of alpha energy emitted by the aggregates.

SOURCE: Kaul and Noffz.[22]

Correcting for the alpha-particle self-absorption within Thorotrast aggregates, the mean radiation dose to a standard 70-kg man at 30 yr after the intravascular injection of 25 ml of Thorotrast was estimated to be 750 rad to the liver, 2,100 rad to the spleen, 270 rad to the red bone marrow, 60–620 rad into various parts of the lung, and 13 rad to the kidney. Based on the tissue distribution of ^{232}Th in Thorotrast-exposed patients and the mean concentration of ^{232}Th in various organs of Thorotrast-exposed patients, Figure 5-3 (from the study of Kaul and Noffz[22]) illustrates the mean steady-state alpha-radiation dose rates in the liver, spleen, and red bone marrow. These are plotted against the volume of Thorotrast injected to give values of dose rate for any volume of intravascularly injected Thorotrast between 10 and 100 ml. A typical injection of 25 ml of Thorotrast administered for angiography would result in an estimated dose rate of about 25 rad/yr in the liver and an average of about 16 rad/yr to the endosteal cells of bone.[50] Dose rates to various parts of bone tissue (bone surface, compact and cancellous bone) were estimated by applying the International Commission on Radiological Protection (ICRP) model[18] on alkaline earth metabolism to the continuous translocation of thorium daughters to bone and to the formation of thorium daughters by decay within bone tissue. The average dose to calcified bone from translocated ^{224}Ra with its daughters was estimated to be 19 rad at 30 yr after the injection of 25 ml of Thorotrast.

Both the steady-state activity ratio of thorium daughters to ^{232}Th and the self-absorption of alpha particles in $^{232}ThO_2$ aggregates are important for estimating the absorbed dose in tissues due

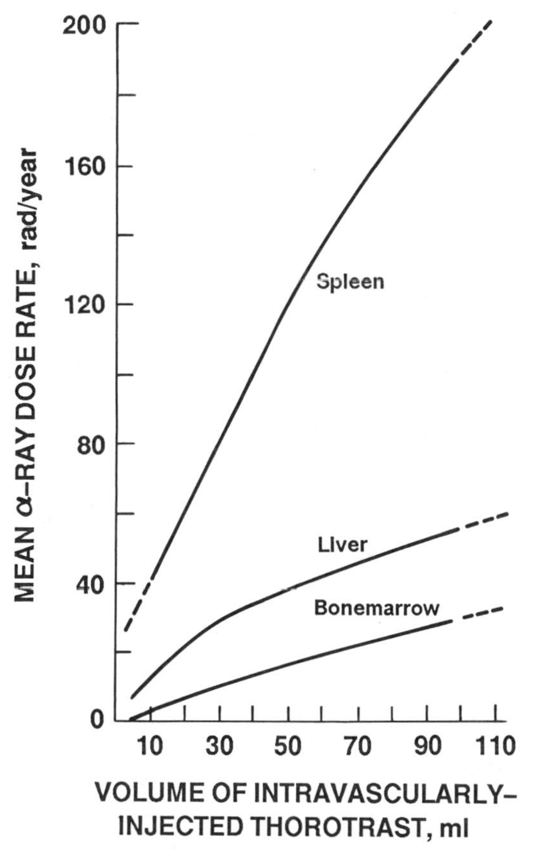

FIGURE 5-3 Mean alpha-radiation dose rates to the liver, spleen, and red bone marrow verses volume of intravascularly injected Thorotrast. SOURCE: Kaul and Noffz.[22]

to Thorotrast. Gamma rays from ^{228}Ac, ^{212}Pb, and ^{208}Tl and alpha rays from ^{232}Th and ^{228}Th emitted from autopsy samples make it possible to estimate the steady-state activity ratio of thorium daughters to ^{232}Th. The steady-state activity ratio of ^{228}Th to ^{232}Th can be determined from an alpha-ray energy spectrum and that of ^{224}Ra to ^{228}Th and ^{228}Ra can be determined from a gamma-ray energy spectrum.[21] For estimation of average absorbed dose in an organ, the distribution of Thorotrast aggregate sizes must be assumed. Examination of Thorotrast-exposed patients and results of laboratory animal experiments demonstrate that the concentration of Thorotrast throughout the liver varies considerably, perhaps by a factor of 100 in Thorotrast-exposed patients.[12,13] Large paravascular injections, together with the heterogeneous distribution in the liver, may be sources of error for the calculation of the tissue dose to the organs of the reticuloendothelial system. Moreover, the estimated tissue dose is dependent on the total injected volume of Thorotrast, the gross organ distribution of the ^{232}Th and its daughter products, the average size of the ThO_2 aggregates, and the alpha-particle self-absorption within the aggregates. At the cellular level, there may be dose rate differences up to a factor of 10,000. Currently, only estimates of the mean organ dose are available.

Kato et al.[20] estimated the absorbed dose in the liver, spleen, and bone marrow in 30 Japanese Thorotrast-exposed patients who died of liver cancer, liver cirrhosis, and other Thorotrast-associated conditions. In the liver, a mean dose rate of 36 rad/yr and a total absorbed dose of 939 rad was calculated; for the spleen the doses were 200 rad/yr and 5,760 rad, respectively; for the bone marrow the doses were 99 rad/yr and 3,087 rad, respectively. For the Japanese patients with hepatic tumors,[20] the mean latent period was 31 yr, the mean absorbed dose in the liver was 939 rad (range, 145–3,234 rad), self-absorption was 0.5, body weight was 50–60 kg, and liver weight was 1,200 g.

Kaul and Noffz[22] estimated the mean dose rates in West German patients based on a 25-ml intravascular injection of Thorotrast in a 70-kg person, to be as follows: liver, 25 rad/yr; spleen, 70 rad/yr; bone marrow, 9 rad/yr; endosteal layer in bone, 16 rad/yr; main pulmonary bronchi, 13 rad/yr, and kidneys, 0.4 rad/yr. The calculations assume the ^{212}Bi activity equals the ^{212}Pb activity in all tissues; if the kidney concentrates ^{212}Bi from the blood plasma, then the kidney dose rate could be much higher than 0.4 rad/yr. The high dose rate to the endosteal layer in bone is due to the thorium dioxide in adjacent bone marrow and translocation of ^{224}Ra from deposits in the reticuloendothelial system to bone surfaces. For the West German patients, the mean latent period was 30 yr, the mean absorbed dose in the liver was 824 rad (range, 384–1,391 rad), the self-absorption was 0.15–0.48, the body weight was about 70 kg, and the liver weight was about 1,800 g. The mean absorbed dose in the liver in the Japanese data was 14% higher than that in the German data, in part, because of the less massive livers in the Japanese patients.[20]

ANIMAL STUDIES

LIVER AND SPLEEN TUMORS

Several animal studies provide a better understanding of the carcinogenic potency of Thorotrast in humans. Early reports discussed whether, in addition to radiation, a foreign body effect or the chemical properties of Thorotrast should be taken into consideration as potential causal factors in tumor induction. Bensted[2] examined the effects of zirconium dioxide aquasol (Zirconotrast) and conventional and ^{230}Th-enriched Thorotrast in mice, and found no clear evidence of an increased incidence of Thorotrast-specific tumors compared with Zirconotrast. Faber[6] injected rabbits with various amounts of ^{230}Th-enriched Thorotrast and found a shortened latency period for hemangioendotheliomas when compared with that caused by commercial Thorotrast. Riedel et al.[35,36] examined the distribution of colloidal thorium, zirconium, and hafnium dioxides and found that the organ distribution of Thorotrast and the kinetics of thorium daughters demonstrated comparable biological behavior in mice, rats, dogs, rabbits, and humans. The other colloids studied failed to show any significantly different effects due to their distribution from those of the thorium dioxide sol.

The investigations by Wesch et al.[54,55] are of particular interest since their objectives were to test for a dose response for carcinogenesis and to determine whether a foreign body effect was involved. In these experiments, ^{232}Th was enriched with different fractions of ^{230}Th to allow variation in dose rate for constant volumes of Thorotrast injected or varying volumes for a constant burden of radioactivity. They found that the frequency of liver and spleen tumors following a single injection of Thorotrast followed a linear dependence on radiation dose rate, but was not correlated with the volume of Thorotrast injected. At a constant dose rate, an increase in the volume of Thorotrast did not increase the tumor risk but did decrease the mean latent period. For a constant activity injected, a factor of 10 increase in the mass injected resulted in further life-shortening. A linear dose-response relationship for liver cancer was found; $I(D) = 3.3 + 0.79D$, where $I(D)$ is the crude incidence (I) of all liver tumors and D is the dose rate. The correlation between dose and incidence was 0.97. The value of I at $D = 0$ did not differ from the observed control incidence of 2.7%.

In later studies, Wesch et al.[56] studied rats injected with Zirconotrast (colloidal ZrO_2) in which ^{228}Th was incorporated. The liver cell carcinomas, intrahepatic bile duct carcinomas, and hemangiosarcomas induced were similar to Thorotrast tumors in humans. The number of hepatic or splenic tumors increased by a factor of 15 compared to controls; the frequency was dose rate dependent but did not correlate with the number of injected particles. The inactive colloid (without ^{228}Th) did not induce primary hepatic or splenic tumors in excess, nor did it increase the tumor incidence at a constant dose rate.

Taylor et al.[43] examined the liver carcinogenicity of ^{241}Am and Thorotrast in mice and found that at comparable doses, in rad, of ^{241}Am and Thorotrast it was approximately equal. The toxicity ratio (^{241}Am/Thorotrast) for liver cancer induction approximated 1.2, with a range of 0.6–1.6. This further suggests that nonradiation factors of Thorotrast were not significant in liver tumor induction.

Brooks et al.[3] injected hamsters with Thorotrast and found that the chromosome aberration frequency in liver cells increased linearly as a function of time and radiation dose. The slope of the dose-response relationship was estimated to be 0.56 aberrations/cell/Gy. This slope can be compared to the value of 0.48 aberrations/cell/Gy observed with injected ^{239}Pu citrate. The data suggest that the dose distribution, chemical effects, or particle loading in the liver do not increase the frequency of chromosome aberrations induced by Thorotrast above that predicted for the more homogeneously distributed alpha radiation from ^{239}Pu citrate. This provides some evidence that the data from Thorotrast-exposed patients may not overestimate the risk for primary liver damage from internally deposited alpha-emitting radionuclides.

Wegener and Hasenöhrl[53] examined rats injected intravenously with different quantities and different alpha doses of Thorotrast. The total frequency of liver and spleen tumors in animals receiving ^{230}Th-enriched Thorotrast was dependent on the dose given. The relationship between dose and effect was almost linear. The volume of injected Thorotrast, given a constant dose rate, had only a slight influence on the number of tumors induced.

The experimental evidence from studies on laboratory animals suggests that Thorotrast-induced tumors appear to arise in large measure from the effects of radiation, and that the carcinogenic effect may not be directly related to the physical presence of the particulate material in the tissues, to the chemical properties of thorium, or to the fibrotic tissue formed by cells killed by the radiation. Tissue destruction of significance does not precede the development of liver neoplasia in rats and mice, even when the radiation dose is very high.[36]

Fibroblast proliferation may occur at injection sites in the subcutaneous tissues in a large proportion of Thorotrast-induced tumors in rats, and this may suggest that neoplasia could develop in those animals that have a vigorous inflammatory reaction to the presence of Thorotrast. However, no difference has been found in the incidence of hepatic tumors in a comparison in mice and rats of the late effects of Thorotrast and the nonradioactive colloidal contrast medium Zirconotrast.[2,56] Studies on the effects of radioactive colloidal gold in laboratory animals demonstrated that it was not necessarily the colloidal state of the material that rendered it carcinogenic for the liver. Whereas a number of investigations have emphasized the possible role of cirrhosis and related biochemical factors in the livers of Thorotrast-exposed patients, no recent experimental evidence is available to indicate induction of liver damage and cirrhosis in rats and mice following the administration of Thorotrast.[12]

Bone Tumors

For ^{228}Th, the tissues that are of importance in the neoplastic response in bone are the osteogenic tissues at the surface of bones and possibly in or near zones of endochondral bone formation, especially endosteal tissue.[4] Mays et al.[26] found that ^{228}Th is 8 times as effective as ^{226}Ra for the induction of osteosarcomas from injected bone-seeking alpha-emitting radionuclides in beagle dogs. The higher effectiveness of ^{228}Th is apparently due to the surface deposition of the thorium closer to the osteogenic tissue, in contrast to the distribution of the radium isotope throughout the bone tissue volume.

Studies of the bones of beagle dogs receiving single intravenous injections of ^{228}Th have shown that the histopathological changes preceding the development of osteosarcomas are similar to those caused by ^{239}Pu and ^{226}Ra and those in radium-bearing humans. High radiation doses altered the vasculature and circulation and caused bone necrosis, bone resorption, reduced bone formation, and marrow fibrosis.[19,26]

Lloyd et al.[24] determined toxicity ratios for bone sarcoma induction at low dose rates and at low total doses in life-span observations of beagles, injected as young adults, for incorporated ^{228}Th relative to that for ^{226}Ra. For equal incidence of bone sarcoma, ^{228}Th was about 8.5 ± 2.3 times as effective as ^{226}Ra on the basis of cumulative average skeletal dose at 1 yr after death. ^{228}Th was about 9.1 ± 2.5 times as effective as ^{226}Ra when skeletal doses were compared at the time of death.

HUMAN STUDIES

The studies in humans are almost all done following administration of Thorotrast. When Thorotrast is injected intravenously, the particles are taken up by the macrophages of the reticuloendothelial system; and the organs that show the greatest concentrations of aggregates of crystals are the liver, spleen, bone marrow, and lymph nodes. Hematological studies of Thorotrast-exposed patients demonstrated that it is common to find anemia, with an increase in the early forms of the myeloid series.[52] Because of Thorotrast deposition in the bone marrow, destruction of erythropoietic and myelopoietic tissues and the subsequent appearance of circulating immature blood cells would be expected.

The use of Thorotrast for hepatolienography and angiography in order to examine the reticuloendothelial system resulted in the induction of primary sarcomas, carcinomas, and mixed neoplasms in the liver.[52] Since hepatic carcinomas are associated with other pathologic conditions of the liver, for example, cirrhosis, and since in some of these patients Thorotrast was administered to diagnose and evaluate liver disease, it is difficult to assess the role of precancerous conditions that may have existed at the time of the administration of Thorotrast and the extent to which the radioactive colloid may have accelerated the induction of malignancy.

Many Thorotrast-exposed patients have been reported to have hepatocellular and cholangiocellular carcinoma of the liver.[4] While histologically similar, these neoplasms were classified as hepatosarcomas, hemangioepitheliomas, endothelial cell sarcomas, and hemangioendotheliomas. On the basis of the limited clinical and experimental material available, it has been suggested that the hemangioendotheliomas may very well be almost a Thorotrast-specific tumor.[1]

There are five epidemiological follow-up studies of Thorotrast-exposed patients, namely, the German Thorotrast study,[46–51] the Japanese Thorotrast cases,[20,21,29–32] the Thorotrast exposed patients in Portugal,[1,5,17] the Danish Thorotrast study,[6–11] and the American study.[14]

THE GERMAN THOROTRAST STUDY

The German Thorotrast study[46–51] now consists of a follow-up of 5,159 Thorotrast-exposed patients and 5,151 controls followed since 1933 and 1935, respectively. The Thorotrast-exposed patients underwent diagnostic x-ray examination during the period 1930 to 1951; these were primarily intravascular injections of x-ray contrast medium for cerebral angiography and angiography of the lower and upper limbs. There were 2,334 Thorotrast-exposed patients and 1,912 control patients who survived 3 yr or more after treatment and could be traced. Follow-up to 1984 has been performed on 894 Thorotrast-exposed patients and 662 control patients; 1,964 Thorotrast-exposed patients and 1,409 control patients have died.

The causes of death in the latter group are listed in Table 5-3.[51] Most evident in the Thorotrast-exposed patients was 347 cases of liver cancer, compared with 2 cases in the control group; the liver tumors were carcinomas, primarily cholangiocellular and hemangiosarcomas. Cirrhosis was present in many of the patients with liver tumors. The shortest latency period was 16 yr, and some now range to latency intervals of more than 40 yr. The accumulated alpha-radiation tissue dose to the liver is estimated to range from 200 to 1,500 rad.[50]

Myeloproliferative disorders occurred in 35 Thorotrast-exposed patients and 3 controls; the diseases included acute myeloid leukemia and erythroleukemia, monocytic leukemia, and chronic myeloid leukemia. The shortest latency interval for leukemia was 5 yr. The estimated accumulated dose to the red bone marrow is estimated

TABLE 5-3 The German Thorotrast Study: Causes of Death in Examined and Nonexamined Patients (Combined)

Disease	Thorotrast (n = 1, 964 of 2,334)	Control (n = 1, 409 of 1,912)
Liver cancer	347	2
Myeloproliferative disease	35	3
Chronic lymphatic leukemia	3	2
Non-Hodgkins lymphoma	16	7
Bone sarcoma	4	1
Lung cancer	46	40
Pleural mesothelioma	4	0
Kidney cancer	4	2
Liver cirrhosis	292	42
Bone marrow failure	20	1
Cardiovascular diseases	587	468

SOURCE: van Kaick et al.[24]

to range from 50 to 400 rad. Non-Hodgkins lymphoma occurred in 16 Thorotrast-exposed patients and in 7 controls. Only four bone sarcomas have appeared in the 2,334 Thorotrast-exposed patients; the accumulated dose to the bone surface was estimated to be about 200–470 rad. Bone marrow failure due to aplastic anemia, agranulocytosis, or thrombocytopenia occurred in 20 Thorotrast-exposed patients and in 1 control patient; it is possible that some aplastic anemias were misdiagnosed aleukemic leukemias.

The results of the German Thorotrast study,[51] when compared with those of the Portuguese,[5] Danish,[11] and Japanese[32] studies, show similar excess rates of liver cancers and leukemia.[27,28] Dose-effect relationships for liver cancers and leukemias have been observed in the West German study (Figure 5-4).[51] However, the influence of the dose rate to bone marrow on the leukemia incidence cannot, as yet, be established. The cumulative incidence of liver cancers and leukemias plotted against time after Thorotrast injection (Figure 5-4) assumes an average volume of 25 ml of Thorotrast per injection, which corresponds to a tissue dose rate of 25 rad/yr in the liver and 9 rad/yr in the bone marrow. Leukemias appeared 5 yr after injection and continued to increase subsequently, while liver cancers did not appear until almost 20 yr after injection and then increased very rapidly. The proportion of cumulative leukemia incidence to cumulative liver cancer incidence is 1.2 to 12% by 40 yr, or 1 to 10. Time after injection is an important factor in the incidence of liver cancers but is much less so in the case of leukemias.

Figure 5-5 illustrates the cumulative incidence of liver tumors in examined German Thorotrast-exposed patients with different liver dose rates.[50] Three groups of patients were studied: those receiving more than 20 ml of Thorotrast (dose rate, approximately 30 rad/yr), those receiving approximately 11–20 ml (dose rate, approximately 18 rad/yr), and those receiving less than 10 ml (dose rate, approximately 10 rad/yr). The dose and dose rate dependence are indicated in Figure 5-5, both with the shortening of the latent interval with increasing dose and dose rate and with the increased frequency of liver cancers in these patients with increasing dose and dose rate. In this study, the cumulative incidence of liver tumors was not influenced by the age at injection.

There is an apparent lack of excess lung cancers in Thorotrast-exposed patients, (46 [2.3%] observed versus 40 [2.8%] in controls), even though the bronchi are exposed to chronic alpha radiation from

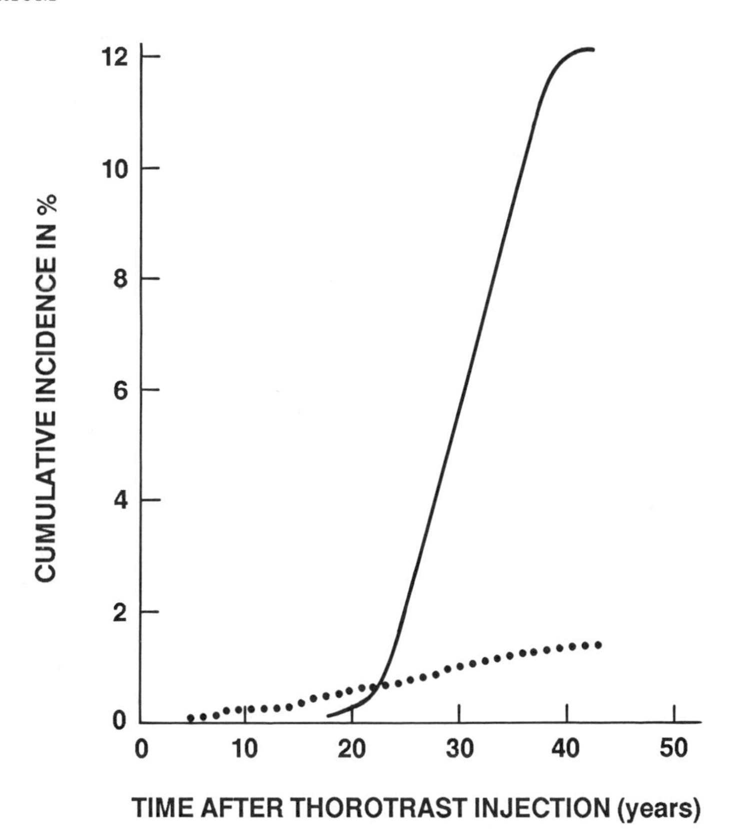

FIGURE 5-4 The German Thorotrast study. Cumulative incidence of liver tumors (solid line) and leukemia (dotted line) in examined and nonexamined patients (n = 2,135). SOURCE: van Kaick et al.[50]

^{220}Rn, which is exhaled with the breath. Initial estimates of the accumulated radiation dose to the bronchial airways of these patients suggested doses as high as 1,000 rad, based on the injection of 1 μCi of ^{232}Th and about 45 ml of Thorotrast deposited in the reticuloendothelial system.[15] At this dose level it might be expected that as many as 50 excess lung cancers might occur in the Thorotrast-exposed patients surveyed; however, no excess occurred. Reevaluation of lung doses[16] by using a Weibel model and calculating doses for bronchial stem cells at generation-specific depths resulted in a decrease in the mean bronchial dose estimates by a factor of 4.3 and in the segmental bronchi dose estimates by a factor of 3.3; this brings the estimates of excess lung cancer in Thorotrast-exposed patients closer to the level observed in the control group.

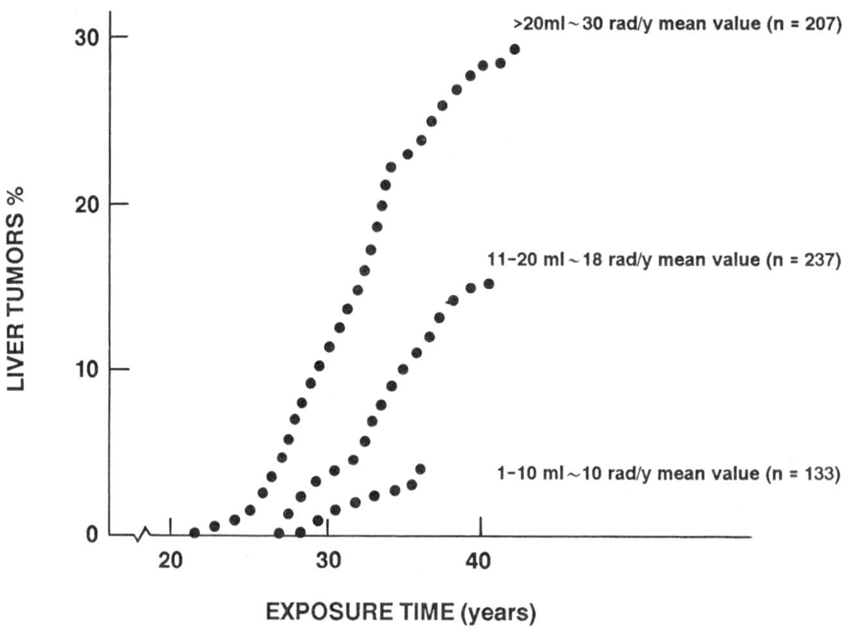

FIGURE 5-5 The German Thorotrast study. Cumulative incidence of liver tumors in examined Thorotrast-treated patients with different liver dose rates. SOURCE: van Kaick et al.[50]

The Portuguese Thorotrast Study

The epidemiological study of Thorotrast-exposed patients in Portugal[1,5,17] represents a 30-yr follow-up of about 2,500 patients exposed mainly between 1929 and 1955 and approximately 2,000 controls. Some 60% of the patients were given Thorotrast for cerebral angiography; the remainder were given Thorotrast for reasons that included limb arteriography and venography, aortography, hepatosplenography, and examination of the paranasal sinuses. By the end of 1976, 955 of the 1,244 traced Thorotrast-exposed patients and 656 of the control cases had died; 137 of the patients died from malignant tumors, 87 of which were primary liver cancers. Of the 32 liver cancers with confirmed histological classification, 18 were hemangioendotheliomas and 4 were biliary duct carcinomas. There were eight carcinomas of the stomach, five carcinomas of the lung, two carcinomas of the larynx, and five primary bone tumors. A total of 23 patients died of blood disorders (12 from leukemias, mostly acute and myeloid) and 27 died of cirrhosis of the liver. There was only 1 case of liver cancer in the 656 deaths in the control group, no cases of leukemia, and 6 cases of liver cirrhosis. Analysis of the data has shown that the number of observed deaths from malignancies—liver, bone, bronchus, larynx, and leukemias—and from liver cirrhosis in the Thorotrast-exposed patients was significantly higher than the expected corresponding numbers in the general Portuguese population.

The majority of the liver tumors were hemangioendotheliomas, one-third were cholangiocarcinomas, two were hepatomas, and one was reticulosarcoma. There were four cases of multicentric tumors of the reticuloendothelial organs, including the liver. Estimates of the amount of Thorotrast sol injected into the patients who had died ranged from 18.0 to 38.9 ml, with an average of about 26 ml. The latency periods varied with the main causes of death. The highest average latent periods were found among those who died from malignancies (greater than 27 yr); for liver cancers, the range was 29–34 yr; for leukemias it was about 20 yr.[5,17]

THE JAPANESE THOROTRAST STUDY

An epidemiological study is being conducted in Japan of 282 patients who were given Thorotrast for angiography and hepatolienography during World War II.[20,21,29–32] Their follow-up now extends to 38–46 yr post-Thorotrast administration. The amount of Thorotrast injected intravascularly in 159 cases ranged from 1.0 to 139 ml (0.02 to 2.78 μCi); the mean injection volume per patient was 17.1 ml (0.3 μCi). In the 261 cases with intravascular injection, there have been 50 cases of liver cancer, 4 cases of blood disease, 3 cases of lung cancer, 1 case of osteosarcoma, 22 other malignant tumors, and 16 cases of liver cirrhosis. These data are summarized together with the data on the non-Thorotrast-exposed control group in Table 5-4. The mortality rates due to hepatic and other malignant tumors, blood diseases, and cirrhosis of the liver and the overall mortality rate were significantly higher in the group treated with Thorotrast intravascularly than in the controls. Figure 5-6 shows the cumulative indices of malignant hepatic tumors, liver cirrhosis, and blood disease. The first case of liver cancer occurred 21 yr after Thorotrast injection, and the number increased rapidly thereafter, reaching 15% of the total number of cases at 40 yr postinjection. Deaths due to liver cirrhosis and blood diseases were first observed at 18 and 16 yr after injection, respectively, and increased to 5.6 and 1.3%, respectively, of the total number of cases at 38 and 36 yr postinjection. These observations concerning malignant hepatic tumors, liver cirrhosis,

TABLE 5-4 The Japanese Thorotrast Study: Causes of Death in Intravascular Thorotrast Group and in Controls

	Thorotrast Group		Non-Thorotrast Group	
Cause of Death	No. of cases	%	No. of cases	%
Malignant tumors	75	28.8[a]	94	7.3
Hepatic tumors	50	19.2[a]	6	0.5
Other	25	9.6	88	6.8
Blood diseases	4	1.5[a]	2	0.2
Liver cirrhosis	16	6.1[a]	17	1.3
Other diseases	67	25.7	243	18.8
Total dead cases	180	69.0[a]	446	34.6
Total living cases	74	28.3	844	65.4
Untraced cases	7	2.7	—	—
Total cases	261	100.0	1,290	100.0

[a] $P < 0.001$.

SOURCE: Modified from Mori et al.[32]

and blood diseases are in accord with the findings of the German Thorotrast study.[49,50]

In the Japanese Thorotrast study,[31] the absorbed dose rate in the liver, spleen, and bone marrow was estimated for 71 autopsy cases of Thorotrast-treated patients who died from cholangiocarcinoma, hemangioendothelioma, liver cell carcinoma, liver cirrhosis, and blood and other diseases. The mean dose rate in the liver, spleen, and bone marrow, classified by cause of death, was estimated to be 22.2–34.7, 67.8–137.8, and 15.9–36.6 rad/yr, respectively. Mean latent periods for the different causes of deaths in patients exposed to Thorotrast (liver cancers, liver cirrhosis, and blood diseases) ranged from about 30–37 years and decreased with increasing dose rate.[21]

Cumulative indices of malignant hepatic tumors, liver cirrhosis, and blood diseases reached 19.2, 6.1, and 1.5%, respectively (Figure 5-6), of the total number of patients exposed to Thorotrast intravascularly at 43 yr after injection. This is in accord with those values reported in the West German Thorotrast study.[51]

Of the 21 patients given Thorotrast nonintravascularly, 6 were alive, 14 were dead, and 1 was untraceable. The causes of death were 1 carcinoma of the pancreas, 2 liver cirrhoses, 10 other diseases, and 1 accident. No significant relationship was found between the causes of death and Thorotrast injection when compared with controls.

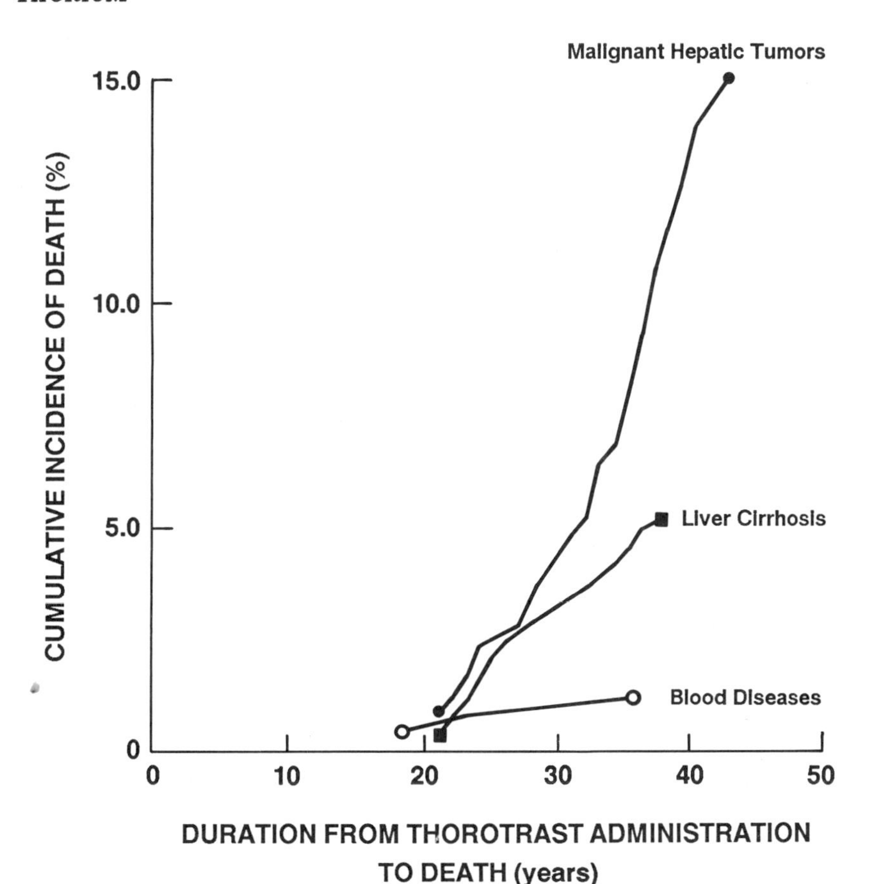

FIGURE 5-6 The Japanese Thorotrast study. Duration from Thorotrast administration to death due to hepatic malignant tumors, liver cirrhosis, and blood diseases in the group exposed to Thorotrast intravascularly. SOURCE: Mori et al.[31]

In the patients exposed to Thorotrast intravascularly, the dose rates to the liver estimated in 96 cases ranged from 2 to 69 rad/yr; the mean absorbed dose was 919.6 rad (standard deviation [SD], 409.0 rad) for 67 malignant hepatic tumors, 958.6 rad (SD, 251.6 rad) for 8 liver cirrhoses, and 757.3 rad (SD, 334.5 rad) for 21 other tumors and diseases. The dose rates to the spleen, estimated in 82 cases, ranged from 8 to 743 rad/yr. The dose rates to the bone marrow in 63 cases ranged from 1 to 157 rad/yr.

In a Japanese series of 120 autopsy cases of patients who died of Thorotrast-associated conditions, there were reported[23] 36 cases of cholangiocarcinoma, 25 cases of angiosarcoma, 10 cases of hepatocellular carcinoma, and 4 cases of multiple hepatic malignancies. The latent periods were as follows: cholangiocarcinoma, mean, 34.1 ± 6.6 yr (range, 23–45 yr); angiosarcoma, mean, 36.4 ± 5.4 yr (range, 27–49 yr); hepatocellular carcinoma, mean, 35.3 ± 5.8 yr (range, 23–41 yr). No unusual histological features were recorded in liver cancers in the Thorotrast- and non-Thorotrast-exposed patients. The coexistence of two or three different malignant neoplasms of the liver was found in 4 (5.3%) of the 75 Thorotrast-induced hepatic malignancies. In 55 Japanese patients who received Thorotrast intravascularly 29–50 yr previously, significant dose-dependent changes were found both in the appearance of Howell-Jolly bodies in the erythrocytes, which increased significantly with thorium body burden, as was an increase in osmotic resistance of erythrocytes with an increase in thorium deposition.[42]

THE DANISH THOROTRAST STUDY

A follow-up study of Danish neurosurgical patients injected with Thorotrast during the years 1935–1946,[6–11] was begun a few years after the cessation of the radiological use of Thorotrast. The control population used is derived from the Danish Cancer Registry. The malignant tumors found in excess in 1979 were as follows: cancers of the digestive tract, 71 observed versus 21 expected; liver tumors, 50 versus 0.75; lung cancers, 14 versus 7.5; and leukemias 14 versus 1.6. In the 1986 report of results to the end of 1983,[11] 1,169 patients had died and 150 were alive. Cancer types have shown little difference over time, and only liver tumors and leukemias show great divergence from expected rates. Liver tumors were the largest single cause of death from 1980 to 1983. There have been 93 liver cancers versus 0.89 expected, and 23 leukemias versus 3.12 expected. There also appeared to be an excess of lung cancer (19 observed versus 9.1 expected). This apparent difference is unexplained.

THE AMERICAN THOROTRAST STUDY

Falk et al.[14] carried out a preliminary epidemiological investigation of Thorotrast-exposed patients in the United States covering the years 1964–1974 and found 26 cases of Thorotrast-induced hepatic angiosarcoma. All patients had undergone either hepatolienography or cerebral angiography. This hepatic tumor incidence was still increasing in the early 1970s, and a larger proportion of the more recent cases had undergone relatively low-dose Thorotrast radiological procedures and prolonged latent periods, ranging from 19.8 to 28.0 yr, and 1 case was as long as 40 yr.

OTHER HUMAN STUDIES

Toohey et al.[44] have measured the activity of thorium daughters (^{228}Ac, ^{212}Pb, ^{212}Bi) in vivo in studies of the health effects of thorium exposure on 133 former workers in a thorium refinery; in addition, the exhalation rate of ^{220}Rn (from ^{224}Ra) was determined for each subject. The values observed were elevated and appeared to be representative of the given individual only. No correlation was made concerning health outcomes.

Xing-an et al.[57] have examined exhaled thoron activity and ^{228}Th lung burden in 20 miners inhaling thorium dust in iron mines; the ^{228}Th lung burden was approximately four times higher than that in nonexposed controls. The thoron concentrations in the breath of miners were 3 to 4 times higher than those in controls. They also found that the ^{228}Th body burden in 20 persons living in a high-background area (Dong-anling region) was 3 times greater than that in controls. No health effects were examined.

ESTIMATION OF EXCESS RISK FOLLOWING THOROTRAST ADMINISTRATION

The primary sources for determining the risks for tumor induction after exposure to thorium are the epidemiological studies of Thorotrast-exposed patients. Although these can now provide estimates for the risks of liver cancer and possibly leukemia, these risk estimates are applicable only to intravascular Thorotrast exposure. Animal studies indicate that it is primarily the alpha radiation from $^{232}ThO_2$ that causes the tumors. Other forms of thorium would be subject to different pharmacodynamics, and thus, the dose distribution and health effects would be different.

In order to calculate the risk of dying by liver cancer after Thorotrast injection, it is necessary to know the size of the Thorotrast population cohort, the average dose to the liver per year, the number of persons dead at time t, the number of liver cancers at time t, and finally the number of liver cancers in the control group at time t. In addition, it is necessary to assume a death rate at time t (to estimate total liver cancers when the entire cohort is dead) and latent period.

TABLE 5-5 Liver Cancer Data from Thorotrast Studies

Thorotrast Study	No. of Years Followed (date)	Average Liver Dose (rad/yr)	Total Cohort Size	No. of Deceased	Liver Cancers	No. of Liver Cancers/ No. of Controls
German	40 (1984)	25	2,334	1,964	347	2/1,409
Japanese	40 (1984)	36	254[a]	180	50	6/446
Portuguese	30 (1976)	25	1,244	955	87	1/656

[a]Of 261 cases, 7 are untraced.

Both the German and the Japanese Thorotrast cases have been followed for about 40 yr, and in the Portuguese study, results are available as of 1976, at which time those cases had been followed for about 30 yr. Table 5-5 lists the information from these three studies needed to make approximate estimates of the liver-cancer risk.

The major assumptions in this calculation are (1) the rate at which the study group is dying (which determines the total lifetime of the study) and (2) the latency period. Figures 5-3 and 5-5 appear to provide evidence that the latent period is about 20 yr. Estimation of the rate of dying is more difficult from the information available. The rate is expected to increase with age, and the simple linear model following liver-cancer deaths should be a rough approximation to what will actually occur. An example calculation of the risk is given in the box entitled "Example Risk Estimate for Liver Cancer in the German Thorotrast Study."

Using these assumptions, excess lifetime risks have been calculated for liver cancer for the three different Thorotrast studies, namely, the German, the Japanese, and the Portuguese studies. These risks are shown in Table 5-6.

An assumption of a shorter latent period, for example 10 yr as in the Biological Effects of Ionizing Radiation (BEIR) III report,[34] will reduce these risk values because the effective dose will have increased due to the longer time at risk. A 10-yr assumed latent period will reduce the risk estimates by about one third. The 1980 BEIR III report[34] based its projections on an assumed minimal latent period of 10 yr and observed mortality to the end of life for the total population in the three studies still alive and at risk; it estimated approximately 300 excess liver cancers/10^6 person-rad of alpha radiation to the liver.

It must be remembered that these estimates are for Thorotrast, not thorium. The dosimetry of thorium in other forms will likely be

EXAMPLE RISK ESTIMATE FOR LIVER CANCER IN THE GERMAN THOROTRAST STUDY

Assumptions:

1. Total death rate after 20 yr parallels liver-cancer death rate and is linear during the last 20 yr.
2. The latency period is 20 yr.

Assumed cohort average annual death rate = 1964/20 = 98.2 deaths/yr

Estimated remaining mean time to death of cohort = (2,334 − 1,964)/98,2 = 4 years

Total expected number of liver cancers = 347[(20+4)]/20 = 416

Person-rad–wasted dose = 25(56,016 × 25) = 1,380,950

Total excess number of liver cancers = 416 − (2/1,409) × 2,334 = 413

Risk per 10^6 person-rad = 413/1,380,950 = ≈ 300/10^6 person-rad

TABLE 5-6 Estimated Liver Cancer Risks from Thorotrast

Thorotrast Study	Expected Excess Liver Cancers	Person-Rad-Wasted Dose	Risk/10^6 Person-Rad
German	413	1.30 × 10^6	300
Japanese	67	0.256 × 10^6	260
Portuguese	111	0.40 × 10^6	280

quite different from the dose distributions associated with Thorotrast aggregates, and the risk values will also be different.

Faber[6,8,9] estimated the excess rate of liver cancer in adults as 4.2 cases/year/10^6 person-rad. For a 40-yr follow-up, this would correspond to about 170 cases/10^6 person-rad. Such estimates, however, are not based on modeling the pattern of risk over time and must be considered provisional until more complete data are available.

Deaths from leukemia in the Thorotrast surveys in Germany, Portugal, and Denmark are in excess of the national rates of death from leukemia. Two categories of malignant disease exist, namely, (1) malignant disease originating in the bone marrow, that is, leukemia (including acute myeloid and chronic myeloid leukemia), multiple myeloma, and hemangiosarcoma confined to the bone marrow; and (2) malignant disease arising in the lymphoid tissues, that is, malignant disease that includes thymoma, reticulosarcoma, and acute lymphoid leukemia. By 1978, the total of the former category in the combined surveys exceeded 40 cases, which is a combined rate of about 12 cases/1,000 persons.[27] The expected number of cases would depend on the age distribution of the population of Thorotrast-exposed patients. If an expected value of 2/1,000 patients is assumed, the excess due to Thorotrast would be 10/1,000.[27,28] The average dose to bone marrow was about 150–200 rad.[27] This would result in an estimated lifetime linear risk coefficient of 50–60 excess leukemia cases/10^6 person-rad.[27]

In the second category, a total of 11 cases have been recorded,[27] which is a combined incidence rate of about 3/1,000 patients. If the expected rate were 1.5/1,000, this excess would be significant. However, no risk coefficient can be estimated since diseases as uncommon as those listed are difficult to distinguish in national registries, and there are no reliable data on the dose to the lymphoid tissues in the Thorotrast-exposed patients.[27]

Mole[28] reported that by 1979, of 3,772 Thorotrast-exposed patients in the German, Danish, and Portuguese Thorotrast surveys, 26 died from bone marrow failure, that is, 6.9/1,000. If the expected control value were approximately 1.6/1,000 and the bone marrow dose is taken as 270 rad over 30 yr for a 25-ml injection, then a lifetime linear risk coefficient of 20 excess cases/10^6 person-rad can be estimated. However, the risk coefficient may be nearer to 30/10^6 person-rad since the deaths in the Danish subjects occurred at 7–24 yr (mean, 16 yr)[4] and in the Portuguese subjects at 8–37 yr (mean, 25 yr)[5] after Thorotrast administration.

Mays and Spiess[25] have estimated the risk of bone-tumor induction in Thorotrast-exposed patients. In Germany, Portugal, and Denmark, 3,000 patients followed for more than 10 yr had contributed about 45,000 person-yr at risk beyond the first 10 yr by 1979; 3–6 bone sarcomas had occurred, compared with 0.5 expected cases. Rowland and Rundo[37] have calculated that a typical intravascular injection of 25 ml of Thorotrast gave an average dose rate from translocated ^{224}Ra of about 1 rad/yr to the marrow-free skeleton of an adult. Assuming that translocated ^{224}Ra is the source of exposure, the risk coefficient estimated is 55–120 excess bone sarcomas/10^6 person-rad (average dose to the skeleton without bone marrow). For comparison, the risk coefficient for protracted injections of ^{224}Ra is estimated to be about 200 excess bone sarcomas/10^6 person-rad, based on 54 cases of bone sarcoma.[25] The effect of age at the time of Thorotrast administration on the induction of neoplasia is poorly understood; patients receiving Thorotrast at younger ages appear to have an excess of bone sarcomas, whereas patients receiving Thorotrast at older ages do not.[25]

Liver tumors arising from hepatic parenchymal or bile duct cells, or hemangioendotheliomas, have not been recorded in excess in humans exposed to external low linear energy transfer radiations, although leukemia is commonly induced from such exposures.[34] This is in contrast to the Thorotrast-exposed patients where the linear risk coefficient for liver tumors is considerably higher than that for leukemia.[27,28] This may be due, in part, to the practice of averaging the dose in the liver; local deposits of Thorotrast provide sufficiently high local alpha-radiation doses to induce cycles of necrosis and regeneration. While radiation plays an important role, it has been suggested that it may be only the hepatocellular tumors and not the hemangioendotheliomas that are associated with cycles of liver necrosis and regeneration in the absence of radiation.[6]

The wide local variation of Thorotrast dose distribution in the liver also occurs in the bone marrow, lymph nodes, and spleen; Mole[27,28] speculated that local radiation levels from Thorotrast deposits are much greater than dose averages throughout the tissue and that this could be responsible for the high incidence of leukemias in Thorotrast-exposed patients, and perhaps also for the apparent excess of multiple myeloma and lymph node neoplasms. The inhomogeneous radiation produced by alpha-emitters and the nonuniform and patchy anatomic distribution of Thorotrast complicate any attempt to calculate radiation dosage to the tissues of these patients. Correlation with histopathological findings based on terminal burdens is difficult, since the uneven and irregular distribution with increasing aggregation and flocculation of Thorotrast granules and migration and redistribution of thorium constantly change the levels of radiation dose. Further, some of the decay products of the complicated thorium series are soluble, translocate, and are bone seekers. Thus, average dose to the tissues may be an inappropriate parameter, and calculations based on terminal burdens do not necessarily represent the radiation dose that may be responsible for initiating malignant processes.

In summary, the combined epidemiological studies of Thorotrast-exposed patients provide estimates for the cancer risks and are listed in Table 5-7.

TABLE 5-7 Lifetime Excess Cancer Risks from Thorotrast

Tissue	Risk Coefficient/ 10^6 Person-rad	Latent Period (yr) for Colloidal $^{232}ThO_2$
Liver	260–300	20
Leukemia	50–60	5
Bone	55–120	10

The extent to which these risk numbers apply to other thorium radionuclides in other forms is unknown.

REFERENCES

1. Abbatt, J. D. 1973. Human leukemic risk data derived from Portuguese Thorotrast experience. Pp. 451–464 in Radionuclide Carcinogenesis, CONF-72055, C. D. Sanders, R. H. Busch, J. E. Ballou, and J. E. Mahlum, eds. Washington, D.C.: U.S. Atomic Energy Commission.
2. Bensted, J. P. M. 1967. Experimental studies in mice on the late effects of radioactive and nonradioactive contrast media. Am. N.Y. Acad. Sci. 145:728–733.
3. Brooks, A. L., R. A. Guilmette, M. J. Evans, and J. H. Diel. 1986. The induction of chromosome aberrations in the livers of Chinese hamsters by injected Thorotrast. Strahlentherapie 80(Suppl.):197–201.
4. Casarett, G. W. 1973. Pathogenesis of radionuclide-induced tumors. Pp. 1–14 in Radionuclide Carcinogenesis, CONF-720505, C. L. Sanders, R. H. Busch, J. E. Ballou, and D. D. Mahlum, eds. Springfield, Va.: U.S. Atomic Energy Commission.
5. da Motta, L. C., J. da Silva Horta, and M. H. Tavares. 1979. Prospective epidemiological study of Thorotrast-exposed patients in Portugal. Environ. Res. 18:152–172.
6. Faber, M. 1973. Pp. 137–147 in Proceedings of the Third International Meeting on the Toxicity of Thorotrast, Riso Report, M. Faber, ed. Copenhagen: Danish Atomic Energy Commission.
7. Faber, M. 1977. Epidemiology of Thorotrast Malignancies in Man. Review paper prepared for the World Health Organization Scientific Group on the Long Term Effects of Radium and Thorium in Man. WHO Working Paper 12. Geneva: World Health Organization. (Copy obtainable from Prof. Mogens Faber, Finsen Institute, Strandboulevard 49, Copenhagen, Denmark.)
8. Faber, M. 1978. Malignancies in Danish Thorotrast patients. Health Phys. 35:154–158.
9. Faber, M. 1979. Twenty-eight years of continuous followup of patients injected with Thorotrast for cerebral angiography. Environ. Res. 18:37–43.
10. Faber, M. 1983. Current (1981) status of the Danish Thorotrast study. Health Phys. 44(Suppl. 1):259-260.
11. Faber, M. 1986. Observations on the Danish Thorotrast patients. Strahlentherapie 80(Suppl.):140–142.
12. Fabrikant, J. I. 1972. Radiobiology. Chicago: Year Book Medical Publishers.

13. Fabrikant, J. I., R. J. Dickson, and B. F. Fetter. 1964. Mechanisms of radiation carcinogenesis at the clinical level. Br. J. Cancer 18:458–477.
14. Falk, H., N. C. Telles, K. G. Ishak, L. B. Thomas, and H. Popper. 1979. Epidemiology of Thorotrast-induced hepatic angiosarcoma in the United States. Environ. Res. 18:65–73.
15. Grillmaier, R., and H. Muth. 1971. Radiation dose distribution in lungs of Thorotrast patients. Health Phys. 20:409–419.
16. Hoffman, W., and F. Dasehil. 1986. Dose distribution and lung cancer incidence in Thorotrast patients. Strahlentherapie 80 (Suppl.):143–146.
17. Horta, J. da Silva, M. E. Horta, L. C. da Motta, and M. H. Tavares. 1978. Malignancies in Portuguese Thorotrast patients. Health Phys. 35:137–152.
18. International Commission on Radiological Protection (ICRP). 1972. Alkaline Earth Metabolism in Adult Man. ICRP Publication 20. Oxford: Pergamon.
19. Jee, W. S. S., M. H. Bartley, N. L. Dockum, J. Yee, and G. H. Kenner. 1969. Vascular changes in bones following bone-seeking radionuclides. Pp. 437–456 in Delayed Effects of Bone-Seeking Radionuclides, C. W. Mays, W. Jee, R. Lloyd, B. Stover, J. Dougherty, and G. Taylor, eds. Salt Lake City: University of Utah Press.
20. Kato, Y., T. Mori, and T. Kumatori. 1979. Thorotrast dosimetric study in Japan. Environ. Res. 18:32–36.
21. Kato, Y., T. Mori, and T. Kumatori. 1983. Estimated absorbed dose in tissues and radiation effects in Japanese Thorotrast patients. Health Phys. 44(Suppl. 1):273–279.
22. Kaul, A., and W. Noffz. 1978. Tissue dose in Thorotrast patients. Health Phys. 35:113–122.
23. Kojiro, M., T. Nakashima, Y. Ito, and H. Ikezaki. 1986. Pathomorphological study on Thorotrast-induced hepatic malignancies. Strahlentherapie 80(Suppl.):119–122.
24. Lloyd, R. D., M. E. Wrenn, G. N. Taylor., C. W. Mays, W. S. S. Jee, F. W. Bruenger, S. C. Miller, and A. S. Paschoa. 1986. Toxicity of ^{228}Ra and ^{228}Th relative to ^{226}Ra for bone sarcoma induction in beagles. Strahlentherapie 80(Suppl.):65–69.
25. Mays, C. W., and H. Spiess. 1979. Bone tumors in Thorotrast patients. Environ. Res. 18:88–93.
26. Mays, C. W., T. F. Dougherty, G. N. Taylor, R. D. Lloyd, B. J. Stover, W. S. S. Jee, W. R. Christensen, J. H. Dougherty, and D. R. Atherton. 1969. Radiation-induced bone cancer in beagles. Pp. 387–408 in Delayed Effects of Bone-Seeking Radionuclides, C. W. Mays, W. S. S. Jee, R. D. Lloyd, B. J. Stover, J. H. Dougerty, and G. N. Taylor, eds. Salt Lake City: University of Utah Press.
27. Mole, R. H. 1978. The radiobiological significance of the studies with ^{224}Ra and Thorotrast. Health Phys. 35:167–174.
28. Mole, R. H. 1979. Carcinogenesis by Thorotrast and other sources of irradiation, especially other α-emitters. Environ. Res. 18:192–215.
29. Mori, T., Y. Kato, T. Shimamine, and S. Watanabe. 1979. Statistical analysis of Japanese Thorotrast-administered autopsy cases. International Meeting on the Toxicity of Thorotrast and Other Alpha-Emitting Heavy Elements, Lisbon. Environ. Res. 18:231–244.
30. Mori, T., T. Maruyama, Y. Kato, and S. Takahashi. 1979. Epidemiological followup study of Japanese Thorotrast cases. Environ. Res. 18:44–54.
31. Mori, T., Y. Kato, T. Kumatori, T. Maruyama, and S. Hatakeyama. 1983. Epidemiological followup study of Japanese Thorotrast cases—1980. Health Phys. 44(Suppl. 1):261–272.

32. Mori, T., T. Kumatori, Y. Kato, S. Hatakeyama, R. Kamiyama, W. Mori, H. Irie, T. Maruyama, and S. Iwata. 1986. Present status of medical study on Thorotrast-administered patients in Japan. Strahlentherapie 80(Suppl.):123–134.
34. National Research Council, Committee on the Biological Effects of Ionizing Radiations (BEIR). 1980. The Effects on Populations of Exposure to Low Levels of Ionizing Radiation. Washington, D.C.: National Academy Press. 524 pp.
33. National Research Council. 1976. The Report of the Ad Hoc Committee on Hot Particles of the Advisory Committee on the Biological Effects of Ionizing Radiations (BEIR). Washington, D.C.: National Academy of Sciences.
35. Riedel, W., R. Hirschberg, A. Kaul, H. Schmier, and U. Walter. 1979. Comparative investigations on the biokinetics of colloidal thorium, zirconium, and nafnium dioxides in animals. Environ. Res. 18:127–139.
36. Riedel, W., W. Dalheimer, A. Said, and U. Walter. 1983. Recent results of the physical and biological properties of Thorotrast equivalent colloids. Health Phys. 44(Suppl. 1):293–298.
37. Rowland, R. E., and J. Rundo. 1973. The skeletal dose from ^{224}Ra following the intravascular administration of Thorotrast. Pp. 95–102 in Proceedings of the Third International Meeting on the Toxicity of Thorotrast, Riso Report 294, M. Faber, ed. Copenhagen: Danish Atomic Energy Commission.
38. Rundo, J. 1978. The radioactive properties and biological behavior of ^{224}Ra (ThX) and its daughters. Health Phys. 35:13–20.
39. Spiers, F. W. 1968. Radioisotopes in the human body. New York: Academic Press.
40. Spiess, H., and C. W. Mays. 1970. Bone cancers induced by ^{224}Ra (ThX) in children and adults. Health Phys. 19:713–729.
41. Spiess, H., and C. W. Mays. 1979. Liver disease in patient injected with ^{224}Ra. Environ. Res. 18:55–60.
42. Sugiyama, H., Y. Kato, T. Ishihara, K. Hirashima, and T. Kumatori. 1986. Late effects of Thorotrast administration: Clinical and pathophysiological studies. Strahlentherapie 80(Suppl.):136–139.
43. Taylor, G. N., C. W. Mays, R. D. Lloyd, C. W. Jones, J. Rojas, M. E. Wrenn, G. Ayorou, A. Kaul, and W. Riedel. 1986. Liver cancer induction by ^{241}Am and Thorotrast in deer mice and grasshopper mice. Strahlentherapie 80(Suppl.):172–177.
44. Toohey, R. E., J. Rundo, J. Y. Sha, A. Essling, C. Pedersen, and J. M. Slane. 1986. Activity ratios of thorium daughters in vivo. Strahlentherapie 80(Suppl.):147–150.
45. United Nations Scientific Committee on the Effects of Atomic Radiation (UNSCEAR). 1982. Ionizing Radiation: Sources and Biological Effects. Report E. 82. IX. 8. New York: United Nations. 773 pp.
46. van Kaick, G., A. Kaul, D. Lorenz, H. Muth, K. Wegener, and H. Wesch. 1978. Late effects and tissue dose in Thorotrast patients. Recent results of the German Thorotrast study. Pp. 263–276 in Late Biological Effects of Ionizing Radiation, Vol. 1. Vienna: International Atomic Energy Agency.
47. van Kaick, G., D. Lorenz, H. Muth, and A. Kaul. 1978. Malignancies in German Thorotrast patients and estimated tissue dose. Health Phys. 35:127–136.
48. van Kaick, G., H. Muth, A. Kaul, H. Immich, D. Liebermann, D. Lorenz, W. J. Lorenz, H. Luhrs, K. E. Scheer, G. Wagner, K. Wegener, and

H. Wesch. 1983. Recent results of the German Thorotrast study—epidemiological results and dose effect relationships in Thorotrast patients. Health Phys. 44(Suppl. 1):299–306.

49. van Kaick, G., H. Muth, A. Kaul, H. Immich, D. Liebermann, D. Lorenz, W. J. Lorenz, W. J. Luhrs, K. E. Scheer, G. Wagner, K. Wegener, and H. Wesch. 1984. Results of the German Thorotrast study. In Radiation Carcinogenesis. Epidemiology and Biological Significance, J. D. Boice, Jr. and J. F. Fraumeni, Jr., eds. New York: Raven.
50. van Kaick, G., H. Muth, and A. Kaul. 1984. The German Thorotrast Study. Results of Epidemiological, Clinical and Biophysical Examinations on Radiation-Induced Late Effects in Man Caused by Incorporated Colloidal Thorium Dioxide (Thorotrast). Report No. EUR 9504 EN. Luxembourg: Commission of the European Communities.
51. van Kaick, G., H. Muth, A. Kaul, H. Wesch, H. Immich, D. Liebermann, W. J. Lorenz, H. Luhrs, K. E. Scheer, G. Wagner, and K. Wegener. 1986. Report on the German Thorotrast study. Strahlentherapie 80(Suppl.):114–118.
52. Vaughan, J. 1986. Carcinogenic effects of radiation on the human skeleton and supporting structures. Pp. 311–344 in Radiation Carcinogenesis, A. C. Upton, R. E. Albert, F. J. Burns, and R. E. Shore, eds. New York: Elsevier.
53. Wegener, K., and K. Hasenöhrl. 1983. Recent results of the German Thorotrast study—pathoanatomical changes in annual experiments and comparison to human Thorotrastosis. Health Phys. 44(Suppl. 1):307–316.
54. Wesch, H., H. Kampmann, and K. Wegener. 1973. Assessment of organ distribution of thorium by neutron-activation-analysis. Pp. 52–60 in Proceedings of the Third International Meeting on the Toxicity of Thorotrast, Riso Report 294, M. Faber, ed. Copenhagen: Danish Atomic Energy Commission.
55. Wesch, H., W. Riedel, K. Wegener, A. Kaul, H. Immich, K. Hasenöhrl, H. Muth, and G. van Kaick. 1983. Recent results of the German Thorotrast study—statistical evaluation of animal experiments with regard to the nonradiation effect in human Thorotrastosis. Health Phys. 44(Suppl. 1):317–321.
56. Wesch, H., U. W. Reidel, K. Hasenöhrl, K. Wegener, A. Kaul, H. Muth, and G. van Kaick. 1986. German Thorotrast study: Results of the long-term animal studies on the effect of incorporated radioactive and nonradioactive particles. Strahlentherapie 80(Suppl.):186–188.
57. Xing-an, C., H. Qingmei, D. Zhiva, L. Wenyuan, W. Yidien, C. Yongru, T. Genhong, H. Weihui, C. Maozhau, F. Runlin, L. Liangjin, and L. Rongbo. 1986. Activity concentration of exhaled ^{220}Rn and burden of ^{228}Th in workers working at the Bai Yuan Iron Mine in Innermongolia and in inhabitants living in the high background radiation area in China. Strahlentherapie 80(Suppl.):157–162.

6
Uranium

INTRODUCTION

Minerals containing uranium are widely distributed in the surface areas of the earth's crust. Some are of commercial value and contain various oxides of uranium, including uraninite, pitchblende, carnotite, and brannerite. Uranium is also found in phosphate rock, lignite, and monazite sands. The potential health effects of uranium in mining or in refining operations are complicated by the presence of other alpha-emitters in the ore, such as radium and radon. Natural uranium contains about 99.283% of ^{238}U by weight, 0.711% ^{235}U, and 0.0054% ^{234}U. ^{238}U has a very long half-life of 4.5×10^9 yr so that this isotope, although accounting for the largest fraction, by weight, of natural uranium in the soil, accounts for only half of the radioactivity. The remainder is derived from ^{235}U and ^{234}U.

Uranium is a dense metal (19.07 g/cm^3 at 25°C) that is chemically reactive and combines with most elements. Its chemistry has been studied in great detail. In the crystalline state it can have valences ranging from +3 to +6. Only the uranic compounds, U(IV), and the hexavalent uranyl compounds, UO_2^{2+}, are sufficiently stable, both thermodynamically and kinetically, in aqueous solution to be of biological importance.

Uranium forms a highly complex series of oxides, including UO_2, U_3O_8, and UO_3. Uranates are obtained when uranium is fused with alkaline earth carbonates. Uranium hexafluoride, UF_6, is an important industrial compound since it is readily volatile (melting point, 64.1°C).

Uranium has assumed enormous importance as a result of its use in nuclear fuel. Previously, however, its industrial use was limited, and most uranium that was recovered as a by-product of vanadium mining was discarded. It has some utility as an intensifier in photography, in dry copying ink, as a colorant in ceramics or glass, in the production of armor-piercing projectiles, and for use as ballast.

ABSORPTION AND DISTRIBUTION OF NATURAL URANIUM

Uranium is ubiquitous in soil and, following uptake into crops, is a trace constituent of food, particularly cereals. Food is the principal origin of the natural uranium content of the body for most populations, although water may be an important source in certain areas. Gastrointestinal uptake is generally low—about 1% of soluble salts and less than 1% of insoluble compounds.[65] Absorption is reasonably independent of the mass of uranium ingested.[12] As expected, there appears to be no preferential biological uptake of ^{234}U compared to ^{238}U, and their ratio in the body is similar to that ingested.[12] There are considerable regional variations, particularly in the concentrations in water supplies, and these are reflected in variations in the body content of uranium of populations in those areas.[13]

The distribution of uranium found in human postmortem studies has been reviewed by Wrenn et al,[65] who noted a range in total body content of 2 to 62 μg. The skeleton is a major storage depot, and the differences between the concentrations observed in various populations is considerable. The concentrations of uranium in skeletal tissue from Nepal was about six times that observed in the United States.[13] There appears to be no increase in skeletal content with increasing age,[12] and this has been interpreted as suggesting that equilibrium is established between intake and excretion during life and that the biological half-life of uranium in bone is fairly short. However, the natural uranium content of lung, liver, kidney, and vertebra was found to be age dependent.[11] Fisenne and Welford[11] also noted that the skeletal content in their material was only one-tenth of that predicted by the International Commission on Radiological Protection (ICRP).[25]

PHARMACOKINETICS AND TOXICOLOGY

The uranyl ion, UO_2^{2+}, is central to an understanding of the toxicology of uranium, because it forms stable complexes with carbonate and phosphate ligands in biological fluids and, to a smaller degree, with carboxyl and hydroxyl ligands. The stability of the UO_2HCO^{3+} complex depends on the pH of the solution, and the pH at different anatomic sites affects its distribution in the body.

Early toxicological studies were reviewed by Hodge,[20] who noted that the toxic effects of uranium were first examined after uranium was extracted from pitchblende, followed by the preparation of uranyl nitrate, sulfate, and chloride. In 1853 Leconte (cited by Hodge[20]) discovered that uranium acetate could induce glycosuria in dogs. This discovery resulted in its widespread use as a homeopathic remedy in the treatment of diabetes mellitus because of the mistaken conclusion that the effect was an abnormality of carbohydrate metabolism. However, extensive research soon showed that it was the result of renal damage, and uranium compounds became the agent of choice in the production of chronic renal lesions in experimental animals. Chronic nephritis was an important and widespread disease at the time, and the discovery of an experimental model that appeared to mimic the human disease attracted much interest. The renal lesion was found to be unique, and was later assumed to be the only toxic effect of uranium important to man, in that uranium was not thought to accumulate in bone or to constitute a radiation hazard. However, the enrichment of natural uranium and the production of artificial isotopes with high specific activities stimulated intensive research, which has been reviewed by various authors who used data drawn largely from the original Manhattan Engineering District studies.[20,48,51,55]

A classic handbook of experimental pharmacology comprehensively treating uranium, plutonium, and transplutonic elements was edited by Hodge et al.[22] It contained seven chapters devoted solely to uranium and uranium mining. Chapter 1[22] summarized the early history of uranium poisoning (1824–1942), Chapter 3[66] the results of toxicologic experiments in animals since 1942, and Chapter 4[23] the direct information on the metabolism of uranium in humans. Chapter 5[46] described the development of criteria for the protection of humans against uranium intoxication. This book was followed by a conference[62] on occupational health experience with uranium up to 1975. The proceedings included a comprehensive review of the metabolism and effects of uranium in animals.[9] In 1975, a Brazilian symposium reviewed the environmental transport and accumulation of all naturally occurring alpha-emitters in areas of high natural background radiation.[64] A 1979 symposium on actinides in humans and animals[63] brought together researchers working on environmental transport, occupational exposure, and metabolism and effects. More recently, a colloquium on biokinetics and analysis of uranium in man,[34] sponsored by the U.S. Uranium Registry, brought information and progress in several fields up to date. It reviewed the biokinetics and toxicology of uranium in animals; updated metabolic models of uranium with special attention to circulatory transport, deposition, and retention in kidney and bone; and evaluated biokinetic models and internal radiation dosimetry. It also included a historical review of uranium in humans, a discussion of human epidemiology, and sessions on analytical methods. The implication of recent findings regarding toxicity, radiation effects, and extrapolation of results to humans were summarized.

As previously described, the uptake of uranium from food is the principal source of natural uranium in the general population. In the industrial environment, the respiratory tract is the most important route of entry. Soluble salts of uranium can be absorbed through the skin, but there are no data on the rate of such absorption in humans.[57]

The disposition of inhaled uranium aerosols in the body is determined by the size of the inhaled particles and their solubility. Insoluble particles with an aerodynamic diameter (ADD) smaller than about 5 μm may be deposited in the alveoli. After some months they are removed from the alveoli to the cilia lining the upper airways and are swallowed so that they are excreted through the gastrointestinal tract. A proportion of the deposited particles remains permanently in the lung tissue or is stored in the hilar lymph nodes. Larger aerosol particles (≥ 5 μm ADD) are deposited on the upper airways and cannot penetrate as far as the alveoli. These particles are removed by the cilia within a few hours or days. They are swallowed and excreted by the gastrointestinal tract. The retention of an insoluble uranium aerosol in the lung is thus dependent on whether it is of a size such that it is deposited mainly in the alveoli or on the ciliated airways.

In addition to the site of deposition in the lung, the solubility of the aerosol is also an important determinant of the biological half-life. In fact, ICRP[25] classified the biological half-life of uranium aerosols according only to their estimated solubility and took no account of particle size. The stable complexes formed between the uranyl ion and bicarbonate anions emphasize the importance of distinguishing between solubility in water and solubility in biological fluids.

Results of animal studies[60] have suggested that uranium in the tetravalent state, U(IV), is oxidized to U(VI) before absorption, unless given intravenously; the metabolism of U(IV) is therefore less relevant to occupational or environmental exposure of man and will not be reviewed here. After absorption from the lungs or after intravenous injection in experimental animals, soluble salts of U(VI) are partitioned in the bloodstream—approximately 60% in a form generally accepted to be the bicarbonate complex and the remaining 40% bound to plasma proteins, primarily transferrin.[15,47] The low-molecular-weight bicarbonate complex is filtered out of the bloodstream in the renal glomeruli and passed into the tubules. The fraction bound to plasma protein cannot penetrate the glomerular lining, but because equilibrium between the two fractions in the bloodstream is maintained, in due course all plasma uranium passes to the renal tubules, except for that which is diverted to the skeletal tissue. Between 10 and 30% is reversibly bound to the surface of the bony structures. During glomerular excretion of uranium, the stores of uranium in the extracellular fluid and skeletal system are mobilized, and the renal tubules continue to excrete uranium; over 60% of a single intravenous dose is excreted in the first 24 h.[15,23,46,47]

Within the renal tubules, water and bicarbonate are absorbed. The resulting decrease in pH causes the uranyl-bicarbonate complex to dissociate and release the highly reactive uranyl ion. This forms complexes with phosphate ligands on the luminal surface of cells lining the tubules. The complexing probably suppresses cellular respiration due to enzyme inhibition and results in slow cell death.[20] The accumulation of uranium in the kidneys accounts for about 20% of an intravenous dose during the first 24 h.[47] The rate of excretion depends on the urinary bicarbonate content and acidity. Very alkaline urine limits dissociation of the uranyl complex and increases uranium excretion.

The deposition of uranium in bone was originally considered not to be important, because of the low specific radioactivity of natural uranium and in view of the obvious severity of the renal effects. However, uranium behaves chemically like calcium, and the skeletal system might be the target organ in the case of enriched uranium. The use of autoradiography with isotopes with high specific activities has provided much experimental information on the mechanism of deposition of uranium in bone. It has shown that uranium shares characteristics with radium, which is distributed throughout bone (a volume seeker), and with plutonium, which remains on the surface of the bone.[39,42] The initial deposition is nonuniform and is at the sites of active calcification.[37,51] Subsequently, there is a slow relocation throughout the volume of the bone. The mechanisms are uncertain.

Many inhalation experiments using uranium have been carried out in animals. In general, soluble salts, such as UF_6, UO_2F_2, and $UO_2(NO_3)_2 \cdot 6H_2O$, are rapidly cleared to the kidneys and bones, with none remaining in the lungs after 30 days. In contrast, less soluble compounds such as UO_2 are largely cleared from the pulmonary tissue to the ciliated airways and eliminated by the gastrointestinal tract. A fraction is retained in the lung tissue or hilar lymph nodes, and little accumulates in tissues outside the thorax. Recent studies have shown that the distinction between compounds which are soluble in water and those which are insoluble is not a reliable indicator of clearance rates because of the rapidity of formation of the bicarbonate-uranyl complex in biological fluids. For example, using UO_3 which was classified by ICRP[26] as a class W compound on the basis of solubility, Stradley et al.[49] exposed rats to an aerosol and by direct injection of an aqueous suspension of the dust into the pulmonary region of the lung. The aerosol had an activity mean diameter of 1.4 μm (σ 4.0). Retention of uranium in the lung was represented as the sum of two exponential terms with half-times of 0.9 (96%) and 60 (4%) days. This unexpectedly rapid clearance of uranium from the lungs resulted from its solubility in biological fluids with translocation in the bloodstream and later excretion in urine. The size of the aerosol particles was such that they had a high probability of deposition in the alveoli where clearance to the ciliated airways would take months in the case of insoluble particles.

Data derived from human experience are sparse. Human intravenous exposures for experimental purposes were carried out in six hospital patients (Rochester, N.Y.), in eight terminally ill brain tumor patients (Boston, Mass.), in studies to investigate bone metabolism in seven patients with different bone diseases, and in healthy control subjects.[21] These experiments provided information on the distribution and renal effects of acute exposures, but are of less value in predicting the effects of chronic low-level exposure. In an inhalation experiment reported by Harris and Davies,[16] a subject inhaled approximately 30 mg of UO_3 or UF_4 over a period of 24 days. The particle size of the aerosol was not stated. Retention was estimated by measuring total excretion of uranium from the body. The

TABLE 6-1 Distribution of Uranium in Postmortem Tissues of a Single Uranium Worker

Tissue	Wet Weight (g)	Uranium Concentration (μg/g wet weight)	Uranium Weight in Sample (μg)	Uranium Weight in Whole Organ (μg)	Normal Content of Whole Organ in Reference Man (μg)
Lung	1,041	1.2	1,249	1,250	1.1
Lymph nodes	12	1.8	22	35	0.4
Bone	114	0.09	10.3	1,000	25.0
Kidney	217	0.15	32.6	30	0.14
Liver	177	0.02	3.5	35	0.4
Other	—	—	—	—	12.5

SOURCE: Data from Donoghue et al.[8] (occupational) and Wrenn et al.[65] (general public).

results showed that 60% of the material that had been inhaled was still present after 32 days, with a period of rapid urinary elimination during the first 20 h after inhalation and slower excretion during the next 20–200 h.

In an alternative approach for obtaining data on the kinetics of uranium in human subjects, Donoghue et al.[8] examined, for uranium content, postmortem tissues of workers who had been exposed when alive. They reported autopsy data on a worker who had been employed in a uranium workshop for 10 yr and who died suddenly from natural causes. Dust exposure had been to uranium mostly in the form of U_3O_8 (85%), with the remainder being in the form of UO_2. Those estimates were based on analysis of settled dust and might not have reflected the composition of inhaled dust. The distribution of uranium in the postmortem tissues is shown in Table 6-1.

Extensive information is available on the content and distribution of uranium in the human body under different circumstances. The purpose of Table 6-1 is to contrast the distribution associated with the inhalation of U_3O_8 derived from occupational sources with that typically found in the body of nonoccupationally exposed individuals. The total amounts present in the latter would depend on the uranium content of food and water during life and thus on the geographical area from which the specimens were obtained.

Donoghue et al.[8] compared the body content of uranium in the case that they reported with that calculated from estimates of occupational exposure during life and the retention values predicted from ICRP models.[25] Uranium in the lungs and thoracic lymph nodes were less than 1% of the predicted values. These discrepancies could be due to errors in the exposure estimates or in the assumptions contained in the model and emphasize the large uncertainties in estimating risks.

NEPHROTOXIC EFFECTS

In the case of natural uranium, the toxicological data identify the kidney as the target organ, although specific acute toxic effects are associated with the inhalation of certain compounds such as UF_4 and UCl_4. Studies done after the acute administration of soluble uranium to animals[23,55] indicated that substantial renal damage occurred when the concentration of uranium in the kidney exceeded 3 μg/g of renal tissue. If the same concentrations were relevant to human experience and renal tissue of a standard man[24] of 310 g is used, this corresponds to a total uranium content in the two kidneys of 930 μg.

Wrenn et al.[65] have recently reviewed metabolic models of the accumulation of uranium in the human kidney under conditions of equilibrium where the intake rate equals the excretion rate. The several models used[26,46,65] similar equilibrium values of uranium in kidney if the same daily input to blood is assumed for each. The ICRP model[25] uses a two-component exponential retention function in kidney with half-times of 6 days associated with 12% of the dose to blood and 1,500 days associated with 0.05% of the dose to blood. The model of kidney retention adopted by Wrenn et al.[65] is that of Spoor and Hursh.[46] Spoor and Hursh used a 15-day half-time in the kidney with 11% transfer from blood to kidney. Durbin[34] has recently evaluated the information on retention of uranium in the kidney in seven mammalian species and concluded that the rate of elimination of uranium deposited in kidney has two phases, with the first (95% of the dosage) having a half-time from 2 to 6 days and the second having a half-time from 30 to 340 days. Under equilibrium conditions of intake and excretion, the daily urinary excretion predicted by these models associated with the buildup of 3 μg/g in human kidney would be about 400 to 500 μg of uranium, of which only a part is contributed from the release of uranium deposited in kidney. In man about 70% of injected uranyl nitrate was excreted in urine in 24 h.[46] For a rapid acute exposure of man to soluble uranium compounds, if 11% were deposited in kidney, an acute intake of uranium to blood of 8,400 μg would be required to produce an initial concentration of 3 μg/g kidney, and about 70% (5,900 μg) would be excreted in urine during the subsequent 24 h.

The mechanism of the renal lesions has been extensively investigated.[14,15] The kidney responds to toxic levels of uranium within the first 1 to 2 days after a single injection[35] but, unlike mercury poisoning, the changes are progressively severe over the first 5 days; whether this is due to the continuous release of further uranium by mobilization from the skeleton or the progression of the initial lesion is not known. The renal deposits formed after a single administration are also excreted rapidly, either because of reversal of the phosphate-binding sites in the tubular cells or because of the sloughing and excretion of dead cells that contain uranium. This acute phase of damage can be followed by repair with modified epithelium consisting of flattened, imperfectly differentiated cells that are remarkably resistant to the toxic effects of further injections of uranium. Young animals have considerably more resistance than older animals, and there are important species differences in sensitivity, possible owing to differences in urinary pH.[19] Rats are much more resistant than dogs. Humans are probably close to dogs in sensitivity (P. Morrow, personal communication, 1986); this is considered below in the discussion of thresholds of toxicity.

The lesions in the proximal convoluted tubules result in the appearance of glucose, low-molecular-weight proteins, and amino acids in the urine. These substances are normally reabsorbed from the tubular fluid, and their appearance in urine is due mainly to malabsorption by the damaged tubular cells and partly to tubular excretion, but not to increased glomerular permeability. They are of practical relevance, because they make it possible to detect early renal damage in industrial workers through routine monitoring of urine for abnormal proteins.

The work of Morrow et al.[36] has cast some doubt on the validity of the conclusion that 3 $\mu g/g$ is the threshold level of nephrotoxicity. In the dog kidney, they concluded that ≤ 1 $\mu g/g$ was associated with histological and transient biochemical abnormalities for injected doses of 0.01 mg/kg UO_2F_2. In the female Wistar rat, Bentley et al.[5] showed that renal concentrations were 10 $\mu g/g$ at 0.05 mg/kg of injected uranyl nitrate at 24 h postinjection, and that 0.01 mg/kg produced a transient proteinuria.

Morrow et al.[36] report half-times of 16 days in rat kidney after single exposures to UO_2F_2, whereas Bentley et al.[5] report on a two-phase retention, with the first component having a half-time of 2.2 days and the second a half-time of 13 days, consistent with the general mammalian metabolic model of Durbin.[34] Morrow et al.[36]

find a rapid (unspecified rate) and slower (half-life, 9.3 days) release rate from dog kidney.

Whether the rat or dog is more sensitive is not clear. Morrow et al.[36] believe the rat is more resistant than the dog, but experiments in which the 50% lethal dose (LD_{50}) is determined suggest that the dog is more resistant based on lethality.[9] The question of which animal is a better model for uranium effects in man deserves further investigation, as does the pharmacokinetics of early uranium deposition in kidney up to 10 days postinjection and its relationship to biochemical and histological indicators of response and damage.

In the absence of reliable epidemiological evidence, the interpretation of animal toxicological data has assumed considerable importance as a source of inference about the likely effects of human exposure. However, it is clear from the preceding discussion that there are important species differences in sensitivity and in the effects of single versus chronic exposure. Both single and chronic exposures are relevant to human populations. The laboratory experiment of single acute exposure mimics accidental exposures, but chronic exposure experiments are more important in estimating environmental effects or those in working populations. The major difficulty is in choosing the animal species that is the best model for predicting human toxicity.

The literature contains a sufficient number of investigations in which injected $UO_2(NO_3)_2$ was used to compare relative species sensitivity. Durbin and Wrenn,[9] in their review of toxicity in the rabbit, guinea pig, dog, cat, and mouse noted a variation in sensitivity of 2 orders of magnitude, from 0.1 mg/kg in rabbit to 20–25 mg/kg in C3H mice. The greater sensitivity of the rabbit and guinea pig may be due to the greater acidity of urine in herbivores. Information on humans is inadequate to estimate an accurate LD_{50} for man, but there is information on levels that are not acutely toxic, and systemic doses of 0.1 to 0.3 mg/kg of soluble uranium have not produced mortality due to kidney lesions in man.[6,30,46] The LD_{50} for man is probably similar to that in the rat or dog. Generally one-tenth the LD_{50} produces no or few acute deaths in experimental animals. Thus, the LD_{50} in man should be no lower than that for the rat or dog, about 1 to 2 mg/kg. Parentally administered $UO_2(NO_3)_2$ and UO_2F_2 are equally toxic in mice and rats and twice as toxic as less soluble UCl_4.[17]

In the adult male Wistar rat, there is a gender sensitivity, with the LD_{50} for the male being about 2.5 mg/kg and that for the female being about 1 mg/kg, for intraperitoneal injection of UO_2F_2.[17]

Thirty-day feeding experiments in rats with compounds of uranium that are insoluble in water (UO_2, U_3O_5, and UF_4) were found to be nontoxic, whereas water-soluble compounds were relatively more toxic. For the toxic compounds, 2 to 10% in the diet produced 100% mortality, and at 0.1 to 1% it produced growth depression. The relative species susceptibility in oral 30-day feeding studies is rabbit > dog > rat. For interspecies scaling for nephrotoxicity, the relative absorption from the gastrointestinal tract is important. The recent article by Wrenn et al.[65] summarizes what is known about mammalian metabolism of uranium, including gastrointestinal absorption.

Tolerance has sometimes been used in the literature to mean the ability of animals to resist the toxic action of uranium more effectively when a dose is given repeatedly rather than acutely. Although the experiments of Haven[17] in the rat showed that survival to an LD_{50} could be increased two- to fourfold by administering conditioning doses up to one-third that of an LD_{50}, the tolerance induced was temporary; it was accompanied by histological alterations in the kidney, with some cells excreting atypical amounts of citric acid; and there were gross alterations of kidney appearance at autopsy. Tolerance does not develop at lower doses. Therefore, it seems to be a laboratory phenomenon of little practical importance, especially as applied to occupational protection of workers.

There is some evidence in mice and dogs that the combination of alpha radiation and chemical toxicity produces a greater nephrotoxic effect than either does seperately.[19] The relevance of this to human risk estimates is difficult to determine.

In one of the few studies to show an effect of uranium on humans, Thun et al.[53] evaluated kidney function among uranium mill workers and found a statistically significant excretion of beta-2-microglobulin and five amino acids. Although the levels of tubular proteinuria were low, a correlation existed between the clearance of beta-2-microglobulin, relative to that of creatinine, and the length of time that the uranium workers had spent in work areas with the highest exposures to soluble uranium. Thun et al.[53] believe that these data suggest reduced renal proximal tubular reabsorption, which is consistent with uranium nephrotoxicity.

RADIATION EFFECTS

Because of the difficulty of obtaining good epidemiological evidence concerning uranium-induced radiation effects in humans, it is necessary to make use of both experimental work in animals and the epidemiological investigations which have been made of uranium miners and millers.

BONE

The radiation effects seen in both animals and humans in association with radium isotopes are primarily bone sarcomas and head carcinomas. The head carcinomas are believed to be due to ^{222}Rn and its daughters, and apparently also ^{226}Ra. Bone sarcomas have been induced in mice with high-specific-activity ^{232}U and ^{233}U, and a retrospective dosimetric analysis[9] showed that, with respect to a linear dose-response relationship, the effectiveness of both uranium isotopes was equal to and not distinguishable from that of ^{226}Ra in the same strain of mouse. Investigators from the USSR have shown that highly enriched uranium can induce bone sarcomas in rats. Accordingly, it is reasonable to believe that high-specific-activity uranium can produce bone sarcomas in humans. If there is a linear dose-response relationship[32] one could speculate that natural or slightly enriched uranium could also induce bone sarcomas in humans, but the likelihood would be very small because of the low specific activity of natural uranium.

Nonetheless, Mays et al.[32] have estimated, using toxicity ratios, the risk of bone sarcoma induction from the chronic ingestion of natural uranium, on the basis of an assumed linear dose-response function for ^{226}Ra. The proportionality of response to average energy deposition in the skeleton by alpha particles is a reasonable assumption, because there is metabolic evidence that uranium is a skeletal volume seeker similar to, although not identical with, radium. Stevens et al [47] have shown with neutron-induced autoradiographs that $^{233}U(VI)$ acutely administered to dogs in citrate buffer deposited initially nonuniformly on bone surfaces, with areas of intense hot spots adjacent to other areas with intermediate and low concentrations. Later (after 1 yr), the redistribution or diffusion of uranium in bone had produced a more volume-like distribution similar to, but not identical with, that of ^{226}Ra. Rowland and Farnham[42] also reported volume-like distributions of uranium in dogs. Bruenger (personal communication, 1986) has shown that naturally occurring

^{235}U in dogs is uniformly distributed throughout the bone volume. Schlenker and Oltman[44] have found diffuse fission-induced tracks in bone volume obtained at autopsy from a person given ^{239}Pu by injection: They interpreted the effect as being due to naturally occurring ^{235}U. Thus, the most reasonable assumption is that the alpha-emitting uranium isotopes are as effective as ^{226}Ra in inducing bone sarcomas. This assumption should be experimentally investigated in an animal with appropriate skeletal remodeling properties, such as the dog, with ^{233}U and ^{232}U.

For ingestion in water or food at a constant daily rate of 1 pCi/day, Mays et al.[32] estimated that the risk of bone-sarcoma induction over a lifetime is 1.5 bone sarcomas/million persons for ^{233}U, ^{234}U, ^{235}U, ^{236}U, or ^{238}U, if the dose response is linear. In a million people in the United States, the naturally occurring bone sarcomas would number about 750.[32] If the dose-response relationship is quadratic, virtually no effect is expected at environmental natural uranium levels. If ^{233}U contains ^{232}U, as is normal with reactor-produced ^{233}U from *n*, 2*n* reactions, the ^{232}U should be considered separately. It should be noted that the isotopes ^{232}U, ^{233}U, and ^{236}U are special cases without direct relevance to the general population.

The work of Finkle, with retrospective dose estimates by Durbin and Wrenn,[9] suggested that ^{232}U and ^{233}U are equally effective in inducing bone sarcomas in mice, on the basis of average dose to the skeleton. But longer-lived animals such as dogs and humans afford a greater opportunity for the buildup of the first daughter, ^{228}Th, in the skeleton, and the equal efficiency of bone-sarcoma induction by ^{233}U and ^{232}U in mice might not exist in longer-lived animals and humans. In an experiment comparing the efficacy of bone-sarcoma induction in dogs, injected ^{228}Ra was about 2.5 times as effective as ^{226}Ra (per average rad delivered to the skeleton).[32] Because ^{228}Ra decays by beta emission to ^{228}Th (the same first daughter of ^{232}U), these results in animals are the most relevant empirical results for use in the assessment of human risk.

Bone Marrow

Mays et al.[32] have summarized the relevant epidemiological experience with alpha emitters in human bone and concluded that "the risk from radiation-induced leukemias has been insignificant relative to that for bone sarcoma." They noted that, in 2,940 located workers, mostly women, who were radium workers before 1970, 63 cases of bone sarcoma, compared with 1 expected, and 10 cases of leukemia, compared with 9 expected, were found.[44,46] Thus, the excess of bone sarcomas[66] greatly exceeded the excess of leukemias (1 case, which in this case is not significantly different from 0). Clearly, the risk (if any) in humans of leukemia induction from alpha irradiation by a volume seeker in mineral bone, such as ^{226}Ra or uranium, is less than that of bone-sarcoma induction by more than an order of magnitude.

LUNG

The late effects of inhaled insoluble compounds (primarily UO_2) have been studied in rats, dogs, and monkeys by observing animals given brief exposures and after prolonged exposure.[27,28] For the insoluble uranium compounds, the lungs and pulmonary lymph nodes were the only organs affected.

In monkeys, dogs, and rats, the chronic inhalation of insoluble UO_2 at 25 mg/m^3 eventually produced fibrosis of lung tissue and malignant lung tumors. On the basis of analogy with ^{239}Pu inhalation experiments in dogs and baboons, this appears to be a direct result of the alpha-particle irradiation of the lung.[28,66]

Durbin and Wrenn[9] reanalyzed the data of Leach et al.[28] and concluded that "neoplastic or metaplastic changes in lung epithelial cells occurred in 21% of dogs after accumulation of 160 rad." By fitting to a linear dose-response curve, a risk coefficient of 0.12%/rad for atypical epithelial cell forms were inferred. Radiation-induced fibrosis occurred in lungs of monkeys exposed to more than 1,000 rad.[66]

Ballou et al.[4] exposed groups of about 60 male Wistar rats to two doses of ^{232}U- and ^{233}U-labeled uranyl nitrate in "nose only" experiments. They developed significant increases, primarily of malignant lung tumors (mostly pulmonary adenocarcinomas) and osteosarcomas, with a frequency of an order of magnitude lower. The osteosarcomas were of doubtful significance as a result of their appearance in rats that inhaled control aerosols. ^{232}U appeared to be more effective in inducing osteosarcomas, although this result was not statistically significant. In humans, sufficiently high doses of high-specific-activity ^{232}U- or ^{233}U-labeled uranyl nitrate should be capable of producing malignant lung tumors and skeletal osteosarcomas. The relative proportions of the two end effects in humans cannot be predicted from this experiment in rats. This is because of the longer human life span and the increased time for skeletal irradiation with these long-lived isotopes in humans afforded by the much longer expected residence time (and consequently larger dose) in the human skeleton than in the lung. This experiment did not provide enough information for confident extrapolation of either risk coefficients or osteosarcoma-to-lung tumor ratios to humans. Inhalation studies in a longer-lived animal such as the dog are required for a more confident extrapolation to humans.

EPIDEMIOLOGICAL EVIDENCE

Workers engaged in the extraction of uranium have long been known to be at risk of increased mortality from cancer. More than one etiologic agent might be involved,[29] but the dominant source of radiation damage to the respiratory system in miners is generally considered to be the inhalation of radon daughters, rather than the uranium content of the ore. Miners are also exposed to quartz (silica), arsenic, and vanadium at substantial concentrations. The mines in the Colorado Plateau were, in fact, started for the purpose of obtaining vanadium rather than uranium. Even exposure to nonspecific underground dust, regardless of its composition, is known to have adverse respiratory health effects on the basis of studies of coal miners.[41] These multiple exposures constitute confounding factors that mitigate against attempts to identify human health effects due specifically to uranium in mining populations.

Enrichment of ore during milling is also associated with mixed exposures, except in the final stages of purification. In the initial stages, airborne dust in the mills can contain arsenic; vanadium; ^{226}Ra; and equal amounts of ^{234}U, ^{238}U, and ^{230}Th.[2] Concern that millers, as well as miners, might be at risk resulted in their inclusion in the epidemiological surveys that began in the United States in 1950.[33] That such concern was justified was supported by the first reports of increased cancer risk in uranium millers in Europe.[40] These results were at variance with other claims[10] that, despite the large numbers of uranium millers studied for years in the United States, there had been little evidence of ill health associated with this occupation. The validity of the latter claim must be assessed in relation to the power of surveys to detect an effect if one were present.

In the design of epidemiological studies to investigate health risks due specifically to uranium, and not to concomitant exposure to other substances, it would be reasonable to search for bone, hematopoietic, or lung tumors as an expression of the radiation effect and for the renal pharmacological effects that have been demonstrated in experimental studies. The distribution of pathological effects in an exposed population would be expected to be a function of the solubility of the uranium compound, with renal effects being dominant in association with more soluble compounds or ores.

In an early mortality study by the U.S. Public Health Service of uranium miners and millers in the Colorado Plateau, Wagoner et al.[56] reported no apparent increase in mortality in a cohort of 611 white uranium millers who had no apparent previous uranium-mining exposure. Archer et al.[2] presented a more detailed analysis with a longer follow-up of the same population. The study included white millers who were available for medical examinations in 1950, 1951, and 1953. The total number studied was 662, but it is not clear whether all millers were included in the medical examinations, nor whether nonwhite millers were excluded. Follow-up through 1967 was achieved by various sources, and more than 99% of the cohort was traced. In a life-table analysis, expected deaths were calculated from the populations of the areas whose mills were studied (Colorado, Utah, New Mexico, and Arizona). The report stated neither what precautions were taken to exclude men with previous underground mining exposure nor the number with such exposure.

A total of 104 deaths were available for analysis—almost identical with the number expected (105.11). An apparent excess of malignant neoplasms of the type in 6th ICD code 200-203, 205 (lymphatic and hematopoietic other than leukemia) was noted; four cases were observed, instead of the 1.02 expected, and the excess (standardized mortality ratio [SMR], 392) was unlikely to be due to chance ($P \leq 0.05$). Interpretation of this finding is difficult, because the four cases, as noted below, included three diagnostic categories.

Exposure data were not available, nor was the analysis undertaken in relation to duration of employment. In general, employment appears to have been fairly brief, between 2 and 13 yr. The four cases noted above (one of giant follicular lymphoma, one of multiple myeloma, and two of lymphosarcoma) were in millers who had not worked in the furnace areas where uranium exposure and vanadium exposure were highest. However, they had all worked in 1943–1953, when hygiene was probably poor; it was concluded that the excess mortality was due to occupational exposure. The specific cause of the hazard was less easy to identify.

In considering this problem, Archer et al.[2] noted that the airborne dust in the plant contained almost equal amounts (in curies) of ^{238}U, ^{234}U, and ^{230}Th. But animal exposure experiments[50] have shown that, after exposures to uranium ore dusts, the radioactivity in tracheobronchial lymph nodes due to ^{230}Th was 20–50 times that due to ^{234}U or ^{238}U. This difference led Archer et al.[2] to conclude that the excess mortality observed in uranium workers was a manifestation of ^{230}Th exposure. It should be noted that they observed no excess mortality due to tumors of bone or liver, and there was no excess of renal lesions. However, the epidemiological power of the study indicates that it constitutes fairly weak evidence against a specific toxic effect of uranium. For example, malignant disease of the digestive system (ICD 6th revision codes 150-159) was associated with an expected number of deaths of 5. The power to detect a 50% increase in cause-specific mortality was only about 25%. In the case of other cardiovascular and renal diseases (ICD codes 330-334, 444-468, and 592-594), the expected number of deaths was 9.15, and the corresponding power was about 40%.

A direct approach to the question of specific uranium toxicity is provided by surveys of workers exposed to enriched uranium. Such an investigation was published by Polednak and Frome[38] in connection with a cohort of 18,869 white men employed between 1943 and 1947 at a uranium conversion and enrichment plant in Oak Ridge, Tennessee. The plant was engaged in the enrichment of uranium with an electromagnetic separation process. Workers were exposed to uranium dust, including uranium oxide and uranium tetrachloride. Airborne uranium concentrations decreased over a number of years, but this was probably not associated with a concomitant decrease in the radiation hazard, because the product was more highly enriched in ^{234}U and ^{235}U, which have much higher specific activities (radioactivity per unit mass) than ^{238}U. The type and solubility of the uranium compounds also changed over time. In the earlier years, insoluble oxides were present with the more soluble chloride (UCl_4). Later, UF_6 was received directly from another facility, so that insoluble oxides were partly replaced by more soluble compounds (UF_6 and UF_2O_2). However, the UF_6 gas was immediately converted to oxides and then to green salt (UF_4). Even though the process was mostly enclosed, exposures to oxides and UF_4 could have occurred. Radium (the source of radon gas) had been largely removed from the uranium before it entered the plant. As a result, the pulmonary radiation hazard to these workers was related only to the inhalation of dust containing the various uranium isotopes and compounds. Ascertainment of deaths was obtained through the Social Security

TABLE 6-2 Causes of Death after Uranium Dust Exposure

	Length of Employment			
	<1 yr (n = 4.337)		≥1 yr (n = 4.008)	
Cause of Death	Observed	SMR	Observed	SMR
Cancer of lung	50	0.92	66	1.06
Cancer of bone	1	0.82	1	0.74
Leukemia	2	0.25	9	1.02
Diseases of respiratory system	61	1.19	57	0.93
Chronic nephritis	5	0.69	4	0.51

Administration, and causes of death were determined from death certificates. Duration of follow up extended to 25 or 30 yr after first employment. Average air concentrations of uranium, according to surveys carried out by industrial hygienists employed by the company, ranged from 25 to 500 $\mu g/m^3$. It should be noted that this method of expressing the concentration of uranium in the air may not be appropriate in the case of enriched uranium. In general, the mortality experience of the cohort did not show increased SMRs for lung or bone cancer or for renal disorders, such as chronic nephritis. The summary results for selected causes of death among white men who worked in departments involving uranium dust exposure are shown in Table 6-2.

Although there was no overall increase in lung-cancer deaths, a more detailed analysis provided evidence of a greater risk among those who were first exposed when older. Among workers first hired (and thus assumed first exposed) over the age of 45, the odds ratio was 1.51 (95% confidence interval, 1.01–2.31).

A weakness of this study is that the duration of exposure was only 1 or 2 yr in most instances. This limits any inferences regarding effects of long-term uranium exposure.

In a further analysis of the same cohort, Cookfair et al.[7] undertook a case-control study nested within the cohort to examine the risk of lung-cancer death among men who received radiation exposure to the lungs as a result of inhaling uranium dust or the dust of uranium compounds.

Cases consisted of the 330 cohort members who died of lung cancer. They were matched on year of birth with two sets of controls, one consisting of men who died of diseases other than cancer and the other of men known to be alive. The cumulative lung radiation dose

TABLE 6-3 Cumulative Lung Dose Odds Ratios (OR)

Cases versus Deceased Exposure[a]	Cases versus Deceased Controls OR (95% confidence interval; $P = 0.05$)	Cases versus Controls OR (95% confidence interval)
Hire age ≤ 45 yr		
Low	0.84 (0.50-1.41)	0.5 (0.34-0.92)
Medium	0.49 (0.26-0.92)	0.31 (0.17-0.56)
High	0.65 (0.36-1.17)	0.54 (0.31-0.97)
Hire age ≥ 45 yr		
Low	2.27 (0.85-6.07)	1.03 (0.44-2.42)
Medium	2.86 (0.70-11.54)	1.81 (0.510-6.50)
High	4.48 (1.20-16.85)	3.79 (1.01-134.3)

[a]Low, 0.001-5 rads; medium, 5.001-20 rads; high, 20.001-75 rads.

resulting from the inhalation of uranium compound dust and other potential carcinogenic agents was calculated for each member of the study population. Data were subjected to Mantel-Haenszel stratified analysis and logistic regression. The results of the logistic model are shown in Table 6-3.

The data show that, among the older group, the relative risk increased with exposure. The pattern was reflected in subgroups of known smoking status. The data support the hypothesis that radiation exposure of the lungs resulting from the inhalation of uranium dust and the dust of uranium compounds is a risk factor for lung cancer among white men who were 45 yr old or older when first exposed. No increased risk was noted in those who received less than 5 rad to the lungs. Two hypotheses might help to explain this apparent interaction between age of hire and cumulative lung dose: Either older workers are more susceptible to uranium-induced lung cancer or the latent period is longer for younger workers than for older workers. Increased susceptibility in older workers could have many explanations, including delayed clearance of uranium dust and previous tissue damage.

Another retrospective study of mortality patterns among a cohort of uranium mill workers was reported by Waxweiler et al.[58] Records from seven uranium mills throughout the Colorado Plateau were obtained, and 2,002 men who had worked for at least 1 yr in the mills were selected for study. Only those who stated on their job applications that they had never worked in uranium mining were included; the purpose of this restriction was to examine the health risks of uranium exposure in the absence of uranium mining. The cohort was traced with standard mechanisms. Risk of mortality was analyzed with a modified life-table system, with expected deaths based on U.S. death rates specific for cause, age, race, sex, and calendar period. Follow-up was 98% complete, and 533 deaths were observed, compared with 605 expected. Labor turnover was high: Only a small fraction worked for more than 5 yr. Mortality from all causes combined, and particularly from stroke and cardiovascular disease, was well below expected. This was interpreted as representing a healthy worker effect. A prior hypothesis included an excess of lung cancer (SMR, 83; 95% confidence interval, 54–121), malignancies due to lymphatic and hematopoietic tissue other than leukemia (7th revision ICD, 200-203, 205; 7 deaths observed versus 5.6 expected), and chronic renal disease (7th revision ICD, 592-594; SMR, 167; 95% confidence interval, 60–353). Detailed analysis of the small excess of deaths due to lymphatic malignancies (other than leukemia) indicated that the excess risk was limited to the induction latency period beyond 20 yr (6 observed verses 2.6 expected). This corroborated the excess of deaths due to the same malignancies reported in another cohort of uranium millers.[2]

An occupational etiology was considered plausible toxicologically, because yellow-cake dust and insoluble yellow-cake uranium compounds accumulate in the tracheobronchial lymphatic system after inhalation,[8,28,52] so they constitute a potential source of radiation to the lymphatic glands. However, the findings should be treated with caution since there was a paucity of observed and expected deaths beyond 5 yr of employment, and the findings were not statistically significant. There appeared to be no increase in lung-cancer deaths even when latency and the low regional rate of death due to malignant lung disease was allowed for. Of the six deaths associated with chronic renal diseases, three were probably due to prostatic obstruction, and the three due to glomerulonephritis were all in men who were exposed only briefly. A clear relationship to occupation was not established.

The previous studies identified in this section referred to mortality. Morbidity due to nonmalignant respiratory disease in uranium miners has been claimed to result from radiation exposure.[1] Archer et al.[3] discussed the implications of their earlier findings of pulmonary impairment in working miners. Their mortality data indicated that death due to pulmonary insufficiency was as common in miners with radiologic evidence of pneumoconiosis as in those with no x-ray evidence of pneumoconiosis. The dominant causative factor in both groups was presumed to be radiation exposure or cigarette smoking. However, these exposures are not related to the specific question of uranium inhalation but rather to radon-gas inhalation and exposure to mixed mine dust.

Wilson[61] investigated this problem in a cohort of white men hired between 1952 and 1972 in a uranium mill. A few of the workers had worked elsewhere in the company before the uranium mill began operating. The study population consisted of 4,101 workers who had worked for at least 3 months. Respiratory morbidity events were abstracted from medical insurance claims in which a diagnosis was given by a physician. Cohort and Mantel-Haenszel stratified analyses were used after estimated cumulative radiation exposures were categorized into low, moderate, and high. These exposure estimates were made on the basis of the toxicity, frequency of use, and amounts of substances used in the various departments. They were not based on measurement of airborne radioactivity. An increased relative risk of respiratory disease was found with increasing cumulative uranium exposure, even when age at diagnosis, duration of employment, and, to some extent, smoking habits were controlled for. Smoking histories were available only for workers employed after 1968. The smoking data included 17.4% of the study population, both current and former workers. Because of these limitations, it was possible only to examine the distribution of smoking in the various uranium-exposure categories, rather than to make detailed adjustments. The assumption was made that the estimates correctly indicated the smoking habits of other members of each category. Wilson[61] noted that 80% of the employees had no respiratory symptoms but that estimated uranium exposure appeared to be the principal occupational exposure that contributed to the development of nonmalignant respiratory disorders. It was not possible to determine whether the effect of exposure was due to some chemical attribute of the uranium or to its radioactivity. A critical review of the validity of this study must take account of the inherent weakness of morbidity data based on diagnoses obtained from insurance claims, even when given by a physician. There were no hard data on the extent of exposure to chemical agents or uranium or on smoking histories. In view of these limitations, the study results cannot be accepted as convincing evidence of an excess of nonmalignant respiratory disease due to uranium exposure.

Indirect evidence of uranium exposure and mortality was obtained in a geographic survey by Wilkinson,[59] who found higher rates of mortality from gastric cancer in several northern New Mexico counties with substantial deposits of uranium, compared with counties without such deposits. The differences persisted after socioeconomic status or ethnic group was accounted for. A number of etiologic possibilities were considered, including exposure to radon or radon daughters and trace elements, such as arsenic, cadmium, selenium, and lead. These are all commonly found in areas with uranium deposits, and the study did not provide evidence of an effect of uranium itself.

In summary, these investigations have provided suggestive but not convincing evidence of deleterious human effects of chronic exposure to uranium dust. There has not yet been any clear indication of renal disease in man due to low-specific-activity uranium, and the only positive finding involved the relative risk of lung cancer in the case-control study of Cookfair et al.[7] Caution is required in the interpretation of these results as an indication of the absence of any effect. The surveys generally included a large number of workers who were exposed for only a short time, and environmental exposure estimates were poor. Clearly, long-term follow-up of workers with adequate documentation is required.

RISK ESTIMATES

Uranium presents two separate potential risks due to its nephrotoxic action and as a result of alpha radiation. At present there is little convincing epidemiological evidence that serious renal disease has occurred in human populations as a result of chronic low-level exposure nor of increased rates of malignant tumors. However, this does not constitute reliable evidence of the absence of important health effects in occupationally exposed groups since the available epidemiological studies had limited power to detect increased rates of disease if these were present. It is for this reason that much weight has been given to inferences drawn from the results of animal studies and from tumor rates in human populations exposed to other alpha-emitting elements.

The renal threshold concentration of uranium that results in significant damage is a matter of controversy, and estimates range from 3 to ≤ 1 $\mu g/g$. The appropriate level depends, to an extent, on the choice of animal species and metabolic model. Adoption of a specific toxic threshold level and its relation to an airborne concentration for control purposes in the occupational environment requires further assumptions which are being reevaluated by the International Commission on Radiological Protection. At present it is premature to attempt a risk estimate for the probability of developing renal damage, and there is an evident need for well-controlled epidemiological studies. The importance of such information is emphasized by the likelihood that control levels in industry may need to be guided more by the potential for nephrotoxic effects than by the effects of alpha radiation. Quantification of the risks associated with alpha emission during chronic exposure to uranium cannot be determined from published epidemiological studies because of confounding factors and because of the limited power of the surveys to detect increased rates of tumor incidence or mortality. For this reason estimates have been based, by analogy, on the effects of other alpha-emitting elements in human populations and from experiments using uranium in animals. The most probable effect, if any, of exposure to uranium would be expected to be an increase in bone sarcomas. It is certainly reasonable to believe that this can result from high-specific-activity uranium. The likelihood of sarcomas resulting from population exposure to natural exposure is exceedingly low and is only demonstrable if a linear dose-response relationship is assumed.[32] If the dose-response relationship is quadratic, then virtually no effect would be expected as a result of exposure to natural uranium. Assuming a linear relationship and a constant nonoccupational intake of 1 pCi/day, Mays et al.[32] estimate that the risk of bone-sarcoma induction over a lifetime is 1.5 bone sarcomas/million persons. In the United States this may be contrasted with the naturally occurring incidence of bone sarcomas of about 750.

This evidence suggests that exposure to natural uranium is unlikely to be a significant health risk in the population and may well have no measurable effect.

REFERENCES

1. Archer, V. E., H. P. Brinton, and J. K. Wagoner. 1964. Pulmonary function in uranium miners. Health Phys. 40:1183–1194.
2. Archer, V. E., J. K. Wagoner, and F. E. Lundin. 1973. Cancer mortality among uranium mill workers. J. Occup. Med. 15:11–14.
3. Archer, V. E., J. D. Gillam, and J. K. Wagoner. 1975. Respiratory disease mortality among uranium miners. Ann. N.Y. Acad. Sci. 271:280–293.
4. Ballou, J. E., R. A. Gies, G. E. Dagle, M. D. Tolley, and A. C. Case. 1980. P. 142ff in Desposition and Late Effects of Inhaled $^{232}UO_2$ (NO_3) in Rats. PNL Annual Report 1980 (Biomedical Sciences) PNL-3700 P & 1. Richland, Wash.

5. Bentley, K. W., D. R. Stockwell, K. A. Britt, and C. B. Kerr. 1985. Transient proteinuria and aminoacidurea in rodents following uranium intoxication. Bull. Environ. Contam. Toxicol. 34:407–416.
6. Boback, M. W. 1975. A review of uranium excretion and clinical urinalysis data in accidental exposure cases. Pp. 226–231 in the Conference on Occupational Health Experience with Uranium. ERDA-93. Washington, D.C.: U.S. Energy Research and Development Administration.
7. Cookfair, D. L., W. L. Beck, C. Shy, C. C. Lushbaugh, and C. L. Sowder. 1983. Lung cancer among workers at a uranium processing plant. Presented at Epidemiology Applied to Health Physics. Pp. 398–406 in Proceedings of the Health Physics Society.
8. Donoghue, J. K., E. D. Dyson, J. S. Hislop, A. M. Leach, and N. L. Spoor. 1972. Human exposure to natural uranium. Br. J. Ind. Med. 29:81–89.
9. Durbin, P. W., and M. E. Wrenn. 1975. Metabolism and effects of uranium in animals. In M. E. Wrenn, ed. Pp. 67–129 in the Conference on Occupational Health Experience with Uranium. ERDA-93. Washington, D.C.: U.S. Energy Research and Development Administration.
10. Ely, T. S. 1959. Medical findings summary. Symposiums on Occupational Health Experience and Practices in the Uranium Industry. HASL-58. Washington, D.C.: Office of Technical Services, Department of Commerce.
11. Fisenne, I. M., G. A. Welford. 1986. Natural U concentrations in soft tissues and bone of New York city residents. Health Phys. 50:739–746.
12. Fisenne, I. M., P. Perry, N. Y. Chu, and N. H. Harley. 1983. Measured ^{234}, ^{238}U and fallout ^{239,240}Pu in human bone ash from Nepal and Australia. Health Phys. 44 (Suppl 1):457–467.
13. Fisenne, I. M., H. W. Keller, and P. Perry. 1984. Uranium and ^{226}Ra in human bone ash from Russia. Health Phys. 46:438–440.
14. Haley, D. P. 1982. Morphologic changes in uranyl nitrate-induced acute renal failure in saline- and water-drinking rats. Lab. Invest. 46:196–208.
15. Haley, D. P., R. E. Bulger, and D. C. Dobyan. 1982. The long-term effects of uranyl nitrate on the structure and function of the rat kidney. Virchows Arch. (Cell Pathol.) 41:181–192.
16. Harris, W. B., and E. Davies. 1961. Experimental clearance of uranium dust from the human body. Inhaled particles and vapours. London: Pergamon.
17. Haven, F. L. 1949. Tolerance to uranium compounds. Pp. 729–758 in The Pharmacology and Toxicology of Uranium Compounds. New York: McGraw-Hill.
18. Haven, F. L., and H. C. Hodge. 1953. Toxicity following parenteral administration of certain soluble uranium compounds. Pp. 281–308 in The Pharmacology and Toxicology of Uranium Compounds. New York: McGraw-Hill.
19. Haven, F. L., and H. C. Hodge. 1953. Toxicity following the parenteral administration of certain soluble uranium compounds. Pp. 281–308 in The Pharmacology and Toxicology of Uranium Compounds. New York: McGraw-Hill.
20. Hodge, H. C. 1956. Mechanism of uranium poisoning. Arch. Ind. Health 14:43–47.

21. Hodge, H. C. 1973. A history of uranium poisoning (1824–1942). Pp. 5–68 in Uranium, Plutonium, and the Transplutonic Elements. Handbook of Experimental Pharmacology, Vol. 36, H. C. Hodge, J. N. Stannard, and J. B. Hursh, eds. New York: Springer-Verlag.
22. Hodge, H. C., J. N. Stannard, and J. B. Hursh. 1973. Uranium, Plutonium, and the Transplutonic Elements. Handbook of Experimental Pharmacology, Vol. 36. New York: Springer-Verlag.
23. Hursh, J. B., and N. L. Spoor. 1973. Chapter 36 in Uranium, Plutonium, and the Transuranic Elements. Handbook of Experimental Pharmacology, Vol. 36, H. C. Hodge, J. N. Stannard, and J. B. Hursch, eds. New York: Springer-Verlag.
24. International Commission on Radiological Protection (ICRP). 1959. Permissible dose for internal radiation. ICRP Publication 2. Oxford: Pergamon.
25. International Commission on Radiological Protection (ICRP). 1975. Report of the Task Group on Reference Man. ICRP Publication 23. Oxford: Pergamon.
26. International Commission on Radiological Protection (ICRP). 1979. Limits of Intakes of Radionuclides by Workers. ICRP Publication 30, Part I. Oxford: Pergamon.
27. Leach, L. J., E. A. Maynard, H. C. Hodge, J. K. Scott, C. L. Yuile, G. E. Sylvester, and H. B. Wilson. 1970. A five-year inhalation study with natural uranium dioxide (UO_2) dust—I. Retention and biologic effect in the monkey, dog, and rat. Health Phys. 18:599–612.
28. Leach, L. J., C. L. Yuile, H. C. Hodge, G. E. Sylvester, and H. B. Wilson. 1973. A five-year inhalation study with natural uranium dioxide (UO_2) dust—II. Postexposure retention and biologic effects in the monkey, dog, and rat. Health Phys. 25:239–258.
29. Lorenz, E. R. 1944. Radioactivity and lung cancer: A critical review of lung cancer in the miners of Schneeburg and Joachimsthal. J. Natl. Cancer Inst. 51:1–15.
30. Luessenhop, J., J. C. Gallimore, W. H. Sweet, E. G. Strucness, and J. Robinson. 1958. The toxicity in man of hexavalent uranium following intravenous administration. Am. J. Roentgenol. 79:83–100.
31. Lundin, F. E., J. W. Lloyd, E. M. Smith, V. E. Archer, and D. A. Holaday. 1969. Mortality of uranium miners in relation to radiation exposure, hard rock mining and cigarette smoking—1950 through 1967. Health Phys. 16:571–578.
32. Mays, C. W., R. E. Rowland, and A. F. Stehney. 1985. Cancer risk from the lifetime intake of radium and uranium isotopes. Health Phys. 48:635–647.
33. Miller, S. E., D. A. Holiday, and H. N. Doyle. 1956. Health protection of uranium miners and millers. Arch. Ind. Health 15:48–55.
34. Moore, R. H., ed. 1984. Biokinetics and Analysis of Uranium in Man. USUR-65 HEHF-47. Richland, Wash.
35. Morrow, P., R. Gelein, H. Beiter, J. Scott, J. Picano, and C. Yuile. 1982. Inhalation and intravenous studies of UF_6/UO_2F_2 in dogs. Health Phys. 43:859–873.
36. Morrow, P. E., L. J. Leach, F. A. Smith, R. M. Gelein, J. B. Scott, H. D. Beiter, F. J. Amato, J. J. Picano, C. L. Yuile, and T. G. Consler. 1982. Metabolic fate and evaluation of injury in rats and dogs following

exposure to the hydrolysis products of uranium hexafluoride, July 1979–October 1981. Report NUREG/CR-2268. Washington, D.C.: U.S. Nuclear Regulatory Commission December.
37. Neuman, W. F., C. Voegtlin, and H. C. Hodge. 1949. Deposition of uranium in bone. Pp. 1911–1991 in National Nuclear Energy Series, Part 1 of Div. VI, Vol 1, W. F. Neuman, ed. New York: McGraw-Hill.
38. Polednak, A. P., and E. L. Frome. 1986. Mortality among men employed between 1943 and 1947 at a uranium processing plant. J. Occup. Med. 23:169–178.
39. Priest, N. D., G. R. Howells, D. Green, and J. W. Haines. 1982. Uranium in bone; metabolic and autoradiographic studies in the rat. Human Toxicol. 1:97–114.
40. Rockstroh, H. 1959. Zur Atiologie des Bronchialkrebses in arsen verarbeitenden Nickelhutten. Beitrag zur Syncarcinogenese des Berufkrebes. Arch. Geschwulstforsch. 14:151–162.
41. Rogan, J. M., M. D. Attfield, M. Jacobsen, S. Rae, D. D. Walker, and W. H. Walton. 1973. Role of dust in the working environment in the development of chronic bronchitis in British coal miners. Br. J. Ind. Med. 30:217–226.
42. Rowland, R. E., and J. E. Farnham. 1969. The deposition of uranium in bone. Health Phys. 17:139–144.
43. Rowland, R. E., A. F. Stehney, and H. F. Lucas. 1983. Dose-response relationships for radium-induced bone sarcomas. Health Phys. 44(Suppl. 1):15–31.
44. Schlenker, R. A., and B. G. Oltman. 1981. Pp. 473–475 in Actinides in Man and Animals, M. E. Wrenn, ed. Salt Lake City: RD Press.
45. Spiers, F. W., H. F. Lucas, J. Rundo, and G. A. Anast. 1983. Leukemia incidence in the U.S. dial workers. Health Phys. 44(Suppl. 1):65–72.
46. Spoor, N. L., and J. B. Hursh. 1973. Protection criteria. Pp. 241–270 in Uranium, Plutonium, and the Transplutonic Elements. Handbook of Experimental Pharmacology, Vol. 36, H. C. Hodge, J. N. Stannard, and H. B. Hursh eds. New York: Springer-Verlag.
47. Stevens, W., F. W. Bruenger, D. R. Atherton, J. M. Smith, and G. N. Taylor. 1980. The distribution and retention of hexavalent ^{233}U in the beagle. Radiat. Res. 83:109–126.
48. Stockinger, H. E. 1963. In Metals Excluding Lead in Industrial Hygiene and Toxicology, Vol. II, D. W. Fassett and D. D. Irish, eds. New York: Interscience.
49. Stradley, G. N., J. W. Stather, M. Ellender, S. A. Sumner, J. C. Moody, C. G. Towndrow, A. Hodgson, D. Sedgwick, and N. Cook. 1985. Metabolism of an industrial uranium trioxide dust after deposition in the rat lung. Human Toxicol. 4:563–572.
50. Stuart, B. O., and T. M. Beasley. 1964. Selection tissue accumulation of uranium and thorium in rats after inhalation of uranium ore dust. Pp. 22–24 in Hanford Biology Research Annual Report for 1964. BNWL-122. Richland, Wash.: U.S. Atomic Energy Commission.
51. Tannebaum, A., ed. 1951. Toxicology of uranium. National Nuclear Energy Series, Div. IV, Vol. 23. New York: McGraw-Hill.
52. Thomas, R. G. 1968. Transport of relatively insoluble materials from lung to lymph nodes. Health Phys. 14:111–117.

53. Thun, M. J., D. B. Baker, K. Steenland, A. B. Smith, W. Halperin, and T. Berl. 1985. Renal toxicity in uranium mill workers. Scand. J. Work Environ. Health 11:83–90.
54. Voegtlin, C., and H. C. Hodge, eds. 1949. In The Pharmacology and Toxicology of Uranium Compounds. National Nuclear Energy Series, Div. IV, Vol. 1. New York: McGraw-Hill.
55. Voegtlin, C., and H. C. Hodge, eds. 1953. Pharmacology and Toxicology of Uranium Compounds. New York: McGraw-Hill.
56. Wagoner, J. K., V. E. Archer, B. E. Carroll, and D. A. Holaday. 1964. Cancer mortality patterns amongst U.S. uranium miners and millers, 1950 through 1962. J. Natl. Cancer Inst. 32:787–801.
57. Walinder, G., B. Fries, and B. Billaudelle. 1967. Incorporation of uranium. Distribution of uranium absorbed through the lungs and the skin. Br. J. Ind. Med. 24:313–319.
58. Waxweiler, R. J., V. E. Archer, R. J. Roscoe, A. Watanabe, and J. J. Thun. 1983. Mortality patterns among a retrospective cohort of uranium mill workers. Presented at Epidemiology Applied to Health Physics. Pp. 428–435 in Proceedings of the Health Physics Society.
59. Wilkinson, G. S. 1985. Gastric cancer in New Mexico counties with significant deposits of uranium. Arch. Environ. Health. 40:307–312.
60. Wilson, H. B., H. E. Stockinger, and G. E. Sylvester. 1953. Acute toxicity of carnotite ore dust. Arch. Ind. Hyg. Occup. Med. 7:301.
61. Wilson, J. 1983. An Epidemiologic Investigation of Nonmalignant Respiratory Disease Among Workers at a Uranium Mill. Ph.D. thesis. Chapel Hill: University of North Carolina.
62. Wrenn, M. E., ed. 1975. Conference on Occupational Health Experience with Uranium. ERDA-93. Washington, D.C.: Energy Research and Development Administration.
63. Wrenn, M. E. 1986. Pp. 131–157 in Symposium on Areas of High Natural Radioactivity. Internal Dose Estimates.
64. Wrenn, M. E., ed. 1986. Actinides in man and animals. Proceedings of the Snowbird Actinide Workshop. Salt Lake City: RD Press.
65. Wrenn, M. E., P. W. Durbin, B. Howard, J. Lipsztein, J. Rundo, E. T. Still, and D. L. Willis. 1985. Metabolism of ingested U and Ra. Health Phys. 48(5):601–633.
66. Yuile, C. L.. 1973. Pp. 165–196 in Uranium, Plutonium, and the Transplutonic Elements. Handbook of Experimental Pharmacology, Vol. 36, H. C. Hodge, J. N. Stannard, and J. B. Hursh, eds. New York: Springer-Verlag.

7

Transuranic Elements

INTRODUCTION

Transuranic elements are members of the actinide series beyond uranium, beginning with neptunium (atomic number 93). The last in the series is element 103 (lawrencium). All are artificially produced in nuclear reactors, accelerators, or explosions of nuclear weapons, and all have several isotopes that emit alpha rays. The energies of the alpha particles emitted from the transuranic elements range from about 5 to well over 8 MeV, with the higher-energy alphas coming largely from the isotopes with the shortest half-lives. Berkelium (element 97), einsteinium (element 99), fermium, mendelevium, nobelium, and lawrencium are produced in such small amounts, mostly for research purposes; and most of the isotopes produced have such short half-lives, a few seconds or minutes, that they are an unlikely health concern. Californium (element 98), a useful neutron radiation source, is available in slightly larger amounts. Neptunium, plutonium, americium, and curium, elements 93 to 96, respectively, are the most abundant and the most extensively used of these man-made actinide series elements. All are produced in nuclear reactors and, because of the alpha-emitting isotopes with very long half-lives, for example, 2.1×10^6 yr for ^{237}Np, 24,400 yr for ^{239}Pu, 458 yr for ^{241}Am, and 17.6 yr for ^{245}Cm, comprise a major radioactive waste disposal concern. Table 7-1 lists the principal transuranic elements which constitute potential health hazards.

TABLE 7-1 Transuranium Nuclides of Potential Biological Significance

Element	Isotope	Half-Life (yr)	Mean α Energy (MeV)
Neptunium	^{237}Np	2.1×10^6	4.7
Plutonium	^{238}Pu	86	5.6
	^{239}Pu	24,400	5.2
	^{240}Pu	6,580	5.3
	^{241}Pu	13	5.1
	^{242}Pu	379,000	5.0
Americium	^{241}Am	458	5.6
	^{243}Am	7,370	5.4
Curium	^{242}Cm	0.45	6.0
	^{244}Cm	17.6	5.9
Californium	^{252}Cf	2.7	6.2[a]
Einsteinium	^{252}Es	0.5	6.7

[a]Includes energy of fission fragments, neutrons, and gamma rays.

Plutonium-239 is a constituent of nuclear weapons and, since 5 metric tons were dispersed into the atmosphere and the environment by the nuclear weapons tests of the 1950s and 1960s, trace amounts can be found almost everywhere. Since relatively large amounts are present in nuclear power reactors, the potential release of plutonium in a reactor accident is a concern, although none was released in the Three Mile Island accident and only small amounts were released at Chernobyl.[42] Plutonium-238, with a half-life of 86.4 yr, is 280 times more radioactive per unit mass than ^{239}Pu. Because of this high specific activity, it is used as a heat source to power thermoelectric devices used in cardiac pacemakers and space vehicles. The use of very small amounts of ^{238}Pu in pacemakers has not caused concern, but the potential for the reentry and destruction of space vehicles, dispersing kilogram quantities of ^{238}Pu into the environment, is a concern. Americium-241 is a contaminant of plutonium in nuclear weapons and thus has been distributed throughout the environment. Very small amounts have become a dependable source of ionizing radiation required in battery-powered smoke detectors. This use has not caused public health concern. Relatively large quantities of ^{237}Np are produced in fission reactors and, with plutonium, americium, and curium isotopes, must be dealt with as a contaminant in cooling water and as a long-lived component of nuclear reactor waste. For all of the transuranic elements, occupational exposures pose a greater potential for causing detectable health effects than environmental exposures, but there is a greater potential that much larger populations will be exposed to it by environmental exposures, but only to trace amounts. Occupational exposures at nuclear materials production facilities have resulted from inhalation of airborne transuranic elements accidentally released from containment equipment and from entry through wounds occurring in the hands of persons handling these materials in glove boxes.

The biological effects of plutonium and other alpha-emitting transuranic elements, unlike gamma-emitting radionuclides, are primarily dependent upon their entering the body and being deposited in radiosensitive tissues. Further, the presence of transuranic elements in the environment does not necessarily infer their deposition in human tissues. In the following discussion it will be seen that transuranic elements are not readily absorbed from the gastrointestinal tract and are even less readily absorbed through intact skin. If environmental conditions lead to transuranic elements becoming airborne, there is a chance of their being inhaled and deposited in the respiratory tract. Deposition in the respiratory tract represents the highest probability for eventual health effects.

This chapter describes the disposition of transuranic elements that enter the body and the biological effects that may result; it also discusses methods for estimating risks and suggests estimates of risk derived from other sources that might be applied to transuranic elements in human beings. Nearly all of the information on health effects has come from laboratory experiments since there are few human data. However, there are human data to supplement extensive animal data on the distribution of transuranic elements in the tissues of the body. For example, since the beginning of the Manhattan Project in 1943, from 5,000 to 10,000 persons have been employed in positions in the United States involving risk of plutonium exposure. Follow-up of the distribution of plutonium in tissues has been accomplished by obtaining tissue samples at autopsy, or infrequently, from surgical specimens from persons who received such exposure. By 1986, the U.S. Transuranium Registry has collected data from about 200 autopsies on exposed workers whose tissues showed increased concentrations of plutonium.[111] Elevated concentrations of plutonium in tissues of the general population are attributed to fallout from atmospheric nuclear weapons testing during the period from 1945 to 1963.[119]

Since this is not intended as an exhaustive or definitive review of the subject, no attempt is made to ensure comprehensive or specific documentation of the information presented. It is intended, however, that all information can be traced to its source through the literature cited, especially the several reviews and symposia publications from which much of the information was obtained.[34,36,41,76,80,117,118,126] Although the committee intended to use only information published in the open literature, reference is made to several recent highly relevant laboratory annual reports.

ROUTES OF INTAKE AND DEPOSITION IN THE BODY

PERCUTANEOUS

Because of the relatively short range of alpha particles in tissues, the radiation-sensitive cells of the basal layer of the skin are not irradiated unless the alpha-emitting radionuclides penetrate the stratum corneum or horny layer of the epidermis. The unbroken skin has been shown to be an effective barrier to the penetration of transuranic elements. This has been observed in skin contamination incidents in nuclear industries and in animal experiments.[41,46,47,81,82]

Insoluble forms such as oxides are easily removed from intact skin by washing. Soluble transuranic compounds, such as nitrates, citrates, chlorides, and complexes with organic solvents, have a greater potential for absorption, even though it is very small. The most common human skin exposures involve plutonium nitrate in nitric acid solutions, plutonium tributylphosphate in carbon tetrachloride, and plutonium in hydrochloric acid. These can be effectively washed from skin with chelating compounds or detergents.

If plutonium, americium, neptunium, or einsteinium is deposited on human or animal skin in a wide range of nitric acid concentrations (0.1 to 10 N) for 1 h, about 5×10^{-4} is absorbed. Americium nitrate in tributylphosphate exhibits nearly a factor of 10 greater percutaneous absorption, about 3×10^{-3}.[46] While data for transuranic oxides are sparse, one can predict that percutaneous absorption through intact skin would be less than 1×10^{-5} during the first hour after deposition. An approximate 10-fold increase in absorption was seen when soluble transuranic compounds remained on the skin for 3–5 days; over this prolonged time, there was evidence that higher concentrations of nitric-acid-enhanced percutaneous absorption. Under the most extreme conditions, such as plutonium nitrate in 10 N nitric acid for 4 days or americium nitrate in 8 N nitric acid for 3 days, maximum percutaneous absorption was only 2%.

Damage to skin by trauma, wounds, acid, or thermal burns facilitates a more rapid transfer of soluble transuranic compounds into the subcutaneous tissue and blood. Insoluble particles and metal slivers deposited below the level of the epidermis are slowly cleared to regional lymph nodes.

There are three mechanisms for transport of transuranic elements from skin: (1) transfer into the subcutaneous microcirculation and then into the blood and lymphatic systems, (2) transfer onto the skin surface with sudoriferous and sebaceous secretions, and (3) loss from body with desquamation of skin. Autoradiographic studies of cutaneously deposited plutonium, americium, and neptunium show a decreasing concentration with increasing depth in skin, but with focal concentrations of activity in the upper epidermis, hair follicles, sebaceous glands, and microvasculature.[47]

In the event of accidental occupational exposure to plutonium through skin wounds in the hands and other sites of the body, plutonium may be retained at the wound site and removed by surgery, sloughed from the surface, solubilized and translocated to internal organs, or transported to regional lymph nodes. Wounds penetrating the horny layer of skin lead to more rapid absorption and translocation of soluble transuranic compounds to bone, liver, and other tissues.

The most serious of these accidents, in terms of quantity of activity deposited in the body, involved an explosion that resulted in deposition of a total of 1–5 mCi of ^{241}Am on the face and, by inhalation, in the respiratory tract.[12] About 1 mCi remained in the body, mostly in wounds on the face, after initial emergency decontamination. Chelation therapy by intravenous injection of diethylenetriaminepentaacetic acid (DTPA) facilitated the excretion of 900 μCi.

Investigators gave beagle dogs subcutaneous implants of 9.5 μCi $^{239}PuO_2$ or 1.3 μCi $^{239}Pu(NO_3)_4$ in their forepaws to mimic hand wounds received by plutonium workers. At 5 and 8 yr following exposure, the injected paws still retained 21 and 16%, respectively, of deposited plutonium; in both cases the highest concentrations of activity were found in regional lymph nodes, with the liver showing the next highest concentration of plutonium. The skeleton retained a greater amount of plutonium from the nitrate than from the dioxide compound.[25] Similar results were observed with subcutaneously injected monomeric and polymeric ^{239}Pu in mice, with the monomer behaving like plutonium nitrate and the polymer like plutonium dioxide.[44]

Gastrointestinal Tract

The gastrointestinal tract, provides a substantial barrier to the uptake of transuranic elements ingested with food or water. Although the fraction absorbed is usually low, continuous ingestion of contaminated food and water may lead to the presence of measurable amounts in the body. Gastrointestinal absorption is also a consideration in assessing the risk from inhaled transuranic elements because of clearance from the lungs to the gastrointestinal tract, but is small compared to direct absorption from the lungs.

The importance of chemical form and oxidation state in the absorption of transuranic elements has been verified in several laboratories. These data have been tabulated by the International Commission on Radiological Protection (ICRP).[41] Relatively high concentrations of chlorine in some domestic water supplies could oxidize Pu^{+4} to Pu^{+6}.[52] However, subsequent animal studies failed to provide convincing evidence that changes in valence state for ingested plutonium had a significant effect on gastrointestinal absorption. Fasting may increase the absorption of plutonium by about 1 order of magnitude. Both calcium- and iron-deficient diets tend to enhance absorption of plutonium.[87,112] The absorption of transuranic elements incorporated in food fed to experimental animals may be 2 to 10 times greater than the absorption of chemical forms such as citrate and nitrate.

In adult animals, $<0.01\%$ of plutonium and most other transuranic elements is absorbed from the intestines.[41] Overall, there is little variation in absorption of plutonium and americium nitrates in adult rats, guinea pigs, or dogs.[110] Swine exhibit a greater plutonium absorption than rats or dogs.[111] The absorption of ^{237}Np can be increased 10 to 100 times by increasing the mass ingested; however, at occupationally and environmentally relevant levels the absorption is more like that of the other transuranic elements. With few exceptions, the absorption of transuranic compounds from the gastrointestinal tract in adult experimental animals varies over 3 orders of magnitude 10^{-5} to 10^{-2}. This led the ICRP[41] to adopt values of 0.1×10^{-4} for plutonium oxides and 1×10^{-4} for plutonium nitrate for application to occupational exposures. These values would apply to inhaled material cleared from the respiratory tract to the gastrointestinal tract, as well as to ingested material. For all other plutonium compounds and compounds of all other transuranic elements, including those incorporated in food products and drinking water, the ICRP adopted an absorption factor of 10×10^{-4} for application to exposures of both workers and the public.*

Ingestion of alpha-emitters is not considered a radiological hazard to the gastrointestinal tract since the range of alpha particles is insufficient to penetrate the mucus and intestinal contents and reach the crypt cells.

The neonatal rat and guinea pig absorb about 100 times more plutonium than the adult, while newborn swine absorb 20 times more plutonium than either the newborn rat or guinea pig. In addition to the increased fraction absorbed, a substantial fraction of transuranic elements given in soluble form was retained for several days within the mucosa of the small bowel.[109] The ICRP has adopted an absorption value of 100×10^{-4} for the first year of life in contrast to the value of 10×10^{-4} cited above for all later years.[41]

RESPIRATORY TRACT

Inhalation is probably the most common pathway by which transuranic elements cross the barriers of the body and penetrate into and across living cells. The aerodynamic particle size of the aerosol, which accounts for not only the sizes of particles but also their density and shape, determines the fractional deposition and sites of deposition in the respiratory tract. The subsequent rates and routes of clearance; the translocation to, deposition in, and rate of clearance from other tissues; and the excretion in urine and feces of inhaled transuranic compounds depend on particle size, solubility, density, shape, and other physicochemical characteristics of the aerosol. In this way the physical and radiological properties of the transuranic compound, and the physiological characteristics of the exposed individual determine the amount deposited and thus, the radiation dose rates and total doses delivered to the tissues of the

---

*These are revised from those used to calculate annual limits on intake by the ICRP.[41] The f_1 values (fraction transferred to blood) used were 0.1×10^{-4} for plutonium oxides; 1×10^{-4} for nitrates and other plutonium compounds; 5×10^{-4} for all americium, curium, and californium compounds; and 100×10^{-4} for neptunium compounds.

respiratory tract and other organs of the body. Aerodynamic particle diameter is a useful predictive characteristic of an aerosol for the estimation of deposition in regions of the respiratory tract. Several dosimetric models have been developed for describing particle deposition and clearance in the human respiratory tract. These models provide a basis for estimating deposition, distribution, and retention of inhaled radioactive aerosols, taking into account particle size and chemical form of the aerosol. Mathematical models were developed to describe the deposition and clearance of inhaled materials from the several compartments of the nasal passage, the trachea and bronchial tree, the pulmonary parenchyma, and the thoracic lymph nodes. The models, when used for radiation protection purposes, apply to a reference man, a 70-kg male worker. Thus, they can be expected to only approximate the deposition, distribution, and retention of inhaled radionuclides in any given individual.

For inhalation of an aerosol with an activity median aerodynamic diameter (AMAD) of 1 μm, according to the ICRP model (based largely on experiments with nonradioactive aerosols in human subjects), 30% of the particles are deposited in the nasal passages, 8% in the trachea and bronchial tree, and 25% in the pulmonary parenchyma, for a total deposition in the respiratory tract of 63% of the amount inhaled. The amount exhaled, not deposited, is 37%. These deposition fractions will vary with particle size and the breathing rate and volume.[39] The ICRP has devised metabolic models to describe the retention and translocation of transuranic compounds from these sites of deposition, based largely on data from experimental animals. The following summarizes this information for several transuranic compounds.

Following deposition in the lungs, particles are quickly phagocytized by alveolar macrophages. The attenuated cytoplasm of type I alveolar epithelial cells may also phagocytize particles.[95] Up to 1% of particles, including transuranic oxides, deposited in the lung may also be taken up by tracheobronchial and bronchiolar epithelia.[13] Particles penetrating the respiratory epithelium may be phagocytized in interstitial areas and, if insoluble, eventually cleared to regional lymph nodes of the thoracic cavity. Since relatively soluble transuranic compounds, such as nitrates, citrates, and the oxides of americium and curium, are rapidly cleared into the blood, only small fractions are cleared to lymph nodes. However, plutonium inhaled as relatively insoluble plutonium oxide particles is very slowly cleared into the blood. Thus, within a few hours after inhalation, about half of inhaled $^{239}PuO_2$ deposited in the alveoli can be removed from experimental animal lungs by multiple lavage,[71,92] with most particles having been phagocytized by pulmonary macrophages. This phagocytosis of plutonium particles may facilitate their transport from the lungs by the mucociliary epithelium and possibly contributes to their transport to lymphatic tissues.

The distribution of transuranic elements within the lungs is relatively uniform after inhalation, more uniform for the most soluble forms. Following the initial clearance process and especially the absorption of the more transportable material by the blood circulating through the lungs, the distribution within the lungs becomes much less uniform. The material retained in the lungs for long times is localized primarily in bronchiolar, alveolar, and lymphatic structures of the lung parenchyma, frequently in regions of fibrosis and scar tissue.[24] This pattern appears to be consistent among all experimental animals studied, and while there are few observations, there is no evidence to the contrary for human lungs.

Respiratory tract clearance of inhaled plutonium in human accidental exposure cases is similar to that seen in PuO_2 studies with large animals (dog,[5,70] sheep,[108] baboon,[8] burro,[108] and rhesus monkey[50]) with half-times for three exponential phases of approximately 1, 30, and 300 to over 1,500 days, respectively.[124] The second phase is not always distinguishable. Early clearance of plutonium is from the nasal passages and upper tracheobronchial regions, while clearance with longer half-times is from the bronchiolar and alveolar or pulmonary regions. High fecal to urine ^{239}Pu ratios (between 50 and 500), indicative of a high insolubility, are observed in humans for long periods following inhalation of $^{239}PuO_2$. A large fraction, as much as 50%, of inhaled and deposited insoluble $^{239}PuO_2$ and up to about 25% of $^{238}PuO_2$ may eventually be transported to the thoracic lymph nodes of dogs.[83]

In contrast to $^{239}PuO_2$, inhaled ^{241}Am and ^{244}Cm dioxides, as well as plutonium nitrates, in humans and animals are relatively soluble, with about half of the amounts deposited in the bronchiolar and alveolar regions cleared with a half-life of 10 to 40 days and the remainder cleared with half-lives generally ranging from 200 to 500 days. Less than 1% of these relatively soluble transuranic compounds deposit in thoracic lymph nodes.[6]

While plutonium oxide particles are generally quite insoluble in the respiratory tract, there are some exceptions. For example, it has been demonstrated in several animal species that the conditions under which plutonium is oxidized may affect the fate of particles deposited in the respiratory tract.[5] Electron micrographs suggest that plutonium particles oxidized at high temperatures have less surface area than those oxidized at much lower temperatures and, thus, could have lower dissolution rates in body fluids. This was verified in studies in which plutonium oxide particles formed at high temperatures (over 1,000°C) tended to have lower translocation rates from the lungs than plutonium oxidized in air at ambient temperatures or calcined at relatively low temperatures. Also, alveolar clearance and translocation of $^{238}PuO_2$ to other tissues such as liver and bone are nearly always more rapid than those for comparably prepared $^{239}PuO_2$.[5,6,83] Plutonium-238 is 280 times more radioactive than an equal mass of ^{239}Pu. Radiolysis may cause these high-specific-activity $^{238}PuO_2$ particles to fragment within the lungs, greatly increasing the surface area of the ^{238}Pu particles, and thus their dissolution rate.

Nanometer-diameter plutonium oxide particles have been found to be cleared from the respiratory tract very rapidly and appear to be excreted in the urine as particles.[105] If transuranic elements are inhaled simultaneously with other materials, their disposition may depend on how the transuranic element is combined with the other material in the aerosol. For example, calcining $^{239}PuO_2$ with a relatively large amount of sodium, potassium, calcium, aluminum, or uranium increases the solubility of ^{239}Pu in the lung.[3,106] Increasing the ratio of plutonium to sodium in laser-vaporized aerosols of PuO_2-UO_2 and sodium from 0 to 1:1 and to $>$ 1:10 increased the rate of clearance from the lungs and translocation to extrapulmonary tissues from 0.5 to 5.0 and 24%, respectively.[58] After inhalation of an aerosol of $^{239}PuO_2$ and $^{244}CmO_2$ calcined as a mixture, both plutonium and curium remained in the lung somewhat longer than when calcined and inhaled separately.[101] The translocation of curium to extrapulmonary tissues was largely prevented by incorporation into the much greater mass of the PuO_2 matrix. However, in rats the rate of alveolar clearance and translocation of ^{169}Yb and ^{239}Pu inhaled as an oxide, prepared by calcining ^{169}Yb mixed with ^{239}Pu, were not significantly different from the rates of clearance and translocation of $^{169}Yb_2O_3$ or $^{239}PuO_2$ inhaled separately.[102]

The high rate of accumulation of inhaled insoluble plutonium in lymph nodes has stimulated considerable interest. Lymph nodes draining the lungs attain concentrations of inhaled plutonium many times higher than those of any other tissue. The particles are preferentially localized along sinusoids in the paracortical area and in medullary cords and less so in the lymphoid germinal centers.[40] More than 10% of $^{239}PuO_2$ deposited in the alveoli was taken up by thoracic lymph nodes of dogs by 1 yr postexposure, increasing to 15% by 2 yr and 30 to 50% by 5 to 15 yr. Accumulation of inhaled $^{238}PuO_2$ in thoracic lymph nodes was less than $^{239}PuO_2$; it reached a maximum of 20 to 24% and gradually declined to $<10\%$ after 10 to 15 yr.[83] No significant differences in uptake by thoracic lymph nodes were noted for monodisperse $^{239}PuO_2$ at AMAD values of 0.72, 1.4, and 2.8 μm.[32] Less than 1% of alveolar-deposited $^{238}Pu(NO_3)_4$, $^{239}Pu(NO_3)_4$, ^{241}Am oxide, or ^{244}Cm oxide was found in thoracic lymph nodes of dogs at >1 yr postexposure.[21–23,64] Uptake of inhaled $^{239}PuO_2$ in lymph nodes of baboons appears to be similar to that in lymph nodes of beagle dogs.[8]

Liver and Bone

In addition to the respiratory tract, a considerable research effort has focused on the deposition and retention of transuranic elements in liver and bone. Animal experiments and analysis of human tissues confirm that liver and skeleton are the principal receptors of transuranic elements that enter the blood. The distribution of transuranic elements between these two tissues varies depending on the form of the transuranic element taken into the body. Concentrations of fallout plutonium in human liver and bone range between about 0.5 and 1.5 pCi/kg. In most cases the concentrations are higher in liver,[18] but higher concentrations in bone have been reported.[45] Although the skeleton is about 4 times the mass of the liver, the liver is generally found to contain as much or more of the total plutonium in the body than is in the skeleton. Following occupational exposures to plutonium, depositions in the liver range from being about equal to about twice those in the skeleton.

Transuranic elements within the liver are uniformly distributed throughout the hepatic epithelium only for a short time after intravenous injection. At long times after injection and following other routes of intake, transuranic elements localize in the phagocytic lining cells of the sinusoids, the Kuppfer cells of the reticuloendothelial system.

In the skeleton transuranic elements tend to concentrate on trabecular and cortical bone surfaces, with the endosteal cells being the principal recipients of emitted alpha radiation. Transuranic elements may be found in bone marrow soon after intravenous injection, but the levels decrease as the levels in the circulating blood decrease. At long times after exposure, slightly increasing amounts may be seen in bone marrow, possibly resulting from redistribution of transuranic elements by bone resorption processes.

Over 20 cases of accidental exposure to ^{241}Am, mostly by inhalation, have been reported in the literature, with total body burdens ranging from 10 nCi to 25 μCi.[127] The depositions in skeleton generally exceeded those in lung. However, the distribution of ^{241}Am in one accidental inhalation exposure case treated with DTPA at about 1 yr postexposure was as follows: 41% of body burden in lung, 47% in liver, and 12% in bone.[31] This suggests that DTPA selectively enhances the excretion of systemic americium and is consistent with the results from animal experiments that show that DTPA is ineffective in removing transuranic elements from lungs.[6] In six accidental human exposures to curium oxide, retention and excretion were similar to those expected for soluble plutonium compounds.[86]

While deposition of transuranic elements in liver and bone are qualitatively similar among mammalian species, there are quantitative differences. There are also differences in retention, especially in the liver. The initial fraction of injected ^{241}Am and ^{244}Cm found in rat liver was three times that of injected ^{239}Pu, although clearance from the liver was rapid in all cases.[104] Deposition and retention of neptunium in liver appears to be appreciably less than those in bone, but most of the experiments were done with rats which have been shown to lose actinides from their livers much more rapidly than other species, including man.[116] In studies of dogs,[83] following the inhalation of $^{239}PuO_2$, up to about 20% may accumulate in liver (over 1% of the body content) but only 2%a accumulates in skeleton (less than 1% of the body content after 15 yr. However, in dogs that inhaled $^{238}PuO_2$ or $^{239}Pu(NO_3)_4$, liver and skeleton accumulated comparable fractions of the amount deposited after 6 yr (20%) and had comparable fractions of the body content (about 40 to 45%).[30,83] After intravenous injection as a citrate, dog liver accumulated about 30% of plutonium, 50% of americium, and 35% of curium after 1 week.[54] The retention half-time for both plutonium and americium in dog liver is about 10 yr. In rats, a large fraction of the plutonium is lost through bile secretion.[9] Less ^{239}Pu was taken up by liver and bone in monkeys than in dogs following inhalation of $^{239}Pu(NO_3)_4$, with the liver retention time being much less in monkeys than in dogs.[15] In cynomolgus monkeys about 16% of the initial lung burden of $^{241}AmO_2$ was was taken up by the liver and 8% by the skeleton, with the retention half-time for ^{241}Am in liver being considerably shorter than that in the liver of the dog.[65] After the first week postexposure, ^{241}Am excreted in the feces was eliminated mostly from the liver.[20].

The ICRP, after reviewing all the experimental data and the results from human autopsies, concluded that the liver can be considered to receive about 30% and the skeleton about 50% of the amount of plutonium that enters the blood, with retention half-times of 20 yr and 50 yr, respectively. The same applies to americium and curium, except that the retention times in liver may be less. For californium, berkelium, and einsteinium, deposition was taken to be 25% in liver and 65% in skeleton. For neptunium liver deposition was taken to be about 15% and skeleton deposition is about 65%.[41]

HEALTH EFFECTS STUDIES IN ANIMALS

Tissues of interest with respect to potential health effects following intake of a transuranic element are lungs, liver, bone, bone marrow, and lymph nodes, and to a lesser degree thyroid gland, gonads, and kidney. By far the greatest emphasis has been placed on lungs and bone since these two tissues have been the predominant sites of neoplasia in experimental animals.

Life-span studies in animals for the purpose of examining the carcinogenicity of orally or percutaneously deposited transuranic compounds have not attracted much interest because of the very low rate of absorption of transuranic compounds. No completed life-span study of transuranic compounds taken into the body by any route has indicated a significant increase of gastrointestinal tract tumors.

RESPIRATORY TRACT

Inhalation of comparatively large amounts of transuranic compounds in experimental animals results in radiation pneumonitis and fibrosis with histological features similar to those observed after external radiotherapy.[19,24] Respiratory insufficiency, caused by diffuse fibrosis in the lungs and characterized by increased respiratory rate, decreased arterial oxygen, and increased carbon dioxide, can lead to death within a month or 2 after deposition of 1 to 2 μCi/g of lung tissue of alpha-emitting transuranic elements in rodents, dogs, and primates.[38] An initial lung deposition of 200 nCi of $^{239}PuO_2$ in rats, which is in the range that yields a maximum response, leads to a replacement of 12% of the lung volume with fibrotic tissue at about 1.5 yr after exposure.

Alveolar deposition of 0.1 μCi/g of lung of $^{239}PuO_2$ in dogs may lead to respiratory failure within 10 months. It is characterized by pulmonary edema, severe vascular damage, fibrinous accumulations in bronchioles and alveoli, and pulmonary fibrosis. Alveolar deposition of about 0.05 μCi/g of lung induced pulmonary fibrosis, bronchioloalveolar hyperplasia and metaplasia, alveolar histiocytic proliferation, pleural fibrosis, and early tumor formation within 5 yr after exposure. In comparable studies with inhaled ^{238}Pu oxides, these lung lesions were found to be more closely related to total cumulative radiation lung dose than to dose rate.[40] Inhaled PuO_2 particles in dog lung are associated with areas of pulmonary fibrosis, as is the case in the rat and the Syrian hamster. Migration of particles over time results in focal concentrations of the particles in peripheral regions of the lungs.

A more homogeneous spatial-temporal dose distribution pattern was seen in the lungs of baboons than in the lungs of dogs or rats following inhalation of $^{239}PuO_2$, leading to a more diffuse interstitial pneumonitis and fibrosis at high doses.[60] The acute mortality doses for baboons with inhaled $^{239}PuO_2$ were similar to those seen in dogs.[8] The lesions in the baboon lung were described as more homogeneous than those in dogs and consisted of interstitial pneumonitis with fibrosis rich in elastic fibers, hyalinized arteries, and intense proliferation of type II alveolar epithelial cells and foci of giant cell interstitial pneumonia.

At levels below those that cause acute radiation pneumonitis, chronic alpha irradiation produces a progressive interstitial fibrosis. Both pneumonitis and fibrosis interfere with respiratory function by increasing the barrier distance in alveolar septal interstitial tissue to the diffusion of gases. The nadir of the pneumonitis reaction is typically seen at 60–200 days after the deposition of transuranic elements. After about 200 days, the acute pneumonitis either repairs or slowly progresses to a chronic inflammatory condition associated with interstitial fibrosis. The terminal phase of radiation pneumonitis/fibrosis is characterized by an increased respiratory rate, depressed carbon dioxide consumption, decreased CO diffusion, and decreased pulmonary compliance.[24]

Pulmonary fibrosis from inhaled PuO_2 particles is seen most frequently at sites of particle concentration in subpleural regions. Epidermoid carcinoma and adenocarcinoma in rats and bronchiol-alveolar carcinoma in dogs, usually preceded by associated metaplastic changes, often arise from these areas of intense fibrosis. The origin and mode of development of these tumors closely parallel the development of certain lung carcinomas in the human lung periphery that can be traced to old peripheral tubercular or trauma-induced scars.[55] Metaplastic lesions occupy only a small part of the lung volume when compared with tumor volume. Yet, cell turnover times for adenomatous metaplasia, adenocarcinoma, squamous cell metaplasia, and squamous cell carcinoma of the rat following inhalation of PuO_2 were similar.[90] Induction of nasal or paranasal tumors has not been seen in any experimental animal or human population exposed to transuranic elements by inhalation or any other route, nor have induced primary tumors been seen in the oral-pharyngeal cavity, larynx, or trachea.

Clinical pathological changes following exposure to transuranic elements reflect the quantities deposited and the amounts of cellular damage or induction of primary tumors in extrapulmonary organs, particularly the liver and bone. These changes are dose related and usually occur late in the animal's life span. They reflect nonspecific hepatic or skeletal changes and are more significant in dogs inhaling soluble than insoluble transuranic compounds.[85]

The biological effects of inhaled plutonium have been studied for 30 yr in over 1,000 beagle dogs. At the Pacific Northwest Laboratory this included 116 exposed to $^{239}PuO_2$,[83] 116 exposed to $^{238}PuO_2$,[83] and 105 exposed to inhaled $^{239}Pu(NO_3)_4$;[23] there were also 66 unexposed control dogs. The aerosols were polydispersed with an AMAD of 2.3 μm for $^{239}PuO_2$ and 1.8 μm for $^{238}PuO_2$. A total of 172 of these dogs were alive as of September 30, 1986. Deaths from radiation pneumonitis resulted from high doses of inhaled ^{239}Pu compounds ($\sim$0.5 μCi/kg or 50 nCi/g of lung and a total lung dose of about 4,000 rad), but not from comparable levels of $^{238}PuO_2$, which clears the lungs more rapidly.

Of 98 deaths that occurred 3 to 15 yr after exposure to $^{239}PuO_2$, there were 48 lung cancers. There were three bone tumors, all in the lowest two dose groups, which resulted in an uncertain relationship between bone tumor formation and plutonium exposure. Other tumors in the exposed animals were also seen among the controls. Of 72 deaths that occurred 3 to 12.5 yr after exposure to $^{238}PuO_2$, 4 were related to lung tumors and 33 to bone tumors (13 of the dogs also had lung tumors unrelated to death).[83] Of 38 deaths that occurred 3 to 9 yr after exposure to $^{239}Pu(NO_3)_4$, 4 were related to lung tumors (2 also had radiation pneumonitis) and 21 to bone tumors (11 of the dogs also had lung tumors unrelated to death).[23] There is a high probability that these deaths were attributed to the plutonium exposures. The relationship of other causes of death to plutonium exposures is very uncertain. For example, a number of dogs had malignant lymphoma, but they were distributed among both control and plutonium-exposed animals. Further analysis of morbidity and mortality in these dogs would be premature until the experiments are completed. As a preliminary estimate, the risk of developing a lung tumor ranged from about 450 to 650 lung tumors/10^6 rad to the lung for ^{238}Pu and ^{239}Pu, but time and competing causes of death, that is, radiation pneumonitis and bone cancer, were not adequately accounted for.[84] The lowest lung doses at which lung tumors have been observed in this incomplete study were 30 to 120 rad to the lung for dogs that inhaled $^{239}PuO_2$ and 100 rad to the lung in dogs that inhaled $^{238}PuO_2$; the tumors occurred after 166 to 175 months and after 134 months, respectively. Because of the number of lung tumors in unexposed control dogs (4/20, or 25%) and the limited number of dogs receiving each dose, it may not be possible to evaluate the lung-tumor risk at doses approaching those in human exposure cases or at doses comparable to the limits for occupational exposure.[39]

Another major life-span inhalation study in dogs is in progress at the Inhalation Toxicology Research Institute with largely monodisperse $^{238}PuO_2$ and $^{239}PuO_2$ aerosols with AMADs of 0.75, 1.5, or 3.0 μm.[75] Initial lung burdens in 576 dogs ranged from 0.0002 to 2 μCi/kg of body weight with 12 dogs in each dose and particle-size group and a total of 96 unexposed control dogs. An initial lung burden of 0.00023 μCi/kg of body weight in a dog is approximately equivalent to 0.016 μCi of plutonium in the lungs of a 70-kg human.

A total of 61 of 72 young adults dogs exposed to 1.5-μm and 64 of 72 dogs exposed to 3.0-μm $^{238}PuO_2$ were dead after 11 to 13 yr. Osteosarcoma of the skeleton was the most commonly observed tumor in over half the animals; few primary lung tumors were present.[68] Of 216 dogs exposed to $^{239}PuO_2$, 135 were dead after 8 to 10 yr. Radiation pneumonitis and pulmonary fibrosis were the causes of death in the highest dose groups, and lung carcinoma was the most frequent cause of death and the only fatal cancers in dogs at lower dose levels.[73] No clear pattern of death based on particle size of inhaled plutonium has been seen from available data. Studies of immature (3-month-old) or aged (8- to 10-yr-old) dogs exposed to monodisperse $^{239}PuO_2$ have not yet indicated significant differences in lung-tumor induction as a function of age at exposure.[74] Dogs exposed repeatedly by inhalation to $^{239}PuO_2$ at levels high enough to cause fatal pulmonary pneumonitis and fibrosis died after about the same cumulative radiation dose to the lung as those exposed only once to the aerosol.[26]

The bronchioloalveolar junction appears to be the site of lung tumor formation following inhalation of plutonium compounds. Lung tumors arise from peripheral areas of the lung, typically in proximity to areas of interstitial fibrosis or from small cavities communicating with bronchioles. The cells of origin are considered to be undifferentiated, nonciliated precursor epithelial cells, with various phenotypes developing in tumor cells giving different histological patterns. These were classified as bronchioloalveolar carcinoma, combined epidermoid and adenocarcinoma, adenocarcinoma, epidermoid carcinoma, and mixed sarcoma and carcinoma. Multiple tumors are frequently present in the same lung, occasionally with more than one histological type.[24]

A large body of experimental data exists for carcinogenic effects of inhaled transuranic compounds in rats.[40] Spontaneous lung tumors are rare in control rats, occurring in $<0.1\%$ of unexposed Wistar rats and up to 1–2% in other strains, such as in Fischer rats. Among the inhaled alpha-emitters shown to induce lung tumors in rats are ^{238}Pu, ^{239}Pu, ^{241}Am, ^{244}Cm, and ^{253}Es. A statistically significant increase in lung tumors was seen in experimental animals at alpha-particle lung doses above 10 rad.

In most life-span studies of inhaled transuranic elements in rats, average doses are calculated for groups of animals that inhaled roughly the same amounts of radionuclides. Since large variability could occur within a dose group, it was probable that induced lung tumors in a group would be skewed to the individual rats given the highest doses. A recent life-span study of inhaled $^{239}PuO_2$ with 3,192 rats (including 1,058 sham-exposed controls) at individually measured initial lung deposition levels ranging from about 0.5 to 180 nCi, indicates a possible threshold dose of 100–200 rad to the lungs for lung-tumor formation.[103] This result contrasts with the appearance of tumors at doses as low as 10 rad in an earlier study.[99] Both studies were carried out in young, adult, female, SPF, Wistar rats. Possible, as yet undetected, genetic differences, as well as improved dosimetry at the lower dose levels, may account for these differences in lung-tumor response between the two groups studied in the same laboratory. Also, the incidence of premalignant metaplastic lesions (squamous cell and adenomatous lesions) in the lung was significantly increased only at doses exceeding 100–200 rad.[103] Even though the analyses are incomplete, it could be concluded that the tumor response over the dose range addressed in this study was nonlinear.

Experimentally induced mesotheliomas have been described in rats following intraperitoneal instillation of $^{239}PuO_2$ with morphogenesis sequelae similar to those observed for intraperitoneally instilled chrysotile asbestos. Pleural mesotheliomas have also been seen in rats and dogs following inhalation of transuranic elements.[24]

In rats, most of the nasal, laryngeal, tracheal, bronchial, and bronchiolar branches of the respiratory tract are lined by pseudostratified, ciliated and mucus-secreting goblet, and columnar epithelial cells, all of which, along with the alveolar epithelium, are relatively radioresistant. Respiratory epithelium has a relatively large capacity to repair sublethal radiation damage. The important progenitor cells in renewing bronchiolo-bronchial epithelium may be small mucus-containing cells rather than basal cells. These cells differentiate and proliferate during regeneration and preneoplastic, metaplastic renewal. The probable target cell for carcinoma formation in the rat lung is the peripheral terminal bronchiolar epithelium, possibly the nonciliated Clara cell. The alveolar endothelial cell is the target cell for hemangiosarcoma formation, the fibroblast for fibrosarcoma formation, and the pleural mesothelial cell for mesothelioma formation.[24]

Mice appear to be less susceptible to PuO_2-induced lung tumors than do rats.[51] Pulmonary fibrosis was increased at >6 months after an initial alveolar deposition of >4 Bq. There was also a decrease in total lung cellularity; the latter was partially compensated for by hyperplasia in less affected areas of the lung. Plutonium dioxide particles were markedly concentrated within fibrotic nodules in the lungs. In the mouse, most spontaneous lung tumors are either papillary adenomas or adenocarcinomas consisting of either cuboidal alveolar cells or columnar bronchiolar cells which occur multicentrically in the periphery of the lung late in life. The highest lung-tumor incidence was seen with the smallest $^{239}PuO_2$ particle size, indicating that the most homogeneous dose-distribution pattern is the most carcinogenic.[51]

Protraction of $^{239}PuO_2$ inhalation exposures in mice (bimonthly intervals for 1 yr to lung burdens of 0.5, 2.5, or 12.5 nCi) resulted in a greater incidence of pulmonary adenoma and adenocarcinoma than was observed with a similar radiation dose delivered after a single inhalation exposure.[57] Protraction in mice was calculated to increase the volume of exposed lung by threefold as compared to a single exposure. In contrast, protraction of $^{239}PuO_2$ exposure in Syrian hamsters[56] or rats[97] had no significant effect on lung-tumor incidence. However, the Syrian golden hamster appears to be resistant to the induction of lung tumors from inhaled radionuclides, although the hamster lung is relatively sensitive to lung-tumor induction by intratracheally instilled ^{210}Po in saline.[53] Although there are substantial spatial-temporal dose-distribution pattern differences for alpha irradiation from intratracheally instilled ^{210}Po as compared to inhaled PuO_2, this does not adequately explain the relatively few lung tumors induced in hamsters with comparable or higher doses of inhaled $^{238}PuO_2$ and $^{239}PuO_2$. Hamsters exposed to radon and radon decay products were also resistant to lung-tumor induction. Only two lung tumors were seen in 600 hamsters following inhalation of PuO_2, and these occurred only at lung doses of $>$1,000 rad.[94] No malignant lung tumors were seen in about 1,000 hamsters exposed to inhaled $^{238}PuO_2$, $^{239}PuO_2$, or $^{241}AmO_2$.[56,66] Intravenous injections of highly radioactive ^{238}Pu microspheres retained in the capillaries of the lungs resulted in very few lung tumors in hamsters.[4]

Available published data do not indicate that inhaled transuranic elements are associated with as high an incidence of respiratory carcinoma in nonhuman primates as that seen in rats and dogs. Although the study is not completed, there have been a few lung tumors in baboons more than 10 yr after inhalation of $^{239}PuO_2$. These were generally small lesions associated with regions of severe radiation pneumonitis. One baboon died because of a large epidermoid carcinoma at 2,528 days postexposure.[8] A single pulmonary fibrosarcoma has been found in a rhesus monkey 9 yr after inhalation of $^{239}PuO_2$.[33]

None of the results of the many animal experiments with inhaled transuranic elements have suggested an enhanced risk when the material is deposited in the form of discrete particles (hot particles) rather than dispersed throughout the tissue. The experimental results tend to support the concept that, at relevant levels of occupational and environmental exposures, a slightly greater risk may be associated with alpha-emitting transuranic elements dispersed throughout a tissue than concentrated in a few particles. An explanation for this observation is that the few cells containing or adjacent to the particles are more likely to receive killing doses than transforming doses of radiation, whereas the opposite would occur with a more diffuse distribution of the radioactivity among a much larger population of cells.[40,78]

The risk of cancer formation in the lungs or elsewhere following exposure to multiple or combined agents, including radionuclides, is poorly understood. Combined interactions may behave in an antagonistic, independent, additive, or synergistic manner. Cigarette smoking and alpha-irradiation interactions in uranium miners are examined in Appendix VII. Cigarette smoke depressed lung clearance of inhaled $^{239}PuO_2$ in rats and dogs,[28,29] but the effect on pulmonary carcinogenesis has not been studied. An additive tumor response was seen in rats that showed induction of abdominal sarcomas following intraperitoneal injection of $^{239}PuO_2$ and benzo(a)pyrene (BaP), a common carcinogenic hydrocarbon in cigarette smoke.[93] Intratracheally instilled BaP appeared to act synergistically with inhaled $^{239}PuO_2$ in causing lung cancer in rats.[63] Studies of the possible interaction of plutonium with other inhaled toxic substances such as asbestos, urethan, cadmium oxide, beryllium oxide, and nitrogen dioxide have produced results that are equivocal with respect to enhancing plutonium lung-cancer risks.

LIVER

Hepatocytes and the biliary epithelium are relatively radioresistant, although as a whole organ the liver is moderately radiosensitive. Early hepatocyte injury is not due to depressed hepatocyte proliferation since cell turnover in the intact liver is low. Chromosome damage in hepatocytes persists for long periods of time following irradiation and is seen only when hepatocytes are stimulated to proliferate, as by partial hepatectomy.

The liver retains transuranic elements with long biological half-times in some species, resulting in substantial radiation doses to liver over a normal life span. Liver tumors have been observed in some life-span studies of inhaled radionuclides in dogs and Chinese hamsters. The incidence of liver tumors in Chinese hamsters injected with ^{239}Pu was 39–47% at liver doses of 1,400–4,500 rad and 26–32% at 270–720 rad; liver tumor incidence was similar for injected ^{239}Pu citrate and $^{239}PuO_2$-labeled particles.[16] Intraperitoneally injected ^{241}Am citrate in deer and grasshopper mice resulted in an increased liver-tumor incidence; grasshopper mice were more sensitive than deer mice to tumor formation.[61] The longer life spans for these cricetid rodents than for the laboratory mouse or rat and the higher retention rate of transuranic elements in liver combined to give a relatively high liver-tumor yield. Liver-tumor risks were calculated to be 765 tumors/10^6 mice/rad to the liver for deer mice and 1,390 tumors/10^6 mice/rad to the liver for grasshopper mice.

Liver-tumor risk is much less following injection of transuranic compounds in rats or dogs and even rarer following inhalation exposure. The rapid loss of transuranic activity from the liver of rats may explain the low liver-tumor rates seen in this species. A significant increase in liver tumors has not been observed in life-span studies of beagle dogs that have inhaled any plutonium compound. However, bile duct carcinoma and a lesser number of sarcomas and fibrosarcomas have occurred in beagle dogs given ^{239}Pu or ^{241}Am citrate intravenously.[115] The primary tumors occurred after long latent periods and thus were seen only in the dogs that received doses of plutonium and americium that were sufficiently low to allow a long life span. Radiation doses to the liver of dogs that developed tumors were as low as 10 rad. There were 9 bile duct carcinomas and 2 hepatic cell carcinomas in 219 dogs given plutonium and 11 bile duct carcinomas and 3 hepatic cell carcinomas in 128 dogs given americium. The fact that liver tumors are rarely induced by inhaled transuranic elements in experimental animals does not negate a potential liver-tumor risk in humans.

BONE

The initial deposition of transuranic elements on bone surfaces is uneven. For example, in the femur of the dog the relative distribution of injected ^{239}Pu is 1.0 for the periosteum, 1.2 for haversian canals, 1.5 for epiphyseal areas, 2.6 for endosteal areas, and 3.0 for metaphyseal areas.[43] In humans, the ratio of periosteal to endosteal surface area is 8:100, which implies that, based on area, the endosteum is the most likely site of malignant change. Plutonium is deposited at higher concentrations in the vertebrae than in long bones. The amount and type of a transuranic elements on the bone surface, its residence time on the bone surface during which it is irradiating osteoblasts, the number of osteoblasts exposed to alpha particles, osteoblast migration and proliferation rate, and bone remodeling by osteoclastic and osteoblastic activities all alter the spatial-temporal dose-distribution pattern in bone and influence subsequent bone-tumor formation.

High doses of transuranic elements deposited in bone can result in pathological fractures, most frequently in the ribs.[79] A moderate but generalized osteoporosis is seen along with cortical thickening in the long bones. Growth stunting is seen in long bones of young animals given >3 μCi ^{239}Pu/kg.

A significantly increased incidence of bone tumors was estimated at 0.38%/rad to the bone in beagle dogs, 0.10%/rad to the bone in mice, and 0.06%/rad to the bone in rats following intravenous administration of ^{239}Pu citrate.[62] Spontaneous bone tumors occur so rarely in the rat that they cannot be taken into account.[30] In mice, the monomeric form of ^{239}Pu is about twice as effective as the polymeric form in producing bone sarcomas.[91] The St. Bernard dog is about 5 times more sensitive than the beagle dog to ^{239}Pu-induced bone sarcoma formation, but also has a higher spontaneous incidence.[114]

Inhaled $^{238}PuO_2$ and $^{239}Pu(NO_3)_4$ but not $^{239}PuO_2$, are potent inducers of bone tumors in dogs. After 15 yr postexposure, the skeletons of dogs had taken up only 1% of the initial alveolar-deposited ^{239}Pu from inhaled $^{239}PuO_2$, in contrast to 20% of initially deposited ^{238}Pu from inhaled $^{238}PuO_2$ after 12 yr[39] and 25% of ^{239}Pu from inhaled $^{239}Pu(NO_3)_4$ after 9 yr.[23] Twelve years after inhalation of $^{238}PuO_2$, a total of 31 dogs had osteosarcomas of 116 exposed dogs at cumulative skeletal doses ranging from 50 to 480 rad; 13 of the tumors originated in the vertebrae.[83] After 11 to 13 yr, bone sarcoma was the primary cause of death in 84 dogs; an initial lung burden of 0.02 μCi of ^{238}Pu/kg of body weight is the lowest dose at which fatal bone sarcoma has occurred in this, as yet, incomplete study.[68] Skeletal radiation doses for these dogs have not been reported. Inhalation of monodisperse 1.5-μm particles did not cause a bone tumor rate different from that of inhalation of monodisperse 3.0-μm particles.[66] Bone tumors were seen at skeletal doses ranging from 50 to 480 rad in dogs exposed to polydisperse $^{238}PuO_2$.[83] Bone tumors are not caused by inhaled insoluble $^{239}PuO_2$ because of its long retention time in respiratory tract tissues and low rate of translocation to bone.

Inhaled transuranic elements are not as carcinogenic in bones of rats as in those of dogs. The fractionation of inhaled $^{244}CmO_2$ over 10 exposures at 3-week intervals, starting at 70 days of age, resulted in an increase in the bone-tumor incidence of 27%, compared with 12%

in rats given a single exposure at that age.[98] Intratracheal instillation of $^{253}EsCl_3$ appeared to cause a higher bone-tumor incidence than that observed with inhaled $^{253}Es(NO_3)_3$.[10,11] Intratracheally instilled ^{239}Pu sodium triacetate resulted in a higher bone-tumor incidence (20%) than that observed with inhaled $Es(NO_3)_3$ (4%). Inhaled air-oxidized $^{239}PuO_2$,[96] high-fired $^{238}PuO_2$,[100] and single or protracted high-fired $^{238}PuO_2$[97,99] in rats and inhaled high-fired $^{238}PuO_2$ and $^{239}PuO_2$ in hamsters[94] did not induce bone tumors.

BONE MARROW

Although radiation leukemogenesis occurs in humans and experimental animals after and irradiation with x rays and gamma rays, it is not a significant finding after the internal desposition of alpha-emitting transuranic compounds, which concentrate more in bone than in bone marrow. The evidence from either experimental or epidemiological studies that plutonium or any other transuranic compound can induce leukemia is scanty.[120] Myeloid leukemia has been induced in CBA mice following injection of ^{239}Pu, but with a much greater yield of osteosarcoma.[35] Currently, on the basis of the experimental animal studies, no case can be made that transuranic elements are leukemogenic.

LYMPHOCYTES AND LYMPH NODES

The hematological effects of transuranic element deposition reflect irradiation of hematopoietic tissue associated with organs that concentrate transuranic elements, as well as direct irradiation of blood cells circulating through the lung, liver, and lymph nodes. Leukopenia occurs after inhalation of relatively large quantities of transuranic elements, for example, after inhalation of 4 to 10 μCi of ^{241}Am in dogs.[49] However, a reduction in the absolute number of lymphocytes in the circulating blood is the most sensitive hematological response to the deposition of transuranic elements in the respiratory tract. This has been an especially notable observation in dogs exposed to PuO_2. The time of onset and the degree of lymphocytopenia is dose-related following inhalation of plutonium dioxide.[89] Lymphocytopenia can be detected after pulmonary depositions of >0.7 nCi of PuO_2/g of lung. In contrast, the minimum lung burden required to produce a significant lymphocytopenia in dogs inhaling $^{239}Pu(NO_3)_4$ was 2 nCi/g of lung. A lung deposition of

>500 nCi of $^{239}PuO_2$ is required to cause a mean lymphocyte reduction of 50% in rats,[88] while a significant lymphocytopenia is seen in the rhesus monkey only at lung burdens of 900 to 1,800 μCi.[15] Lymphadenitis and replacement of parenchymal cells with scar tissue are common findings in regional lymph nodes nearest sites containing PuO_2 in dogs. A significant risk of primary tumors in lymph nodes containing very low to high concentrations of plutonium has not been demonstrated in rats, dogs, or humans.

Lymphocytes are among the most radiosensitive cells in the body, while reticular cells that act as a source of regeneration lymphocytes in lymph nodes along with macrophages, plasma cells, and antigen-stimulated lymphocytes are radioresistant. Alpha particles exhibit a relative biological effectiveness (RBE) of about 20 when compared to x rays in the production of dicentric aberrations in lymphocytes.[14] Chromosome aberrations have been quantified in blood lymphocytes obtained from monkeys that have inhaled $^{239}PuO_2$ and $^{239}Pu(NO_3)_4$; significant results were seen only at cumulative lung doses of >1,000 rad.[17,50] This suggests that the chromosome aberration frequency of lymphocytes of the monkey is an insensitive indicator of transuranic damage in the lung.

OTHER TISSUES

The deposition of inhaled or injected plutonium compounds in tissues other than lung, lymph nodes, liver, and bone is relatively small. In the relatively small mass of the mammary tissue of rats plutonium increased the incidence of mammary tumors;[59] however, it is not reported in other species. Damage to spermatogenic elements was observed at 5 months in rabbits injected with plutonium at testicular doses of 735 rem.[48]

Electrolyte imbalances, including hyperkalemia, hyponitremia, and hypochloremia, have been seen in dogs exposed to $^{238}PuO_2$-labeled aerosols.[83] These changes have been associated with hypoadrenocorticism (Addison's disease) in six dogs following inhalation of >4.5 nCi PuO_2/g of lung. The pathogenesis of the syndrome is not known.

Relatively high concentrations of americium were observed in the thyroids of beagle dogs.[107] Autoradiography showed that the ^{241}Am deposited primarily in the interfollicular areas. No adverse effects on thyroid function or on the incidence of thyroid disease were observed.

Effects in these other tissues are generally accompanied by much more severe effects in lung, lymph nodes, liver, or bone.

HUMAN EPIDEMIOLOGICAL STUDIES

Persons residing in the Northern Hemisphere have been exposed to very low levels of ^{239}Pu from atmospheric nuclear weapons testing in the 1950s and 1960s and to ^{238}Pu from an accidental disintegration of power sources after aborted spaceflights. However, the levels of plutonium and other transuranic elements deposited in the general population are well below those that might cause detectable health effects. Persons working with nuclear material have also been exposed to transuranic elements. But these, too, have been relatively small exposures; only a few accident victims received relatively high exposures. Since the beginning of the Manhattan Project in 1943, from 5,000 to 10,000 persons have been employed in positions in the United States involving risk of plutonium exposure. In a survey of 203 U.S. government contractor personnel who incurred internal deposition of plutonium between 1957 and 1970, 131 cases were contaminated by inhalation, 48 through wounds in the skin, 8 by both routes, and 16 through an unidentified route.[123]

Studies of employees of Rocky Flats Nuclear Weapons Plant and Los Alamos National Laboratory have been reported.[1,2,121–123,125] The most extensive report on the Rocky Flats employees was a mortality study of a cohort of 5,413 white males who were employed there for at least 2 yr and followed through 1979.[125] Individual radiation exposures were documented from health physics records based on periodic urine bioassays for plutonium and annual summaries of film badge readings for external radiation (gamma, neutron, beta, and x rays). Because systemic depositions of less than 2 nCi of plutonium are not measured reliably, only those workers with exposures of ≥ 2 nCi were considered exposed. Follow-up investigations identified the status of 98.9% of the cohort and located the death certificates for 99.9% of the deceased. Mortality from specific causes was evaluated in two ways. First, standardized mortality ratios were used to compare the observed deaths among the entire cohort versus the expected deaths based on U.S. rates. Second, the authors compared exposed with monitored unexposed workers (unmonitored workers were excluded) by stratifying on age and calendar period of death.

Analyses were also conducted separately by 2-, 5-, and 10-year periods of latency from the date that a worker reached 2 nCi of plutonium exposure or 1 rem of external radiation exposure.

The average external dose for the entire cohort was 4.13 rem, and the average plutonium burden was 1.75 nCi. Approximately 25% of the cohort was exposed to both 2 nCi or more and 1 rem or more. The mortality experience of the entire cohort was less than that expected based on U.S. mortality rates, with a standardized mortality ratio of 62 for all causes of death and 71 for all causes of cancer. The only significant excess risk was for the category of benign and unspecified neoplasms, with a standardized mortality ratio of 376.

To minimize biases, such as the healthy worker effect, comparisons of exposed to unexposed workers within the cohort were carried out. After plutonium exposure was lagged for 5 yr, total mortality and all lymphopoietic malignancy rates were slightly elevated.

No significant linearly increasing dose-response trends in risk with plutonium dose were found with a 2-, 5-, or 10-yr induction period for any causes of death or total mortality. Nevertheless, the authors concluded that this study suggested that plutonium-burdened individuals may experience increased risk of lymphopoietic neoplasms. This increased risk was based on four deaths, one each from lymphosarcoma/reticulosarcoma, non-Hodgkin's lymphoma, multiple myeloma, and myeloid leukemia (the last two are not usually categorized as lymphopoietic neoplasms). Lymphopoietic neoplasms have not been a common observation in the many studies of thousands of experimental animals treated with plutonium over a wide range of doses. The analysis showed no elevated risks for cancer of the tissues that show the highest concentration of plutonium in human autopsy cases and experimental animals, for example, lung, bone, and liver.

A smaller cohort of 26 former Los Alamos workers with the highest known plutonium concentrations at that facility in its early period of operation has been followed for 37 yr and repeatedly evaluated medically.[123] No increased risks attributable to plutonium exposure have been noted in this cohort. An investigation of cancer incidence among Los Alamos workers employed from 1969 through 1978 found no significant excess risks.[2]

RISK ESTIMATES

The limited human epidemiological studies of transuranic element deposition fail to demonstrate any unequivocal association of exposure with cancer formation at any anatomical location. Although clearly identified in experimental animals given plutonium, no significant lung-, bone-, or liver-cancer risk has been found in plutonium workers exposed 30 yr ago or more. Thus, these limited epidemiological studies do not indicate a cancer risk appreciably higher than that estimated from previous calculations made by United Nations Scientific Committee on the Effects of Atomic Radiation (UNSCEAR) or the Committee on the Biological Effects of Ionizing Radiations (BEIR). In the absence of adequate human epidemiological data, cancer risk for transuranic elements is usually estimated on the basis of human studies of other alpha-emitting radionuclides (e.g., uranium miners exposed to radon and its progeny, radium-dial painters, patients undergoing treatment with radium, or thorotrast-exposed patients) and of low linear energy transfer (LET) radiation exposures.

LUNG CANCER

In this report, risk estimates for lung cancer resulting from exposure to radon and radon daughters were obtained from analyses of data on occupationally exposed miners. The BEIR III Committee[77] also used human data to estimate risks from low-LET radiation. Data on humans exposed to transuranic elements are far too limited to permit useful quantification of risks. These data have shown no unequivocal evidence of risk resulting from such exposure, but these negative findings possibly resulted from small sample sizes and the limited magnitude of the exposures.

In the absence of directly relevant human data, there are at least two approaches that can be used to estimate risks. The first involves the use of estimated lifetime risks obtained from laboratory animal experiments. Difficulties with this approach relate to the many differences between animals and humans, including differences in histological types of cancers, differences in confounding exposures (e.g., smoking), differences in spontaneous risks, and differences in life span. The second approach involves expressing risks obtained from humans exposed to alpha radiation from radon decay products or to low-LET x and gamma radiation in terms of dose (or dose equivalent) to the lung or other relevant tissues, and then applying these risk estimates to the doses resulting from exposure to high-LET alpha radiation from transuranic elements. A difficulty with this approach is that there may be characteristics of specific exposures that are not fully reflected in a single dose estimate but that may affect resulting health effect risks. In particular, risks may depend on the specific cells of the lung that are irradiated. This may be quite different for transuranic element exposure than for exposure to radon decay products or low-LET radiation. In a report by the ICRP,[40] both approaches were utilized and the results were compared.

In Chapter 2 of this report, a model for estimating risks resulting from radon exposure was provided, based on analyses of data from four groups of miners. This model allows estimation of the lifetime risk resulting from exposure, expressed in working-level months (WLM), at any particular period in a lifetime. Risks resulting from exposures received during different periods of time throughout life can be summed to obtain an overall risk estimate for any specified sequence of exposures. The model specifically incorporates observed patterns of risk over time in miners, and also uses life-table methods to account for attrition of the population from death for reasons unrelated to radiation exposure.

A possible approach for estimating risks of exposure to inhaled transuranic elements would be to apply the model developed for radon exposure. This would require, as a minimum, conversion of the WLM to an appropriate measure of dose to the lung and would also require determination of the dose and its distribution over time resulting from any transuranic element exposure for which risk estimates were desired. Before such an approach can be applied, its validity needs to be confirmed by evaluating available laboratory animal data. Instances in which experiments involving both radon decay products and transuranic element exposure have been conducted in the same animal species are especially relevant for this purpose.

To evaluate the adequacy of the BEIR IV (Chapter 2, this volume) radon model for predicting risks in animals exposed to transuranics elements, data from relevant experiments need to be analyzed by methods that are comparable to those employed in analyzing data from epidemiological cohorts. In particular, it is not adequate to base analyses only on the proportion of animals that have developed lung tumors. Instead, the pattern of risk over time needs to be explicitly examined by modeling risk as a function of the exposure history as well as factors such as age at risk, age at exposure, and time since exposure. Such an approach allows explicit consideration of the time distribution of dose and also minimizes problems related to competing risks from bone tumors or other diseases resulting from the exposure.

The particular findings observed in the analyses of miners described in Chapter 2 of this volume with respect to the effect of age at exposure, time since exposure, and age at risk need to be checked using available laboratory animal data. Such patterns could be compared for different species and for the radionuclide involved in the exposure, whether it is radon or various transuranic elements. It must be recognized, however, that there are difficulties in examining time-related effects in animals in the same manner as in humans. These difficulties are related to the short life span, small numbers of animals (especially in canine laboratory experiments), and the lack of adequate data on time of occurrence in instances in which lung tumors are incidental findings and not the cause of death. Finally, it is unclear whether lung cancers induced in man by inhaled transuranic elements would occur in the lung periphery, as in rats and dogs, or in the bronchi, a tumor location rarely found in experimental animal studies with inhaled transuranic elements but a frequent site of cancer in human lungs.

It is also important to conduct analyses that allow quantitative comparison of risks resulting from different types of exposure in the same species. A method of analysis is needed that accounts for competing risks as well as different temporal and spatial patterns of dose. This can be accomplished by modeling the age- or time-specific relative risk as a function of the estimated cumulative dose to lungs. Such analyses could be useful for comparing risks due to radon with those due to exposures to various transuranic elements and might indicate ways in which the BEIR IV radon model (Chapter 2, this volume) would need to be modified to predict risks from transuranic element exposure.

In addition, analyses could be conducted that allow quantitative comparison of risks resulting from similar types of exposure, but in different animal species. Such comparisons should provide insights that are relevant to the use of laboratory animal data for estimating risks in humans. Comparisons of this type have been made and have generally been based on the proportion of animals that developed tumors. This approach may not be adequate if competing risks differ substantially in the species being compared; certainly, competing risks in humans are quite different from those in experimental animals. Another approach for making such comparisons is to examine the hazard or risk per unit of time. This hazard could be expressed either as a relative or absolute excess, and could also be used to estimate the probability of developing a lung tumor by time *t*, given survival to time *t*. Time might be expressed as a fraction of life span that would be similar for the species being compared. The choice of a measurement of risk that will provide the best comparison across species and thus be most appropriate for the extrapolation of risks from animals to man is required. Appropriate analyses of available experimental data could provide insights with regard to this issue.

Since the above methods have not yet been applied to experimental animal data for inhaled radionuclides, it is necessary to rely on risk estimates obtained by other methods, such as those used by the ICRP.[40] The available data on lung-tumor induction by alpha- and beta-gamma-emitters were nearly all from worldwide rat studies. The studies were conducted under different protocols and were complicated by varying methods of dose estimation, exposure (inhalation or intubation), and diagnosis of malignant tumors. While other factors were not constant (e.g., age at exposure), the lack of data on time and cause of death for individual animals precluded the use of much better models that incorporate this information. Because of these deficiencies, a data-selection scheme was devised, and the probit and weighted linear models were selected as two possible models to describe the available dose-response data, recognizing that both were probably inappropriate. Both the linear and probit models gave a reasonable description of the alpha-emitter dose-response data, while neither model was useful in mimicking the extremely variable beta-gamma-emitter data over the range of observed doses.

Risk estimates based on the Mantel-Bryan procedure were stated by the ICRP[40] in terms of the dose that causes 1 cancer/million animals. These estimates were 52, 14, 40, and 1,190 mrad to the lung for soluble alpha-, insoluble alpha-, all alpha-, and beta-gamma-emitting radionuclides, respectively. In contrast, an extrapolation of the linear model used by the ICRP[40] gave a dose estimate of about 3 mrad to the lung for all alpha-emitters, about 13 times higher than that resulting from use of the Mantel-Bryan procedure.

The risk estimate provided in this report (BEIR IV) from analyses of miners exposed to radon and radon daughters is 350 lung-cancer deaths/million persons/WLM. If expressed per rad, using a nominal value of 0.5 rad/per WLM, this estimate would be 700 cancer deaths/million person-rad and would be equivalent to a dose of 1.4 mrad causing 1 cancer/million persons. The estimated risk based on human radon data is within about a factor of 2 of the estimate obtained through linear extrapolation from animals exposed to transuranic elements but is considerably larger than the estimates obtained by using the Mantel-Bryan procedure. However, it should be recalled that the Committee's estimate for radon projects that most of the cancers occur in smokers. For nonsmokers, the risks are about a factor of 10 less.

BONE CANCER

Extensive human data on bone cancer from alpha irradiation are available from studies of about 1,700 people exposed to radium from 1910 to 1930 with a follow-up period of more than 55 yr; 54 bone cancers and 27 cancers of the paranasal sinuses and mastoids were found in this group by 1974.[77] Also, a large number of experimental animal studies with radium and other alpha-emitting radionuclides including transuranic elements have produced substantial data on bone cancer.

Data from several studies on the effect of internal deposition of two isotopes of radium and two isotopes of plutonium on bone-cancer death rates have been collected and summarized by the committee in an easily compared form. Annex 7A describes the 15 different data sets of quantities: n, the number of bone cancer deaths; N, the number of individuals; D, the total cumulative dose to the skeleton received by these individuals; and T, the total animal- or person-years of observation of the individuals by dose group within each study. These summary statistics are often available in the published papers that describe each study and are the minimum needed for each of 5 to 10 well-spaced dose-rate groups within each study. It is also necessary to assume that the dose rate is roughly constant over time and over animals within each dose-rate group. The summary of the radium-dial painter data contained only three broad dose-rate groups. Because of this, the analyses included here are intended more as examples of the proposed methodology than as definitive results.

The committee has applied a Bayesian methodology developed by DuMouchel and Harris[27] to estimate the bone-cancer risk in humans due to exposure to plutonium (see Annex 7A). Using the summary data tables, the committee fitted a linear dose-response model to the data from each study. This produced an estimate of the bone cancers per rad observed in each study, with an estimate of the within-study sampling variation attached to each slope. This approach allows the use of a Bayesian components of variance model to estimate how ratios of slopes from different studies differ by more than can be explained by the within-study sampling variation. However, there are indications that there may be no hope of extrapolating dose-response slopes more accurately than to a factor of 2 or 4. This would be true even if very good data on the effects of other isotopes on human bone-cancer rates and of plutonium on several animal systems were available. This question cannot be settled without gathering more data from other combinations of isotopes that act on biological systems. The Bayesian methodology employed here allows quantification and adjustment for prior uncertainty that is impossible to achieve when an approach to statistical inference based on frequencies is used.

In this regard, it is necessary to consider how each of the studies fits into the matrix of other studies already performed so that analyses of all the studies can be most informative. For example, one crucial hole in the array of studies available was that there were no measures of the effect of radium on bone cancer in rats. This gap prevented the analysis from making effective use of the several plutonium studies on rats. Similarly, the fact that all the radium studies on beagle dogs used the injected mode of dose administration, while most of the plutonium studies on beagles used inhalation as the mode of dose administration, introduced a prior uncertainty that lessened the accuracy of the analysis.

To summarize the tentative conclusions of the Bayesian analysis presented here, the potency of plutonium deposition in human bone is estimated to be 300 bone-cancer deaths/million person-rad received beyond a latency period of relatively little increased risk. The 95% confidence interval includes the range from 80 to 1,100 bone-cancer deaths/million person-rad. These values are 5 to 10 times higher than the corresponding estimates of the effects of two isotopes of radium. The chief contribution of this analysis is that it provides a more realistic appraisal of the interval of uncertainty.

Finally, published data on a few humans injected with plutonium were reanalyzed and integrated into the larger analysis. The analysis showed that these data are too meager to provide any important information on the bone-cancer effects of plutonium deposition.

LIVER CANCER

Although liver tumors have not been associated with any human exposures to transuranic elements, they have occurred in populations given Thorotrast (colloidal $^{232}ThO_2$) as a contrast medium in diagnostic radiology (see Chapter 5). Liver cancers have also been observed in experimental animal studies of transuranic elements, particularly those in which the animals were given transuranic compounds by intravenous injection. Because liver cancers appear to have a long latency period, the only animals at risk are those that have not succumbed to lung and bone cancers (which have a strong association with exposure to transuranic elements) or died of other causes.

Studies of dogs given alpha-emitting transuranic elements by intravenous injection have led to estimates of liver-cancer mortality risk in dogs of 920/million rad.[72] In Chapter 5, a risk estimate is derived for internally deposited Thorotrast of 260–300 fatal liver cancers/million person-rad. This suggests that either the effective dose from Thorotrast aggregates is less than the calculated value or that dogs may be about 3 times more sensitive to radiation-induced liver cancer than Thorotrast-exposed patients. Since there are no human data for transuranic elements and an acceptable method has not been developed for extrapolating the results from animal experiments to humans, it might be possible to apply the same risk estimate to transuranic elements in liver. Before this is done, careful consideration should be given to the differences between Thorotrast aggregates and deposits of transuranic elements, as well as to the uncertainties that are involved.

OTHER TISSUES

Among tissues irradiated by transuranic elements deposited in the body, only lymph nodes that drain regions containing deposits of transuranic particles are likely to receive radiation doses approaching or exceeding those received by lungs, liver, and bone. In spite of the large radiation doses received by thoracic, abdominal, and regional lymph nodes in thousands of experimental animals, there is little evidence of primary neoplasia. A few lymphatic vessel tumors and hemangiosarcomas have been observed in lymph nodes. Lymph nodes are relatively resistant to radiation carcinogenesis, and the committee has not attempted to derive a risk estimate for lymphatic tissue.

Risk estimates for transuranic elements are frequently applied to cancers known to originate in other tissues following irradiation from other sources such as external gamma and x radiation. For example, a risk of 400 leukemia deaths/million/rad of alpha radiation to the bone marrow and 400 deaths due to gastrointestinal tract cancer/million/rad of alpha radiation have been estimated for exposures to transuranic elements.[4] These tumors have not been identified as likely causes of death in animal experiments with transuranic elements or observed following human exposures. Thus, the validity of such risk estimates for transuranic element exposures is highly uncertain. In applying these or other risk estimates to transuranic elements, however, the most uncertainty may be in the calculation of the doses to the tissues. Dose calculations that may be appropriate for radiation protection purposes, for example, those by the ICRP[39] may be entirely misleading for projecting risks of cancer mortality from transuranic element exposures.

SUMMARY

The transuranic elements, which are produced in nuclear reactors, accelerators, and explosions of nuclear weapons and which are characterized by a predominance of isotopes emitting alpha radiations with energies ranging from 5 to over 8 MeV, are dominated quantitatively by plutonium, neptunium, and americium. The transuranic elements are not readily absorbed from the skin ($<5 \times 10^{-4}$). Absorption of transuranic compounds from the gastrointestinal tract at less than 1×10^{-4} may be increased to a level of 1×10^{-3} if incorporated into food products. Because of the short range of alpha radiation in tissues, the alpha-emitting transuranic elements are not a health concern unless they enter the body and deposit in radiation-sensitive tissues through wounds or the respiratory tract.

Insoluble transuranic compounds, primarily plutonium dioxide, are avidly retained in the lungs and the thoracic lymph nodes. Other plutonium compounds and essentially all compounds of other transuranic elements are more mobile when taken into the body through the respiratory tract or through wounds and are deposited in bone, liver, and, to a lesser extent, other tissues. Transuranic elements deposited in lungs, lymph nodes, bone, and liver are generally retained for a long time, frequently with half-times of many months or years. Distribution of transuranic elements within the tissues may be diffuse at first, but they often accumulate or form aggregates within cells or cellular structures. Particles and aggregates of transuranic elements, possibly mobilized by macrophages, may be deposited eventually in lymphatic or fibrotic tissues. In lungs, transuranic elements tend to accumulate in bronchiolar-alveolar and lymphatic structures in the parenchyma, frequently in regions of fibrosis. Transuranic particles are preferentially localized in paracortical and medullary regions of lymph nodes, which are also associated with fibrotic tissue. In liver transuranic elements localize in reticuloendothelial cells and in bone, primarily on the endosteal surfaces.

It is clear that transuranic elements are not homogeneously distributed throughout the body or throughout the tissues in which they are deposited. Further, since the range of alpha radiation in tissues is short, less than 100 μm, tissues in which they are deposited will be very nonuniformly irradiated. Only under conditions of very high deposition would more than a few percentage of the total cells in a tissue be exposed to alpha radiations, and many of these would receive doses more likely to kill than initiate neoplastic transformation. Thus, it is likely that, under most conditions, only a very small fraction of the alpha energy would be available for cancer induction. Nevertheless, the association of cancers in lungs, bone, and liver with the deposition of transuranic elements in these tissues in several animal species under experimental conditions but has not been demonstrated in several thousand human beings who have been accidentally exposed predominantly to low levels of transuranic elements.

Therefore, estimates of risk for transuranic elements cannot be derived from human epidemiological studies. Although risk estimates have been derived from experimental animals studies, they cannot readily be extrapolated to human. Until problems associated with this extrapolation are resolved, the only acceptable alternative is to apply risk estimates derived from studies of human populations exposed to other alpha-emitting radionuclides. For lung cancer the risk estimate is 700 lung-cancer deaths/million person-rad, based on the estimate for radon and its progeny. This value is about one-third larger than those that can be derived from current incomplete studies of plutonium in dogs. For bone cancer, the risk estimate is 80 to 1,100 bone cancer deaths/million person-rad from Bayesian analysis of human radium and animal transuranic and radium data. For liver, the risk estimate is 300 cancer deaths/million person-rad, based on human Thorotrast data. In applying these risk estimates to transuranic elements, their origin as well as the great uncertainties associated with their calculation should be remembered.

REFERENCES

1. Acquavella, J. F., G. L. Tietjen, and G. S. Wilkinson. 1981. Malignant melanoma incidence at the Los Alamos National Laboratory. Lancet i:883–884.
2. Acquavella, J. F., G. S. Wilkinson, and L. D. Wiggs. 1983. An evaluation of cancer incidence among employees at the Los Alamos National Laboratory. Pp. 338–345 in Proceedings of the 16th midyear topical Meeting of the Health Physics Society. CONF-830101, UC-41. Springfield, Va.: National Technical Information Service.
3. Allen, M. D., J. F. Briant, O. R. Moss, E. J. Rossignol, D. D. Mahlum, L. G. Morgan, J. L. Ryan, R. P. Turcotte. 1981. Dissolution characteristics of LMFBR fuel-sodium aerosols. Health Phys. 40:183–193.
4. Anderson, E. C. L. M. Holland, J. R. Prine, D. M. Smith, and R. G. Thomas. 1978. Current summary of intravenous microsphere experiments. In Biomedical and Environmental Research Program of the LASL Health Division. LA-7254-PR. Los Alamos, N.M.: Los Alamos Scientific Laboratory.
5. Bair, W. J. 1976. Recent animal studies on the deposition, retention and translocation of plutonium and other transuranic compounds. Pp. 51–83 in Proceedings of an IAEA and WHO Seminar, Diagnosis and Treatment of Incorporated Radionuclides. Vienna: International Atomic Energy Agency.
6. Bair, W. J. 1979. Metabolism and biological effects of alpha-emitting radionuclides. Pp. 908–912 in Proceedings of the Sixth International Congress of Radiation Research. Tokyo: Toppan Printing Co. Ltd.
7. Bair, W. J., and V. H. Smith. 1969. Radionuclide contamination and removal. Pp. 157–223 in Progress in Nuclear Energy, Series XII. Health Physics, Vol. II. New York: Pergamon.
8. Bair, W. J., H. Metivier, and J. F. Park. 1980. Comparison of early mortality in baboons and dogs after inhalation of $^{239}PuO_2$. Radiat. Res. 82:588–610.
9. Ballou, J. E., and J. O. Hess. Biliary plutonium exretion in the rat. Health Phys. 22:369-372, 1972.
10. Ballou, J. E., G. E. Dagle, and W. G. Morrow. 1975. The long-term effects of intratracheally instilled $^{253}EsCl_3$ in rats. Health Phys. 29:267–272.
11. Ballou, J. E., G. E. Dagle, R. A. Gies, and L. G. Smith. 1979. Late effects of inhaled $^{253}Es(NO_3)_3$ in rats. Health Phys. 37:301–309.
12. Breitenstein, B. D. 1983. Hanford americium exposure incident: Medical management and chelation therapy. Health Phys. 45:855–866.
13. Briant, J., and C. L. Sanders. In press. Inhalation deposition and retention patterns of U-Pu chain aggregate aerosol. Health Phys.
14. Brooks, A. L., R. J. LaBauve, R. O. McClellan, and D. A. Jensen. 1976. Chromosome aberration frequency in blood lymphocytes of animals with ^{239}Pu lung burdens. Pp. 106–112 in Radiation and the Lymphatic System, CONF-740930. J. E. Ballou, ed. Springfield, Va.: National Technical Information Service.
15. Brooks, A. L., R. J. LaBauve, H. C. Redman, J. L. Mauderly, W. H. Halliwell, and R. O. McClellan. 1976. Biological effects of $^{239}PuO_2$ inhalation

in the rhesus monkey. In Inhalation Toxicology Research Institute Annual Report for 1975–1976. LF-56. Albuquerque, N.M.: Lovelace Biomedical and Environmental Research Institute.

16. Brooks, A. L., S. A. Benjamin, F. F. Hahn, D. G. Brownstein, W. C. Griffith, and R. O. McClellan. 1983. The induction of liver tumors by ^{239}Pu citrate or $^{239}PuO_2$ particles in the Chinese hamster. Radiat. Res. 96:135–151.
17. Brooks, A. L., H. C. Redman, F. F. Hahn, J. A. Mewhinney, J. M. Smith, and R. O. McClellan. 1983. The retention, distribution, and cytogenetic effects of inhaled $^{239}Pu(NO_3)_4$ in the cynomolgus monkey. Pp. 283–287 in Inhalation Toxicology Research Institute Annual Report for 1982–1983. LMF-107. Albuquerque, N.M.: Lovelace Biomedical and Environmental Research Institute.
18. Bunzl, K., and W. Kracke. 1983. Fallout of $^{239/240}Pu$ and ^{238}Pu in human tissues from the Federal Republic of Germany. Health Phys. 44:441–448.
19. Casarett, G. W. 1980. Radiation histopathology, Vol. II. Boca Raton, Fla.: CRC Press.
20. Cohen, N., M. E. Wrenn, R. A. Guilmette, T. Lo Sasso. 1976. Enhancement of ^{241}Am excretion by intravenous administration of Na_3 (Ca-DTPA) in man and baboon. Pp. 461–475 in Proceedings of an IAEA and WHO Seminar, Diagnosis and Treatment of Incorporated Radionuclides. Vienna: International Atomic Energy Agency.
21. Craig, D. K., J. E. Ballou, G. E. Dagle, D. D. Mahlum, J. F. Park, C. L. Sanders, M. R. Sikov, and B. O. Stuart. 1978. Deposition, translocation, and effects of transuranic particles inhaled by experimental animals. Pp. 191–121 in Airborne Radioactivity. No. 710001. La Grange Park, Ill.: American Nuclear Society.
22. Graig, D. K., J. F. Park, G. J. Powers, and D. L. Catt. 1979. The disposition of americium-241 oxide following inhalation by beagles. Radiat. Res 78:455–473.
23. Dagle, G. E. 1987. Inhaled plutonium nitrate in dogs. Pp. 21–25 in Pacific Northwest Laboratory Annual Report for 1986 to the Department of Energy Office of Energy Research, Part. 1. PNL-6100. Richland, Wash.: Battelle Pacific Northwest Laboratory.
24. Dagle, G. E., and C. L. Sanders. 1984. Radionculide injury to the lung. Environ. Health Perspect. 55:129–137.
25. Dagle, G. E., R. W. Bistline, J. L. Level, and R. L. Watters. 1984. Plutonium-induced wounds in beagles. Health Phys. 47:73–84.
26. Diel, J. H., F. F. Hahn, and B. A. Muggenburg. 1986. Repeated inhalation exposure of beagle dogs to aerosols of $^{239}PuO_2$. X. Pp. 243–246 in Inhalation of Toxicology Research Institute Annual Report for 1985–1986. LMF-115. Albuquerque, N.M.: Lovelace Biomedical and Environmental Research Institute.
27. DuMouchel, W. H., and J. E. Harris. 1983. Bayes methods for combining the results of cancer studies in humans and other species, with discussion. J. Am. Stat. Assoc. 78:293-315.
28. Filipy, R. E.. 1982. Cigarette smoke and plutonium. Pp. 93–97 in Pacific Northwest Laboratory Annual Report for 1981 to the Department of Energy Office of Energy Research, Part 1. PNL-4100. Richland, Wash: Battelle Pacific Northwest Laboratory.

29. Filipy, R. E., J. L. Pappin, D. L. Stevens, and W. J. Bair. 1981. The impairment of pulmonary clearance of $^{239}PuO_2$ in rats by prolonged exposure to cigarette smoke. Pp. 110–111 in Pacific Northwest Laboratory Annual Report for 1980 to the Department of Energy Office of Energy Research, Part 1. PNL-3700. Richland, Wash: Battelle Pacific Northwest Laboratory.
30. Flitinov, N. N., and J. N. Soloviev. 1973. Tumors of the bone. Pp. 169 in Pathology of Tumors in Laboratory Animals, Vol. I. Tumors of the Rat, Part 1. Lyon, France: International Agency for Research on Cancer.
31. Fry, F. A. 1976. Long term retention of americium-241 following accidental inhalation. Health Phys. 31:13–20.
32. Guilmette, R. A., J. H. Diel, B. A. Muggenburg, J. A. Mewhinney, B. B. Boecker, and R. O. McCllelan. 1984. Biokinetics of inhaled $^{239}PuO_2$ in the beagle dog: Effect of aerosol particle size. Intern. J. Radiat. Biol. 45:563–581.
33. Hahn, F. F., A. L. Brooks, and J. A. Mewhinney. 1984. A pulmonary sarcoma in a rhesus monkey after inhalation of plutonium dioxide. Pp. 267-271 in Inhalation Toxicology Research Institute Annual Report for 1983–1984. LMF-113. Albuquerque, N.M.: Lovelace Biomedical and Environmental Research Institute.
34. Hodge, H. C., J. N. Stannard, and J. B. Hursh, eds. 1973. Uranium, plutonium, and transplutonic elements. Handbook of Experimental Pharmacology, Vol. 36. New York: Springer-Verlag.
35. Humphreys, E. R., J. F. Loutit, and V. A. Stones. 1985. The induction of myeloid leukemia and osteosarcoma in male CBA mice. Pp. 343–351 in Metals in Bone, N. D. Priest, ed Lancaster, England: MTP Press.
36. International Atomic Energy Agency (IAEA). 1979. Proceedings of an International Symposium on Biological Implications of Radionuclides Released from Nuclear Industries. Organized by the International Atomic Energy Agency, Vols. 1 and 2. Vienna: International Atomic Energy.
37. International Atomic Energy Agency (IAEA), United States Energy Research and Development Administration (ERDA). 1976. Proceedings of the Symposium on Transuranium Nuclides in the Environment. Organized by the U.S. Energy Research and Development Administration and the International Atomic Energy Agency. Vienna: International Atomic Energy Agency.
38. International Commission on Radiological Protection (ICRP). 1977. Recommendations of the International Commission on Radiological Protection. ICRP Publication 26. Oxford: Pergamon.
39. International Commission on Radiological Protection (ICRP). 1979. Limits for Intakes of Radionuclides by Workers. Part 1. ICRP Publication 30. Oxford: Pergamon.
40. International Commission on Radiological Protection (ICRP). 1980. Biological Effects of Inhaled Radionuclides. ICRP Publication 31. Oxford: Pergamon.
41. International Commission on Radiological Protection (ICRP). 1986. The Metabolism of Plutonium and Related Elements. ICRP Publication 48. Oxford: Pergamon Press.
42. International Nuclear Safety Advisory Group. 1986. Summary Report on the Post-Accident Review Meeting on the Chernobyl Accident. Safety Series No. 75-INSAG-1. Vienna: International Atomic Energy Agency.

43. Jee, W. S. S. 1971. Bone-seeking radionuclides and bones. Pp. 186 in Pathology of Irradiation. Baltimore: The Williams & Wilkins Co.
44. Kashima, M., H. Joshima, and O. Matsuoka. 1978. Relationship between physico-chemical form of plutonium and its behavior in tissues and effects on reticuloendotherial system in mice. Distribution patterns of monomeric and polymeric Pu after subcutaneous injection in mice. Nippon Acta Radiol. 38:65, 992.
45. Kawamura, H., and G. Tanaka. 1983. Actinide concentrations in human tissues. Health Phys. 44:451–456.
46. Khodyreva, M. A., R. Y. Sitko, G. M. Parkhomenko, and V. A. Sarychev. 1976. The effect of organic solvents on ^{241}Am penetration into the organism via the skin. Gig Sanit. 9:45–49.
47. Khodyreva, M. A., R. Y. Sitko, A. V. Simakov, and N. A. Andreeva. 1977. Distribution of alpha radiators in the skin and body in various types of contamination. Gig. Sanit. 8:57–61.
48. Koshurnikova, N. A. 1961. The histopathology of sex glands of rabbits under the action of incorporated plutonium. Pp. 164–173 in Biological Effects of Radiation and Problems of Radioactive Isotope Distribution (AEC Translation AEC-tr-5265), A. V. Tebedinskii and Y. I. Moskalev, eds. Moscow: Atomizdat.
49. Kudasheva, N. P. 1972. Changes in the dog's blood system following damage due to inhalation of americium-241. Pp. 455–459 in Biological Effects of Radiation from External and Internal Sources (AEC-tr-7457, 1974), Y. I. Moskalev and V.S. Kalistratova, eds. Moscow: Meditsina.
50. LaBauve, R. J., A. L. Brooks, J. L. Mauderly, F. F. Hahn, H. C. Redman, C. Macken, D. O. Slavson, J. A. Mewhinney, and R. O. McClellan. Cytogenetic and other biological effects of $^{239}PuO_2$ inhaled by the rhesus monkey. Radiat. Res. 82:310.
51. Lambert, B. E., M. I. Phipps, P. J. Lindop, A. Black, and S. R. Moores. 1982. Induction of lung tumors in mice following the inhalation of $^{239}PuO_2$. In Radiological Protection—Advances in Theory and Practice. Third SRP International Symposium, Vol. 1. Berkeley, England: Society for Radiation Protection.
52. Larsen, R. P., and R. D. Oldham. 1977. Plutonium in drinking water: Effects of chlorination on its maximum permissible concentration. Science 201:1008–1009.
53. Little, J. B., A. R. Kennedy, and R. B. McGandy. 1975. Lung cancer induced in hamsters by low doses of alpha radiation. Science 188:737–738.
54. Lloyd, R. D., D. R. Atherton, C. W. Mays, S. S. McFarland, and J. L. Williams. 1974. The early excretion, retention and distribution of injected curium citrate in beagles. Health Phys. 27:61–68.
55. Luders, C.J., and K.G. Themel. 1954. Die narbenkrebse der lungen als beitrag zur pathogenese des peripheren lungenkarzinoms. Virchows Arch. Pathol. Anat. 325:499.
56. Lundgren, D. L., F. F. Hahn, A. H. Rebar, and R. O. McClellan. 1983. Effects of the single or repeated inhalation exposure of Syrian hamsters to aerosols of $^{239}PuO_2$. Int. J. Radiat. Biol. 43:1–18.
57. Lundgren D. L., N. A. Gillett, F. F. Hahn, and R. O. McClellan. 1985. Repeated inhalation exposure of mice to aerosols of $^{239}PuO_2$. Pp. 259–264 in Inhalation Toxicology Research Institute Annual Report for 1984–1985.

LMF-114. Albuquerque, N.M.: Lovelace Biomedical and Environmental Research Institute.

58. Mahlum, D. D., J. O. Hess, and M. D. Allen. 1978. Tranlocation of mixed LMFBR fuel-sodium aerosols from the lung following inhalation by rodents. In Pacific Northwest Laboratory Annual Report for 1977, Part. 1. PNL 2500. Richland, Wash.: Battelle Pacific Northwest Laboratory.
59. Mahlum, D. D., M. R. Sirov, J. O. Hess, G. M. Zwicker, and D. B. Carr. 1979. Age and carcinogenesis of ^{239}Pu. Pp. 43–60 in Proceedings of an International Symposium on Biological Implications of Radionuclides Released from Nuclear Industries, Organized by the International Atomic Energy Agency, Vol. 1. Vienna: International Atomic Energy Agency.
60. Masse, R., D. Nolibe, P. Fritsch, H. Metivier, J. Lafuma, and J. Chretien. 1975. Chronic interstitial pneumonitis induced by internal-irradiation of the lung; value of the experimental model. In Alveolar Interstitium of the Lung, Pathological and Physiological Aspects, International Meeting. Prog. Respir. Res. 8:74.
61. Mays, C. W. 1982. Risk estimates for liver. Pp. 182–196 in Critical Issues in Setting Radiation Dose Limits, Proceedings of the 17th Annual Meeting of the National Council on Radiation Protection and Measurements. Bethesda, Md.: National Council on Radiation Protection and Measurements.
62. Mays, C. W., and R. D. Lloyd. 1972. Bone sarcoma incidence vs. alpha particle dose. Pp. 409–430 in Radiobiology of Plutonium. B. J. Stover and W. S. S. Jee, eds. University of Utah, Salt Lake City: J.W. Press.
63. Metevier, H., R. Masse, J. Wahrendorf, and J. Lafuma. 1986. Combined effects of inhaled plutonium oxide and benzo[a]pyrene on lung carcinogenesis in rats. Pp. 413–428 in Life-Span Radiation Effects Studies in Animals: What Can They Tell Us? Proceedings of the 22nd Hanford Life Sciences Symposium. CONF-830951. Springfield, Va.: Office of Science and Technical Information, U.S. Department of Energy.
64. Mewhinney, J. A., and J. H. Diel. 1983. Retention of inhaled $^{238}PuO_2$ in beagles: A mechanistic approach to description. Health Phys. 45:39–60.
65. Mewhinney, J. A., and B. A. Muggenburg. 1985. Comparison of Retention of ^{241}Am in immature young adult, and aged dogs and in monkeys after inhalation of $^{241}AmO_2$. Pp. 348–353 in Inhalation Toxicology Research Institute Annual Report for 1984–1985. LMF-114. Albuquerque, N.M.: Lovelace Biomedical and Environmental Research Institute.
66. Mewhinney, J. A., C. H. Hobbs, and R. O. McClellan. 1976. Toxicity of inhaled polydisperse or monodisperse aerosols of $^{238}PuO_2$ in Syrian hamsters. IV. Pp. 238–244; J. A. Mewhinney, C. H. Hobbs, and T. Mo, Toxicity of inhaled polydisperse or monodisperse aerosols of $^{241}AmO_2$ in Syrian hamsters. III. Pp. 251–258; and J. A. Mewhinney and C. H. Hobbs, Toxicity of Inhaled polydisperse aerosols of $^{239}Pu(NO_3)_4$ in Syrian hamsters. II. Pp. 259–262. All in Inhalation of Toxicology Research Institute Annual Report for 1975–1976. LF-56. Albuquerque, N.M.: Lovelace Biomedical and Environmental Research Institute.
67. Mewhinney, J. A., F. F. Hahn, B. A. Muggenburg, N. A. Gillette, J. H. Diel, J. L. Manderly, B. B. Boecker, and R. O. McClellan. 1985. Toxicity of inhaled $^{239}PuO_2$ in beagle dogs. A. Monodisperse 1.5 μm AMAD particles. B. Monodisperse 3.0 μm particles. XII. Pp. 226–235 in Inhalation Toxicology Research Institute Annual Report for 1984–1985.

LMF-114. Albuquerque, N.M.: Lovelace Biomedical and Environmental Research Institute.

68. Mewhinney, J. A., F. F. Hahn, B. A. Muggenburg, N. A. Gillett, J. H. Diel, J. L. Mauderly, B. B. Boecker, and R. O. McClellan. 1986. Toxicity of inhaled $^{238}PuO_2$ in beagle dogs: A. Monodisperse 1.5 μm AMAD particles. B. Monodisperse 3.0 μm particles. XIII. Pp. 215–225 in Inhalation Toxicology Research Institute Annual Report for 1985–1986. LMF-115. Albuquerque, N.M.: Lovelace Biomedical and Environmental Research Institute.
69. Morgan, A., A. Black, and S. R. Moores. 1984. Retention of ^{239}Pu in the mouse lung and estimation of consequence dose following inhalation of sized $^{239}PuO_2$. Radiat. Res. 99:272.
70. Morrow, P. E., F. R. Gibb, H. Davies, J. Mitola, D. Wood, N. Wraight, and H. S. Campbell. 1967. The retention and fate of inhaled plutonium dioxide in dogs. Health Phys. 13:113.
71. Muggenburg, B.A., S.A. Felicetti, and S.A. Silbough. 1977. Removal of inhaled radioactive particles by lung lavage—a review. Health Phys. 33:213–220.
72. Muggenburg, B. A., B. B. Boecker, F. F. Hahn, W. E. Griffith, and R. O. McClellan. 1986. The risk of liver tumors in dogs and man from radioactive aerosols. Pp. 556–563 in Life-Span Radiation Effects Studies in Animals: What Can They Tell Us? Proceedings of the 22nd Hanford Life Sciences Symposium. CONF-830951. Springfield, Va.: Office of Science and Technical Information, U.S. Department of Energy.
73. Muggenburg, B. A., R. A. Guilmette, F. F. Hahn, B. B. Boecker, and R. O. McClellan. 1986. Toxicity of inhaled $^{239}PuO_2$ in beagle dogs. A. Monodisperse 0.75 μm AMAD particles. B. Monodisperse 1.5 μm AMAD Particles. C. Monodisperse 3.0 μm AMAD particles. IX. Pp. 226–238 in Inhalation Toxicology Research Institute Annual Report for 1985–1986. LMF-115. Albuquerque, N.M.: Lovelace Biomedical and Environmental Research Institute.
74. Muggenburg, B. A., F. F. Hahn, N. A. Gillett, R. A. Guilmette, B. B. Boecker, and R. O. McClellan. 1986. Toxicity of $^{239}PuO_2$ inhaled by aged beagle dogs. VIII. Pp. 239–242 in Inhalation Toxicology Research Institute Annual Report for 1985–1986. LMF-115. Albuquerque, N.M.: Lovelace Biomedical and Environmental Research Institute.
75. Muggenburg, B. A., J. A. Mewhinney, R. A. Guilmette, D. L. Lundgren, F. F. Hahn, B. B. Boecker, and R. O. McClellan. 1986. Toxicity of inhaled alpha-emitting radionuclides—status report. Pp. 208–214 in Inhalation Toxicology Research Institute Annual Report for 1985–1986. LMF-115. Albuquerque, N.M.: Lovelace Biomedical and Environmental Research Institute.
76. National Council on Radiation Protection and Measurements (NCRP). Critical Issues in Setting Radiation Dose Limits. Proceedings of the 17th Annual Meeting of the National Council on Radiation Protection and Measurements. Bethesda, Md.: National Council on Radiation Protection and Measurements.
77. National Research Council, Committee on the Biological Effects of Ionizing Radiations (BEIR). 1980. The Effects on Populations of Exposure to Low Levels of Ionizing Radiation. Washington, D.C.: National Academy Press. 524 pp.

78. National Research Council, National Academy of Sciences. 1976. Health effects of alpha-emitting particles in the respiratory tract. EPA 520/4-76-013. Washington, D.C.: Office of Radiation Programs, Enviromental Protections Agency.
79. Nenot, J. C., and J. W. Stather. 1979. The toxicity of plutonium, americium and curium. Commission of the European Communities. Oxford: Pergamon.
80. Okada, S., M. Imamura, T. Terashima, and H. Yamaguchi, eds. 1979. Radiation Research. Proceedings of the Sixth International Congress of Radiation Research. Tokyo: Toppan Printing Co. Ltd.
81. Osanov, D. P. 1983. Dosimetry and Radiation Biophysics of the Skin. Moscow: Gosenergoizdat Publishers.
82. Osanov, D. P., E. B. Ershov, O. V. Klykov, and V. A. Rakova. 1971. Kinetics of dose distribution in structural layers of skin contaminated with radioactive materials. Health Phys. 20:559–566.
83. Park, J. F. 1987. Inhaled plutonium oxide in dogs. Pp. 5–20 in Pacific Northwest Laboratory Annual Report for 1986 to the Office of Energy Research, U.S. Department of Energy, Part 1. PNL-6100. Richland, Wash.: Battelle Pacific Northwest Laboratory.
84. Park, J. F., G. E. Dagle, H. A. Ragan, R. E. Weller, and D. L. Stevens. 1986. Current status of life-span studies with inhaled plutonium in beagles at Pacific Northwest Laboratory. Pp. 455–476 in Life-Span Radiation Effects Studies in Animals; What Can They Tell Us? Proceedings of the 22nd Hanford Life Sciences Symposium. CONF-830951. Springfield, Va.: Office of Scientific and Technical Information, U.S. Department of Energy.
85. Park, J. F. 1986. Inhaled plutonium oxide in dogs. Pp. 3–17 in Pacific Northwest Laboratory Annual Report for 1985 to the Department of Energy Office of Energy Research, Part 1. PNL-5750. Richland, Wash.: Battelle Pacific Northwest Laboratory.
86. Parker, H. G., A. de G. Low-Beer, and E. L. Saac. 1962. Comparison of retention and organ distribution of americium-241 and californium-252 in mice; the effect of in vivo DTPA chelation. Health Phys. 8:679–684.
87. Ragan, H. A. 1975. Enhanced plutonium absorption in iron-deficient mice. Proc. Soc. Exp. Biol. Med. 150:36–39.
88. Ragan, H. A. 1976. Hematologic effects of $^{239}PuO_2$ inhalation in rats. Pp. 107–108 in Pacific Northwest Annual Report for 1975, Part 1, to Division of Biomedical and Environmental Research, Environmental Research and Development Agency. BNW-2000. Richland, Wash.: Battelle Pacific Northwest Laboratory.
89. Ragan, H. A., R. L. Buschbom, J. F. Park, G. E. Dagle, and R. E. Weller. 1986. Hematologic effects of inhaled plutonium in beagles. Pp. 427–487 in Life-Span Radiation Effects Studies in Animals: What Can They Tell Us? Proceedings of the 22nd Hanford Life Sciences Symposium. CONF-830951. Springfield, Va.: Office of Scientific and Technical Information, U.S. Department of Energy.
90. Rhoads, K., J. A. Mahaffey, and C. L. Sanders. Dosimetry and response in rat pulmonary epithelium following inhalation of $^{239}PuO_2$. Pp. 59–65 in Current Concepts in Lung Dosimetry, Part I. CONF-802492. Springfield, Va.: National Technical Information Service.
91. Rosenthal, M. W., and A. Lindenbaum. 1969. Osteosarcomas as related to tissue distribution of monomeric and polymeric plutonium in mice. Pp.

371 in Delayed Effects of Bone-Seeking Radionuclides. Salt Lake City: University of Utah Press.

92. Sanders, C. L. 1969. The distrbution of inhaled $^{239}PuO_2$ particles within pulmonary macrophases. Arch. Environ. Health 18:904–912.
93. Sanders, C. L. 1973. Cocarcinogenesis of $^{239}PuO_2$ with chrysotile asbestos or benzopyrene in the rat abdominal cavity. Pp. 138–153 in Radionuclide Carcinogenesis. CONF-720505. Springfield, Va.: National Technical Information Service.
94. Sanders, C. L. 1977. Inhalation toxicology of $^{238}PuO_2$ in syrian golden hamsters. Radiat. Res. 70:334–344.
95. Sanders, C. L., and R. R. Adee. 1970. Ultrastructural localization of inhaled $^{239}PuO_2$ in alveolar epithelium and macrophages. Health Phys. 18:293–295.
96. Sanders, C. L., and J. A. Mahaffey. 1979. Carcinogeneicity of inhaled air-oxidized $^{239}PuO_2$ in rats. Int. J. Radiat. Biol. 35:95–98.
97. Sanders, C. L., and J. A. Mahaffey. 1981. Inhalation carcinogenesis of repeated exposures to high-fired $^{239}PuO_2$ in rats. Health Phys. 41:629–644.
98. Sanders, C. L., and J. A. Mahaffey. 1981. Influence of dose protraction and age on lung and bone tumorigenesis from inhaled $^{244}CmO_2$. Pp. 105–106 in Pacific Northwest Laboratory Annual Report for 1980, Part 1. PNL-3700. Richland, Wash.: Battelle Pacific Northwest Laboratory.
99. Sanders, C. L., G. E. Dagle, W. C. Cannon, D. K. Craig, G. J. Powers, and D. M. Meier. 1976. Inhalation carcinogenesis of high-fired $^{239}PuO_2$ in rats. Radiat. Res. 68:349–360.
100. Sanders, C. L., G. E. Dagle, W. C. Cannon, G. J. Powers, and D. M Meier. 1977. Inhalation carcinogenesis of high-fired $^{238}PuO_2$ in rats. Radiat. Res. 71:528–546.
101. Sanders, C. L., J. Mahaffey, J. M. Morris, and K. Rhoads. 1983. Inhaled transuranics in rodents. Pp. 65–67 in Pacific Northwest Laboratory Annual Report for 1982 to the Department of Energy Office of Research. Part 1. PNL-4600. Richland, Wash.: Battelle Pacific Northwest Laboratory.
102. Sanders, C. L., K. E. McDonald, B. W. Kelland, J. A. Mahaffey, and W. C. Cannon. 1986. Low-level inhaled $^{239}PuO_2$ life-studies in animals: What can they tell us? Pp. 429–449 in Life-Span Radiation Effects Studies in Animals: What They Tell Us? Proceedings of the 22nd Hanford Life Sciences Symposium. CONF-830951. Springfield, Va.: Office of Scientific and Technical Information, U.S. Department of Energy.
103. Sanders, C. L., E. S. Gilbert, K. E. Lauhala, J. A. Mahaffey, and K. E. McDonald. 1987. Low-level $^{239}PuO_2$ lifespan studies. Pp. 31–35 in Pacific Northwest Laboratory Annual Report for 1986, Part 1. PNO-6100. Richland, Wash.: Battelle Pacific Northwest Laboratory
104. Seidel, A., and V. Volf. 1972. Removal of internally deposited transuranium elements by Zn-DTPA. Health Phys. 22:779.
105. Smith, H., G. N. Stradling, B. W. Loveless, and G. J. Ham. 1977. The *in vivo* solubility of plutonium-239 dioxide in the rat lung. Health Phys. 33:539–551.
106. Stather, J. W., A. C. James, J. Brightwell, and P. Rodwell. 1979. Clearance of Pu and Am from the respiratory system of rodents after the inhalation of oxide aerosols of these actinides either alone or in combination with other metals. Pp. 3–25 in Proceedings of a Symposium on Biological Implications

of Radionuclides Released from Nuclear Industries. IEAE-SM-237. Vienna: International Atomic Energy Agency.

107. Stevens, W., B. J. Stover, F. W. Bruenger, and G. N. Taylor. 1969. Some observations on the deposition of americium-241 in the thyroid gland of the beagle. Radiat. Res. 39:201–206.
108. Stewart, K., D. M. C. Thomas, J. L. Terry, and R. H. Wilson. 1965. A preliminary evaluation of the biological measurements on operation roller coaster. AWRE-O-29/65. Atomic Weapons Research Establishment, Aldermaston, England.
109. Sullivan, M. F. 1980. Absorption of actinide elements from the gastrointestinal tract of neonatal animals. Health Phys. 38:173–185.
110. Sullivan, M. F. 1980. Absorption of actinide elements from the gastrointestinal tract of rats, guinea pigs and dogs. Health Phys. 38:159–171.
111. Sullivan, M. F., and L. S. Gorham. 1982. Further studies on the absorption of actinide elements from the gastrointestinal tract of neonatal animals. Health Phys. 43:509–519.
112. Sullivan, M. F., B. M. Miller, and L. S. Gorham. 1983. Nutritional influences on plutonium absorption from the gastrointestinal tract of the rat. Radiat. Res. 96:580–591.
113. Swint, M. J., and R. L. Kathren. 1986. In U.S. Transuranium Registry Annual Report, October 1, 1985–September 30, 1986. HEHF 54-86. Richland, Wash.: Hanford Environmental Health Foundation.
114. Taylor, G. N., G. B. Thurman, C. W. Mays, L. Shabesturi, W. Angus, and D. R. Atherton. 1981. Plutonium-induced osteosarcomas in the St. Bernard. Radiat. Res. 88:180–186.
115. Taylor, G. N., C. W. Mays, M. E. Wrenn, L. Shabestari, and R. D. Lloyd. 1986. Incidence of liver tumors in beagles with body burdens of ^{239}Pu or ^{241}Am. Pp. 268–285 in Life-Span Radiation Effects Studies in Animals: What Can They Tell Us? Proceedings of 22nd Hanford Life Sciences Symposium. CONF-830951. Springfield, Va.: Office of Scientific and Technical Information U.S. Department of Energy.
116. Thompson, R. C. 1982. Neptunium—the neglected actinide: A review of the biological and environmental literature. Radiat. Res. 90:1–32.
117. Thompson, R. C., and W. J. Bair, eds. 1972. The biological implications of the transuranium elements. Proceedings of the 11th Hanford Biology Symposium. Health Phys. 22
118. Thompson, R. C., and J. A. Mahaffey, eds. 1986. Life-Span Radiation Effects Studies in Animals: What Can They Tell Us? Proceedings of the 22nd Hanford Life Sciences Symposium. CONF-830951. Springfield, Va.: Office of Scientific and Technical Information, U.S. Department of Energy.
119. United Nations Scientific Committee on the Effects of Atomic Radiation (UNSLEAR). 1982. Ionizing Radiation: Sources and Biological Effects. Report E.82.IX.8. New York: United Nations.
120. Vaughan, J. Plutonium—a possible leukamic risk. Pp. 691–705 in The Health Effects of Plutonium and Radium. Salt Lake City: J. W. Press.
121. Voelz, G. L., G. S. Wilkinson, J. F. Acquavella G. L. Tietjen, R. N. Brackbill, M. Reyes, and L. D. Wiggs. 1983. An update of epidemiologic studies of plutonium workers. In Proceedings of the International Meeting on the Radiobiology of Radium and the Actinides in Man. Health Phys. 44(Suppl 1):493–503.

122. Voelz, G. L., G. S. Wilkinson, and J. W. Healy et al. 1983. Mortality study of Los Alamos workers with high exposures to plutonium.. Pp. 318–327 in Proceedings of the 16th Midyear Topical Meeting of the Health Physics Society. CONF-830101, UC-41. Springfield, Va.: National Technical Information Service.
123. Voelz, G. L., R. S. Grier, and L. H. Hempelmann. 1985. A 37-year medical followup of Manhattan project plutonium workers. Health Phys. 48:249–259.
124. Watts, L. 1974. Clearance rates of insoluble plutonium-239 compounds from the lung. Health Phys. 29:53–59.
125. Wilkinson, G. S., G. L. Tietjen, L. D. Wiggs, W. A. Galke, J. F. Acquavella, M. Reyes, G. L. Volez, and R. J. Waxweiler. 1987. Mortality among plutonium and other radiation workers at a plutonium weapons facility. Am. J. Epidemiol. 125:231–250.
126. Wrenn, M.E., ed. 1979. Actinides in man and animals. Proceedings of the Snowbird Actinide Workshop. Salt Lake City: RD Press.
127. Wrenn, M. E., and R. L. Roswell. 1981. A review of ^{241}Am accumulation by man and an estimation of the carcinogenic risks. Pp. 443–453 in Actinides in Man and Animals. University of Utah, Salt Lake City: RD Press.

ANNEX 7A

A Bayesian Methodology for Combining Radiation Studies

INTRODUCTION

This study reviews and integrates several other studies in which the effect of high linear energy transfer (LET) radiation on the risk of bone cancer has been measured. The methods used are very similar to those described by DuMouchel and Harris.[1] The general goal is to enable the quantitative use of the results of animal studies for the estimation of human cancer risks from exposure to ionizing radiation, especially plutonium. The choice of bone cancer as an endpoint and of plutonium as the source of exposure for this study was made partially because of its inherent interest and because of issues of data availability and suitability.

DATA SETS USED

Since very little data exist on the long-term effects of plutonium deposition in humans, for purposes of risk estimation it becomes necessary to use data from different animal species exposed to different isotopes and chemical forms of plutonium and other internal alpha-emitters. Animal studies have been designed and carried out at several laboratories in various countries over the last few decades in an attempt to fill this information gap. In addition, it was anticipated that an epidemiological follow-up study of the radium-dial painters could provide a calibration point that could be used to scale the bone-cancer risk observed in animals exposed to internal alpha-emitters, yielding a calibrated bone-cancer risk estimate for humans exposed to plutonium. However, until now no formal statistical methods for integrating the data from all these studies have been proposed or applied in the literature. The problem is that the many different studies whose results need to be combined into a "meta-analysis" have differing data collection designs, differing sample sizes, and thus, differing probable sampling errors. Most importantly, they have differing degrees of relevance to the problem of estimating the risk from plutonium deposition in humans.

This section gives a brief description of all the data the committee used to obtain the bone-cancer risk estimate for plutonium deposition. The endpoint in all studies considered here is bone cancer. The data sets were obtained from different publications or, if unpublished, directly from the investigators. For some experiments, information on skeletal dose and survival time after exposure for each individual or individual animal was available. For other studies, only rough summary statistics for groups of individuals could be obtained. This inhomogeneity of available data posed some difficulties for the data analysis.

A total of 15 sets of data were assembled, reanalyzed as individual data sets, and then integrated into one meta-analysis. These 15 data sets came from fewer than 15 studies, because we separated data according to the isotope of radium or plutonium involved, even though in some studies they were published together. Table 7A-1 lists the biological system, the isotope, and the source of data for each of the 15 data sets.

A large group of data came from the studies with beagles at the University of Utah[7] and the Inhalation Toxicology Research Institute (ITRI) of the Lovelace Biomedical and Environmental Research Institute in Albuquerque.[2] These data are available in detail. The Utah studies included in this analysis used beagles injected with ^{239}Pu, ^{228}Ra, and ^{226}Ra. For most of the animals, we were able to obtain the mean skeletal dose from published reports.[7] For a few beagles, this information was not available. In these cases, we approximated the dose by scaling injected doses to that of other beagles with similar injected activities. The report[7] listed amount of injected activities, time to death, and cause of death for all dead beagles. For most beagles that were still alive at the data reporting date; the report listed the cumulative dose up to this point, and the time since injection.

TABLE 7A-1 Data Sets Used

Data Set	Biological System	Isotope	Data Source
1	Human (ingestion)	^{226}Ra	R. Schlenker, personal communication, 1986
2	Human (ingestion)	^{228}Ra	R. Schlenker, personal communication, 1986
3	Human (injection)	^{238}Pu	Rowland and Durbin[5,6]
4	Human (injection)	^{239}Pu	Rowland and Durbin[5,6]
5	Beagle dog (injection)	^{226}Ra	U. Utah, Radiobiol. Div.[4]
6	Beagle dog (injection)	^{228}Ra	U. Utah, Radiobiol. Div.[4]
7	Beagle dog (injection)	^{239}Pu	U. Utah, Radiobiol. Div.[4]
8	Beagle dog (inhalation, 1.5 μm)	^{238}Pu	ITRI[2]
9	Beagle dog (inhalation, 3.0 μm)	^{238}Pu	ITRI[2]
10	Beagle dog (inhalation)	^{238}Pu	J. Park, personal communication, 1986
11	Rat (inhalation)	^{238}Pu	ICRP, Table 6[3]; C. L. Sanders, personal communication, 1986
12	Rat (inhalation)	^{238}Pu	ICRP, Table 6[3]
13	Rat (inhalation)	^{239}Pu	ICRP, Table 6[3]
14	Rat (inhalation)	^{239}Pu	ICRP, Table 6[3]
15	Rat (inhalation)	^{239}Pu	ICRP, Table 6[3]

Identical information was available for the inhalation studies with beagles at ITRI.[2] We used the data on beagles exposed to $^{238}PuO_2$ (1.5 μm [AMAD]) and $^{238}PuO_2$ activity median aerodynamic diameter (3.0 μm AMAD) monodisperse aerosols. We did not use data in beagles exposed to ^{239}Pu aerosols since no osteosarcomas were observed in these animals. The absence of bone sarcomas in these beagles is due to the lower specific activity of ^{239}Pu, which did not cause a fragmentation of the aerosol particles, resulting in virtually no delivered dose to the skeleton.

From the Pacific Northwest Laboratory (PNL), we used data (J. Park, personal communication) on beagles exposed to $^{238}PuO_2$ aerosols. Individual skeletal doses, survival time, and cause of death were also available for each animal in this study.

From Table 6 in ICRP Publication 31 from the International Commission on Radiological Protection (ICRP),[3] we obtained summary statistics on studies of rats exposed to ^{239}Pu citrate, ^{239}Pu ammonium pentacarbonate, ^{239}Pu nitrate, ^{238}Pu nitrate, and $^{238}PuO_2$. Additional dose information on the $^{238}PuO_2$ study was provided by C. L. Sanders (personal communication).

The data on the ^{226}Ra- and ^{228}Ra-dial painters were provided by R. Schlenker (personal communication). We used tabular information on the $^{226}Ra/^{228}Ra$ dose ratio, approximate dose rates, number of osteosarcomas, and number of person-years of follow-up. Only the data with a large or small ^{226}Ra to ^{228}Ra cumulative-dose ratio were used for this analysis.

The limited information on ^{238}Pu and ^{239}Pu in man was taken from reports by Rowland and Durbin.[4,5] We used survival time since exposure and cumulative alpha-ray doses to bone for all plutonium injection cases who lived more than 5 yr after injection. As discussed later in this annex, the limited nature of these data required that they be handled differently than the other data sets.

SEPARATE ANALYSES OF EACH DATA SET

STATISTICAL MODEL

For comparative purposes, it is necessary to perform parallel analyses of the data from each of the studies. Because data from the different studies are often only available in summarized form, the choice of analysis method is somewhat restricted. The data from each study are aggregated by dose group. In each study, all individuals within a given dose group are assumed to have been exposed to approximately the same dose rate (rads per day), which is assumed to be approximately constant over the period of observation of each individual. For each dose group in each study, four totals are collected: N is the number of individuals in the dose group, n is the number of deaths from bone cancer, T is the total person-days or animal-days of observation, and D is the sum of the cumulative doses in rads up to death, or time of last contact of each individual in the dose group.

A simple linear-effect model is used to relate the dependence of n on N, T, and D. For each individual, the hazard rate for bone-cancer death is assumed to be approximately equal to λ D/T, where the dose rate is set at D/T for each member of the group, and the parameter λ is the potency of the particular isotope when applied to the particular biological system and is measured in cancers per rad. Within each dose group, the number of cancer deaths, n, would then be approximately distributed as a Poisson variable with expectation:

$$E(n) = \lambda D(1 - \tau N/T). \qquad (7A\text{-}1)$$

The quantities λ and τ are possibly different for each experiment. The time period τ is the latency of bone sarcomas for a particular isotope within the given biological system and is measured in years. The fraction $(1 - \tau N/T)$ represents the approximate proportion of the years at risk, T, which fell before the beginning of the minimum latency period. Thus, the effective total dose to the dose group is D times this proportion, which is then multiplied by λ to produce the expected number of cancer deaths. (In all of the species considered here, the natural mortality rate from bone cancer is very low and will be assumed to be zero.)

For each experiment, the values of λ and τ are estimated by a Poisson regression analysis, which also produces approximate standard errors for these estimates. Although this statistical model is presumably only an approximation to reality, it is possible to compute a measure of goodness of fit of the model by comparing the observed and fitted cancer counts by using the usual likelihood ratio or Pearson chi-squared statistics. If there are K dose groups in an experiment, the chi-squared value has $K - 2$ degrees of freedom. Unfortunately, in most of the studies under consideration there were too few observed bone-cancer deaths to give the test of fit much power.

ADEQUACY OF THE MODEL

It is very unlikely that such a simple model holds exactly in each of the biological systems under consideration. The dose effect may not be linear in some or all species, the proper concept of latency is certainly more complicated than Equation 7A-1 indicates, metabolic differences between species produce different retention times and patterns of deposition, and the age-specific susceptibility to bone cancer may differ between species, among many other possibilities. However, the available data and the available scientific theory do not permit use of a more detailed statistical model. The chief merit of the proposed model is that it allows an assessment of how consistent is a natural and simple measure of carcinogenic potency, namely, the cancers per rad for each combination of an isotope acting on a biological system. Similarly, it is clear that bone-cancer deaths never appear immediately after exposure, so it is necessary to make some adjustment for latency. The model in Equation 7A-1 is simple, yet it is about as realistic as possible, given that only N, n, T, and D are available for each dose group. Use of this model does not deny that the mechanisms of carcinogenesis are different for different species; on the contrary, comparisons of estimated values of λ provide a way of assessing the magnitudes and patterns of these differences.

The model assumes that each individual receives a constant dose rate during the period of observation. This is a reasonable assumption in studies in which the dose comes from a one-time internal deposition of isotopes with long half-lives and clearance times. All the animal experiments were of this type.

The proper model to use for the radium-dial painter data is especially uncertain. It is noteworthy that Rowland et al.[6] report that a quadratic-exponential model fits these data much better than does a linear-effects model. However, the versions of the linear-effects model that Rowland et al.[6] fit differed in several respects from that which we proposed above in Equation 7A-1, so the question of appropriateness of our model is still unanswered until the radium-dial painter data can be analyzed in greater detail. (The data on the radium-dial painters have only three very broad dose groups, and are too broad for the assumptions of our analysis.) With these caveats in mind, the radium-dial painter data are included in the analysis.

Analysis of the data on the effects of plutonium in man is even more problematic. These data, presented and described by Rowland and Durbin,[4,5] consists of records on 18 individuals who were injected with one of the two isotopes of plutonium under consideration. Since none of these individuals has so far contracted bone cancer, the data by themselves can only provide a rough upper bound on the potency of internal deposition of plutonium in human bone. In addition, one statistical approximation that occurs in our methodology for combining the results of many studies requires that each included study have some bone-cancer cases. Therefore, the data on plutonium in man is not included in the initial analysis in which all studies are summarized. Instead, those data are used at a later stage of the analysis (see below).

The assumption that each individual in a dose group received exactly the average dose given to individuals in the group is not critical to the analysis. If individual doses were available, the analysis would gain estimating power. So long as there are at least five or six well-spaced dose groups, the loss of power is not appreciable.

The model also assumes that the minimum latency period, τ, is approximately independent of the dose level. There is some evidence of this based on examination of the data from the radium-dial

TABLE 7A-2 Results of Individual Analyses

Study	λ (tumors/1,000 rad)	Standard Error	τ (yr)	Chi-Squared	Degrees of Freedom
1	0.037	0.012	5[a]	3.4	—
2	0.088	0.063	11	0.7	1
3	—	—	—	—	—
4	—	—	—	—	—
5	0.42	0.09	2.6	1.8	6
6	0.73	0.12	1.8	5.8	5
7	5.4	0.9	3.5	20.3	8
8	4.5	1.1	2.7	1.8	6
9	3.1	0.8	2.4	5.7	7
10	8.2	2.3	3.6	2.7	4
11	0.81	0.47	0.5[a]	0.8	—
12	2.9	1.2	0.7	5.4	3
13	0.87	0.63	0.5	14.3	4
14	1.5	0.4	0.6	4.0	4
15	0.87	0.65	0.5	9.2	4

[a]The available data from this study did not permit a sensible estimate of the latency parameter τ. The value in the table was used instead as being more consistent with those from other studies. In these cases the goodness of fit statistic will not have an asymptotic chi-squared distribution and so the degrees of freedom column has been left blank.

painters, where the first cancers begin appearing at about the same time lag after the first exposure, regardless of the level of exposure.

Results of the Individual Analyses

Poisson regressions were performed on 13 of the data sets discussed earlier in this annex (excluding only the data on plutonium in man). Table 7A-2 presents the estimates of the parameter λ, their estimated standard errors, the estimate of τ, and the likelihood ratio goodness of fit statistics, with their degrees of freedom.

THE BAYESIAN MODEL COMBINING ALL STUDIES

The reanalyses of the individual data sets are only a necessary preliminary step to the real goal of providing a unified method of interpreting the entire ensemble of studies. To do this, a Bayesian framework, developed in detail by DuMouchel and Harris,[1] is applied. In that report they introduced a theoretical model and illustrated it with an extensive analysis combining 37 different studies on the carcinogenic and mutagenic potency of 10 different polyaromatic hydrocarbons in five different biological systems. The present analysis of radiation-induced bone-cancer studies follows closely the model of the earlier study. Readers interested in the theoretical development of the Bayesian model for combining studies should refer to that paper and its accompanying discussion.

COMPONENTS OF VARIATION

The principal goal of the model is to impose a formal theoretical structure on the previously ill-defined problem of interspecies extrapolation. In this effort, three ideas are central. First, there is a crucial distinction between the error of measurement within each dose-response study and the error of imperfect relevance among the studies. Second, the uncertain relevance between experiments can be formalized by reference to a hypothetical superpopulation model that generates all the experimental data. Relevance is then roughly quantifiable by the fit of the data to the underlying model. Third, the available biological and physical information about species differences, characteristics of isotopes, disease mechanisms, and the like enters the analysis in the form of prior assumptions about the parameters of the underlying model relating the experiments.

SUMMARY STATISTICS FOR EACH STUDY

As discussed by DuMouchel and Harris,[1] each of the separate dose-response studies is summarized by a single number, together with its estimated standard error. This number is the natural logarithm of the estimated dose-response slope. The slope was denoted by λ in the previous section; we now define $\theta = \log \lambda$. Let y denote the estimate of θ from a particular study. The standard error of y, denoted by c, is taken to be the coefficient of variation of the estimate of λ. That is, θ is log λ, y is the estimate of θ from a single experiment and equals the log (estimate of λ), and c is the estimate of the standard error of y and equals (standard error of estimate of λ)/(estimate of λ). The values of y and c for each of the separate studies can be computed directly from the first two columns of Table 7A-2. They are shown in Table 7A-3.

FORMAL MODEL

Let y_{ij} be the estimated log slope as taken from the ith row and jth column of Table 7A-3, which is assumed to be an unbiased estimate of θ_{ij}, the true log slope for the particular animal

TABLE 7A-3 Summaries of the Individual Studies

Biological System	Isotope[a] ^{226}Ra	^{228}Ra	^{238}Pu	^{239}Pu
Human	−3.30 (0.32)	−2.43 (0.72)	—	—
Beagle dog (injection)	−0.87 (0.21)	−0.32 (0.16)	—	1.69 (0.17)
Beagle dog (inhalation)	—	—	1.55 (0.15)	—
Rat	—	—	0.64 (0.34)	0.29 (0.24)

NOTE: Dashes in this table correspond to combinations of isotope and biological system with no available study. In case data from multiple studies were available for a cell, estimates were averaged, as if they were unbiased estimates of the same quantity, to produce one value of y and c in each such cell. The weight given to each y was inversely proportional to the square of its value of c, and values of c^{-2} were summed to obtain the value of c^{-2} for the average.

[a]Data are arranged as follows: y_{ij}, in units of log tumors/1,000 rads (c_{ij}, as estimated standard error of y_{ij}).

system-isotope combination. Conditional on θ_{ij}, y_{ij} is assumed to be normally distributed with standard deviation c_{ij}. Formally:

$$y_{ij}|\theta_{ij} \sim N(\theta_{ij}, c_{ij}{}^2). \tag{7A-2}$$

The values of θ have normal prior distributions that are independent, conditional on the values of further parameters:

$$\alpha_1, \ldots \alpha_4, \gamma_1, \ldots \gamma_4, \text{ and } \sigma:$$

$$\theta_{ij}|(\alpha, \gamma, \sigma) \sim N(\alpha_i + \gamma_j, \sigma^2). \tag{7A-3}$$

Finally, the parameters α_i, γ_j, and σ all have prior distributions as discussed below. The specification of these prior distributions, together with the values of y and c given in Table 7A-3, completes the specification of the formal Bayesian model. Equations given by DuMouchel and Harris[1] then provide final estimates and standard deviations for each of the θ_{ij} values, including those for which no corresponding y_{ij} value is available.

The crux of the Bayesian analysis is the specification of the prior distributions for α, γ, and σ. In order to do this, it is necessary to have a good understanding of the meanings of these parameters and their conceptual role in the analysis. These parameters, which are only used to specify the prior distribution of the θ_{ij} term, the parameters of direct interest, are called hyperparameters.

INTERPRETATION OF THE HYPERPARAMETERS

The prior mean of θ_{ij}, the log of the slope of bone-cancer risk versus dose, is, by Equation 7A-3, the sum of an average for the ith biological system, α_i, and an effect due to the jth isotope, γ_j. This additive model translates to a multiplicative model on the original scale. This means that the prior expectation is that the ratio of the carcinogenic potency of any two isotopes is preserved across species. The hyperparameter σ measures how well the actual θ_{ij} values conform to this prior expectation. A belief that σ is very near zero implies a belief that the relative potency of isotopes is almost exactly the same for every species. Larger values of σ imply more probability that some of the species systems have isotope-specific reactions to radiation.

The fact that the mean value of each θ_{ij} has an additive representation implies that we cannot identify a priori which biological system is most likely to exhibit a distinctive reaction to any particular isotope. The specific values of α_i measure the sensitivity of the ith biological system to the average isotope, while the specific values of the γ_j measure the average potency of the jth isotope across biological systems.

PRIOR DISTRIBUTIONS USED IN THE ANALYSES

DIFFERENCES BETWEEN BIOLOGICAL SYSTEMS

The values of α_1, α_2, α_3, α_4, respectively, represent the average log potency of the isotopes being considered, when dose is measured as rads to the skeleton, in the four biological systems. Except for the data now being analyzed, there is very little knowledge about these quantities. Therefore, it seems appropriate to choose prior distributions for the α_1 term that are very broad. These parameters are all assumed to have identical normal distributions, with a mean of 0 and a standard deviation of 10. (Note that all of the values of y are between -4 and $+2$. This shows that the data are much more precise than the assumed marginal prior distributions of α_i.) Although there is little prior information about the individual α_i, a predominant fact concerning them is that two of the four biological systems are in the beagle, the only difference being the mode of administration of the dose. As discussed above in this annex, the studies varied as to whether the dose was delivered by ingestion, injection, and inhalation. The inhalation method is a quite different pathway to the skeleton than is either ingestion or injection. The ingestion and injection experiments are designed to be as comparable as possible, and theoretically, their bone-cancer effects should be the same if the dose to the skeleton is the same. However, considering the difficulty in determining dose to the skeleton, metabolic differences, and other differences between the two groups of beagles, one cannot rule out the possibility that all of these differences together result in a systematic difference in the estimated potency of all isotopes. This possibility will be described by following the prior probability statement:

$$P(|\alpha_2 - \alpha_3| > 0.2) = 0.05. \tag{7A-4}$$

This states that there is only a 5% probability that the ratio of average potencies (tumors per rad to the skeleton) from the two modes of administration is greater than $e^{0.2}$ in favor of either mode. In terms of the assumption of normal distributions, this translates to an assumption that the standard deviation of $\alpha_2 - \alpha_3$ is 0.1. Since the variances of each α_i have been assumed to equal 100, this implies that the covariance of α_2 and α_3 is 99.995. To summarize, the prior distributions of the α_i are assumed to be normal with means of 0 and covariance matrix:

$$\begin{matrix} 100 & 0 & 0 & 0 \\ 0 & 100 & 99.995 & 0 \\ 0 & 99.995 & 100 & 0 \\ 0 & 0 & 0 & 100 \end{matrix}$$

The fact that the variances and covariances are chosen to be exactly 100 and 99.995 is not crucial here. The only important feature is that the standard deviation of $\alpha_2 - \alpha_3$ is 0.1, and the standard deviations of all other linear combinations of α_i are assumed to be very large. Any other prior distributions of α_i that have these features lead to almost exactly the same results.

DIFFERENCES BETWEEN ISOTOPES

Next, the prior distributions of γ_j are considered. These distributions represent the average differences (across species), on a log scale, of the potencies of the four isotopes under consideration. Here there is some scientific knowledge. If one really believed "a rad is a rad is a rad," then one would assume that every $\gamma_j = 0$. However, the possibility that the different isotopes have different potencies per rad to the skeleton will be assumed here. The hyperparameters γ_1 and γ_2 correspond to the isotopes ^{226}Ra and ^{228}Ra, respectively. The latter isotope has a much shorter half-life, a different decay chain with daughters that emit different radiations of different energies, and this may interact with the phenomenon of carcinogenesis in unpredictable ways. It is barely possible that either of the two radium isotopes is as much as twice as potent as the other. This will be stated probabilistically as:

$$P(|\gamma_1 - \gamma_2| > \log 2) = 0.05. \tag{7A-5}$$

Similarly, γ_3 and γ_4 correspond to the isotopes ^{238}Pu and ^{239}Pu, respectively. These two isotopes each have very long half-lives, and it is harder to find a rationale for the possibility of a consistent difference between these two isotopes. Accordingly, a ratio of potencies of 1.5 is barely possible here. Probabilistically,

$$P(|\gamma_3 - \gamma_4| > \log 1.5) = 0.05. \tag{7A-6}$$

Finally, compare the potencies of radium and plutonium. Here there is also scientific knowledge. Because it is known that plutonium concentrates more in the outer layers of bone cells than does radium, and because osteosarcomas also tend to originate in these layers of cells, the same dose to the skeleton of plutonium will tend to produce more tumors in all species than will radium. The relative potency of either isotope of plutonium to either isotope of radium is judged to be almost surely greater than 1 but less than 10. Probabilistically,

$$P(O < \gamma_j - \gamma_k < \log 10) = 0.95; j = 3, 4 \text{ and } k = 1, 2. \tag{7A-7}$$

If one uses the assumed normality of the prior distributions of γ_j, the above probabilities can be used to derive the means and covariance matrix of γ_j. The means are (−0.25 log 10, −0.25 log 10, 0.25 log 10, 0.25 log 10). The prior covariance matrix for the γ_j is:

0.166	0.106	0.0	0.0
0.106	0.166	0.0	0.0
0.0	0.0	0.166	0.145
0.0	0.0	0.145	0.166

Using the terminology of DuMouchel and Harris,[1] the values of Y, C, X, b, and V are now specified for the Bayesian analysis. The values of Y, C, and X are given in Table 7A-4.

TABLE 7A-4 Values of Y, C, and X

Biological System	Isotope	Y	C	X
Human	^{226}Ra	−3.30	0.32	1 0 0 0 1 0 0 0
Human	^{228}Ra	−2.43	0.72	1 0 0 0 0 1 0 0
Beagle dog (injection)	^{226}Ra	−0.87	0.21	0 1 0 0 1 0 0 0
Beagle dog (injection)	^{228}Ra	−0.32	0.16	0 1 0 0 0 1 0 0
Beagle dog (injection)	^{239}Pu	1.69	0.17	0 1 0 0 0 0 0 1
Beagle dog (inhalation)	^{238}Pu	1.55	0.15	0 0 1 0 0 0 1 0
Rat	^{238}Pu	0.64	0.34	0 0 0 1 0 0 1 0
Rat	^{239}Pu	0.29	0.24	0 0 0 1 0 0 0 1

TABLE 7A-5 Values of b and V

Hyperparameter	Prior Mean b	Prior Covariance Matrix V							
Human	0.0	100.0	0.0	0.0	0.0	0.0	0.0	0.0	0.0
Beagle dog (injection)	0.0	0.0	100.0	99.995	0.0	0.0	0.0	0.0	0.0
Beagle dog (inhalation)	0.0	0.0	99.995	100.0	0.0	0.0	0.0	0.0	0.0
Rat	0.0	0.0	0.0	0.0	100.0	0.0	0.0	0.0	0.0
^{226}Ra	−0.58	0.0	0.0	0.0	0.0	0.166	0.106	0.0	0.0
^{228}Ra	−0.58	0.0	0.0	0.0	0.0	0.106	0.166	0.0	0.0
^{238}Pu	0.58	0.0	0.0	0.0	0.0	0.0	0.0	0.166	0.145
^{239}Pu	0.58	0.0	0.0	0.0	0.0	0.0	0.0	0.145	0.166

The values of C in Table 7A-4 are estimated standard errors. They would be squared and then represented as a diagonal matrix to conform to the notation of DuMouchel and Harris.[1] The first four columns of X identify the four biological systems, while the last four columns of X identify the four isotopes. The corresponding values of b and V are given in Table 7A-5.

PRIOR DISTRIBUTION FOR σ

The value of σ determines how reliable the interspecies extrapolation is expected to be. From Equation 7A-3 each log potency, θ_{ij}, has prior mean $\alpha_i + \gamma_j$ and prior standard deviation σ. For any two biological systems, i and i', and any two isotopes, j and j', the linear combination $\Delta = \theta_{ij} - \theta_{i'j} - \theta_{ij'} + \theta_{i'j'} = \log\,(\lambda_{ij}/\lambda_{i'j})/(\lambda_{ij'}/\lambda_{i'j'})$, is assumed to be normally distributed with mean of 0 and a standard deviation of 2σ, conditional on σ. The interpretation of Δ is that e^{Δ} is the ratio by which the extrapolation of potency fails when the isotopes j and j' are compared for the pair of biological systems i and i', if individual potencies were perfectly measured. We judge that this extrapolation is highly unlikely to fail by more than a factor of 10. Probabilistically,

$$P(|\Delta| > \log 10) \approx 0.05. \tag{7A-8}$$

Now, conditional on σ:

$$P(|\Delta| > \log 10|\sigma) = 2[1 - \Phi(\log 10/2\sigma)], \tag{7A-9}$$

where Φ is the standard normal distribution function. Therefore,

$$P(|\Delta| > \log 10) = E\{2[1 - \Phi(\log 10/2\sigma)]\}, \tag{7A-10}$$

where E{} refers to the expectation with respect to the prior distribution of σ. If we assume that, a priori, σ takes each of the 10 values 0.05, 0.15, . . ., 0.95 with a probability of one-tenth, then expectation in Equation 7A-10 is, in fact, about 0.06, which is in agreement with the subjective assessment of P(| Δ | > log 10). Therefore, this prior distribution for σ is used in the Bayesian analysis.

RESULTS OF THE BAYESIAN ANALYSIS

Having defined the quantities Y, C, X, b, V, and the prior distribution of σ, it is now straightforward to use the procedures given by DuMouchel and Harris[1] to compute the posterior distributions of σ and θ_{ij}. When this is done, the posterior distribution of σ is $\pi(\sigma\,|Y)$, given by:

$\sigma =$ 0.05	0.15	0.25	0.35	0.45	0.55	0.65	0.75	0.85	0.95
$\pi(\sigma\|Y) =$ 0.241	0.212	0.166	0.122	0.087	0.062	0.043	0.030	0.021	0.015

Thus, although the prior distribution of σ was approximately uniform over the interval (0,1), the posterior mean of σ is 0.25 and $P(|\ \sigma < 0.5 \mid Y) = 0.83$. Roughly, this analysis suggests that σ is about half as large as was supposed a priori. Extrapolation on the basis of the comparison of two isotopes on each of two species is likely to be off by a factor of 3 to 5 rather than by a factor of 10. The posterior probability that a new extrapolation will be off by a factor of 10 or more is $P(\mid \Delta \mid > \log 10 \mid Y) = 0.18$, down from the value of 0.06 computed from the prior distribution.

TABLE 7A-6 Summary of Posterior Distributions after Combining Studies

Parameter of Posterior Distribution	Isotope			
	^{226}Ra	^{228}Ra	^{238}Pu	^{239}Pu
$E\{\theta\}$	−3.22	−2.81	−1.11	−1.12
Standard deviation $\{\theta\}$	0.30	0.46	0.65	0.65
$\lambda_{0.5}$	0.04	0.06	0.33	0.33
$\lambda_{0.025}$	0.02	0.03	0.09	0.09
$\lambda_{0.975}$	0.07	0.15	1.12	1.12

Table 7A-6 displays information on the posterior distributions of the potencies of each of the four isotopes in man. The last three lines of Table 7A-6 show the medians and 95% confidence limits of the potency, in bone cancers per 1,000 rad to the skeleton, of each of the isotopes. These limits were computed by assuming that the posterior distributions of θ ($= \log \lambda$) are Gaussian, as is approximately true. Note that the uncertainty ratios $\lambda_{0.975}/\lambda_{0.025}$ for the potencies of plutonium are greater than those for radium, since no direct data on the effects of plutonium in man have yet been incorporated into the analysis.

USING THE DATA ON HUMAN EXPOSURE TO PLUTONIUM

The posterior distribution for the effects of plutonium on man resulting from the Bayesian analysis described above, in which the data from Rowland and Durbin[4,5] were not used, is now used as the prior distribution for the analysis of those data. This second analysis proceeds as a Bayesian update by using the Poisson likelihood function of the data, namely

$$L(\lambda) = P(n = O|\lambda, N = 1, D, T) = \exp[-\lambda \Sigma_i D_i (1 - \tau/T_i)], \quad \text{(7A-11)}$$

where $D = (D_1, D_2, \ldots)$ and $T = (T_1, T_2, \ldots)$ are, respectively, the doses (in thousands of rads) and observation times (in years) for the individuals included in the studies. Each individual is considered as a group of size $N = 1$, and since no bone cancers were observed in these individuals, every $n = 0$. The value $\tau = 5$ yr was used as the latency parameter for this analysis. When the values of D, T, and τ are substituted into the above formula for $L(\lambda)$, it becomes:

TABLE 7A-7 Converting from Log Normal to Gamma Distributions

Parameter of Posterior Distribution	Isotope			
	^{226}Ra	^{228}Ra	^{238}Pu	^{239}Pu
$E\{\log \lambda\}$	−3.22	−2.81	−1.11	−1.12
Standard deviation $\{\log \lambda\}$	0.30	0.46	0.65	0.65
$E\{\lambda\}$	0.042	0.067	0.406	0.405
Standard deviation $\{\lambda\}$	0.013	0.032	0.294	0.293
Gamma c	10.3	4.3	1.9	1.9
Gamma d	246.1	64.1	4.7	4.7

$$\begin{aligned} L(\lambda) &= \exp(-0.449\lambda) \qquad (^{238}Pu), \text{ and} \\ &= \exp(-0.324\lambda) \qquad (^{239}Pu). \end{aligned} \tag{7A-12}$$

Under our model, the expected number of bone-cancer deaths among the individuals in the studies by Rowland and Durbin[4,5] would be 0.449λ and 0.324λ for those exposed to ^{238}Pu and ^{239}Pu, respectively. Since the previous analysis concluded that λ is probably less than 1, human bone-cancer death/thousand person-rad of plutonium exposure, there cannot be much further information in these data.

The previous analysis approximated the distribution of λ by a log-normal distribution. However, in order to combine this distribution with the exponential likelihood function given above, it is convenient to use a gamma-distribution approximation. For a given mean and variance of λ, the gamma distribution with the same first two moments as the log-normal distribution will be considered equivalent to it. The gamma density is:

$$G(\lambda; c, d) = k(c, d)\lambda^{c-1}\exp(-\lambda d), \tag{7A-13}$$

where $c > 0$ and $d > 0$ are parameters determining the mean and variance of λ, and $k(c,d)$ is a normalizing constant ensuring that the density integrates to unity over the range $\lambda > 0$. The mean of λ is c/d, while the variance of λ is c/d^2. When the distributions from Table 7A-6 are converted from log-normal to gamma representations, the resulting values of c and d are shown in Table 7A-7.

If λ has the $G(\lambda;\ c,d)$ density, the values of c and d have a simple interpretation. Bayesian probability intervals for λ then coincide numerically with the frequentist confidence intervals which would result if c cancers were observed in a population exposed to a total of d (times 1,000) rad (outside the latent period). Thus, the

TABLE 7A-8 Percentiles of λ

Isotope	^{238}Pu	^{239}Pu
$\lambda_{0.5}$	0.30	0.31
$\lambda_{0.025}$	0.08	0.09
$\lambda_{0.975}$	1.07	1.09

conclusions reached by this Bayesian analysis of the data in Table 7A-3, excluding the data on plutonium in man, are numerically very similar to those that might be reached by a non-Bayesian statistician who had observed 1.9 (i.e., about 2) bone-cancer deaths in a human population exposed to a total of 4,700 person-rad from plutonium exposure.

The $G(\lambda;\ c,d)$ prior distribution is mathematically convenient because it is easily updated: On observation of n cancers in a population exposed to a total of d' cumulative (latency-adjusted) thousands of rad, the posterior density of λ is $G(\lambda;\ c+n,\ d+d')$. In incorporating the data from Rowland and Durbin,[4,5] the values of n are 0 for both isotopes, while $d' = 0.449$ for ^{238}Pu, and $d' = 0.324$ for ^{239}Pu. The posterior gamma densities are therefore $G(\lambda;\ 1.9,\ 5.1)$ for ^{238}Pu and $G(\lambda;\ 1.9,\ 5.0)$ for ^{239}Pu. The mean values of λ then become 0.370 and 0.379, respectively. If the gamma distributions are converted back to the log-normal distributions with the same mean and variance, the percentiles of λ are as given in Table 7A-8. As can be seen by comparing the results given in Table 7A-8 with those in Table 7A-6, the use of the human plutonium data has very little effect on the distribution of λ.

CONCLUSION AND SUMMARY

Data from several studies on the effect of internal deposition of two isotopes of radium and two isotopes of plutonium on bone-cancer death rates have been collected and summarized in an easily compared form. The 15 different data tables of the quantities n, the number of bone-cancer deaths; N, the number of individuals; D, the total cumulative dose to the skeleton received by these individuals; and T, the total animal or person-years of observation of the individuals, by dose group, within each study were described. These summary statistics are often available in the published papers that describe each study, and they are the bare minimum needed to make any cross-study meta-analysis possible. The values n, N, D, and T are needed for each of 5 to 10 well-spaced dose-rate groups within each study. It is also necessary to assume that the dose rate is roughly constant over time and over animals within each dose-rate group. The summary of the radium-dial painter data that were available within the time frame of this analysis contained only three broad dose-rate groups. With more detailed data, the analysis could be improved.

Using these summary data tables, a linear-effects dose-response model was fitted to the data from each study. This produced an estimate of the bone cancers per rad observed in each study, with an estimate of the within-study sampling variation attached to each slope. This allowed the use of a Bayesian components of variance model to estimate by how much ratios of slopes from different studies differ more than could be explained by the within-study sampling variation. The posterior distribution of the parameter σ showed that, although the variation in the ratios of estimated slopes for different isotopes deposited in the same species could possibly be explained purely by within-study sampling errors, making extrapolation across species and isotopes potentially accurate, the fact that the value $\sigma = \log 2$ also had nonnegligible probability means that there may be no hope of extrapolating dose-response slopes more accurately than by a factor of 2 or 4, even if very good data on the effects of other isotopes on human bone-cancer rates and of plutonium on several animal systems are available. The question cannot be settled without gathering more data from other combinations of isotopes that act on biological systems.

In this regard, it would be greatly advisable for researchers to consider how their proposed studies fit into the matrix of other studies already performed, so that meta-analyses of all the studies can be most informative. For example, one crucial hole in the array of studies available was that there were no measures of the effect of radium on bone cancer in rats. This prevented the analysis from making effective use of the several plutonium studies on rats. Similarly, the fact that all the radium studies on beagles used the injection mode of dose administration, while most of the plutonium studies on beagles used the inhalation mode of administration, introduced a prior uncertainty, which lessened the accuracy of the Bayesian analysis.

The Bayesian methodology illustrated here allows a quantification and adjustment for prior uncertainty which is impossible to achieve by using the frequentist approach to statistical inference. The particular prior distributions of the hyperparameters $\{\alpha_1\}$, $\{\gamma_j\}$, and σ that were used in the analyses were intended to make use of as much scientific information as possible, in addition to the information contained in the data under analysis.

In summary, the Bayesian analysis presented here gives an estimate of the risk of bone cancer due to internally deposited plutonium of about 300 cancer deaths per million person rad for dose received beyond the latency period of very small increased risk. The 95% confidence interval includes the range from about 80 to 1100 bone cancer deaths per million person rad. These risks are 5 to 10 times larger than the estimated risks for ^{226,228}Ra in humans, but the interval of uncertainty determined here is considered to be realistic.

Finally, the published data on a few humans injected with plutonium were reanalyzed and integrated into the larger analysis. It was determined that these data are too meager to provide any important information on the bone-cancer effects of plutonium deposition.

REFERENCES

1. DuMouchel, W. H., and J. E. Harris. 1983. Bayes methods for combining results of cancer studies in humans and other species, with discussion. J. Am. Stat. Assoc. 78:293–315.
2. Inhalation Toxicology Research Institute (ITRI). 1985. Annual Report LMF-114. Albuquerque, N.M.: Lovelace Biomedical and Environmental Research Institute.
3. International Commission on Radiological Protection (ICRP). 1976. Biological Effects of Inhaled Radionuclides. ICRP Publication 31. Oxford: Pergamon Press.
4. Rowland, R. E., and P. W. Durbin. 1976. Survival, causes of death, and estimated tissue doses in a group of human beings injected with plutonium and radium. Salt Lake City: University of Utah Medical Center: The J.W. Press.
5. Rowland, R. E., and P. W. Durbin. 1978. The plutonium cases: An update to 1977. Oral presentation at the Scientific Group Meeting on Long-term Effects of Radium and Thorium in Man, Geneva: World Health Organization. (A summary can be found in pp. 138–141 in Radiological and Environmental Research Division Annual Report, ANL-78-65, Part II, Argonne, Ill., Argonne National Laboratory, 1978.)
6. Rowland, R. E., A. F. Stehney, and H. F. Lucas, Jr. 1987. Dose-response relationships for female radium dial workers. Radiat. Res. 76:368–383.
7. University of Utah, Radiobiology Division. 1983. Research in Radiobiology. Radiobiology Division Annual Report C00-119-258. Salt Lake City: University of Utah School of Medicine.

8
Genetic, Teratogenic, and Fetal Effects

GENETIC EFFECTS

INTRODUCTION

The health consequences of genetic damage that results from human exposure to low levels of ionizing radiation have been considered in the reports of the Committee on the Biological Effects of Ionizing Radiations (BEIR) [25,26] and in the reports of other national and international groups, such as the United Nations Scientific Committee on the Effects of Atomic Radiation (UNSCEAR).[41] The BEIR III[26] committee's estimates of the risks due to genetic damage were recently updated in a study for the U.S. Nuclear Regulatory Commission.[6] That study incorporated several modifications to the BEIR III estimates, including the adoption of equivalent induced mutation rates for the two sexes, the development of an X-linked mutation rate for humans, and the development of an estimate for induced numerical chromosomal abnormalities of nondisjunctional origin. In addition, rather than simply tabulating estimated increases in genetic effects of various categories, as in the 1980 BEIR III report,[26] the authors developed a computerized health-effects model (HEM) that used existing demographic data to predict health outcomes of radiation accidents through approximately the next five generations. They also incorporated the health impairment concept, developed in the 1982 UNSCEAR report,[41] to estimate the societal impact of radiation-induced genetic effects, as well as their numbers.

A task of the present BEIR committee (BEIR IV) is to estimate the genetic health consequences of human population exposures to alpha-emitting radionuclides. In the absence of any positive empirical human data, or even animal data on many of these radionuclides, estimates of genetic effects due to alpha-emitters must be based largely on estimates of the genetic health consequences of exposure to low linear energy transfer (LET) radiation. Estimates of the genetic consequences of low-LET radiation are based on experimental data (mostly on mice) and must be related to the few animal data on genetic effects due to high-LET alpha-particle radiation.

After some deliberation, the committee decided that attempting to derive new low-LET genetic-effect estimates or to update earlier estimates would be unwarranted. The BEIR III[26] estimates were thought to constitute a logical foundation on which to base the required genetic-effects estimates for alpha-emitting radionuclides. As noted in the BEIR III report, those estimates are numerically not very different from estimates that have appeared elsewhere, for example, in UNSCEAR reports. While the committee believes that updating of the BEIR III genetic-effects risk estimates for low-level human exposure to low-LET radiation might be desirable, the current committee was not constituted with the broad expertise required for such a revision. In addition, some of the issues that would have to be considered are controversial (e.g., the relative sensitivity of males and females to mutation induction and whether low to moderate doses of ionizing radiation induce nondisjunctional events).

Therefore, the committee bases its genetic-effects risks estimates for alpha-emitting radionuclides on the low-LET estimates provided in the BEIR III report.[26] We have, however, noted the influence that the adoption of the HEM[6] would have on the estimates of genetic risk in the BEIR III report. We have also adopted the demographic projection techniques developed for the HEM as a logical extension of the tabulations in the BEIR III report, so that we could project genetic effects into future generations. Because the demographic projections require numerical inputs, whereas the BEIR III estimate for chromosomal abnormalities was stated as "fewer than 10/million liveborn offspring at 1 rem/generation," we have arbitrarily used 9 as the upper limit and 1 as the lower limit for this endpoint.

TYPES OF GENETIC EFFECTS

The term *genetic effects of radiation*, as used here, means stable, heritable changes in the DNA of germ cells or their precursors. (Similar changes that occur in the DNA of somatic cells can be called genetic effects in a broad sense, but they cannot be passed on to future generations and thus do not constitute genetic effects in the sense of our concern.) Genetic effects can be grouped into two broad categories: (1) mutations and (2) chromosomal anomalies. These were once treated as separate classes, but recent work has demonstrated that so-called point mutations and chromosomal aberrations are the extremes of a continuous distribution of changes involving increasingly large portions of the genome. Chromosomal aberrations, usually defined as visible changes in the structure or number of chromosomes, have counterparts that constitute, for example, deletions and exchanges of segments of DNA; these segments are so small that they must be demonstrated by methods other than direct visualization.

Mutations can be grouped according to the mode of their phenotypic expression. If they are recessive, the alleles inherited from both parents must be mutant for them to be expressed. If they are dominant, only one copy of the mutant gene is required for phenotypic expression. Recessive mutations of genes on the X chromosome constitute a special case. Such mutant alleles are expressed in the male, in whom there is only one X chromosome (the hemizygous state), but not in the (heterozygous) female. Such a pattern of inheritance is termed *sex-linked.* Examples of human mutations with these three patterns of inheritance are albinism, inherited as a simple autosomal recessive; achondroplasia, inherited as a simple dominant; and hemophilia, which displays typical X-linked inheritance, being expressed in males but (usually) not in females. In practice, the three categories are not completely distinct. Some recessives have definite, although often different, effects in the heterozygote. Many dominants may have more severe phenotypic effects, or even different ones, in the homozygote than in the heterozygote. And many mutations classified as dominant might fail to be expressed at all in some heterozygous individuals—a phenomenon known as *incomplete penetrance.*

Many mutations induced by ionizing radiation, in both man and experimental animals (such as the laboratory mouse), are now recognized as chromosomal aberrations. In particular, many, perhaps most, apparent point mutations, when examined on the DNA level with the powerful new techniques of molecular biology, are now seen to be deletions of tens, hundreds, or even thousands of base pairs.

Gross chromosomal aberrations are in two distinct categories: chromosomal breaks and rearrangements, which are termed *structural aberrations*; and variations in the number of chromosomes characteristic of the species, a phenomenon resulting from chromosomal nondisjunction during meiosis and termed *aneuploidy*. Structural aberrations include simple deletions, inversions, translocations, and occasional, more bizarre types. In humans, a deletion of a specific segment of one chromosome produces the congenital abnormality known as cri-du-chat syndrome, and a specific translocation between two chromosomes is responsible for a form of renal cancer. Aneuploidy can consist of the absence of a chromosome or the presence of an extra copy of a chromosome. Aneuploidy involving the sex chromosomes is much more easily tolerated than aneuploidy of autosomes, which either are lethal or produce massive congenital defects if they involve any but the smallest autosomes. Down syndrome is a well-known example of the presence of an extra chromosome (trisomy); those affected have 47 chromosomes, including 3 of the small chromosome 21. Examples of sex-chromosome aneuploidy in humans include Turner syndrome, in which those affected have only 45 chromosomes (only 1 X chromosome), and Kleinfelter syndrome, in which those affected have at least 47 chromosomes (including 2 X chromosomes and 1 Y chromosome).

It is important to recognize that mutations of all types—point mutations as well as chromosomal abnormalities—occur spontaneously in humans without any radiation exposure other than the unavoidable ubiquitous background radiation. The most recent edition of McKusick's catalog of human mutations [23] lists 588 definite plus 710 probable recessives, 934 definite plus 893 probable dominants, and 115 definite plus 128 probable X-linked mutations, as well as many chromosomal abnormalities. In addition to these relatively simply inherited conditions, however, much human ill health has some heritable component, even though such conditions are not inherited in any clear-cut simple pattern. Such conditions are said to be irregularly or complexly inherited. Such inherited predispositions to many major human diseases (e.g., diabetes, schizophrenia, and cancer) are well known. Estimation of the numbers of added cases of such diseases that would be caused by irradiation of a human population constitutes one of the most complex and uncertain elements in the consideration of human genetic effects of radiation;

it involves highly subjective approaches to the genetic component of the diseases.

BEIR III ESTIMATES

The BEIR III Subcommittee on Genetic Effects used two somewhat independent methods for estimating human genetic risk: the so-called indirect and direct methods. These methods adhered to the following five principles, which were originally enumerated in the BEIR I report.[25]

> 1. Use relevant data from all sources, but emphasize human data when feasible. In general, when data of comparable accuracy exist, place greater emphasis on organisms closest to man.
> 2. Use data from the lowest doses and dose-rates for which reliable data exist as being more relevant to the usual conditions of human exposure.
> 3. Use simple linear interpolation between the lowest reliable dose data and the spontaneous or zero dose rate. In order to get any kind of precision from experiments of manageable size, it is necessary to use dosages much higher than are expected for the human population. Some mathematical assumption is necessary and the linear model, if not always correct, is likely to err on the safe side. . . .
> 4. If cell stages differ in sensitivity, weight the data in accordance with the duration of the stage.
> 5. If the sexes differ in sensitivity, use the unweighted average of data for the two sexes.

Because there are no positive human data, principle 1 meant that the bulk of the data used for the low-LET genetic effect estimates were on the laboratory mouse, the organism closest to man on which it was deemed practical to accumulate experimental data. Principle 3 provided a reasonable basis for conservative extrapolation, although many geneticists would argue that the mutation-induction curves for acute doses of low-LET radiation might best be fitted with a linear-dose square quadratic model (linear-quadratic), as was in fact done in the 1985 HEM.[6] In mammals, the longest-lasting cell stages during which the largest portion of received radiation would be absorbed are the immature resting oocyte in females and the spermatogonium in males. Many mouse mutation-rate data have been obtained on these stages and applied virtually directly to the derivation of human genetic-effect estimates. This makes principle 4 unnecessary. The degree, if any, of mutational sensitivity difference between males and females has been controversial. The authors of the 1985 HEM[6] decided to adopt equal mutation rates, although the BEIR III report[26] had adopted a female mutation rate (i.e., oocyte sensitivity) that was no more than 0.44 times the male (spermatogonial sensitivity), so application of principle 5 is not straightforward.

The indirect method for estimating genetic risk used by the BEIR III[26] Subcommittee on Genetic Effects was a mutation relative-risk method, as had been used in the BEIR I report.[25] The newer direct method was based on directly observed phenotypic damage induced in a single generation. The relative-risk method is based on the idea that, whatever the contribution of the current incidence to the fraction of genetically related ill health in the population caused by mutation, doubling that incidence ultimately doubles the incidence of genetically related ill health, if the increased mutation rate is maintained over enough generations to reach genetic equilibrium. If the amount of radiation required to double the mutation frequency is known (i.e., the doubling dose), the fractional increase in frequency at equilibrium due to any added radiation exposure can be calculated by using the reciprocal of the doubling dose, the relative risk of mutation per unit exposure. If the current incidence of genetically related ill health is known, the increase in incidence at equilibrium can easily be calculated. From several assumptions (hence, the term indirect), it is possible to extrapolate from the incidence at equilibrium to the incidence anticipated in the first generation after an increase in exposure. The BEIR III[26] Subcommittee on Genetic Effects adopted a range for doubling dose of 50–250 rem (relative risk of mutation, 0.004–0.02/rem) and a current incidence of 107,100/million liveborn offspring, of which 10,000 are thought to be expressions of autosomal dominant and X-linked genes, 1,100 expressions of recessive genes, 6,000 expressions of chromosomal aberrations, and the remaining 90,000 irregularly inherited.

The doubling-dose estimate of 50–250 rem translates into estimates, at genetic equilibrium, of 40–200 autosomal dominant and X-linked effects and 20–900 irregularly inherited effects/million liveborn offspring for a dose of 1 rem/generation until equilibrium is reached. For recessively inherited effects, the BEIR III Subcommittee on Genetic Effects made no numerical estimate, noting only that there would be a very slow increase in such effects. Nor did the subcommittee make a numerical estimate for chromosomal aberrations, noting that the numbers of effects would increase only slightly.

The direct method used by the BEIR III subcommittee depends on observations of phenotypic skeletal anomalies in first-generation offspring of irradiated mice. The subcommittee used estimates of the fraction of all phenotypically expressed anomalies affecting health that might be represented by the skeletal effects and some assumptions, for example, that dose-rate effects and male-female mutation rate differences would apply to this category of mutations, even though it had been measured only for recessive mutation induction. It concluded that between 5 and 65 effects might be expected/million liveborn offspring/rem. No separate estimate was made for irregularly inherited or recessive effects, as the effects in heterozygotes were expected to be included in the 5–65 autosomal dominant and X-linked effects. The numerical estimate for first-generation chromosomal aberrations, also a direct estimate, was based on direct cytogenetic observation of aberrations; fewer than 10 were estimated to result from a 1-rem parental exposure among a million liveborn offspring. As noted by the Subcommittee on Genetic Effects,[26] the estimate of 5–65 added genetic effects/million liveborn offspring/rem of parental radiation is in reasonable agreement with the estimate that would be derived by extrapolation from the equilibrium estimate based on the doubling-dose, or relative-risk, method with the same assumptions as were used by the BEIR I Subcommittee on Genetic Effects in 1972.[25]

It is important to recognize that the BEIR III genetic-effects risk estimates were derived from experiments with laboratory mice in which specific-locus mutations or first-generation skeletal anomalies were determined after exposure of the parental generation to relatively high doses of low-LET radiation, often at high dose rates. The doubling-dose estimates have been corrected for low-dose and low-dose-rate effects, essentially by reducing the slope of the linear downward extrapolation by a factor of 3; that factor was determined experimentally from observations that the spermatogonial specific-locus mutation rate decreased as a function of decreasing dose rate until a plateau was reached at about one-third of the high-dose-rate mutation rate (at about 0.01 rem/min). However, no such observations have been made for the oocyte or for the dominant skeletal mutations that provide the basis for the direct estimates. Whether this dose-rate effect results from the simple loss of a dose-squared component in a quadratic dose-effect curve has been much discussed; the dose-rate factor of 3, however, reflects the empirical observation and is thus independent of the debate over curve shape.

ESTIMATION OF GENETIC EFFECTS OF HIGH-LET RADIATION

A simple way to derive estimates of effects of high-LET radiation from estimates of effects of low-LET radiation is to multiply the low-LET estimate by a factor analogous to relative biological effectiveness (RBE) for the higher-LET radiation. To do this in a valid way, however, the dose-effect curves for both radiation qualities must be essentially linear in the dose range of interest. As detailed in Appendix II, such linearity is reasonable to assume in the case of high-LET radiation. But many geneticists believe that, for mutation induction and the induction of some kinds of chromosomal aberrations, the linear assumption is not a valid interpretation of the experimental high-dose-rate, low-LET results.

Nevertheless, if one considers how the BEIR III Subcommittee on Genetics Effects obtained its low-dose, low-LET genetic-effects estimates from data on experimental mice, it seems reasonable to use the RBE approach in estimating the genetic effects of alpha-emitting radionuclides. The subcommittee used the linearity assumption and simply extrapolated from the lower doses on which empirical mutation-rate data were available down to the 1-rem level. If, as seems likely, the true form of the dose-effect curves for acute doses of low-LET radiation is linear-quadratic—that is, has an important dose squared component—the error thus introduced would lead to overestimation of the genetic effects induced by low-LET radiation. This might raise a question about the validity of the low-LET estimates, but not about the procedure by which high-LET estimates are derived. The subcommittee's estimates were for a very low dose (1 rem) administered at a very low dose rate (accumulated over a 30-yr, one-generation span), and the application of a simple RBE-like proportionality constant still seems appropriate and could easily be used for revisions, if low-LET genetic-effects estimates are revised.

The estimates of the BEIR III Subcommittee on Genetic Effects were for a dose of radiation specified in rems. The rem is meant to make just the sort of allowance for the greater effectiveness of high-LET radiation that is outlined above. However, the factor Q, by which the dose in rads is multiplied to arrive at a number of rems, has two components. One is analogous to RBE; the other is meant to take into account nonuniform distribution of dose within the target organ and is addressed below. Instead of simply taking the BEIR III subcommittee's estimates of doses specified in rems and using them, it seems appropriate to make the adjustments for RBE and dose distribution separately, where possible, for each radionuclide of interest. Because the BEIR III subcommittee's estimates were derived entirely from experiments with low-LET x and gamma rays, we take these estimates to be valid for low-LET radiation specified in rads.

The validity of the application of a simple proportionality constant analogous to RBE to the present calculations depends entirely on the linearity of both dose-effect curves; the proportionality constant is simply the ratio of their slopes. Although, as already discussed, there is some question about the shape of the low-LET curve for the range of low doses and dose rates of interest here, the overwhelmingly predominant contribution to the dose curve would be linear in any case; the dose-squared component, if any, contributes nearly nothing at doses of only a few rads. It is clear (see Appendix II) that, even at much higher doses, the dose-squared component becomes vanishingly small if the dose rate is low enough—below about 0.01 rad/min for mutation induction in mouse spermatogonia. It is also clear that curves for high-LET radiation are essentially linear, at least in the lower-dose range where saturation is not a factor.

RBEs FOR THE INDUCTION OF GENETIC EFFECTS

Mutations

The BEIR III[26] and other estimates of genetic risk have been based mainly on the mutation rates measured by the specific-locus technique or the dominant skeletal-mutation technique in the mouse. It seems appropriate to base our derivation of an appropriate RBE factor on similar studies done with high-LET radiation. Unfortunately, there is not much information on mutation induction by internally deposited alpha-emitters measured with either system. Russell and Lindenbaum[35] used the specific-locus method to determine mutation rates in male mice into which ^{239}Pu was injected. They used what appeared to be an appropriate factor to describe the location of the plutonium and derived an RBE for alpha particles of only about 2.5, compared with low-dose-rate, low-LET radiation. This RBE is surprisingly low when compared with the RBE of about 17 obtained in similar experiments with neutrons[2,37] or with the even higher RBEs obtained for other endpoints. Several explanations may be offered: the location of the plutonium might be inappropriate or the mutations induced by the two radiation qualities might be quite dissimilar. There is considerable support for the latter explanation, both theoretical (see Appendix II) and empirical. Several lines of evidence suggest that mutations produced by ^{239}Pu are, on the average, qualitatively more deleterious than those induced by low-LET radiation: Most, if not all, are lethal when homozygous; many have marked deleterious effects in heterozygotes; and there is a greater loss of mutants induced by ^{239}Pu. Nevertheless, if one reason for the low measured RBE in the mouse specific-locus experiments with ^{239}Pu is indeed the very early loss of more serious mutations, then these would presumably have a smaller impact on human health, as in mice.

No positive information on induction of dominant skeletal mutations by alpha-emitting radionuclides is available. One small experiment in which male mice were given ^{239}Pu failed to yield mutations; thus, it provided some assurance, at least, that the rate of induction of such effects by an internally deposited alpha-emitter is not very large.

Chromosomal Aberrations

Abundant evidence from experiments with alpha-emitters and with neutrons indicate that RBEs for the induction of chromosomal aberrations in plant material or in somatic cells of mammals can be high. The ratio of linear slopes for high-LET radiation when compared with chronic exposure to gamma- or x-radiation tends to lie in the range of 10–20, with some values approaching 100.[5,27,30] Measurements made directly with the radionuclides of interest in this report include those with ^{238}Pu, ^{239}Pu, ^{241}Am, and ^{252}Cf, in comparison with chronic low-LET radiation RBEs were found to range from 10 to 40.[3,5,30]

Determinations of translocation induction in the mouse have been made both by direct observation of cytogenetically detectable translocations in primary spermatocytes and genetically by use of the heritable translocation test after exposure of spermatogonia to internally deposited ^{239}Pu. The results of the two tests are somewhat conflicting. The frequency of translocations as measured cytogenetically appeared to increase during the first several months after injection of the radionuclide, but then to decrease,[10] whereas no such decline was seen in heritable translocation tests.[8] RBEs for ^{239}Pu alpha particles were in the range of 10–20, compared with chronically administered gamma rays, depending on the alpha dose-distribution factor adopted.

Dominant Lethals

Dominant lethals are believed to result largely from chromosomal aberrations. The efficiency of their production has been evaluated for ^{239}Pu.[9,20,21] RBEs in the range of 10–15 were found.

Selection of RBE Values

In the absence of other information, it seems appropriate to adopt an RBE value for mutations (dominant X-linked and recessive mutations and those involved in the production of irregularly inherited genetic effects) of 2.5, as indicated by the mouse spermatogonial specific-locus information for ^{239}Pu. A higher value of 15 seems appropriate for the induction of chromosomal aberrations by ^{239}Pu in spermatogonia. The lack of any substantive information on these effects in females is unfortunate; nevertheless, since the major contributions to the numbers of genetic health effects estimated by the BEIR III Genetic Effects Subcommittee comes from the male, application of the male-derived plutonium-239 RBEs stated above does not seem inappropriate.

There is no direct information on any of the other radionuclides of interest, so there seems little choice but simply to adopt for their alpha particles the same RBE values as those for ^{239}Pu. Our confidence in doing so is reinforced by the somatic-cell observations cited above on ^{238}Pu, ^{241}Am, and ^{252}Cf.

Distribution Factors

The fraction of alpha-emitting radionuclides entering the body that can be expected to end up in the gonads is small, and the distribution within the male gonad in the laboratory mouse is nonuniform. The concentration in the interstitial tissue around the seminiferous tubules is higher than the average testicular concentration, so a dose to the genetically significant spermatogonia is 2–4 times larger than that predicted on the basis of uniform ^{239}Pu distribution in the organ.[4,11,35] Whether this distribution factor is appropriate for the human testis, however, is uncertain, because the primate (including human) testis has a different geometry from the rodent testis, with a greater ratio of interstitial to reproductive tissue and a larger separation between the seminiferous tubules.[4,14] It thus seems appropriate to apply a human distribution factor for ^{239}Pu of 1.0. In the absence of hard data on the other elements with alpha-emitting radionuclides of interest, there seems little choice but to apply the same factor of 1.0.

Nonuniformity of distribution of ^{239}Pu in the mouse ovary has also been observed.[12,13,35,38] Thus, genetic effects induced in females by this radionuclide might be less frequent than would be expected on the basis of uniform distribution of the whole ovary dose. Whether this is true in humans, however, is unknown; in any case, no attempt has been made to determine the ovarian ^{239}Pu distribution factor even for the mouse. It seems unavoidable that we adopt a value of 1.0 be adopted for the ovary and, for the same reasons as in the case of the testis, for all the alpha-emitting radionuclides concerned. Fortunately, any error in this assumption will have only a small effect on the genetic-effects estimates, because of the smaller contribution of female mutation to the estimates in the BEIR III report.[26]

GONADAL RADIONUCLIDE BURDEN

It appears from the work of Richmond and Thomas[32] that between 10^{-5} and 10^{-4} of the initial systemic burden of ^{239}Pu is retained in the mammalian gonad—either testis or ovary. The retention half-life of the ^{239}Pu deposited in the gonad is probably long, on the basis of measurements in mice, rats, Chinese hamsters, and cynomolgus monkeys. We are aware of no corresponding measurements of any of the other radionuclides of interest.

NUMERICAL ESTIMATES

Table 8-1 shows the numerical estimates of genetic effects of an average population exposure of the gonads to 1 rad of alpha particles per 30-yr generation. Table 8-1 is derived from the BEIR III report[26] using RBEs of 2.5 for gene mutations (dominant, X-linked, and recessive) and 15 for chromosomal aberrations. To facilitate calculation of effects in the manner used in the HEM[6] and the multiplication of the basic BEIR III[26] estimates by the appropriate RBE, the committee made several assumptions for cases for which BEIR III gave no numerical estimates. For chromosomal aberrations, BEIR III stated fewer than 10 cases for the first generation; the BEIR IV committee

TABLE 8-1 Genetic Effects of an Average Population Exposure to the Gonads of 1 rad of Alpha Particles per 30-yr Generation[a]

Type of Genetic Disorder	Current Incidence/Million Liveborn Offspring	Effect/Million Offspring	
		First Generation	Equilibrium
Single gene	10,000	13–163	100–500
Autosomal dominant		10–122	75–375
X-linked		3–41	25–125
Irregularly inherited	90,000	—	50–2,250
Recessive	1,100	Very few	Very slow increase
Chromosomal aberrations	6,000	15–135	15–135

[a]Derived from Table IV-2 of the 1980 BEIR III report[26] by application of an RBE of 15 for chromosomal aberrations and of 2.5 for other mutation classes as outlined in the text. Values for dominant and X-linked traits derived by assuming that one-fourth of the single-gene effects are X-linked.[6]

adopted the range of 1–9 additional cases in order to produce numerical estimates. BEIR III did not give separate estimates for autosomal dominant and X-linked conditions. The estimates contained in the HEM[6] review allotted roughly one-fourth of the total for the two classes to the X-linked conditions, and we have similarly apportioned the combined BEIR III estimates in preparing Table 8-1.

The HEM review[6] projected genetic effects into the future with computer programs written to project the incidences of dominant, X-linked conditions, translocations, and aneuploidy separately. No estimates were made for recessive conditions, because cases would not be expected until farther into the future than such projections can reasonably be made. Because BEIR III[26] made no estimate for aneuploidy, we present here only the results for the first three categories. The HEM[6] programs require normal demographic information as input. We have chosen the U.S. population from *Vital Statistics of the United States—1981*, Vol.1, *Natality* (U.S. Department of Health and Human Services, Public Health Service, Hyattsville, Md.) as a basis and derived the age-specific life tables for males and females (all races) in Table 8-2 and maternity by age of mother in Table 8-3.

For simplicity (and following the presentation of the HEM[6] review), we have made a single projection—using the geometric means of the range estimates, given in Table 8-1 $(x_1 x_2)^{1/2}$, to provide the required mutation-rate input—of incidence of the three classes of genetic effects over a 150-yr span. Table 8-4 shows the projections of genetic effects for an average gonadal exposure to 1 rad of alpha

TABLE 8-2 1981 U.S. Age-Specific Population and Births

Age	Males		Females		Births/Million
(yr)	U.S. Population/Million[a]	Stationary Population[b]	U.S. Population/Million[a]	Stationary Population[b]	Females[c]
<1	8,005	98,860	7,735	99,074	0
1–4	29,784	394,131	28,450	395,245	0
5–9	35,784	491,652	34,193	493,313	0
10–14	40,637	490,889	38,906	492,800	43
15–19	45,263	488,921	43,611	491,925	2,298
20–24	47,535	484,935	47,251	490,550	5,283
25–29	43,548	480,331	43,920	489,012	4,919
30–34	40,449	475,883	41,265	487,225	2,534
35–39	30,901	470,873	31,917	383,795	638
40–44	25,725	464,046	26,793	481,080	102
45–49	23,279	453,570	24,626	475,111	5
50–54	24,212	437,186	26,131	465,714	0
55–59	23,833	412,497	26,741	451,657	0
60–64	20,855	377,276	24,216	431,301	0
65–69	17,245	329,470	21,483	402,333	0
70–74	12,832	268,750	17,768	362,005	0
75–79	8,337	198,809	13,329	307,239	0
80–84	4,587	127,118	8,646	234,036	—
>85	3,074	98,167	7,203	255,217	—

[a]Estimated population of United States, by age and sex, all races, July 1, 1981, standardized to a population size of 10^6 from Table 4-2, Section 4, Technical Appendix, Vol. 1, Natality, Vital Statistics of the United States, 1981.

[b]Stationary population by age and sex, all races, for 100,000 male and 100,000 female births each year. From Table 6-1, Section 6, Life Tables, Vol. 2, Mortality, Vital Statistics of the United States, 1981.

[c]Estimated number of births by age group, all races. From Table 1-9, Section 1, Natality, Vol. 1, Natality, Vital Statistics of the United States, 1981.

TABLE 8-3 Live Births by Age of Mother, All Races, United States, 1981

Age of Mother (yr)	No. of Live Births[a]
<15	9,768
15-19	524,589
20-24	1,205,793
25-29	1,122,791
30-34	583,877
35-39	163,615
40-44	23,240
45-49	1,125
All ages	3,629,238

[a]Calculated from Vital Statistics of the United States—1981, Natality (1985) on the assumption that number of live births for mothers of all ages is accurate, whereas the birth rates by maternal age are subject to rounding errors.

particles of each generation (i.e., 170 mrad per 5-yr period for 30 yr) in each 5-yr interval of the entire 150-yr projection span—a situation like that for the equilibrium estimates given in Table 8-1.

FETAL EFFECTS, TERATOGENESIS, AND NEONATAL EFFECTS OF IN UTERO EXPOSURE

By definition, teratogenic effects can be produced only by injury received after conception. Although alpha-emitting radionuclides can be transmitted across a placenta and incorporated into the body of a developing fetus, only the alpha decays that occur during intrauterine life can cause teratogenesis. In the interests of completeness, the committee has included in this consideration of abnormal development both strictly teratogenic effects and those induced postnatally in the developing mammal.

The field of radiation teratogenesis has been reviewed many times.[16,34] Dose-effect curves for the induction of teratogenic effects by low-LET radiation are, with one possible exception, of the threshold type, and the effects themselves are nonstochastic. Effects include early embryonic death, neonatal or early postnatal death, gross malformations, local morphological defects or reductions, central nervous system defects (including mental retardation), and general growth retardation. The little information on human radiation teratogenesis is limited largely to high-dose low-LET radiation, generally delivered acutely. As with genetic effects, therefore, risk estimates

TABLE 8-4 Genetic Effects of Population Exposure to 1 rad of Alpha Particles to the Gonads per 30-yr Generation over Five Generations

	Cumulative Absolute No./~1,000,000 Population			
Generation	Years	Dominants	X-Linked	Translocations
1	5	0	0	0
	10	2	1	2
	15	2	1	2
	20	2	2	4
	25	4	3	4
	30	6	4	6
2	35	6	6	8
	40	8	7	10
	45	10	9	12
	50	10	12	14
	55	12	14	16
	60	14	17	19
3	65	15	19	21
	70	16	22	24
	75	18	25	26
	80	19	28	29
	85	20	31	31
	90	21	35	34
4	95	23	38	37
	100	24	41	39
	105	25	45	43
	110	27	48	45
	115	28	52	47
	120	29	55	50
5	125	31	58	53
	130	33	61	57
	135	34	63	61
	140	35	66	64
	145	37	68	68
	150	39	70	69

NOTE: Values are cumulative totals by 5-yr intervals to an approximately stable population of 1,000,000 persons.

must be based mainly on experimental animal data. Because the gestation periods of experimental animals differ so widely and are so much shorter than that of humans, estimates are based on the times at which comparable developmental events occur in the different species.

EFFECT OF EMBRYONIC DEVELOPMENT STAGE

The teratogenic effects of radiation exposure tend to be closely related to the stage in embryonic development at which the radiation dose is received. Three general periods of development are commonly recognized: preimplantation stage, between conception and the time of implantation of the developing embryo into the uterine lining in humans, generally taken as the period, days 0–9; the stage of major organogenesis, days 14–50; and the fetal period, days 51–280.

Preimplantation is the stage of greatest sensitivity to radiation. Nevertheless, only one effect has been noted in this stage: simple preimplantation loss. During this period, the early embryo is able to develop normally after losing one or more cells, but once cell loss reaches some point, practically any additional effect at all will be lethal. Only animal data are available, but they explain the observation that single doses of less than 10 rad of low-LET radiation produce no detectable effects. It is speculated that the principal reason for the absence of human data showing preimplantation teratogenesis is that the losses occur before the mothers know that they are pregnant and therefore go unnoticed. Theoretical considerations strongly suggest that preimplantation loss must be a nonstochastic effect, with a definite threshold. Exceeding this threshold requires that some minimal number of cells be killed.

During the major organogenesis stage, the embryo appears to be sensitive to all the known teratogenic effects of radiation. But the sensitivity of any one tissue or organ system is limited to a small fraction of the 37-day period in humans. Individual tissue or organ sensitivity appears to be limited to the few days of its critical early-stage development, when the loss of even a few cells cannot easily be remedied. Windows of one to a few days are commonly observed during which a given developmental abnormality can be induced during the major organogenesis stage. Thresholds are expected theoretically and have been observed; single doses below about 10 rad of low-LET radiation appear ineffective.

During the fetal period, only general morphological defects, reductions, and central nervous system effects can be induced. Most such effects are nonstochastic and have thresholds. One prominent effect on which there are human epidemiological data is characterized by a nonthreshold dose-effect curve. Studies of children exposed in utero at Hiroshima and Nagasaki have shown that the incidence of microcephaly, often connected with mental retardation, was much higher among those exposed to doses greater than about 10 rad during weeks 10–19 of fetal development. In a recent study of 1,599 children exposed in utero at Hiroshima and Nagasaki, mental retardation was found to be apparently linearly related to dose during the sensitive period.[28]

When alpha-emitting radionuclides are internally deposited, the radiation exposures they produce are chronic and can affect development in neonates. It is therefore impractical to differentiate completely between teratogenic and neonatal developmental effects, so we include prenatal and postnatal effects, such as decreased growth and survival, physiological and biochemical deficits, and time of appearance and incidence of differences in degenerative processes.

Placental Transfer and Fetal Accretion of Radionuclides

The inherent mechanisms involved in the production of developmental effects by exposure of a conceptus or neonate to radiation from internally deposited radioactive materials do not differ from those associated with exposure to photons from external sources. The several factors to be discussed in detail as pertaining to exposure of adults also lead to differences between the developmental effects produced by internal and external exposures and affect the quantitative relationships between administered dose (or maternal body burden) and effect. It is not feasible to consider effect without simultaneously considering the components of dose. A brief review of these factors will place the subject in context.

It is more difficult to measure the sequentially changing concentrations of internal emitters in a conceptus and its supporting structures than in an adult, so the estimates of absorbed dose are less precise than estimates of prenatal photon exposure or adult exposure to internal emitters. Moreover, some of the factors that influence dose-effect relationships make it more difficult to develop generalized relationships for fetoplacental dosimetry. Several alpha-emitters differ in placental transfer, and transfer is also influenced by the chemical and physicochemical state of the material and by the stage of gestation at exposure. Most radioisotopic materials of concern have specific organ or tissue affinities for localization in the conceptus and throughout the fetoplacental unit; these affinities tend to change during gestation. Estimation of tissue concentrations and associated radiation doses is complicated by the interplay of affinities, translocation, and fetoplacental growth. The resulting exposures occur at low and varying dose rates, and that makes it difficult to establish reliable quality factors. Many of the deleterious effects of internal emitters are displayed as postnatal deficits, and it is necessary to separate the effects produced by prenatal exposure from those produced by postnatal exposure to remaining radioisotope and to material transferred in mother's milk.

There are no comprehensive overviews of prenatal radionuclide dosimetry. But, for some materials—such as radiopharmaceuticals, isotopes used as clinical tracers, and analogs of natural metabolites—there is a useful body of information from animal studies and from human experience. Less information is available on naturally occurring alpha-emitting elements, and there is still less information from animal studies on the radionuclides associated with the nuclear industries; comparative data from human experience are almost absent. The term fetal accretion is used as a general expression of concentration to integrate the complex interplay among maternal biokinetics, placental transfer, and the presence of receptor sites in the conceptus. Many of the materials that are readily or moderately accepted are beta- or gamma-emitters with relatively short biological half-lives and without markedly specific localization during embryonic stages (e.g., during organogenesis). When they are administered during these periods, their effects are dosimetrically compatible with those of photon exposure, if allowance is made for differences in dose rates and protraction. However, specific localization occurs during the histogenesis that characterizes the fetal period and markedly influences the effects after the target tissues have developed.

Most of the alpha-emitting heavy elements—including radium, radon, uranium, and the transuranic elements—are in the low fetal accretion category, although they vary widely in actual values and in fetoplacental distribution. Several elements with higher atomic numbers have been found to be embryotoxic and teratogenic in rodents, but only at high administered doses. This effect is sometimes attributable to chemical toxicity. The associated dosimetric data are incomplete, but the action of some nuclides seems to occur via an effect on the villose visceral splanchnopleure (yolk sac), which has a high affinity for heavy metals and is important in absorption of nutrients by the early rodent embryo. The importance of this structure is not completely paralleled in other mammals, but similar affinities have been demonstrated in nonhuman primates.

It was first demonstrated in the early 1920s that adverse developmental effects resulted from exposure of the mammalian fetus to radioactive materials injected into the pregnant animal. The next reported study, resulting from the efforts of the Manhattan Project, did not appear until 25 yr later. In contrast with the extensive literature dealing with acute exposure to external photon beams, relatively few experiments have been performed with radionuclides, particularly with alpha-emitters, because few laboratories have a knowledge of developmental toxicology and the facilities and expertise to work with alpha-emitters. Performance and interpretation of such experiments is most difficult, in that establishing radiation doses requires sacrifice of the animals and precludes determination of effect, the radiation is not uniformly distributed throughout the fetoplacental unit, and the absorbed radiation is protracted at varying dose rates. As a result, most experiments have been directed either at measurement of placental transfer and fetal content or at determination of effect relative to administered dose, but rarely at both. Thus, the dose-response relationships for internal alpha exposure are less well established than those for external or for internal beta exposure.

Although our information is still limited and there are elements on which there are no data, some progress on the developmental toxicity of alpha-emitters has been made during the last 40 yr. For simplicity, an overall summary of effect patterns relative to dosimetry in order of ascending atomic number is presented here.

Radon and Daughters

The first developmental study with an alpha-emitter apparently was also the first experiment with prenatal administration of an internal emitter.[15] Radon and its daughters were dissolved in isotonic saline and injected subcutaneously into female rats before mating or during gestation in amounts originally equivalent to 5 mCi of radon. Prenatal mortality was increased, and there was macroscopic hemorrhage in the survivors when the rats were exposed at an unstated time of gestation and even if the solution was administered as early as 22 days before mating. Most of the offspring born on the day after injection at 19 days of gestation (dg) had similar hemorrhages, but normal placentas.

The reports did not indicate the extent to which the several nuclides crossed the placenta. Studies have shown that inhaled ^{85}Kr freely crosses the placenta and that its concentration is the same in maternal and fetal blood; it can be inferred that radon would behave in a similar manner. Because of the consistent involvement of the placenta in the effect, the presence of edema and/or hemorrhage, and the mechanical effects in malformation production, it is not clear to what extent the effects were directly on the conceptus or were secondary to placental and/or yolk sac changes.

Radium

Wilkinson and Hoecker[44] dissolved $^{226}RaCl_2$ in saline; the solution was allowed to reach equilibrium and then injected into rats at 15 dg. No radioactivity was detected in the placentas or fetuses at 20 dg, but the availability of the nuclide to the placenta after administration is not clear.

Rajewsky et al.[31] measured the ^{226}Ra content of the bones and soft tissue of approximately 200 human fetuses and 40 additional placentas at various stages of gestation. The specific activity of bone ash (10^{-14} Ci/g) was independent of the stage of gestation and was identical with that measured in adult bone. Fetal soft tissue and placental concentrations were similar (10^{-16} Ci/g) and did not change during gestation, although the total fetal content increased during gestation as a result of the increase in fetal mass. Martland and Martland[22] found less than 10^{-8}g of radium on examination of 17 children from 10 mothers who had been employed as radium-dial painters. Reports regarding the developmental effects of prenatal exposure to radium have not been found, but to the extent that these measurements reflect the amounts transferred, demonstrated effects might not be expected.

Polonium

Lacassagne and Lattes[19] found salts of polonium deposited in the placental syncytium (cells of the chorionic epithelium), but not in the fetal connective tissue or the fetal endothelium.

Uranium

The toxicity of several uranium isotopes has been studied extensively in adult animals of various species, but few data are available on its placental transfer or developmental toxicity, other than those from one experiment[39] that used intravenous exposure to citrated solutions of ^{233}U at 9, 15, or 19 dg at dosages of 1.8, 3.3, 5.75, and 10 μCi/kg. The two highest dosages were toxic to pregnant rats. Exposure at 9 and 15 dg produced dosage-dependent trends toward increased prenatal mortality and decreased fetal and placental weights on evaluation at 20 dg. Injection at 9 dg produced a dosage-dependent increase in the incidence of rib malformation, and the highest dosage resulted in cleft palate. The two highest dosages resulted in fetal edema if injected at 15 dg. Fetoplacental concentrations and partition, measured in other rats, were affected by dosage evaluated late in gestation, but not at earlier times, and there was less selective localization in the yolk sac than was the case with many actinides. The resulting radiation doses throughout the fetoplacental unit were calculated from the distribution data. At the highest dosage, the radiation doses to the conceptus, placenta, and membranes were about 1, 2, and 3 rad, respectively, after injection at 9 dg and 0.2, 1, and 7 rad, respectively, after injection at 15 dg. Comparison of the doses suggests that the early maternal and developmental effects were attributable to chemical toxicity, rather than to radiation.

Transuranic Elements

There are pronounced differences among the transuranic elements relative to their metabolism in pregnant animals, placental transfer, and fetoplacental distribution. The limited data do not suggest marked qualitative differences in the types of responses, but metabolic and dosimetric differences yield substantial differences in toxicity relative to administered dosage. It should be noted that almost all evaluations of effects were performed with administered dosages that were far in excess of those to which people might be exposed. These results are of radiobiological interest, but are not directly applicable to establishing exposure criteria.

Intravenous injection into rats of ^{237}Np as the oxalate at 0.3–5 μCi/kg increased the incidence of preimplantation mortality. Relative to controls, offspring of litters receiving these dosages had greater depression of erythrocyte production after gamma irradiation, prolonged narcosis after hexanol administration, and decreased sexual function.[29]

Plutonium is the most thoroughly studied of the transuranic elements, although some questions remain as to details of its placental transfer, fetoplacental distribution, and developmental toxicity. In the earliest reported study with ^{239}Pu,[7] it was found that mice given plutonium at 0.016, 0.03, or 0.06 μCi/g by injection during gestation had an increased incidence of totally stillborn litters and of stillbirths in viable litters. If injection took place late during gestation, the newborn offspring contained about 1% of the administered radioactivity; the amounts of radioactivity decreased with increasing dosage and with increasing time between injection and measurement. Autoradiographic and radioanalytic studies from several laboratories have extended and quantified those findings, but are in general accord with them. In general, it has been found that small amounts of monomeric (citrated) plutonium cross the placenta. Intravenously or intraperitoneally injected polymeric plutonium has been shown to be less available to the fetoplacental unit.

Studies to examine the partition of ^{239}Pu at various stages of gestation in rabbits have found a difference in distribution from that in rats and mice. The concentrations in the placentas and fetal membranes were not as high as those found in rodents, and the concentration ratio in these two structures and the concentration in the embryo-fetus were lower in the rodents.

To determine whether yolk sac deposition occurs in nonhuman primate species, Andrew et al.[1] injected citrated ^{239}Pu intravenously into baboons at 10 μCi/kg at representative stages of gestation and removed the uteri and their contents 24 h later. The uteri and fetoplacental components were dissected and subjected to radioanalysis; concentration ratios were similar to those found in rodents. The autoradiographic localization of activity was also similar in the two species when allowance was made for morphological differences.

Kelman and Sikov[18] directly examined placental transfer using a system in which the vessels of the fetal side of the near-term guinea pig placenta were cannulated and perfused to eliminate the role of the fetus. Graded dosages of citrated ^{239}Pu and a trace dose of tritiated water were injected into the maternal circulation, and placental transfer was calculated in terms of clearance. Clearance was found to be 2.3 liters/min, a value less than one-fifth of that for inorganic mercury, which had the lowest clearance previously measured with this system. Moreover, on the basis of reduced clearance of tritiated water at the highest doses of plutonium, the maternal blood supply to the placenta was affected; the threshold for this effect was about 5 μCi/kg of body weight.

Administration of ^{239}Pu during early organogenesis (e.g., at 9 dg in rats) results in dose-dependent increases in prenatal mortality and reduced weights of the fetuses, placentas, and fetal membranes. Typically, however, the lowest administered intravenous dosage that consistently produces statistically significant differences from controls is about 10 μCi/kg in rats. Sequential histopathological studies at this and higher dosages demonstrated early shrinking and suppressed development of the villi of the yolk sac, which suggested that the embryotoxic effects might be mediated through changes in this structure. The radiation doses to the embryo-fetus, placenta, and membranes of rats under these general circumstances are approximately 0.45, 1.3, and 2.5 rad, respectively, through 12 dg and about 2, 6, and 33 rad, respectively, through 20 dg. The dose to the yolk sac might be 10–100 times as great as the average membrane dose because it represents only a small portion of the total mass.

Other experiments have failed to detect prenatal mortality or other indicators of prenatal toxicity after exposure of rats to plutonium at dosages as great as 50 μCi (about 150 μCi/kg) at 15 or 19 dg. As indicated above, the pattern of fetoplacental partition is similar to that at earlier stages.

The induction of developmental toxicity involves complicated interactions, as indicated by differences among rat strains in sensitivity to production of embryo lethality and fetal weight reduction. These between-strain differences are incompletely accounted for by distribution differences. Comparable quantitative differences in developmental effects and minor qualitative differences have been observed in other species, including rabbits, but these are partially related to the distribution differences described above.

The primitive cells that ultimately give rise to the gametes and the hematopoietic system are formed in the early yolk sac and migrate into the embryo proper. It can be hypothesized that alpha particles emitted by radionuclides deposited in the yolk sac of the early embryo could produce persistent adverse effects on these primitive cell lines. To test this hypothesis, pregnant rats were evaluated at 14 or 19 dg after intravenous injection with 36 μCi/kg of citrated plutonium at 9 dg. Weight gains of the pregnant rats were reduced, as were reticulocyte and leukocyte counts at both times and erythrocyte concentrations at 19 dg. Exposure increased prenatal mortality but did not significantly affect fetal weights. Fetal hematological changes included a transient decrease in the concentration of circulating non-nucleated erythrocytes and altered distribution of the erythropoietic cell types. These changes were interpreted as a disturbance of the maturation process. The weights of the yolk sac and fetal liver were reduced in exposed litters; their cellularity and that of the spleen were also decreased, but the proportion of cell types was unaffected. Detailed microdosimetry has not been performed, but the radiation dose to the primitive hematopoietic cells might have been as high as 1,000 rad.

Moskalev et al.[24] injected a constant volume of solution containing ^{241}Am at concentrations of 1.2–7.6% intravenously into pregnant rats at 10–19 dg. The resulting ratio of average maternal to fetal concentration at 24 h varied from 6:1 to 2:1 as a function of dose and stage. Although the value is influenced by the interval between injection and evaluation, as well as by sensitivity, they calculated the injection dosages and radiation doses to the fetus that resulted in death of 50% of the fetuses as 0.003, 0.01, and 40 μCi/g and 100, 800, and 1,000 rad respectively, for injection at 10, 14, and 19 dg.

Weiss and Walburg[42,43] reported that the effect of the mass administered was less than they had found with plutonium, but fetal concentration varied only by a factor of 2 between the highest and lowest dosages. These studies and others demonstrated that, on a percentage basis, less americium than plutonium entered the conceptus or fetoplacental unit. Several studies have shown that there was proportionately less deposition of americium than of plutonium in the placenta and membranes.

Results of a contemporaneous study with the two nuclides[33] have confirmed suggestions that the prenatal effects of americium are similar to those of plutonium and include prenatal mortality and rib malformations, but not weight reduction. The effects are smaller if based on intravenous dosages administered to pregnant rats, but there is better correspondence between the effects when they are considered relative to radiation doses to particular components of the fetoplacental unit, especially the yolk sac.

Measurable amounts of ^{253}Es cross the placenta, but, according to the limited data available, the fraction of maternal dose that is deposited in the conceptus is low, approximating that of ^{241}Am. However, einsteinium has a greater tendency than americium to be incorporated in the yolk sac.

CONCLUSIONS

Very recently, a task group of Committee 1 of the International Commission on Radiological Protection completed a study of the effects of radiation on the development of the brain of the embryo and fetus.[17] The task group reported that, within the period of maximum vulnerability, the data it reviewed appeared to be consistent with a linear nonthreshold response. This information was published after this report was prepared and therefore has not been examined by this committee, which reached no conclusions concerning the effects of alpha dose on the developing brain.

With the exception of risks to the developing brain, no national or international expert group has made quantitative risk estimates of purely teratogenic effects of exposures of less than 10 rad of acute low-LET radiation, simply because of the threshold nature of most of the dose-effect curves. For organs other than the brain, the concept of RBE can be used to translate estimates of the effects of acute low-LET exposures to the case of alpha particles. Virtually all other teratogenic effects of radiation are believed to be due to multiple cell killing, and one can simply translate the accepted 10-rad threshold for single-dose low-LET radiation exposures by applying the RBE commonly observed for alpha particles in in vitro cell-killing experiments. RBEs for cell killing by alpha particles are around 10, but could be higher for the very low dose rates expected from internal emitters. Sensitive time windows have been observed, particularly during the stage of major organogenesis, and much (if not all) of the total dose accumulates on either side of this window, which is apparently only a few days long even in man. Thus, most of the total dose accumulated during the entire 280-day gestation period would not be effective. It seems reasonable to conclude that except for brain tissues, high-LET alpha-particle doses below about 1 rad will have no teratogenic effects.

REFERENCES

1. Andrew, F. D., R. L. Bernstine, D. D. Mahlum, and M. R. Sikov. 1977. Distribution of ^{239}Pu in the gravid baboon. Radiat. Res. 70:637–638.
2. Batchelor, A. L., R. J. S. Philips, and A. G. Searle. 1966. A comparison of the mutagenic effectiveness of chronic neutron- and γ-irradiation of mouse spermatogonia. Mutat. Res. 3:218–229.
3. Brooks, A. L. 1975. Chromosome damage in liver cells from low dose rate alpha, beta, and gamma irradiation: Derivation of RBE. Science 190: 1090–1092.

4. Brooks, A. L., J. H. Diel, and R. O. McClellan. 1979. The influence of testicular microanatomy on the potential genetic dose from internally deposited ^{239}Pu citrate in Chinese hamster, mouse and man. Radiat. Res. 77:292.
5. Edwards, A. A., R. J. Purrott, J. S. Prosser, and D. C. Lloyd. 1980. The induction of chromosome aberrations in human lymphocytes by alpha-radiation. Int. J. Radiat. Biol. 38:83–91.
6. Evans, J. S., D. W. Moeller, and D. W. Cooper. 1985. Health effects model for nuclear power plant accident consequence analysis. NUREG/CR-4214. U.S. Nuclear Regulatory Commission, Washington, D.C.
7. Finkel, M. P. 1947. The transmission of radio-strontium and plutonium from mother to offspring in laboratory animals. Physiol. Zool. 20:405–421.
8. Generoso, W., K. T. Cain, N. L. A. Cacheiro, and C. V. Cornett. 1985. ^{239}Pu-induced heritable translocations in male mice. Mutat. Res. 152:49.
9. Grahn, D., B. H. Frystak, C. H. Lee, J. J. Russell, and A. Lindenbaum. 1979. Dominant lethal mutations and chromosome aberrations induced in male mice by incorporated ^{239}Pu and by external fission neutron and gamma irradiation. Pp. 163 in Biological Implications of Radionuclides Released from Nuclear Industries. IAEA-SM-237/50. Geneva: International Atomic Energy Agency.
10. Grahn, D., C. H. Lee, and B. F. Farrington. 1983. Interpretation of cytogenetic damage induced in the germ line of male mice exposed for over 1 year to ^{239}Pu alpha particles, fission neutrons, or ^{60}Co gamma rays. Radiat. Res. 95:566–583.
11. Green, D., G. R., E. R. Howells, and J. Vennart. 1975. Localization of plutonium in mouses testes. Nature 255:77.
12. Green, D., G. R. Howells, J. Vennart, and R. Watts. 1977. The distribution of plutonium in the mouse ovary. Int. J. Appl. Radiat. Isotopes 28:497–501.
13. Green, D., G. R. Howells, and R. Watts. 1979. Plutonium in the tissues of fetal, neonatal and suckling mice after plutonium administration to their dams. Int. J. Radiat. Biol. 35:417–432.
14. Green, D., G. R. Howells, and J. Vennart. 1980. Radiation dose to mouse testes from ^{239}Pu. Health Phys. 38:242–243.
15. Gudernatsch, J. F., and H. J. Bagg. 1920. Disturbances in the development of mammalian embryos caused by radium emanation. Proc. Soc. Exp. Biol. Med. 17:183–187.
16. Hicks, S. P. and C. J. D'Amato. 1966. Effects of ionizing radiations on mammalian development. Adv. Teratology 1:195–266.
17. International Commission on Radiological Protection (ICRP). 1986. Developmental Effects of Inadiation on the Brain of the Embryo and Fetus. ICRP Publication 49. Oxford: Pergamon.
18. Kelman, B. J., and M. R. Sikov. 1981. Plutonium movements across the haemochorial placenta of the guinea pig. Placenta (Suppl. 3):319–326.
19. Lacassagne, A. and J. Lattes. 1924. Compt. Rend. Soc. Biol. 90:485 (cited by Wilkinson and Hoecker[44]).
20. Luning, K. G., and H. Frolen. 1982. Genetic effects of ^{239}Pu salt injections in male mice. Mutat. Res. 92:169.
21. Luning, K. G., H. Frolen and A. Nilsson. 1976. Dominant lethal tests of male mice given ^{239}Pu salt injections. In Biological and Environmental Effects of Low-Level Radiation. IAEA STI/PUB/409. Geneva: International Atomic Energy Commission.

22. Martland, H. S. and H. S. Martland, Jr. 1950. Am. J. Surg. 80:270 (cited by Wilkinson and Hoecker[44]).
23. McKusick, V. A. 1982. Mendelian inheritance in man. Baltimore: The Johns Hopkins University Press.
24. Moskalev, J. I., L. A. Buldakov, A. M. Lyaginskaya, E. P. Ovcharenko, and T. M. Egorova. 1969. Experimental study of radionuclide transfer through the placenta and their biological action on the fetus. Pp. 153–160 in Radiation Biology of the Fetal and Juvenile Mammal, M. R. Sikov, and D. D. Mahlum, eds. U.S. Atomic Energy Commission, Washington, D.C.
25. National Research Council, Committee on the Biological Effects of Ionizing Radiations (BEIR). 1972. The Effects on Populations of Exposure to Low Levels of Ionizing Radiation. Washington, D.C.: National Academy of Sciences. 217 pp.
26. National Research Council, Committee on the Biological Effects of Ionizing Radiations (BEIR). 1980. The Effects on Populations of Exposure to Low Levels of Ionzing Radiation. Washington, D.C.: National Academy Press. 524 pp.
27. Neary, G. J., J. R. K. Savage, H. J. Evans, and J. Whittle. 1963. Ultimate maximum values of the RBE of fast neutrons and gamma rays for chromosome aberrations. Int. J. Radiat. Biol. 6:127–136.
28. Otake, M., and W. J. Schull. 1984. In utero exposure to A-bomb radiation and mental retardation; a reassessment. Br. J. Radiol. 57:409–414.
29. Ovcharenko, E. P., and T. R. Fomina. 1982. The effect of injected ^{237}Np-oxalate on the gonads of rats and their offspring. Radiobiologiya 22:374–379.
30. Purrott, R. J., A. A. Edwards, D. C. Lloyd, and J. W. Stather. 1980. The induction of chromosome aberrations in human lymphocytes by *in vitro* irradiation with alpha-particles from plutonium-239. Int. J. Radiat. Biol. 38:277–284.
31. Rajewsky, B., V. Belloch-Zimmermann, E. Lohr, and W. Stahlhofen. 1965. ^{226}Ra in human embryonic tissue, relationship of activity to the stage of pregnancy, measurement of natural ^{226}Ra occurrence in the human placenta. Health Phys. 11:161–169.
32. Richmond, C. R., and R. L. Thomas. 1975. Plutonium and other actinide elements in gonadal tissue of man and animals. Health Phys. 29:241–250.
33. Rommereim, D. N., and M. R. Sikov. 1986. Relative embryotoxicity of ^{239}Pu and ^{241}Am in rats. Teratology 33:93C.
34. Rugh, R. 1969. The effects of ionizing radiations on the developing embryo and fetus. Seminar Paper 007. Washington, D.C.: Bureau of Radiological Health, U.S. Public Health Service.
35. Russell, J. J., and A. Lindenbaum. 1978. One year study of non-uniformly distributed plutonium in mouse testis as related to spermatogonial irradiation. Health Phys. 36:153–157.
36. Russell, L. B. 1971. Definition of functional units in a small chromosomal segment of the mouse and its use in interpreting the nature of radiation-induced mutations. Mutat. Res. 11:107–123.
37. Searle, A. G., C. V. Beechey, D. Green, and E.R. Humphreys. 1976. Cytogenetic effects of protracted exposures to alpha-particles from plutonium-239 and to gamma rays from cobalt-60 compared in male mice. Mutat. Res. 41:297–310.

38. Searle, A. G., C. V. Beechey, D. Green, and G. R. Howells. 1982. Dominant lethal and ovarian effects of plutonium-239 in female mice. Int. J. Radiat. Biol. 42:235–244.
39. Sikov, M. R., and D. N. Rommereim. 1986. Evaluation of the embryotoxicity of uranium in rats. Teratology 33:41C.
40. Thomas, R. G., J. W. Healy, and J. F. McInroy. 1985. Plutonium in gonads: A summary of the current status. Health Phys. 48:7–17.
41. United Nations Scientific Committee on the Effects of Atomic Radiation (UNSLEAR). 1982. Ionizing Radiation: Sources and Biological Effects. Report E.82.IX.8. New York: United Nations. 773 pp.
42. Weiss, J. F., and J. E. Walburg. 1978. Placental transfer of americium and plutonium in mice. Health Phys. 39:903–911.
43. Weiss, J. F., and J. E. Walburg. 1980. Influence of the mass of administered plutonium on its cross-placental transfer in mice. Health Phys. 35:773–777.
44. Wilkinson, P. N., and F. E. Hoecker. 1953. Selective placental transmission of radioactive alkaline earths and plutonium. Trans. Kans. Acad. Sci. 56:341–363.

APPENDIX I
Dosimetry of Alpha Particles

The name alpha particle was given to the energetic helium nuclei (that is, helium atoms stripped of their two electrons) emitted in radioactive decay. They are emitted with energies in the range of 4 to 9 MeV. Because they are always the same kind of particle, regardless of the nucleus from which they come, all alpha particles of a given energy have the same properties.

PROPERTIES OF ALPHA PARTICLES

SCATTERING

The tracks of alpha particles are nearly straight lines, meaning that they are scattered very little by the material through which they pass. In dosimetric calculations it is usual to assume that the tracks are straight lines.

Alpha particles do undergo scattering, however. One of the earliest developments in nuclear physics was the discovery, by Rutherford, of the alpha-particle scattering law. He found that very small deflections, a degree or so, are frequent and that larger deflections do occur but are quite rare. Consequently, along a typical track, each time an alpha particle scatters it changes direction only slightly and in a random direction, with the result that the track winds back and forth a small amount around a straight line. This multiple scattering contributes to the straggling discussed below.

STOPPING POWER

As alpha particles go through a material, they interact with the material's molecules, losing a little energy in each interaction. Thus, they gradually slow down. At very low energies (less than 1 eV) they acquire two electrons and become atoms of helium gas in thermal equilibrium with the material. Alpha particles of the same energy lose different energies when they go equal distances because of randomness in the number and in the kinds of interactions with the material. The differences in the energy losses are small, however. The actual rates of energy loss are close to the average rate.

The stopping power (S) of a charged particle of specified energy is its average energy loss per unit distance along its path. The mass stopping power (S/ρ) is the quotient of the stopping power by the density of the material.

The experimental and theoretical determinations of the alpha-particle stopping powers are in good agreement.[1,14] Figure I-1 shows the stopping power of alpha particles in soft tissues of unit density. For comparison, the stopping power of electrons is also shown in Figure I-1.[6] (Beta particles, the other common kind of charged particle emitted in radioactive decay, are electrons.) Note that for the energies shown, the stopping powers of the alpha particles are 2 to 3 orders of magnitude larger than those of the electrons. As a consequence, the damage (for example, caused by ionization, excitation, chemical alteration, and biological damage) is generally about that much higher along the alpha-particle tracks.

The mass stopping powers of alpha particles are slightly higher in gases than in liquids or solids.[13] The stopping powers in Figure I-1 are for solid tissue and are about 5% higher than the mass stopping powers for gases.

LINEAR ENERGY TRANSFER

The energy lost by a charged particle produces damage of various kinds in the material with an effectiveness that depends on how densely the energy it loses is spread in the material. The stopping power is one indication of that density, but it is not a complete measure: Some of the energy lost by the particle is transferred to secondary radiations, electrons and photons, that penetrate to distances from the particle track. Several investigators have used various other quantities to characterize the results of this spreading process.[3]

One of the quantities used for this purpose is the linear energy transfer (LET; also referred to as L). While the stopping power gives the rate of energy loss of the particle, the LET gives the rate at which energy is laid down close to the particle track. This is done by excluding the energy

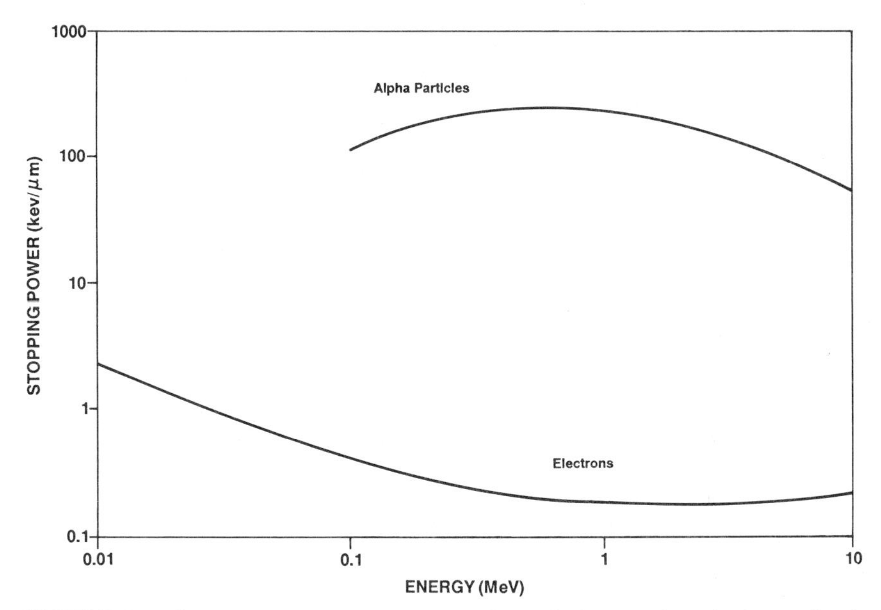

FIGURE I-1 Stopping powers of alpha particles and electrons in soft tissue of unit density.

carried away by photons and energetic electrons from the energy lost by the particle. Two criteria have been used to specify which electrons are to be excluded: (1) electrons with energies above some limit, and (2) electrons with ranges above some limit. In either case it is important to specify the limit; this is usually done by writing it as a subscript to the symbol L. The energy limit is the current preference, because the LET can be calculated from theoretical equations for the stopping power by simply excluding energy losses to secondary electrons above the selected limit.

As an example, a 4-MeV alpha particle has a stopping power of 102 keV μm^{-1}; the LET excluding losses of more than 100 eV is $L_{100} = 56$ keV μm^{-1}; 1,000 eV, $L_{1,000} = 81$ keV μm^{-1}; 10,000 eV, $L_{10,000} = 102$ keV μm^{-1}. Losses of 10 keV are above the maximum energy, about 2,000 eV, that a 4-MeV alpha particle can lose to an electron; therefore, the LET for that cutoff does not differ from the stopping power. It is customary to designate the LET for energy limits larger than the maximum energy transferrable by L_∞. L_∞ is the largest value possible for the LET, and it equals the stopping power S.

RANGE

Alpha particles have a fairly sharply defined range (R), which is the average distance that they travel in coming to rest in a material. Figure

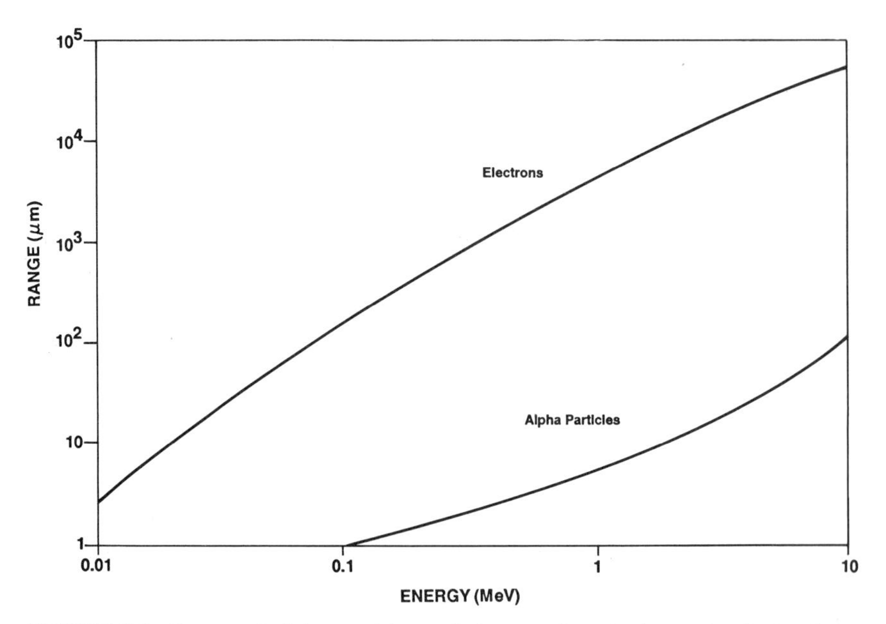

FIGURE I-2 Ranges of alpha particles and electrons in soft tissue of unit density.

I-2 shows the ranges of alpha particles in soft tissue of unit density. For comparison, the average distances traveled by electrons, neglecting energy straggling (see below), are also given. Because of their very much lower stopping powers, the electrons travel 2 or 3 orders of magnitude farther than alpha particles.

The stopping powers and ranges in tissue suffice for most applications to alpha-particle dosimetry. Occasionally, however, it is necessary to consider alpha-particle paths that lie partly in one material, partly in another (e.g., partly in soft tissue and partly in bone). Fortunately, the ranges of an alpha particle of a given energy in different materials are proportional to one another, independently of the energy, to within a few percent.

STRAGGLING

Random variations in the energy lost and in the change in direction in individual interactions with the molecules of the material produce distributions in the actual distances traveled by different alpha particles; this is known as range straggling. Similarly, there is a distribution in the energy remaining after traveling a given distance; this is known as energy straggling.

The probability distribution of the actual ranges is represented fairly accurately by a normal distribution (it neglects the effects of occasional large energy losses in individual collisions). Theoretical derivations of parts of the variance, $\sigma^2(R)$, of the normal distribution have been made, but the observed variances are always larger. Evans[2] recommended a relation that he estimated to be accurate to within 10% for range straggling in air:

$$\sigma(R) = 0.015R, \tag{I-1}$$

where R is the range. This relation shows that, in air, most of the actual distances traveled are within a few percent of the mean distance (the range); theoretical analyses suggest similar narrow distributions for the distances in tissue.

DOSIMETRY

EXPOSURE AND DOSE

A basic step in the study of the risk associated with an agent is the establishment of the relation between the degree of harm it produces and some physically measurable quantity that characterizes its prevalence. The measurable quantity is often referred to as the exposure to the agent. The exposure is seldom the concentration of the agent (or of a product of the agent) in the specific cells or tissues where the harm is thought to arise. The latter, or some closely related quantity, is often called the "dose." Because of the difficulty in making measurements within the body, it is usually hard to determine the dose. The exposure, on the other hand, is usually the concentration outside the body in some material in which it is easier to measure. Usually it is the concentration in the material that is the main carrier of the agent into the body.

The working level used in the study of radon and its daughters is a good example of these general statements; it is treated at length in Annex 2B to Chapter 2. In brief, the working level, an exposure quantity, is a concentration of radon and its daughters in air, and is reasonably easy to measure. The corresponding dose quantity might be the number of atoms of radon and its daughters deposited at some point in the respiratory tract. The relation between the concentration in air and the number of atoms deposited depends on a host of variables (for example, the structure of the respiratory tract and breathing rate; see Chapter 2). This lack of a unique correlation between the exposure and dose quantities is typical.

There is danger of confusion in discussing the exposure and dose concepts: A particular quantity named exposure was introduced early into radiation studies to characterize x-ray and gamma-ray fields; this exposure is the one measured in roentgens.[4] Fortunately, a need to use this particular exposure seldom arises in studying internally deposited alpha emitters; thus, the term can usually be used here in its general sense.

ABSORBED DOSE

Radiation studies employ the dose concept in the quantity called absorbed dose. The determination of absorbed dose is called dosimetry. The International Commission on Radiation Units and Measurements (ICRU) defines absorbed dose to be the mean energy imparted to the irradiated medium, per unit mass, by ionizing radiation. (For definitions of absorbed dose and the other quantities used in dosimetry, see ICRU.[4])

The energy of ionizing radiation is imparted to the medium in a series of individual interactions with it. The number of interactions and the amount of energy lost in each are random variables. The word *mean* used in the definition of absorbed dose requires that the average of the energies imparted be used. In what follows, up to the section "Microdosimetry," it is assumed that the average has been taken; in the section "Microdosimetry" the probability distribution of the energy imparted whose mean is the one required for the absorbed dose will be dealt with.

AVERAGE ABSORBED DOSE

Often one can be satisfied (see the section "Nonequilibrium Doses" below) with the average absorbed dose in some volume. If the range of the alpha particle is much smaller than the dimensions of the volume, most of the alpha particles emitted within that volume are absorbed within it; that is, they impart their energy within it. Only those emitted close to the surface and headed through the bounding surface can escape and impart their energy elsewhere (or particles emitted outside the volume can impart energy to it). In many circumstances, this leakage in and out is negligible, because far more particles are emitted within the volume than are emitted close enough to the surface to leak in or out. In these circumstances the average absorbed dose, $\langle D \rangle$, in the particular tissue equals the product of the number of alpha particles emitted within it and their energy, E, divided by the mass of the tissue. Let $\langle C \rangle$ be the number of particles emitted divided by the mass of the tissue, that is, the mean number emitted per unit mass. Then:

$$\langle D \rangle = \langle C \rangle E. \tag{I-2}$$

CHARGED-PARTICLE EQUILIBRIUM

The leakage is also negligible (actually, zero) in another situation that is representative of many experimental situations. Suppose the tissue and its surroundings are of uniform composition and the number of alpha particles emitted per unit mass (C) is constant throughout the volume of interest and for some distance, greater than the range, into the material on all sides of it (see the next section). The net leakage is then zero, because there is as much leakage into the volume of interest as leakage out of it. This condition is known as charged-particle equilibrium, and the dose (D) is given by:

$$D = CE. \tag{I-3}$$

NONEQUILIBRIUM DOSES

When the average dose does not suffice or when charged-particle equilibrium does not exist, the dose at a point can be calculated from the local density of alpha-particle emission (C) in all elementary volumes (dV) within the alpha-particle range of the point. The number of alpha particles emitted from a particular dV is $C\ \rho\ dV$, where ρ is the density of the medium (assumed constant in the neighborhood of the point). The number of these alpha particles per unit area at the point is $(C\ \rho\ dV)/(4\pi r^2)$, where r is the distance to the point. The number entering an elementary target volume at the point and with area dA facing dV is dA times the number per unit area. Each particle imparts an energy, denoted by $e(r)dx$, to the target, where dx is the thickness of the target. The mass of the target volume is $\rho dAdx$. Thus, the dose to the target from the alpha particles emitted in the particular dV is:

$$[C\rho dV dAe(r)dx/4\pi r^2]/\rho dAdx. \tag{I-4}$$

The total dose is obtained by integrating over all dV within range of the point. Several factors cancel to give, for the dose:

$$D = \int e(r)(CdV/4\pi r^2). \tag{I-5}$$

Different approximations have been used for the kernel $e(r)$. One approximation is to equate it to the stopping power: $e(r) = S$. This expression is approximate for two reasons. First, S gives the energy lost by the alpha particle, not the energy imparted to the medium in the target. Secondary radiations, electrons (called delta rays) and photons, leak energy into and out of the target, as discussed above. Because of the very short ranges of these secondary radiations, leakages in and out tend to compensate each other and make the approximation a good one. Second, because of the straggling described above, there is a spread in the energies of the alpha particles arriving at the target element. This straggling causes a larger error than the leakage just mentioned. An average, $\langle S \rangle$, of the stopping power over the straggling spectrum (σ) would be a better approximation to $e(r)$.

A basic datum, called the Bragg curve, collected for alpha particles and other heavy ions provides another good estimate for $e(r)$. The curve is the ionization as a function of distance in a thin, broad ionization chamber held so that the particles strike the broad face perpendicularly. The ion chamber does the averaging and leakage compensation required for $e(r)$. Furthermore, by being broad, it allows for the multiple scattering discussed above. Use of the stopping power or an average stopping power does not allow for particles emitted in dV and headed for the target element that do not get there because they scatter away from it; it also does not allow for those not headed for it but scattered so that they hit it. The data in a Bragg curve are converted to $e(r)$ by normalizing the curve so it equals the stopping power near the point of emission, where the effects of straggling are smallest.

If one does not require much accuracy, for example, if one is making just a trial or illustrative calculation, the variation of the stopping power with the distance traveled can be neglected; that is, one can approximate $e(r)$ with the average stopping power E/R. Figure I-3 shows the results of a calculation done in the $e(r) = E/R$ approximation to illustrate the doses from nonequilibrium distribution of alpha-particle emitters. In this instance, C alpha particles per unit mass of energy E were emitted from spherical regions of radius a in uniform tissue in which the range of the particles is R. The doses are shown as functions of x, the distance from the center of the sphere. All distances are normalized to the alpha-particle range R. Under charged-particle equilibrium conditions, the density C would produce a uniform dose, CE; all the doses in Figure I-3 are normalized to CE.

Figure I-3 illustrates the conditions required to obtain charged- particle equilibrium. For the two largest spheres, radii of 2 and 3 times the range, the dose equals the equilibrium dose CE in the central region of the sphere. There each point is surrounded by emitters out to a distance at least equal to the range of the alpha particles; emitters farther away have no influence on the dose because the particles cannot reach the point. This example illustrates the requirement discussed above that C must be uniform out to distances equal to the range beyond the edge of the region before charged-particle equilibrium can exist within it.

Inside the largest spheres, at points less than the alpha-particle range from the edge, the dose is less than CE because there are regions within the range that contain no emitters. At the edge of these spheres (that is, for $x = a$), the dose is roughly one-half CE (the dose would be exactly one-half CE at a plane surface in an equilibrium region if all the emitters were removed from one side of the plane; small parts of the surfaces of large spheres are shaped much like planes). Outside a sphere of any size,

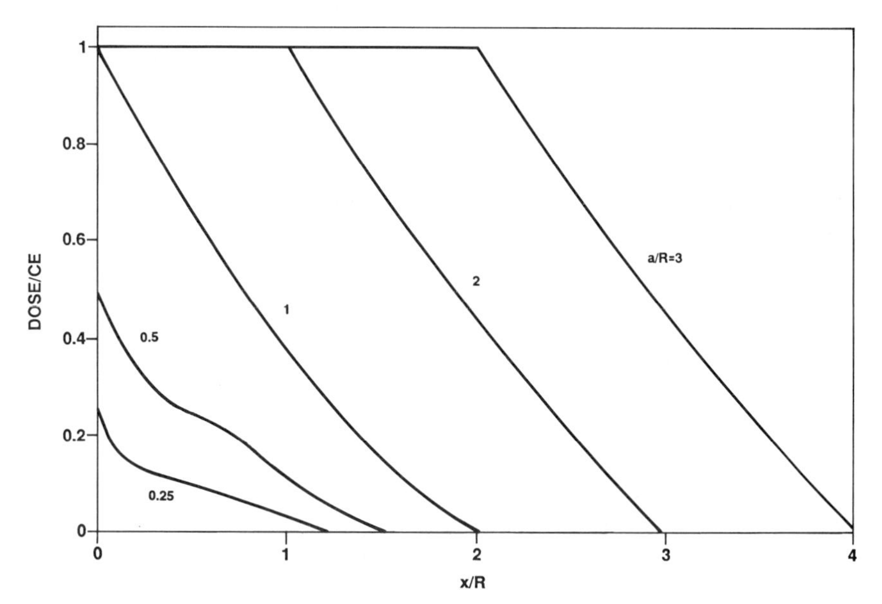

FIGURE I-3 Absorbed dose from alpha particles of energy E and range R as a function of distance x from the centers of spherical regions of radius a that contain uniform concentrations C of the emitter [in the $e(r) = E/R$ approximation].

at distances greater than the alpha-particle range from the edge, the dose drops to zero because particles cannot penetrate that far.

The figure also suggests when the average dose $\langle D \rangle$ discussed above in the sphere is a reasonable estimate of the dose throughout the sphere. Clearly, it is reasonable if the lower doses at the inner edge of the sphere are negligible in the average. Since the mass involved is proportional to the square of the distance from the center, the sphere must be quite large relative to the range of the alpha particles. For about 10% accuracy, the radius of the sphere must be roughly 30 times the range; this usually means that spheres with radii of 1 to 2 mm are needed to give meaningful averages.

When the radius of the sphere equals the range, the dose CE is attained only at the exact center. For smaller spheres, the dose CE is never attained. [In the $e(r) = E/R$ approximation only, the relative dose at the center is a/R.] In general, for uniform distributions in C, the absorbed dose does not exceed CE anywhere. If the sphere (or any other small volume) is very small, then at distances several times its radius:

$$D = [CV\ e(x)]/(4\pi x^2), \tag{I-6}$$

where V is the volume of the sphere and, hence, CV is the number of alpha particles emitted.

UNITS

The ICRU and U.S. National Bureau of Standards[10] recommend the use of the International System of Units (SI). Absorbed dose is a quotient of a quantity with dimensions of energy by one with dimensions of mass; therefore, its unit in the SI is joules per kilogram ($J\ kg^{-1}$). In the radiological sciences, this unit is called the gray (Gy); $1\ Gy = 1\ J\ kg^{-1}$. The gray recently replaced another popular unit, the rad; 1 Gy = 100 rad.

The SI units for the quantities in Equations I-2 and I-3 are the gray, inverse kilograms, and joules. These are seldom convenient units. In particular, radiation energies are universally given in electron volts (eV) or a multiple thereof. Also, C and $\langle C \rangle$ are often given in either becquerels (Bq) times a time per unit mass or curies (Ci) times a time per unit mass. The becquerel is a unit of activity, the rate of radioactive disintegrations; $1\ Bq = 1\ s^{-1}$, that is, 1/s. The curie is an older unit for activity; $1\ Ci = 3.7 \times 10^{10}\ s^{-1}$.

To accommodate mixed systems of units, Equations I-2 and I-3 are rewritten:

$$\langle D \rangle = k \langle C \rangle E, \text{ and} \qquad \text{(I-7)}$$

$$D = kCE, \qquad \text{(I-8)}$$

where k is the same in both and is introduced solely to provide for the different units used. For example, if D or $\langle D \rangle$ is in rads, C or $\langle C \rangle$ is in microcurie hours per gram, and E is million electron volts, then $k = 2.13\ J\ kg^{-1}/\mu Ci\ h\ MeV\ g^{-1}$.

RELATIVE BIOLOGICAL EFFECTIVENESS

Equal doses of different radiations do not always produce the same effect. In radiobiology, therefore, the relative biological effectiveness (RBE) was introduced to compare the effects of different radiations. The RBE of a test radiation with respect to a reference radiation, for a given effect, is defined as the ratio:

$$\text{RBE} = D_{\text{reference}}/D_{\text{test}}, \qquad \text{(I-9)}$$

where D_{test} and $D_{\text{reference}}$ are the doses of the two radiations that produce the same degree of the given effect. If the radiation being tested required less dose than the reference radiation, it would be said to be the more effective one, and its RBE would be greater than 1.

An RBE is a number. But the effect considered can be defined either with numbers (e.g., the number of tumors) or without numbers (e.g., degree of erythema). All that is required is a way of identifying equality of effect (or of identifying one effect as greater than or equal to another).[8]

RBEs are used for comparing radiations. To give a meaningful comparison, everything else that might affect the outcome should be the same during the experimental comparisons. For example, the absorbed dose distribution, the exposure time, the temperature, the atmosphere in which the cells or animals are exposed, and the growth conditions after the exposure should all be the same.

The RBE may depend, in particular, on how long observations are continued after exposure to radiation. In experiments with cells or animals, it is conventional to follow the exposed populations for their lifetime, or at least until new occurrences of the effect cease to appear. In epidemiological studies of human populations, few studies have reached this degree of completion, and caution is required in interpreting the data derived from them.

Making the absorbed dose distributions the same during the irradiations may be difficult. There is seldom difficulty in in vitro cell experiments where the absorbed dose can ordinarily be made uniform throughout the exposed population. In animal experiments, on the other hand, dose uniformity is the exception rather than the rule: Radiations incident from outside the body are subject to different attenuations; for internally deposited radionuclides, the distributions of dose reflect the distributions of the radionuclides and can be very erratic. When one or both of the test and reference dose distributions are nonuniform, the dose to be entered in the definition of RBE is not defined. If the spread in doses is not too great, an average dose can be used, with a consequent uncertainty in the RBE.

DOSE AVERAGING

Because of the nonuniformity of dose typically encountered with alpha particles, the following argument is often made. For low doses the yield is proportional to the dose; thus, if the yield is averaged throughout some tissue, the average yield would be proportional to the average dose (the $\langle D \rangle$ dealt with above). One could then assess RBEs with average doses, which would be a great simplification because the generally difficult determination of nonuniform doses is avoided. But this argument hides several critical assumptions. One is that the cells are equally sensitive throughout the tissue, something that is not obvious in view of the differences in oxygenation and nutrient supply throughout a typical tissue. Probably even more critical is the assumption that the cells are uniformly distributed throughout the tissue (implied by the uniform weighting of the dose in the tissue during the averaging). For example, if the critical cells were the epithelial cells lining small blood vessels and the radionuclide were one that deposited in or near these cells, the dose to them could be very much higher than the average dose in the tissue.

In spite of these criticisms and because of the practical difficulties in determining nonuniform doses, average doses have normally been used in alpha-particle dosimetry. In comparing like situations, the practice of using $\langle D \rangle$ values is a useful, practical expedient. The practice leads to difficulty when data for one radionuclide are applied to another or when data for one species are applied to another. In these applications, the actual doses to the relevant cells should be used in determining RBEs.

MICRODOSIMETRY

If the energy imparted by radiation to the mass in a small volume (usually called a site in the microdosimetric literature) were measured repeatedly under apparently identical conditions, the values obtained would differ. These differences are not experimental errors; the errors can be made much smaller than the differences observed. The differences are inherent; they are due to the randomness in the number of charged particles that impart energy to the site and to the randomness in the energy imparted to the medium in the individual interactions between a particle and the medium. These random features are particularly important for alpha particles and other high-LET radiations where (as can be seen in autoradiographs) the particle density is often so low that many sites are struck by only a few particles and some sites are not struck at all. In ordinary dosimetry, that is, in the determination of absorbed dose, the different values of the energy imparted to the mass would be averaged; information about the extent of the randomness would thus be discarded. In microdosimetry this information is kept and exploited.

While the effect of the randomness is present, no matter what the dose or size of the site studied, the degree of variation encountered will be less the larger the dose or the larger the mass. As a consequence, although microdosimetry applies to sites of all sizes, it generally focuses on low doses and small masses where the differences are larger. Usually the attention is on masses the size of cells or cellular components.

SPECIFIC ENERGY

Dosimetry deals with the absorbed dose, the *mean* energy per unit mass imparted to matter by radiation; microdosimetry deals with the *actual* energy per unit mass. The latter is given another name (specific energy) and symbol (z) to distinguish it from the former. Dose and specific energy have the same units. The definitions of dose and specific energy are framed so that $\langle z \rangle$, the average value of the specific energy over many repetitions of the irradiation, equals the absorbed dose:

$$D = \langle z \rangle. \qquad \text{(I-10)}$$

The specific energy is the result of the energies imparted by individual charged particles. The number of events in which individual particles

TABLE I-1 Representative Values of Microdosimetric Parameters for a 1-μm Sphere

Radiation	$\langle z:1\rangle$ (Gy)	$\frac{\langle z^2:1\rangle}{\langle z:1\rangle}$ (Gy)
65-kVp x rays	0.45	1.05
250-kVp x rays	0.25	0.78
^{60}Co gamma rays	0.08	0.34
0.43-MeV neutrons	10	15
1.8-MeV neutrons	7.1	12.6
14-MeV neutrons	2.5	30
5.3 MeV alpha particles	13	19

impart energy is random with a Poisson distribution. If the mean number for that Poisson distribution is m and if the mean specific energy for single events is denoted by $\langle z : 1\rangle$, then Kellerer and Rossi[9] (see also ICRU[5]) proved that:

$$D = \langle z\rangle = m\langle z : 1\rangle; \tag{I-11}$$

that is, the average specific energy due to all the charged particles (the absorbed dose) equals the product of the average number of events and the average specific energy for a single event.

The mean square specific energy, $\langle z^2\rangle$, is given[9] by:

$$\langle z^2\rangle = m\langle z^2 : 1\rangle + m^2\langle z : 1\rangle^2, \tag{I-12}$$

where $\langle z^2 : 1\rangle$ is the mean square for individual events. The mean square can be used to calculate the variance of the specific energy:

$$\sigma^2 = \langle(z - \langle z\rangle)^2\rangle, \tag{I-13}$$

$$= m\langle z^2 : 1\rangle, \text{ and} \tag{I-14}$$

$$= D\langle z^2 : 1\rangle/\langle z : 1\rangle. \tag{I-15}$$

The variance, the standard deviation (σ), or the coefficient of variation ($\sigma/\langle z\rangle$) can be used to indicate the breadth of the distribution in specific energy. According to Equation I-15, the variance is the product of the absorbed dose and a factor that is independent of dose; the factor depends only on the characteristics of single events. Table I-1 lists values of this factor and of the mean specific energy for single events for a number of radiations.

DISTRIBUTIONS IN SPECIFIC ENERGY

The mean values just discussed are the means and mean squares of probability densities in specific energy. Two basic kinds of densities are of

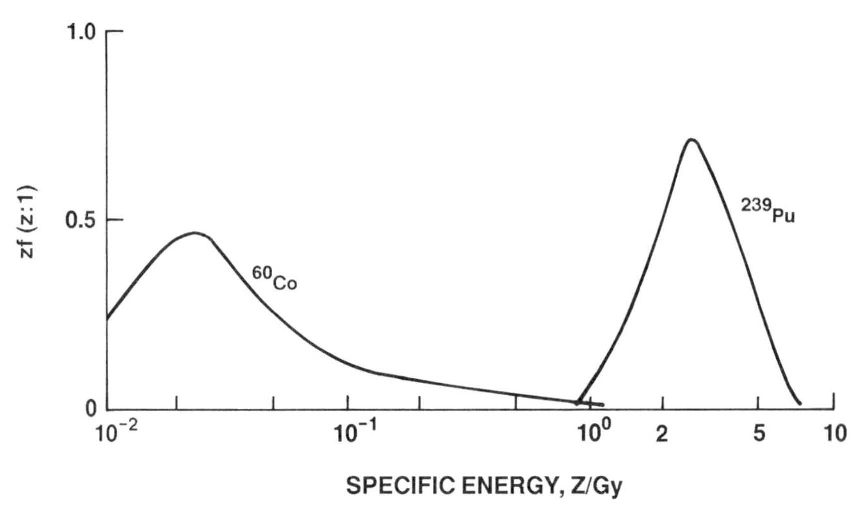

FIGURE I-4 Probability densities in specific energy for single events, $f(z : 1)$ for ^{60}Co and ^{239}Pu.

interest: densities for individual events and densities for a given absorbed dose. The probability density for single events will be denoted by $f(z : 1)$; this means that $f(z : 1)dz$ is the probability that the specific energy due to a single event is in a range dz that includes z. The probability density for a given dose will be denoted by $f(z)$.

In the microdosimetry of alpha particles, each distribution of the alpha-particle emitters can produce a different probability density. Here, only two examples will be given: the charged-particle equilibrium situation considered above, and the situation in which the emitters are agglomerated into particulates from which many alpha particles are emitted. The densities for nonequilibrium distributions of emitters can also be calculated.[11]

Distributions for Single Events

Figure I-4 shows the single-event densities for ^{60}Co gamma rays, a low-LET radiation, and for ^{239}Pu alpha particles, a high-LET radiation, for charged-particle equilibrium. To cover a wide range of the abscissa, the probability density is multiplied by z and then plotted on a logarithmic scale, because equal areas anywhere under such a curve represent equal probabilities of occurrence. On this logarithmic plot the two distributions differ only slightly in shape, but there is a large distance along the abscissa between them due to the difference in the stopping powers of the particles (discussed earlier in this appendix).

The location of the single-event distributions on the specific energy scale also depends strongly on the size of the site considered. The energy imparted to the site by a particle increases in proportion to the diameter of the site, but the mass of the site increases in proportion to the cube of the diameter. The energy per unit mass, therefore, is inversely proportional to the square of the diameter. The following expression relates, approximately, the mean specific energy in single events, $\langle z : 1\rangle$, the mean stopping power of the particles, $\langle S\rangle$, and the diameter, d, of a spherical site in tissue of unit density:

$$\langle z : 1\rangle \simeq 0.2\langle S\rangle/d^2, \tag{I-16}$$

where the units are grays, kiloelectron volts per micrometer, and micrometers, respectively. Thus, for larger sites the distribution is moved to the left in Figure I-4; for smaller sites the distribution is moved to the right. There are also changes in the details of the shapes of the distributions.

The mean number (m) of events in a site is proportional to the cross-sectional area it presents to the charged particles, that is, proportional to the square of the site size. This fact and the inverse dependence of $\langle z : 1\rangle$ on the square of the site size are the reason that the dose $D = m\langle z : 1\rangle$ is independent of the site size.

For alpha particles, the difficulties in measuring such a short-range radiation have forced investigators to use calculations to obtain approximate single-event distributions.[11] The $f\langle z : 1\rangle$ for radiations with longer ranges are determined experimentally with proportional counters.[5,12]

Distributions for a Given Dose

While $f(z : 1)$ is the distribution for a single event, $f(z)$ is the distribution for a number of events. The number of events is random with a Poisson distribution. The $f(z)$ distribution is calculated from $f(z : 1)$ by Fourier-transform methods.[7,11]

Figure I-5 shows the distributions for ^{239}Pu alpha particles for different absorbed doses. For small doses, the chance of a site being hit by more than one alpha particle is very small. The area under $f(z : 1)$ is, by definition, unity; that under $f(z)$ at these doses is approximately equal to the probability of the one event. Consequently, $f(z)$ has the same shape as $f(z : 1)$ but is smaller by a factor equal to the one-event probability.

As the dose increases, the area under $f(z)$ increases because the chance of a site being missed by alpha particles decreases. The increase is seen in two ways: the part similar in shape to $f(z : 1)$ increases, reflecting more single hits. In addition, a small bulge begins to develop on the high-z side due to hits by two alpha particles.

At much higher doses $f(z)$ begins to move to the right. The chance of just one hit has grown small; many now occur and give a higher total z. The area under the curve is close to unity, because the chance of any site being missed is now small. As $f(z)$ moves farther and farther to the right

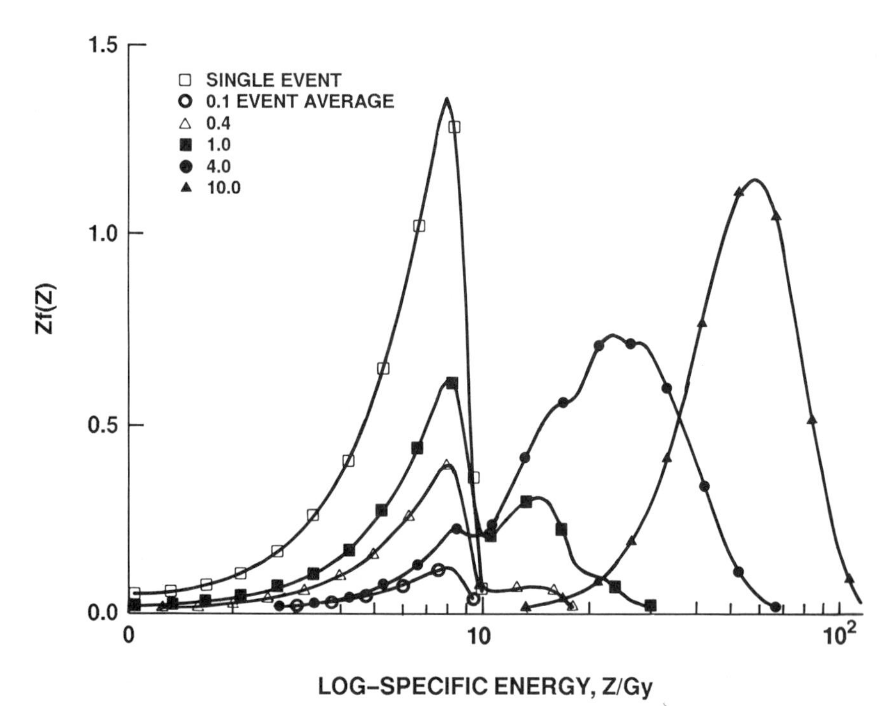

FIGURE I-5 Probability densities $f(z)$ in specific energy for different absorbed doses (i.e., different mean numbers of events) for ^{239}Pu alpha particles compared with the density for single events $f(z : 1)$.

(with increasing dose), the shape of the curve approaches that of a normal distribution.

Distributions for Particulate Sources

The distributions shown in Figures I-4 and I-5 are for radionuclides randomly dispersed in the tissue. But random dispersion does not always occur. Under some circumstances the molecules of a nuclide coalesce with each other (and with other molecules). These agglomerations of radionuclides are called particulates. For a particulate, many alpha particles may emerge from nearly the same point in the medium. Sites near such a particulate stand a larger chance of receiving energy than if the activity were spread more uniformly; sites far away stand a smaller chance.

Figure I-6 illustrates what the agglomeration into particulates can do.[11] It shows the distributions in specific energy, $f(z)$, for the same absorbed dose (0.75 Gy) for different average numbers of alpha particles per particulate. To get the same absorbed dose, the number of particulates per unit volume is changed in inverse proportion to the number of alphas

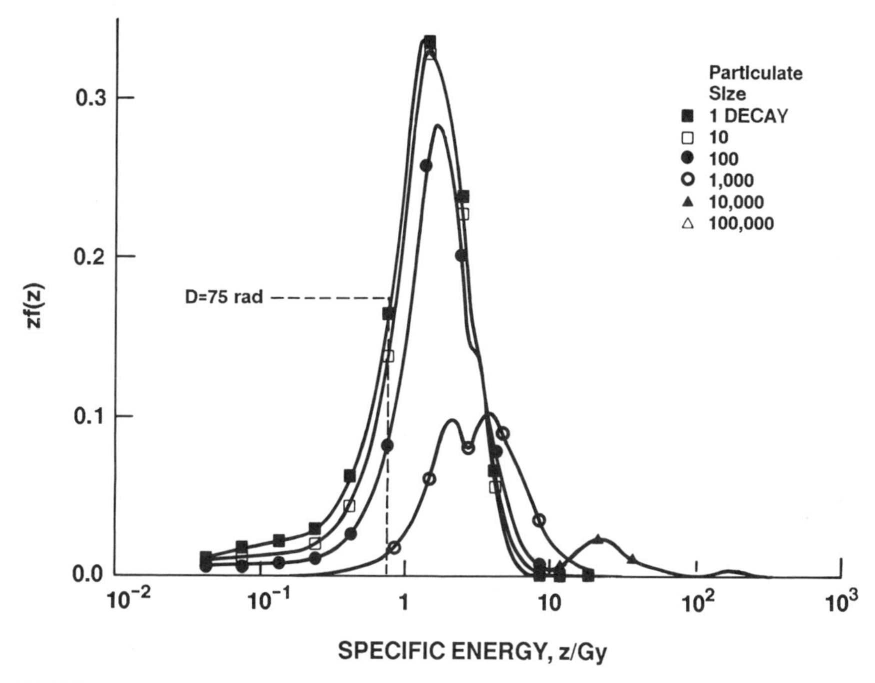

FIGURE I-6 Probability densities in specific energy for particulates of different sizes for the same absorbed dose (0.75 Gy).

per particulate. When the number of alphas per particulate is small, 1 or 10 for this site size, the $f(z)$ values do not differ much; they are actually very close to the $f(z)$ for no agglomeration into particulates. In these circumstances, even though many alpha particles are emitted at a common point, the chance of a significant number of them going in nearly the same direction so as to affect a common site is small. For particulates that emit up to about 100 alpha particles and for this site size, $f(z)$ differs only slightly from $f(z)$ for nonagglomeration, that is, for a uniform distribution of molecules of the radionuclide. For about 1,000 alpha particles and higher, it is distinctly different. A site close to a particulate that emits so many alpha particles stands a good chance of receiving energy from more than one alpha particle, with the result that its $f(z)$ is pushed to higher specific energies. But, when so many alpha particles are emitted from each particulate, there are fewer particulates for a given dose, with the result that there is an increased chance that some sites will not be close enough to any particulate to be hit by any alpha particles. This causes the area beneath $f(z)$ to decrease. At 0.75 Gy, for 1 alpha per particulate and this site size, the chance of being missed is 0.52. For 100,000, it is 0.9978. In other words, only 0.0022 sites get hit at all; but, the sites that are hit are apt to be hit many times.

REFERENCES

1. Bichsel, H. 1972. Passage of charged particles through matter. P. 159 in American Institute of Physics Handbook, D. E. Gray, ed. New York: McGraw-Hill.
2. Evans, R. D. 1955. The Atomic Nucleus. New York: McGraw-Hill.
3. International Commission on Radiation Units and Measurements. 1970. Linear Energy Transfer. Report 16. Bethesda, Md.: International Commission on Radiation Units and Measurements.
4. International Commission on Radiation Units and Measurements. 1980. Radiation Quantities and Units. Report 33. Bethesda, Md.: International Commission on Radiation Units and Measurements.
5. International Commission on Radiation Units and Measurements. 1983. Microdosimetry. Report 36. Bethesda: Md.: International Commission on Radiation Units and Measurements.
6. International Commission on Radiation Units and Measurements. 1984. Stopping Powers for Electrons and Positrons. Report 37. Bethesda, Md.: International Commission on Radiation Units and Measurements.
7. Kellerer, A. M. 1970. Analysis of patterns of energy deposition; a survey of theoretical relations in microdosimetry. Pp. 107–134 in Proceedings, Second Symposium on Microdosimetry, H. G. Ebert, ed. Brussels: Euratom.
8. Kellerer, A. M., and J. Brenot. 1973. Nonparametric determination of modifying factors in radiation action. Radiat. Res. 56:28–39.
9. Kellerer, A. M., and H. H. Rossi. 1970. Summary of the quantities and functions employed in microdosimetry. Pp. 843–853 in Proceedings, Second Symposium on Microdosimetry, H. G. Ebert, ed. Brussels: Euratom.
10. National Bureau of Standards. 1976. Guidelines for Use of the Metric System. Report LC1056. Washington, D.C.: National Bureau of Standards.
11. Roesch, W. C. 1977. Microdosimetry of internal sources. Radiat. Res. 70:494–510.
12. Rossi, H. H. 1968. Microscopic energy distribution in irradiation matter. Pp. 43–92. In Radiation Dosimetry, Vol. I, 2nd ed., F. H. Attix and W. C. Roesch, eds. New York: Academic Press.
13. Thwaites, D. I., and D. E. Watt. 1978. Similarity treatment of phase effects in stopping power for low energy heavy charged particles. Phys. Med. Biol. 23:426–437.
14. Ziegler, J. F. 1980. Handbook of stopping cross-sections for energetic ions in all elements. New York: Pergamon.

APPENDIX II
Cellular Radiobiology

INTRODUCTION

Cellular radiobiology is a well-developed discipline, dating to almost 90 yr ago, when it was first recognized that exposure to ionizing radiation had biological consequences. Much information has since been accumulated from studies with irradiated animals and plants and, over the last 50 yr in particular, from studies of effects at the cellular, physical, and chemical levels. Molecular and cellular radiobiology have contributed greatly both to our understanding of the physical and chemical processes involved in the induction of radiation effects and to our understanding of the responses observed in whole organisms. From the vast store of radiobiological information, it is possible to draw a number of general principles.[26,28,35,36] These principles sometime constitute our only rational basis for making human risk assessments, for example, when direct empirical observations on which to base such assessments are not available.

DOSE-RESPONSE CURVES

Many models have been developed to describe radiobiological responses, but one broad generalization is that empirical dose-response curves for cellular radiobiological effects, whether in vivo or in vitro, take the general form:

$$Y = (a + \alpha_1 D + \beta_1 D^2) \exp(-\alpha_2 D - \beta_2 D^2), \qquad \text{(II-1)}$$

where Y is the response observed in a population, a is the spontaneous incidence, α_1 and β_1 are the coefficients for the induction of the observed effect, D is the dose, and the exponential is a term expressing the loss of observed response due to competing effects, such as cell killing, with α_2 and β_2 as the coefficients for cell killing. For doses well below that at which competing effects are important, the simpler expression, $Y = a + \alpha D + \beta D^2$, suffices. The relationship is often referred to as mixed quadratic or linear-quadratic, although the expression is simply quadratic. Such an expression adequately describes the dose-response relationship at low to moderate doses of low linear energy transfer (LET) radiation for a wide variety of cellular radiobiological endpoints, including induction of mutations and induction of the various classes of chromosomal aberrations. As described in more detail below, the quadratic expression also describes adequately the responses of such systems to high-LET radiation.

Cell killing is often analyzed according to the following expression for survival:

$$S = 1 - (1 - e^{-kD^n}), \tag{II-2}$$

where k is the coefficient for killing, and n is an exponent often called the hit number or extrapolation number. The survival fraction can also be expressed by an $\alpha D + \beta D^2$ model of the form:

$$S = \exp(-\alpha D - \beta D^2), \tag{II-3}$$

the same as the last term in the complete quadratic with saturation model given above in Equation II-1.

Although the quadratic model can be derived entirely empirically from response data, there appears to be a rational biophysical basis. What such dose-effect curves imply is that some of the effects of radiation of any given class are induced by single ionizing events (i.e., the passage of a single photon or of a single particle), whereas others result from the interaction of two or more statistically independent ionizing events. The dose-squared term can be understood in another way: If P is the probability of hitting a target and causing a sublethal amount of damage, then the probability that target will be hit twice (or more), to give an effective hit, is $(P)(P)$, or P^2. The quadratic expression can yield curves ranging from linear (or nearly linear), in the case where β is so small in relation to α that the αD dominates, to essentially dose-squared, in the case where α is negligibly small in relation to β. Mutation induction in simple prokaryotic cells is an example of linear response, and the induction of two-break chromosomal aberrations of the exchange type in higher eukaryotic cells exposed to acute doses of low-LET radiation is an example of quadratic response. Many dose-response curves, however, are somewhere in between.

EFFECT OF DOSE RATE AND FRACTIONATION

The expressions presented above apply specifically to low-LET radiation, such as x or gamma rays, delivered at a high (acute) rate. It was noticed early, however, that when low-LET doses were protracted or appreciable periods passed between successive dose fractions, the effectiveness of the total dose was likely to diminish. In fact, as dose rate decreased, the radiobiological endpoints that at higher dose rates had response curves with an appreciable βD^2 term began to lose that term. Eventually, little was left of the βD^2 term, and the dose-response curve was essentially linear, represented simply by αD. With increasing interfraction time, much the same thing happened, with the βD^2 term for the total dose decreasing until, instead of what is often called complete interaction [i.e., $\beta(D_1 + D_2 + D_3 \ldots)^2$], one observed the sum of βD^2 terms for the individual fractions (i.e., $\beta D_1{}^2 + \beta D_2{}^2 + \beta D_3{}^2 \ldots$). This implied that somehow the subeffective partial lesions left at the end of the first dose fraction were becoming unavailable, or repaired, with time; thus, if enough time elapsed between fractions (a few hours), none were left from the first dose fraction to interact with those from the second. Loss of subeffective partial lesions also accounts for the simple dose-rate effect. Lea[35] understood this early and created a correction factor that he called the G factor to correct for dose-rate effects:

$$G = 2(\tau/T)^2(T/\tau - 1 + e^{-T/\tau}), \tag{II-4}$$

where τ is the average time between breakage and restitution, and T is the duration of treatment.

EFFECT OF LET

As the LET of radiation (usually calculated as track-average LET) increases, two things happen. First, for radiobiological endpoints that have a substantial βD^2 term with acute doses of low-LET radiation, the curves begin to straighten, losing the D^2 component and tending toward linearity. Second, the effectiveness of the αD component also increases, indicating that this is not the same phenomenon as that observed for dose rate or fractionation with low-LET radiation. In sum, the higher-LET radiation seems to be capable, if it deposits energy in a target at all, of causing fully effective events. In other words, the production of subeffective lesions becomes less and less likely, while the production of fully effective lesions becomes more likely as LET increases. This makes sense, because as the ionization per unit track length increases, a target becomes more likely to suffer substantial damage, if it is affected at all. In fact, one might expect that if the LET of radiation becomes high enough, the effectiveness with increasing LET would saturate at some point and then fall off as the LET continued to increase. That is what is observed, particularly in prokaryotic systems. For the induction of radiobiological effects in mammalian cells, the target volume appears to be large enough for the effect to occur at a higher LET, with the maximally effective LET being around 100 keV/μm.

MICROSCOPIC DOSE DISTRIBUTION

With energetic photon irradiation, the quantity dose, used in calculating a dose-effect model, has an easily understood meaning and is easy to measure empirically. However, the situation often is not so simple for particle irradiation from internally deposited radionuclides, particularly when the mean path length of the charged particle is small in relation to the diameter of the average cell or subcellular structure of interest and when the distribution of the radionuclide is very nonuniform. The problem is particularly acute when, as in the case of tritium incorporated into the DNA of a cell in the form of tritiated purines or pyrimidines, only parts of cells have incorporated the radionuclide and only some of the potential target cells have incorporated any radionuclide at all. In such cases, dose expressed in the usual terms of the average amount of energy deposited per unit of tissue mass over the mass of the organs (or parts of organs) becomes meaningless, if not actually misleading. Here it is important to consider the microscopic distribution of dose and the doses accumulated by individual targets (usually taken to be individual cell nuclei).

In the case of the alpha-emitting radionuclides of interest here, we must consider the problem presented by the inadequacy of simply measuring body or even organ burdens of a radionuclide, and try instead to identify the target cells at risk for a particular possible health impact and then to determine the average dose to the nuclei of all these cells. The distribution factor will otherwise be confounded with the relative biological effectiveness (RBE) factor, and that can lead to large uncertainties and ambiguities. In fact, if the RBE for a given particle seems to be reasonably well established, deviations from that RBE can be used to infer the magnitude that must be attributed to the microscopic dose-distribution factor.

MODELS OTHER THAN THE QUADRATIC

Models designed to describe and, thus, to permit prediction of the response of cell populations to radiation with different LETs can be divided broadly into two categories: those derived empirically from observed cellular dose-response curves (observation obviates assumptions about or dependence on underlying molecular or subcellular biological effects or mechanisms), and those built on assumptions about underlying subcellular mechanisms (which might or might not be shown to be applicable). Although the two approaches must ultimately converge, there can be little confidence that this will occur soon.

The first category includes two models that are related to the original target theory:[35] the quadratic model and the model of Bond et al.[8] As noted above, the former is applicable to a wide variety of endpoints in many systems, particularly with low-level exposure to radiation (i.e., either small doses at any dose rate or larger doses at very low dose rates). It also adequately accommodates dose-rate and protraction effects and accurately describes most responses as a function of LET in tissue in terms of an increasing αD component and a decreasing βD^2 contribution as the LET increases.

Thus, with very low doses, the mean absorbed dose to the population of cells increases only because the fraction of cells hit increases; the mean dose (specific energy) delivered stochastically to the hit cells remains constant. However, as the fraction of cells hit approaches unity, the absorbed dose to the cell population can increase only if the number of hits per cell increases.[8] At high doses and high dose rates, where each cell has received many hits, the variance of the mean decreases; so the dose to each cell, the mean dose to the cell population, and the mean dose to the organ or other medium approach equality.

The other model that belongs in this category and that combines some elements of hit theory with those of microdosimetry is under development.[8] The unique addition is an empirically derived hit size effectiveness factor. This factor gives the probability that a cell with a certain amount of energy deposited within it will respond as a function of the energy deposited. Additional testing with different endpoints and biological systems is necessary before the degree of applicability of the model can be ascertained.

The second category includes several models that cover a wide variety of assumptions, some of which are commented on below. The early dual-action model[33] combines microdosimetry with assumptions about the stochastic interactions of microdosimetric events to derive the basic formulation:

$$E = K(\varsigma D + D^2), \qquad \text{(II-5)}$$

where ς is a physical quantity equal to the average specific energy (dose to a subcellular target volume in single events), and K is the sensitivity coefficient. This expression is equivalent to the more general $\alpha D + \beta D^2$ formulation, with $K\varsigma$ equivalent to α and K equivalent to β.

A recently developed thesis is the lethal and potentially lethal model,[24] which, with the earlier repair-nonrepair model of cell survival[46] and other models, is built on the added assumptions of irreparable lethal lesions, repairable potentially lethal lesions, first-order kinetics for correct repair, and second-order kinetics for misrepair. Although the models have been applied only to cell lethality in plateau-phase (stationary-phase) cells in culture, they do take LET and dose rate into account. The principal mathematical formulation is complex and embraces a power expansion, but the first two terms also yield the $\alpha D + \beta D^2$ expression. Such models constitute valuable contributions in attempting to unify the most attractive features of a number of others; the extent of their universality remains to be determined.

Other models have also been developed, for example, the molecular theory of Chadwick and Leenhouts,[21] the kinetic model of Dienes,[25] the cybernetic model of Kappos and Pohlit,[32] the incomplete-repair model of Thames,[45] and the repair model of Braby and Roesch.[19] Many are limited, in that they deal either with only a small fraction of the many radiobiological factors that must be taken into account (e.g., some are limited to high dose rates and others to the question of repair, possible misrepair, and dose rate), or in encompassing only particular endpoints (e.g., some deal only with cell lethality, and others exclude it).

There is evidence that the various models can be reconciled, but it seems unlikely that there will be general agreement or that the mechanistic assumptions behind any of them will soon be proved.

None of the models in this category has the simplicity or the generality of application of the quadratic model, and none is at the point of applicability to the problems of predicting either genetic or carcinogenic responses in mammalian systems. Thus, although they contribute to our general understanding and appreciation of problems remaining to be solved, they cannot yet be applied in practical exercises that require the prediction of risk associated with exposure to principally low-level radiation. Therefore, it appears reasonable that the quadratic model be used.

APPLICATION TO RISK ESTIMATION FOR INCORPORATED ALPHA-EMITTING RADIONUCLIDES

With some notable exceptions—such as Thorotrast-exposed patients, radium-dial painters, or uranium miners—there is generally little direct information on the human health effects of the internally incorporated alpha-emitting radionuclides. However, we can derive several generalizations that can be applied to the problem of extrapolating available empirical evidence of the induction of human health effects by low-LET radiation.

First, even though the question of whether the low-LET effects best fit a simple linear dose-response model or a dose-squared model is unsettled, as far as the epidemiological data on radiation carcinogenesis are concerned (see Committee on the Biological Effects of Ionizing Radiations [BEIR

III][38]), we can be reasonably sure that the dose-effect curves of alpha particles are linear (or essentially linear), at least for the low to moderate doses of interest, for which saturation effects can be ignored. Second, the evidence is that there are no dose-rate effects for high-LET alpha particles (with the one possible exception of in vitro cell transformation as discussed below). Third, we can be sure that the yields of effects per unit dose of alpha particles are greater than those per unit dose for low-LET x or gamma rays, that is, that the RBE for alpha particles is greater than unity. Although the appropriate RBE for a given alpha particle might not have been empirically measured in an appropriate target-cell system, generalizations on the basis of LET seem reasonable, provided that they are based on RBEs determined in cells likely to have similar cell nucleus and target volumes and based on doses that produce similar hit fractions per rad. Thus, in considering the hazard of induction of a malignancy of, say, the liver by deposition of plutonium, it seems reasonable to multiply the low-dose risk coefficients for low-LET radiation available in the BEIR III report[38] by an appropriate RBE factor to derive a new estimate that can be useful in the absence of empirical information on the overall risk associated with the radionuclide of interest.

We present below the empirical evidence from cellular radiobiology on RBE for alpha particles, with necessary background information, on which to base estimates of expected effects of exposure of human populations to alpha-emitting radionuclides.

MAMMALIAN CELL SURVIVAL

Although much work had been done earlier with prokaryotes and unicellular eukaryotes, mammalian cell survival studies became possible only with the development by Puck et al.[39] of a practical clonal assay for the survival of single mammalian cells in culture. The general principles already discussed in this appendix were elucidated mainly with low-LET x and gamma rays and have been reviewed elsewhere.[26,28] Much work has been done with high-LET irradiation, particularly with neutrons, protons, and beams of heavy ions, largely because of interest in their potential application to radiation oncology. Much less has been done on cell killing by alpha particles from the radionuclides of interest here. Therefore, it is necessary to consider the larger body of data from other high-LET radiation.

Extensive studies with fast neutrons of various energies have been done by Broerse et al.,[20] Berry,[6] and by the Radiological Research Laboratory group at Columbia University.[30,34] The last studies are of particular interest because of the essentially monoenergetic neutron beams that the group at Columbia University was able to use. Survival curves for Chinese hamster cells irradiated with neutrons of 0.1–50 MeV have been obtained. As expected, neutron energies below a few million electron volts produced exponential survival curves without the shoulder characteristic of low-LET survival curves. At energies above a few million electron volts, the LET is lower and a small shoulder becomes evident.

RBE varied with neutron energy, exhibiting a rather broad peak at approximately 0.4 MeV. Because of the difference in survival-curve shapes, RBE was a function of the survival fraction at which the comparison was made with the reference curve, in this case for 250-kV x rays, peaking at about 9 for 80% survival and falling to approximately 3 at 0.001% survival.

Blakely et al.[7] recently reviewed extensive mammalian cell-survival studies with heavy ions, ^{12}C to ^{40}Ar, which used a variety of tissue-culture cell types. RBE was found to increase from about 1.0 at a track-average LET of 10 keV/μm to about 2.5 at a track-average LET of around 100 keV/μm. Survival curves lost their shoulders as LET increased, becoming simple exponentials for a higher LET. Thus, the results of both the fast-neutron and the heavy-ion experiments are in general agreement as to the modification of curve shape and increase in RBE with increasing LET. The maximal RBE observed in the neutron experiments, about 5 for 5% survival, is somewhat higher than the maximum observed in the heavy-ion experiments, probably because the heavy ions, with their high velocities, deposit a larger fraction of their energy via lower-LET secondary electrons and are thus less monochromatic with respect to LET.[43] The results of the heavy-ion experiments clearly demonstrate the falloff in efficiency (dose wastage) at average LETs greater than approximately 100 keV/μm.

Barendsen and coworkers[1–4] have studied mammalian cell-survival curves for cyclotron-accelerated alpha particles (and deuterons) and for ^{210}Po alpha particles. Polonium-210 alpha particles induced simple exponential survival curves with an RBE ranging from 2.5 (at high acute doses) to around 6.0 (at low doses). With cyclotron-accelerated alpha particles at LETs of 25–86 keV/μm, survival curves for the higher-LET alpha particles were again simple exponentials, although that for 25-keV/μm, alpha particles did show some evidence of a shoulder. The RBE, in comparison with the curve for 200-kV x rays, was dose dependent but was in the range of 3–5 for survival below about 20%, with the peak for any degree of survival at about 100 keV/μm.

Lloyd et al.[36] also determined mammalian cell-survival curves for accelerator-produced alpha particles. Essentially exponential survival curves were found for alpha particles with LETs of about 85 keV/μm; the slope agreed well with that determined by Barendsen and coworkers.[1–4]

Barnhart and Cox[5] and Thacker et al.[43] have determined survival curves for Chinese hamster cells in tissue culture exposed to alpha particles from ^{238}Pu. Again, the survival curves were exponential. With the irradiation geometry used, the ranges of LET through the cells were 127–165 keV/μm and 100–135 keV/μm, respectively, in the two studies. RBE values of 7–10 were observed for high survival fractions.

Thus, there is general agreement that cell-survival curves for alpha particles, as well as other high-LET irradiation, are exponential with high RBE, peaking at around 100 keV/μm; are dose dependent; and reach maximal calculated values of about 5–10 at low doses. Low-dose-rate, low-LET ionizing radiation seems appropriate for comparison for the purposes of this report (i.e., comparison of the α terms of the quadratic relationship seems appropriate). Therefore, the lowest dose and, consequently, the highest RBE observed seem appropriate as a basis for extrapolating from the BEIR III report[38] and similar risk estimates to alpha-particle estimates when direct human data are not available.

MUTATION IN VITRO

Mutation induction at the hypoxanthine-guanine phosphoribosyl transferase (HGPRT) locus can be studied in vitro in Chinese hamster or human cells by using selection in a thioguanine-containing medium. Dose-effect curves have been determined in this way for both high- and low-LET radiations. Thacker et al.[43] and Cox and Masson[22] determined HGPRT mutation by heavy ions in Chinese hamster cells and human diploid fibroblasts, respectively. Helium, nitrogen, and boron ions with LETs of 28–470 keV/μm in the irradiated cells were used. Mutation-induction curves were essentially linear, giving maximal RBEs of about 6 at 90–200 keV/μm when calculated in terms of mutants per survivor and compared with the initial slope of the curve for gamma rays.

Both Barnhart and Cox[5] and Thacker et al.[43] have reported on the mutagenicity of alpha particles from ^{238}Pu in Chinese hamster cells. The study by Barnhart and Cox,[5] however, appears to have suffered from technical difficulties; as noted by Thacker et al., the latter found an RBE of about 2 for the alpha particles, which had an LET range of 127–165 keV/μm as they traversed the cells, compared with acute exposure to 250-kV x rays.

TRANSFORMATION IN VITRO

A few types of mammalian cells growing in tissue culture can be transformed in vitro from the growth patterns that characterize fairly normal cultured cells to a new phenotype that more closely resembles that of cancer cells. The basic change is from an orderly growth pattern exhibiting contact inhibition on the culture vessel surface to a pattern resulting from the loss of contact inhibition. That loss causes the cells to pile up and overgrow each other and to produce foci of a distinct and recognizable morphology. Examples are shown in Figure II-1.

The development of cell-culture systems has made it possible to study the cellular and molecular mechanisms involved in radiation transformation under defined conditions devoid of host-mediated homeostatic modulating factors, and to assess the underlying mechanisms qualitatively and quantitatively. These systems afford the opportunity to study dose-related and time-dependent interactions of radiation with single cells and to identify factors and conditions that can prevent or increase cellular transformation by radiation. Because such cells transformed in vitro give rise to tumors when injected into hamsters, whereas untreated cells show no spontaneous transformation,[14,15] the system has obvious implications with respect to in vivo carcinogenesis. The details of the system and its utility in cellular radiation biology were recently reviewed by Borek.[11,12]

The role of DNA as a target in radiation transformation was suggested early by the requirement of DNA metabolism for fixation of the transformed state.[15] The ability of genomic high-molecular-weight DNA purified from cells transformed in vitro to transmit the transformed phenotype to normal cells constitutes an important criterion for the neoplastic state of the cells transformed after exposure to a carcinogen,[12,13,17,18,41] indicating that the transformed phenotype of the cells exposed to the carcinogen in vitro is encoded in the DNA. This criterion aids in mechanistic studies of transformation that attempt to analyze the specific transforming genes activated as a result of exposure to the carcinogen and to elucidate genetic changes.[17,18]

For both low- and high-LET radiation, transformants are produced as a function of increasing dose up to approximately 1–5 surviving cells/100 exposed cells, at which point the curves saturate. Transformants are produced more efficiently by fission neutrons than by 250-kVp x rays, with an RBE of approximately 10 in the region below saturation.[16,31] Figure II-2 shows an example for 430-keV neutrons.

Data on alpha-particle-induced in vitro transformation are sparse and often incomplete (there are no data on the tumorigenicity of the transformed cells). The effect of low-energy alpha particles in transforming the C3H/10T-1/2 cells was evaluated by Lloyd et al.[37] Transformation frequency per surviving cell increased as the cube of the dose, peaking at a fluence of 1.5×10^7–2.5×10^7 alpha particles/cm^2 (205–342 rad). Maximal transformation frequency reached 4%. No parallel experiments were carried out with x rays; thus, no RBE for alpha radiation was determined. By taking the dose required to reach the peak transformation region and comparing it with x-ray data determined for the same cell line by Terzaghi and Little,[42] an RBE of approximately 2 is obtained.

A study carried out by Robertson et al.[40] evaluated the effects of ^{238}Pu alpha particles in mouse BALB/3T3 cells. The A31-11 mouse BALB/3T3

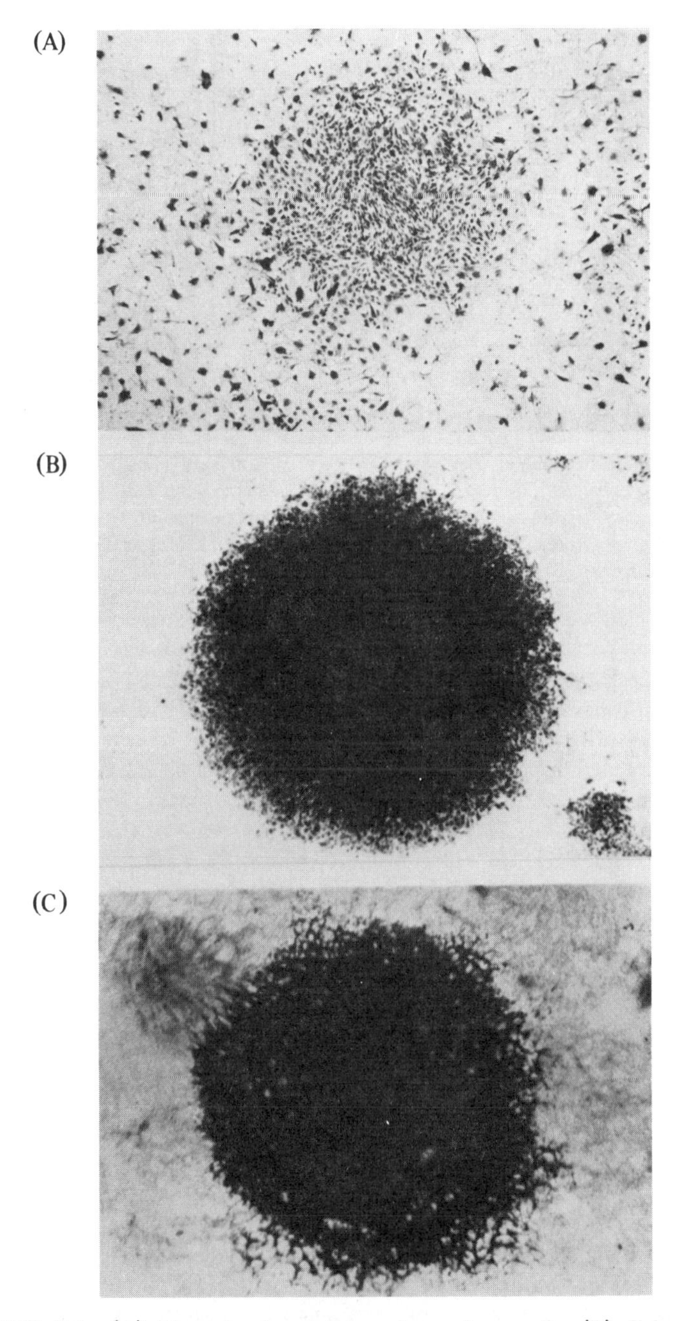

FIGURE II-1 (A) Normal colony of hamster embryo cells. (B) Colony of x-ray-transformed hamster embryo cells. (C) Focus of C3H/10T-1/2 cells transformed by x rays growing over normal cells. Morphology of transformed cells is the same as that after exposure to high-LET radiation. SOURCE: Borek.[12]

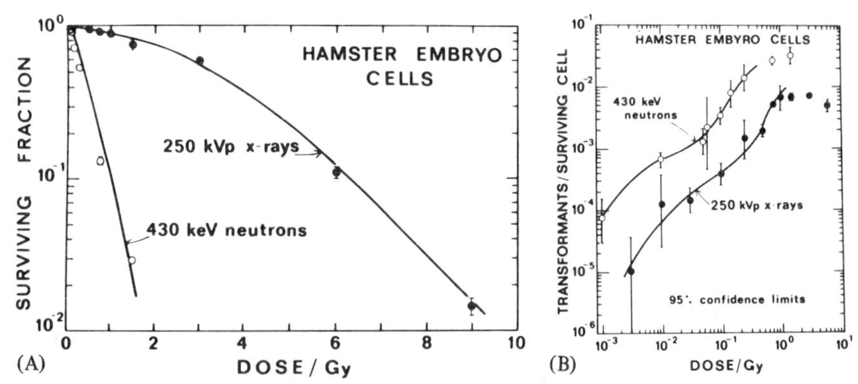

FIGURE II-2 (A) Pooled data on survival of hamster embryo cells irradiated with 250-kVp x rays (solid circles) or 430-keV monoenergetic neutrons (open circles). Error bars show estimated standard deviations. 1 Gy = 100 rad. (B) Pooled data for hamster embryo cells on number of transformants per surviving cell after irradiation with 250-kVp x rays (solid circles) or 430-keV monoenergetic neutrons (open circles) at the Radiological Research Accelerator Facility. Error bars show 95% confidence intervals for estimates. SOURCE: Borek et al.[16]

cell system was used, and in vitro transformation was induced by 5.3-MeV alpha particles from a specially constructed ^{238}Pu source. The biological effects were compared with those of 220-kVp x rays. The alpha-radiation survival curve gave an RBE of 3.5 at 50% survival. The transformation frequency increased exponentially with dose in the range examined (25–250 rad); the maximal RBE for the induction of transformation in growing cells was approximately 3.

However, the RBE for alpha transformation in nonproliferating cells appeared to be much higher; the yield of transformants among x-irradiated cells that were held in the stationary phase of growth for 6–220 h after irradiation declined by a factor of nearly 50, whereas no decrease occurred in alpha-irradiated cells. The findings suggest that carcinogenic damage induced by high-LET radiation in mammalian cells is very inefficiently repaired, compared with that induced by x rays, and that the intracellular carcinogenic effect of exposures to high-LET radiation can be cumulative. They also suggest that the effective RBE for alpha radiation in nonproliferating cell populations in vivo might be much higher than one would predict on the basis of measurements in dividing cells.

Work by Hall and Hei[29] with C3H/10T-1/2 cells compared the transforming action of radiation, from a source of ^{241}Am, delivered at 10 rad/min with that of gamma rays from a ^{137}Cs source with an absorbed dose rate of 137 rad/min at equivalent doses. Alpha particles were substantially more cytotoxic and more efficient than gamma rays in inducing oncogenic transformation. The calculated RBE ranged from 2.3 to 9 for the transformation frequencies examined.

REFERENCES

1. Barendsen, G. W. 1962. Dose-survival curves of human cells in tissue culture irradiated with alpha-, beta-, 20-KV.X- and 200-KV. X-radiation. Nature 193:1153–1155.
2. Barendsen, G. W., and T. L. J. Beusker. 1960. Effects of different ionizing radiations on human cells in tissue culture. I. Irradiation techniques and dosimetry. Radiat. Res. 13:832–840.
3. Barendsen, G. W., T. L. J. Beusker, A. J. Vergroesen, and L. Burke. 1960. Effects of ionizing radiations on human cells in tissue culture. II. Biological experiment. Radiat. Res. 13:841–849.
4. Barendsen, G. W., H. M. D. Walter, J. F. Fowler, and D. K. Bawley. 1963. Effects of different ionizing radiations on human cells in tissue culture. III. Experiments with cyclotron-accelerated alpha-particles and deuterons. Radiat. Res. 18:106–119.
5. Barnhart, B. J., and S. H. Cox. 1979. Mutagenicity and cytotoxicity of 4.4-MeV α-particles emitted by plutonium-238. Radiat. Res. 80:542–548.
6. Berry, R. J. 1974. Modification of neutron effects upon cells by repair, and by physical and chemical means. Pp. 257–271 in Biological Effects of Neutron Irradiation. TI/PUB/352. Vienna: International Atomic Energy Agency.
7. Blakely, E. A., F. Q. H. Ng, S. B. Curtis, and C. A. Tobias. 1984. Heavy-ion radiobiology: Cellular studies. Adv. Radiat. Biol. 11:295–389.
8. Bond, V. P., M. N. Varma, C. A. Sendhaus, and L. E. Feinendagen. 1985. An alternative to absorbed dose, quality, and RBE at low doses. Radiat. Res. 104:S52–S57.
9. Booz, J. 1978. Mapping of fast neutron radiation quality. Proceedings of the Third Symposium on Neutron Dosimetry in Biology and Medicine, G. Burger and H. G. Ebert, eds. EUR 5845 DE/EN/FR:499-514. Brussels: Euratom.
10. Borek, C. 1980. X-ray induced in vitro neoplastic transformation of human diploid cells. Nature 282:776–778.
11. Borek, C. 1982. Radiation oncogenesis in cell culture. Adv. Cancer Res. 37:159–232.
12. Borek, C. 1985. The induction and control of radiogenic transformation in vitro: Cellular and molecular mechanisms. J. Pharmacol. Ther. 27:99–142.
13. Borek, C. 1985. Cellular and molecular mechanisms in malignant transformation of diploid rodent and human cells by radiation. Pp. 365–378 in Mechanisms in Carcinogenesis, Vol. 9, J. Barrett and R. W. Tennant, eds. New York: Raven.
14. Borek, C., and L. Sachs. 1966. In vitro cell transformation by x-irradiation. Nature 210:276–278.
15. Borek, C., and L. Sachs. 1967. Cell susceptibility to transformation by x-irradiation and fixation of the transformed state. Proc. Natl. Acad. Sci. USA 57:1522–1527.
16. Borek, C., E. J. Hall, and H. H. Rossi. 1980. Malignant transformation in cultured hamster embryo cells produced by x rays, 430-KeV monoenergetic neutrons, and heavy ions. Cancer Res. 38:2997–3005.
17. Borek, C., A. Ong, J. Bresser, and D. Gillespie. 1984. Transforming activity of DNA of radiation transformed mouse cells and identification of activated oncogenes in the donor cells. Proc. Am. Assoc. Cancer Res. 25:100.
18. Borek, C., A. Ong, and H. Mason. In press. Distinctive transforming genes in x-ray transformed mammalian cells. Proc. Natl. Acad. Sci. USA.
19. Braby, L. A., and W. C. Roesch. 1978. Testing of dose-rate models with *Chlamydomonas reinhardi*. Radiat. Res. 76:259–270.

20. Broerse, J. J., G. W. Barendsen, and G. R. van Kersen. 1967. Survival of cultured human cells after irradiation with fast neutrons of different energies in hypoxic and oxygenated conditions. Int. J. Radiat. Biol. 13:559–572.
21. Chadwick, K. H., and H. P. Leenhouts. 1973. The molecular target theory. Int. J. Radiat. Biol. 23:185.
22. Cox, R., and W. K. Masson. 1979. Mutation and inactivation of cultured mammalian cells exposed to beams of accelerated heavy ions. III. Human diploid fibroblasts. Int. J. Radiat. Biol. 36:149–160.
23. Cox, R., J. Thacker, D. T. Goodhead, and R. J. Munson. 1977. Mutation and inactivation of mammalian cells by various ionizing radiations. Nature 267:425–427.
24. Curtis, S. B. 1986. Lethal and potentially lethal lesions induced by radiation. A unified repair model. Radiat. Res. 106:252–270.
25. Dienes, G. 1966. A kinetic model of biological radiation response. Radiat. Res. 28:183–202.
26. Elkind, M. M., and G. F. Whitmore. 1967. The radiobiology of cultured mammalian cells. New York: Gordon and Breach.
27. Goodhead, D. T., R. J. Munson, J. Thacker, and R. Cox. 1980. Mutation and inactivation of cultured mammalian cells exposed to beams of accelerated heavy ions. IV. Biophysical interpretation. Int. J. Radiat. Biol. 37:135–167.
28. Hall, E. J. 1978. Radiobiology for the Radiologists, 2nd ed. New York: Harper and Row.
29. Hall, E. J. and T. K. Hei. 1985. Oncogenic transformation *in vitro* by radiations of varying LET. Radiat. Prot. Dosimetry 13:149–151.
30. Hall, E. J., J. K. Novak, A. M. Kellerer, H. H. Rossi, S. Marino, and L. J. Goodman. 1975. RBE as a function of neutron energy. I. Experimental observations. Radiat. Res. 64:245–255.
31. Han, A., and M. M. Elkind. 1979. Transformation of mouse C3H/10T-1/2 cells by single and fractionated doses of x-rays and fission-spectrum neutrons. Cancer Res. 39:123–130.
32. A. Kappos and W. Pohlit. 1972. A cybernetic model for radiation reactions in living cells. I. Sparsely ionizing radiations; stationary cells. Int. J. Radiat. Biol. 22:51–65.
33. Kellerer, A. M., and H. H. Rossi. 1972. The theory of dual radiation action. Curr. Top. Radiat. Res. Q. 8:85–158.
34. Kellerer, A. M., E. J. Hall, H. Rossi, and P. Teedla. 1976. RBE as a function of neutron energy. II. Statistical analysis. Radiat. Res. 65:172–186.
35. Lea, D. E. 1946. Actions of Radiations on Living Cells. Cambridge: Cambridge University Press.
36. Lloyd, E. L., M. A. Gemmell, C. B. Henning, D. S. Gemmell, and B. J. Zabransky. 1979. Cell survival following multiple-track alpha particle irradiation. Int. J. Radiat. Biol. 35:23–31.
37. Lloyd, E. L., M. A. Gemmell, C. B. Henning, D. S. Gemmell, and B. J. Zabransky. 1979. Transformation of mammalian cells by alpha particles. Int. J. Radiat. Biol. 36:467–478.
38. National Research Council, Committee on the Biological Effects of Ionizing Radiations (BEIR). 1980. The Effects on Populations of Exposures to Low Levels of Ionizing Radiation. Washington, D.C.: National Academy Press. 524 pp.

39. Puck, T. T., D. Morkorn, P. I. Marcus, and S. J. Cieciura. 1957. Action of x-rays on mammalian cells. II. Survival curves of cells from normal human tissues. J. Exp. Med. 106:484–500.
40. Robertson, J. B., A. Keehler, J. George, and J. B. Little. 1983. Oncogenic transformation of mouse BALB/3T3 cells by plutonium-238 alpha particles. Radiat. Res. 96:261–274.
41. Shilo, B. Z., and R. A. Weinberg. 1981. Unique transforming gene in carcinogen-transformed mouse cells. Nature 289:607–609.
42. Terzaghi, M., and J. B. Little. 1976. X-radiation-induced transformation in a C3H mouse embryo-derived cell line. Cancer Res. 36:1367–1374.
43. Thacker, J., A. Stretch, and M. A. Stephens. 1979. Mutation and inactivation of cultured mammalian cells exposed to beams of accelerated heavy ions. II. Chinese hamster V79 cells. Int. J. Radiat. Biol. 36:137–148.
44. Thacker, J., A. Stretch, and D. T. Goodhead. 1982. The mutagenicity of particles from plutonium-238. Radiat. Res. 92:343–352.
45. Thames, H. D. 1985. An incomplete repair model for fractionated and continuous irradiations. Int. J. Radiat. Biol. 47:319–339.
46. Tobias, C. A. 1985. The repair-misrepair model in radiobiology: Comparison to other models. Radiat. Res. 104:S77–S95.

APPENDIX III
The Effects of Radon Progeny on Laboratory Animals

Animal studies have been conducted for over 50 yr to examine the respiratory effects of pollutants in the air of mines. This work, emphasizing respiratory cancer, has provided important data on exposure-response relationships and the interactions among the harmful agents to which miners are exposed. Many of the initial studies were concerned with early effects or short-term pathological changes.[21,22,29] In many of the studies, exposures were based primarily on radon-gas concentrations, with little or no consideration of radon-daughter concentrations, which have been shown to contribute the greatest radiation dose to the lung. Two American research centers—the University of Rochester and the Pacific Northwest Laboratory (PNL)—and the Compagnie Generale des Matieres Nucleaires (COGEMA) laboratory in France have contributed most of the experimental data on radon-daughter inhalation by laboratory animals.

INHALATION STUDIES AT THE UNIVERSITY OF ROCHESTER

Beginning in the 1950s, investigators examined the biological and physical behaviors of radon daughters and the dosimetry of radon daughters in the respiratory tract.[1,20,25] Shapiro[31] exposed rats and dogs to radon alone at several concentrations and to radon with radon daughters attached to room-dust aerosols. The degree of attachment of radon daughters to carrier dust particles was shown to be an important determinant of the alpha-radiation dose to the airway epithelium and that more than 95% of the dose to the airway epithelium was due to the short-lived radon daughters radium A (^{218}Po) and radium C′ (^{214}Po), rather than to the parent radon. In 1953, Cohn et al.[9] reported the relative levels of radioactivity found in the nasal passages, the trachea and major bronchi, and the other portions of rat lungs after exposure to radon or radon daughters. The respiratory tracts of animals that inhaled radon with its daughters contained 125 times more activity than those of animals that inhaled radon alone. Beginning in the mid-1950s, Morken,[25–27] and Morken and Scott[28] initiated a series of experiments to evaluate the biological effects of inhaled radon and radon daughters in mice; later experiments also used rats and beagles. The negative results of these studies suggested that alpha irradiation was inefficient in producing tumors in the respiratory system.

These experiments were noteworthy in describing exposure-dose relationships in the whole lung, in regions of the lung, and in other organs. The paucity of pathological effects did not permit examination of exposure-response relationships for carcinogenesis, as demonstrated later by experiments at COGEMA and PNL. In the early experiments, the only apparent late, permanent changes occurred in the alveolar and possibly the bronchiolar regions of the lung. They were observed for a wide range of doses and developed after 3 yr in the dog and 1 and 2 yr in the rat and mouse, respectively. Some of these changes might have been preneoplastic, but the high-level exposures (associated with life-span shortening) and the early termination of experiments precluded further development to neoplasia. The influence of the radon-daughter carrier aerosol (laboratory air) on the results of these experiments is uncertain, but it might have led to more rapid solubilization of the daughters into blood and a resulting decrease in irritation or fibrosis, in comparison with ore-dust and silica aerosols.

INHALATION STUDIES AT COGEMA

The studies by Chameaud and colleagues[2–8] were begun in the late 1960s and early 1970s to determine whether radon and its daughters induced tumors in rats and to provide experimental data to support the epidemiological data on radon-daughter carcinogenesis. Before 1972, rats were exposed to ambient air that was enriched with radon after passage through trays of finely ground ore containing 25% uranium. Resulting radon concentrations were 0.75 μCi/liter; radon-daughter equilibrium factors were about 30%. With filters and electrostatic purifiers, the equilibrium factor was reduced to about 1%. Radon-daughter concentrations were calculated to be around 2,300 and 75 working levels (WL), respectively, for the two radon-daughter equilibrium conditions.

After 1972, animals were exposed to radon derived from underground barrels of radium-rich lead sulfate. Radon was pumped by a closed circuit into a 1-m^3 equilibration container and then to two 10-m^3 metal inhalation chambers. Up to 600 rats could be exposed for as long as 16 h when oxygen was added to the inhalation chambers. The maximum radon concentration was 1.25 μCi/liter, generally at 100% equilibrium with radon daughters. By calculation, the maximum radon-daughter concentration was 12,500 WL. Because of radon-daughter deposition on the cages and the hairs of rats, the disequilibrium of the radon daughters increased as the number of animals in the inhalation chambers increased. Exposure periods ranged from about 1 to 10 months; exposure rates ranged from less than 10 to hundreds of working-level months (WLM)/wk, the majority averaging approximately 200–400 WLM/wk.*

In two major experiments,[2] rats were exposed by inhalation to stable cerium hydroxide or to uranium-ore dust concentrations with and without radon daughters, at 130 mg/m^3, to determine whether the presence of dust altered the carcinogenic effect of radon daughters. Exposure to stable cerium hydroxide before exposure to radon daughters shortened the induction latent period by 2–3 months. Uranium-ore dust (given on days alternating with days of radon-daughter exposure) appeared to have little influence on the tumorigenic process, although too few animals were used to permit a firm conclusion.[5] Radon-daughter exposures varied from 500 to 8,500 WLM. The effect of the radon daughters did not change with the various equilibrium ratios. These experiments confirmed that radon daughters alone induced tumors in rats.

Other changes were observed in these experiments. These are given below.

- After large radon-daughter exposures, large areas of diffuse interstitial pneumonia with hyaline membrane formation and with severe fibrosis of interalveolar septa surrounding capillaries were noted. Death generally occurred within a few weeks to a few months if exposure exceeded 6,000 WLM. No lung cancers were produced.
- Animals lived longer after smaller radon-daughter exposures, with lung carcinomas appearing 12–24 months after the beginning of exposure. The time to appearance of tumors increased with decreasing cumulative radon-daughter exposure. Exposures of 2,000–5,000 WLM, delivered over 300–500 h (during 3–4 months), produced the highest incidence of tumors.
- Bronchiolar metaplasia occurred at the bronchioloalveolar junction and in neighboring alveoli. It consisted of large columnar cells with basal

---

*Some of the exposure values in these French studies have been supplied by COGEMA investigators and might be different from previously published values. (J. Chameaud, personal communication to F. T. Cross, 1986.)

nuclei and light-colored protoplasm that were often ciliated. Alveolar metaplasia of cuboidal cells, with darker protoplasm, appeared in peripheral regions of the lungs.

- Adenomatous lesions of varied size and cell layers covered areas of the alveolar septa. Adenomas consisting of round tumors with cells often clustered together occurred. Some adenomas showed malignant characteristics.
- Malignant tumors of several different types occurred, often in the same animal. These included epidermoid carcinomas, not always clearly differentiated, often keratinized or necrosed, and occasionally extending into the mediastinum; bronchiolar adenocarcinomas, sometimes mucus-producing, containing numerous cellular anomalies, and characterized by a high number of mitoses and invasion of other lung lobes, but seldom metastatic; and bronchioloalveolar adenocarcinomas with few mitoses, but later invading the mediastinum, diaphragm, and thoracic wall.
- The relationship of exposure to tumor incidence, uncorrected for life-span shortening, was not linear over a wide range of exposures; the incidence per unit exposure increased with decreasing high cumulative exposure.

Later experiments, which confirmed these pathological findings, extended the radon-daughter exposures to approximately 20–50 WLM.[5,7,8] Tumor-incidence and survival-time data and lifetime lung-tumor risk coefficients are shown in Table III-1. Although the risk data are uncorrected for life-span shortening, hazard-function analysis demonstrated that when the data are adjusted for competing causes of death, the excess risk of developing pulmonary tumors is approximately linearly related to exposure throughout the range of exposures studied.[19] Further findings are given below.

- The tumor latent period, defined as the interval between the start of radon-daughter exposure and death or killing, of the animal increased with decreasing cumulative WLM. Mean latent periods of tumor-bearing animals were around 750 days for exposures of less than 300 WLM and 650 days for exposures of over 1,000 WLM.
- Lung cancers in rats invaded pulmonary lymph nodes, but metastases to other tissues were rare. Tumor size increased with increasing cumulative WLM.
- No radiation-induced small-cell carcinomas were observed in rats; however, other histological types of lung carcinomas were similar to those observed in humans.

TABLE III-1 Summary of Tumors Primary to Lungs of Rats, Median Survival Times,[a] and Lung-Tumor Risk Coefficients for COGEMA Radon-Daughter Exposures

Group Mean Exposure (WLM)	Nominal Exposure Rate (WLM/wk)	No. of Animals Examined	No. of Animals with Tumors	% Animals with Tumors	Group Median Survival Time (days)	Mean Lifetime Risk Coefficient (10^{-4}/WLM)[b]
20–25	2–4	~1,500	25	1.7	684	7.5[c]
50	2–8	~1,000	30	2.9	687	5.8[c]
290	9	21	2	10	610	3.3
860	370	20	4	20	672	2.8[d]
1,470	370	20	5	25	606	1.7
1,800	200	50	17	34	600	1.9
1,900	310	20	7	35	548	1.8
2,100	220	54	23	43	593	2.0
2,800	310	180	74	41	560	1.5
3,000	370	40	17	43	670	1.4
4,500	370	40	29	73	644	1.6

[a]Data from Chameaud et al.[7-10]
[b]Values are uncorrected for life-span differences from control animals. Lifetime risk coefficients based on raw incidence at very low exposures are considered to define accurately the initial slope of the risk-coefficient curve.
[c]Value is corrected for lung-tumor incidence in control rats of the low-exposure group (0.83%); normal incidence in the absence of appreciable background radon exposure is about 0.1–0.2%. Median survival time of control rats of the two lowest exposure groups was 752 d.
[d]Calculated value at this exposure level is 2.3.

- Cutaneous epitheliomas of the upper lip and cancers of the urinary system were the only two sites other than the lungs where cancers were noted in exposed rats.
- The incidence of lung cancer increased with decreasing high radon-daughter exposure rate. The greatest effect was noted in exposure-fractionation experiments. Rats exposed to radon daughters for approximately 3,000 WLM, at 1,500 WL for 7 h/day or 1 or 5 days/wk (average exposure rates are calculated to be above 50 and 300 WLM/wk) had a nearly fourfold increase in cancer incidence with exposure protraction.
- While the latency period decreased, the lung-cancer incidence did not change with increasing age at first exposure. For 3,000-WLM exposures, the latent periods for ages at first exposure of 150, 280, 400, and 520 days were 640, 510, 450, and 305 days, respectively.
- Synergism was observed between exposure to radon progeny and whole-body cigarette-smoke exposures if the exposure to smoke followed the exposure to radon daughters. However, if the cumulative cigarette-smoke exposure preceded the radon-daughter exposure, no increase in cancer incidence was noted over that produced by radon daughters alone. Thus, the effect of cigarette smoke depended on the sequence of exposures and was attributed to its promoting action.[5] The histological types of cancers observed were not altered by cigarette-smoke exposures. The investigators have not reported whether the latent period for cancer was influenced by smoke exposure; the observation that tumors in the radon-daughter- and smoke-exposed animals were larger and more invasive than those in animals exposed only to radon daughters might be indicative of a shorter latent period for smoking-related tumors.

The COGEMA studies have produced more than 800 lung cancers in about 10,000 rats exposed to radon daughters with ambient aerosols and in mixtures with other pollutants. The exposure-response relationship data shown in Table III-1 therefore constitute only a portion of the data from these experiments. The derived range in mean lifetime risk coefficients, uncorrected for life-span differences from control animals, is about 1.5×10^{-4}–7.5×10^{-4}/WLM for exposures between about 20 and 5,000 WLM. The risk decreases at larger exposures because of life-span shortening. No evidence of a threshold below 20 WLM was apparent.[8]

INHALATION STUDIES AT THE PACIFIC NORTHWEST LABORATORY

Exposures of dogs and rodents to uranium-mine air contaminants were begun in the late 1960s and early 1970s to identify agents and the magnitude of exposures to them that were responsible for producing lesions of the respiratory tract similar to those observed in uranium miners. The early experiments concentrated on lifetime inhalation exposures of hamsters and beagles to mixed aerosols of radon, radon daughters, carnotite uranium-ore dust, diesel-engine exhaust, and cigarette smoke. Most of the final data from these early experiments have been published.[11–13] To provide data that were missing from the earlier dog study, follow-up studies have included exposures of beagles to uranium-ore dust alone (but not to radon daughters alone) and exposures of rats to mixtures of radon, radon daughters, and uranium-ore dust.[10,14–18,33] Because the studies in rats were designed to develop exposure-response relationships, the exposures were truncated rather than extended through the animals' lifetimes. They were also designed to study the roles of carnotite uranium-ore dust concentration and radon-daughter exposure rate, unattachment fraction, and disequilibrium in the production of lung lesions. Histopathological examination, clinical pathological examination, and pulmonary physiology tests were the primary means of measuring response. Urinalyses have recently supplemented serum tests as more sensitive evaluations for kidney damage. Radiometric analyses of tissues have been used to determine mean radon-daughter tissue doses and the body distribution of long-lived radioactivity from the ore dust.

Lifetime exposures of hamsters to radon daughters alone or in combination with uranium-ore dust and diesel-engine exhaust caused no significant ($P > 0.05$) changes in mortality patterns compared with those of controls. The mean radon-daughter exposure in the hamster experiments was about 10,000 WLM. Lifetime exposures of beagles to mixtures of radon daughters, uranium-ore dust, and cigarette smoke caused significant life-span shortening compared with that of controls. Mean survival times of the dogs exposed to mixtures of radon daughters and ore dust, with or without cigarette smoke, were 4–5 yr. Mean survival times of controls and dogs exposed to smoke only were equivalent during the same period. The mean radon-daughter exposure of the dogs was about 13,000 WLM.

Studies in progress show that chronic exposure of rats to mixtures of radon daughters and uranium-ore dust shortens the life span. The data thus far generally show no significant differences in mortality patterns compared with those of controls for exposures up to about 2,500 WLM. Exposures exceeding 5,000 WLM have caused significant life-span shortening, with the effect increasing with exposure. In general, rats that showed life-span shortening also showed weight loss.

Thus far, two life-span-shortening anomalies have been noted in the rat experiments. First, in an interim study to determine any influence of radon-daughter exposure rate, rats exposed to about 640 WLM at the lowest rate (about 44 WLM/wk) died earlier than other animals given comparable cumulative exposures. Second, in a study to determine the influence of unattached radon daughters versus that of attached radon daughters, rats exposed to about 5,100 WLM with the highest unattachment fraction ($f_a = 24\%$) died earlier than other animals given comparable cumulative exposures. Life-table analyses of the survival-time data in the unattachment-fraction study[18] showed that the estimated probabilities that a rat would die with a lung tumor before 600 days were 0.42, 0.65, and 0.75 for 6, 10, and 24% ^{218}Po (radium A) unattachment, respectively. Expressed as percentages of radon concentration, rather than radium A concentration, the unattachment was 1.3, 5.2, and 9.5%. Later experiments at 640 and 53 WLM/wk showed no appreciable life-span shortening.

The mean survival time of tumor-bearing rats (as in the COGEMA data) was always significantly longer than that of non-tumor-bearing rats. The latent period of lung tumors is a large fraction of the rat life span, and tumors must grow to a size sufficient for detection; the shorter-lived animals might have died too soon for tumors, if any, to be detected.

In the life-span studies with dogs, animals with tumors of the respiratory tract generally had cumulative radon-daughter exposures exceeding 13,000 WLM; the exposure rate was 71 WLM/wk. Concomitant exposure to cigarette smoke had a mitigating effect on radon-daughter-induced tumors, possibly because smoking caused thickening of the mucus layer and stimulated mucociliary clearance. The overall incidence of lung primary tumors was 21% for a mean exposure of 13,100 WLM to radon daughters, 37% in the group exposed to radon daughters and uranium-ore dust, but only 5% in the comparable group that was also exposed to cigarette smoke. The overall incidence of nasal carcinoma was 8%. The lung cancers were about 70% bronchogenic carcinomas and 30% bronchioloalveolar carcinomas.[15] The simplified convention used was that squamous cell carcinomas and mucus-staining adenocarcinomas were bronchogenic carcinomas and that tumors of Clara cell or type II alveolar cell origin and non-mucus-staining adenocarcinomas were bronchioloalveolar carcinomas.

Lifetime inhalation exposures of hamsters produced severe radiation pneumonitis but only four squamous cell carcinomas (three in the radon daughters-only group and one in the group exposed to radon daughters and uranium-ore dust) in 306 radon-daughter-exposed animals (1.3%

incidence). Squamous cell carcinoma occurred only in association with squamous metaplasia of the alveolar epithelium, which was found only in hamsters exposed to radon daughters. Thus, it appears that after exposure to radon daughters, the development of squamous metaplasia and the development of carcinoma were related. Because so few lung cancers were produced in these high-exposure experiments, it was concluded that the hamster was an inappropriate surrogate for further study of the carcinogenic potential of inhaled (as opposed to instilled) mine-air pollutants.

Over 4,000 male rats have received chronic exposures to ambient air or to mixtures of radon daughters and uranium-ore dust since 1978. Data are still accumulating, but some general trends can be observed. Lung-cancer risk tended to increase (sometimes significantly) with decreasing radon-daughter exposure rate, increasing unattached fraction of radon daughters, and increasing radon-daughter disequilibrium. The lung cancers induced after exposures of approximately 300–5,000 WLM were about 70% bronchogenic carcinomas and 30% bronchioloalveolar carcinomas. The tumors were most often estimated (by sizing associated bronchi and bronchioles) to be about 50% proximal (bronchus-associated) and 50% distal (bronchiole- and alveolus-associated), in contrast with the greater proportion of proximal lung cancers in humans.[30] The prevalence of squamous metaplasia, and generally carcinoma, of the respiratory tract increased with an increasing unattached fraction of radon daughters.

The PNL data are inadequate for firm conclusions regarding the effect of radon-daughter exposure rate and the magnitude of the lifetime risk coefficient below 100 WLM. However, the data to date indicate an increasing lifetime lung-tumor risk coefficient with decreasing cumulative radon-daughter exposure. Like the COGEMA data, the PNL risk-coefficient data have not been corrected for life-span shortening due to competing causes of death, such as radiation pneumonitis (see Table III-1). It cannot be concluded that the increase in the risk coefficient continues with further decreases in cumulative exposure and exposure rate. The PNL experiments include exposures as low as 20 WLM. The tumor-incidence data, particularly those derived from high-exposure-rate experiments, are similar not only to those from COGEMA but also to present estimated lung-tumor incidence data in humans.

Animal exposure studies show that the tumorigenic efficiency of radon daughters varies with cumulative exposure, exposure rate, unattached fraction, disequilibrium, and concomitant exposures to other pollutants (i.e., cigarette smoke). The COGEMA and PNL data indicate that tumor incidence increases with an increase in radon-daughter cumulative exposure and a decrease in radon-daughter exposure rate. Chameaud et al.[5]

concluded that lung-cancer incidence at comparable cumulative exposures increased as the radon-daughter concentration decreased from 12,000 to less than 3,000 WL. In a related dose-fractionation study with a cumulative exposure of 3,000 WLM and a radon-daughter concentration of 1,500 WL, an approximately fourfold increase in lung cancers was observed when the exposure rate decreased from about 300 to 50 WLM/wk; it is not known whether this exposure-rate dependence persists at the far lower rates. A trend toward increasing a lung-tumor risk with decreasing exposure rate was noted in the earlier PNL rat experiments[14,18] when the rates changed from 180 to 88 and to 44 WLM/wk. Inasmuch as the increase was not significant and results were uncertain at 44 WLM/wk as a result of life-span shortening in that group, the exposure-rate dependence in rats might be lessened at the lower weekly rates of exposure. However, more recent data confirm the increase in lung-tumor risk with decreased exposure rate down to 53 WLM/wk.

Data from the PNL rat experiments also indicate an increase in the risk of lung tumors with increases in radon-daughter unattached fraction and disequilibrium.[18] The risk increase from 1.6 to 10% unattached radium A is significant ($P < 0.05$), but the positive trend reverses at 24% unattachment as a result of life-span shortening in that exposure group. In contrast with the results of the COGEMA experiments, the increase is also significant with radon-daughter disequilibrium (an equilibrium of 10 versus 40%) when the total numbers of lung cancers are compared. However, the trend is of borderline significance ($P = 0.10$) when the total numbers of rats with lung tumors are compared. The data on nasal carcinoma show an increasing trend with increasing unattachment and, as with the neoplastic lesions of the lung, a reverse trend at 24% unattachment. There is no indication that high-disequilibrium radon-daughter exposures, without concomitant high unattachments, produce more nasal carcinomas than do low-disequilibrium exposures.

The role of concomitant exposures to other pollutants depends not only on the nature of those pollutants but also on the sequence of exposures. Simultaneous or same-day exposure to radon daughters and uranium-ore dust, diesel-engine exhaust, or cigarette smoke increased the incidence of preneoplastic lesions but, except for cigarette smoke, did not change the incidence of lung tumors in the PNL experiments. In the COGEMA rat experiments, cigarette smoke was cocarcinogenic with radon daughters if exposure to smoke followed completion of exposure to the radon daughters,[4] but not if smoking preceded the radon-daughter exposures. In the PNL dog experiments, lung-tumor incidence decreased when animals were exposed to radon daughters and cigarette smoke alternately on the same day.

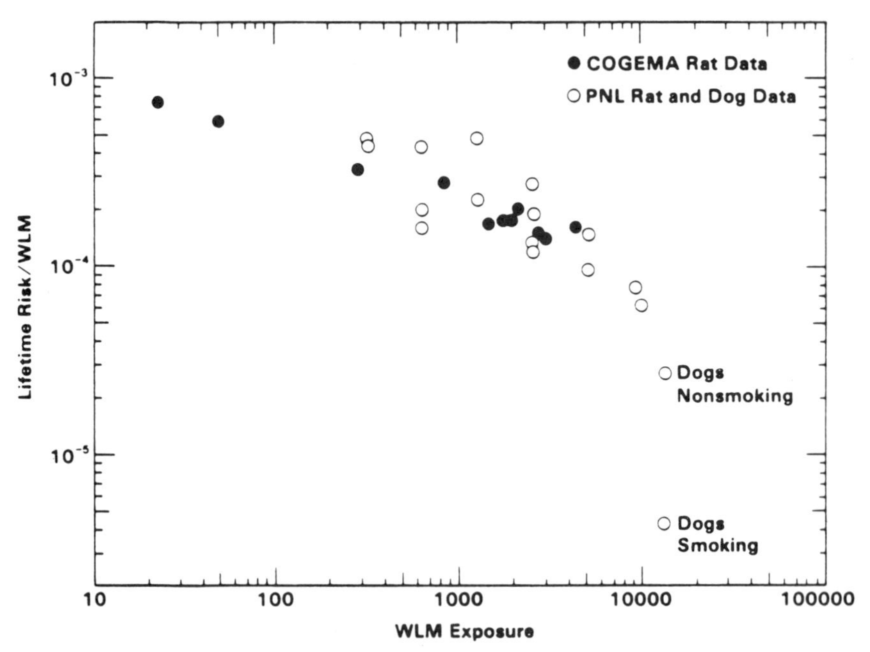

FIGURE III-1 Lifetime risk coefficients for radon-daughter exposure for PNL rat and dog data and COGEMA rat data; error bars are omitted. SOURCE: Personal communication, Dr. F. Cross, Pacific Northwest Laboratories.

LUNG CANCER

In Figure III-1 the mean lifetime lung-tumor risk per WLM (uncorrected for life-span differences from control animals) is plotted against the radon-daughter exposure (WLM) for PNL rats and dogs and COGEMA rats. The higher tumor efficiencies in the PNL studies (in contrast with the COGEMA studies) are probably due to the lower average exposure rates of the PNL experiments.

The uncertainties in the PNL lung-cancer incidence and risk-coefficient data are considered to be due mainly to uncertainties in the exposure data (standard deviations were generally well within ±20% of the means). Whenever PNL exposures were repeated, reproducibility of tumor-incidence data was generally within ±20% of the mean tumor incidence, which included the statistical uncertainties in the exposure data. Because the normal lung-tumor incidence in the absence of appreciable background radon exposures is very low (<0.2%) in the COGEMA and PNL rats, the risk-coefficient data, except for the 20- to 50-WLM COGEMA group of rats, have not been corrected for the incidence in control animals.

Current experiments at PNL, which involve mixtures of radon daughters and uranium-ore dust, will further define the shape of the risk-coefficient curve for very low exposures and exposure rates. For the present, COGEMA data on low exposures and low exposure rates indicate a leveling-off of the risk to a value of 6×10^{-4}–8×10^{-4}/WLM.

Kushneva[23] reported that rats given 50 mg of silica by instillation with inhalation exposures to radon at 8 μCi/liter developed many more pulmonary effects, including both adenomas and carcinomas, than did animals exposed to silica alone; the number of tumors and control animals was small. When silica dusts were included in the exposures, the radon-daughter inhalation studies at COGEMA and PNL showed no increased tumorigenic efficiency over exposures to radon daughters alone if these exposures exceeded a few hundred WLM. However, in contrast with the rat data of Kushneva[23] and the dog data from PNL, Chameaud et al.[5] have not found the silicotic process to be accelerated by the presence of radon daughters.

Little et al.[24,32] have shown in hamsters that when benzo(a)pyrene or saline instillations followed low-dose ^{210}Po instillations, the carcinogenic action of polonium was increased. Because radioactivity appears to be the initiator of the lung cancer, as in all the animal experiments with radon described here, any later exposure to an irritant that stimulates cell proliferation appears to increase the incidence of cancer.

SUMMARY AND CONCLUSIONS

Laboratory animal research programs on the effects of radon-daughter inhalation are being carried out in laboratories in both the United States and France. While much of the early work explored acute effects, more recent experiments involving chronic exposure have resulted in the induction of lung cancer in both rats and dogs. It should be noted, however, that the location and histopathology of such cancers are not analogous to humans, and caution is warranted in extrapolating from experiments with laboratory animals to humans. Nevertheless, substantial information has accumulated that provides insights into radon-daughter carcinogenesis. Table III-2 summarizes recent findings in animal studies of lung-cancer induction by radon decay products.

In rats, lung tumors have been induced at relatively low exposures (20 WLM).[7] As yet, experiments with dogs do not extend to this low-dose range, but tumors have been observed for exposures at the 600-WLM level.[13] It is of interest that lung-cancer incidence in animals increases with a decreasing rate of exposure for fixed cumulative exposure—a finding that has yet to be confirmed in studies of exposed underground miners (Annex 2A). The difficulty of documenting exposure rate for the miners may explain the failure to find a dose-rate effect in the epidemiological studies.

TABLE III-2 Summary of Factors Influencing the Tumorigenic Efficiency of Radon-Daughter Exposure

Factor	Effect on Respiratory Tract Tumor Incidence
Radon-daughter cumulative exposure	Increases approximately linearly with exposure
Radon-daughter exposure rate	Increases with decrease in exposure rate (~200 to 400% increase from about 500 to 50 WLM/wk)
Radon-daughter unattached fraction	Increases with increase in unattached fraction, f_a (~50% increase per WLM exposure from 2 to 10% f_a)
Radon-daughter disequilibrium	Increases with increase in disequilibrium (~30% increase per WLM exposure [borderline significance] from 0.4 to 0.1 equilibrium).
Concomitant exposure to cigarette smoke	Decreases if smoking alternates on same day with radon-daughter exposures
	Increases if smoking follows cumulative radon-daughter exposures
	No effect if smoking precedes cumulative radon-daughter exposures

Large-scale animal studies may become useful for elucidating the interactions between radon daughters and other inhaled pollutants. Information on the extent and duration of smoking is incomplete for human studies, but smoking can be controlled in experiments with animals. It is clear from such experiments that the interactions between smoking and lung cancer induced by radon decay products reflect a complex interplay of these agents in the host. Well thought out experiments with dogs and rats can provide models that aid our understanding of how smoking modulates radiogenic lung cancer. Nevertheless, application to humans is indirect, and confirming experiments with primates may be necessary. However, findings in humans and animals to date are generally parallel for short-half-life radon progeny.

REFERENCES

1. Bale, W. F. 1951 Hazards associated with radon and thoron. Division of Biology and Medicine, U.S. Atomic Energy Commission, Washington, D.C.: March 14. Memorandum. (Also found in Health Phys. 38:1061, 1951.)
2. Chameaud, J., R. Perraud, J. Lafuma, R. Masse, and J. Pradel. 1974. Lesions and lung cancers induced in rats by inhaled radon-222 at various equilibriums with radon daughters. P. 411 in Experimental Lung Cancer. Carcinogenesis and Bioassays, E. Karbe and J. F. Park, eds. New York: Springer-Verlag.
3. Chameaud, J., R. Perraud, J. Chretien, R. Masse, and J. Lafuma. 1976. Lung cancer induced in rats by radon and its daughter nuclides at different concentrations. P. 223 in Biological and Environmental Effects of Low Level Radiation, Vol. II. Publication STI/PUB/409. Vienna: International Atomic Energy Agency.
4. Chameaud, J., R. Perraud, J. Chretien, R. Masse, and J. Lafuma. 1980. Combined effects of inhalation of radon daughter products and tobacco smoke.

P. 551 in Pulmonary Toxicology of Respirable Particles, CONF-791002, C. L. Sanders, F. T. Cross, G. E. Dagle, and J. A. Mahaffey, eds. Springfield, Va.: National Technical Information Service.

5. Chameaud, J., R. Perraud, R. Masse, and J. Lafuma. 1981. Contribution of animal experimentation to the interpretation of human epidemiological data. In Proceedings of the International Conference on Radiation Hazards in Mining: Control, Measurement and Medical Aspects, M. Gomez, ed. Kingsport, Tenn.: Kingsport Press, Inc.
6. Chameaud, J., R. Perraud, and J. Lafuma. 1982. Cancers induced by Rn-222 in the rat. In Proceedings of Specialist Meeting on Assessment of Radon and Daughter Exposure and Related Biological Effects, G. F. Clemente, A. V. Nero, F. Steinhausler, and M. E. Wrenn, eds. Salt Lake City, Utah: RD Press.
7. Chameaud, J., R. Masse, and J. Lafuma. 1984. Influence of radon daughter exposure at low doses on occurrence of lung cancer in rats. Radiat. Prot. Dosimetry 7:385–388.
8. Chameaud, J., R. Masse, M. Morin, and J. Lafuma. 1985. Lung cancer induction by radon daughters in rats. Pp. 350–353 in Proceedings of the International Conference on Occupational Radiation Safety in Mining, Vol. 1, H. Stocker, ed. Toronto: Canadian Nuclear Association.
9. Cohn, S. H., R. K. Skow, and J. K. Gong. 1953. Radon inhalation studies in rats. Arch. Ind. Hyg. Occup. Med. 7:508.
10. Cross, F. T. 1984. A Review of Radon Inhalation Studies in Animals with Reference to Epidemiological Data. Research Report for SENES Consultants, Ltd., Ontario, Canada. Richland, Wash.: Battelle, Pacific Northwest Laboratories.
11. Cross, F. T, R. F. Palmer, R. E. Filipy, R. H. Busch, and B. O. Stuart. 1978. Study of the Combined Effects of Smoking and Inhalation of Uranium Ore Dust, Radon Daughters and Diesel Oil Exhaust Fumes in Hamsters and Dogs. PNL-2744. Springfield, Va.: National Technical Information Service.
12. Cross, F. T., R. F. Palmer, R. H. Busch, R. E. Filipy, and B. O. Stuart. 1981. Development of lesions in Syrian golden hamsters following exposure to radon daughters and uranium ore dust. Health Phys. 41:135–153.
13. Cross, F. T., R. F. Palmer, G. E. Dagle, R. E. Filipy, and B. O. Stuart. 1982. Carcinogenic effects of radon daughters in uranium ore dust and cigarette smoke in beagle dogs. Health Phys. 42:35–52.
14. Cross, F. T., R. F. Palmer, R. H. Busch, and R. L. Buschbom. 1982. Influence of radon daughter exposure rate and uranium ore dust concentration on occurrence of lung tumors. Pp. 189–197 in Proceedings of Specialist Meeting on Assessment of Radon and Daughter Exposure and Related Biological Effects. Salt Lake City: Salt Lake City Press.
15. Cross, F. T., R. F. Palmer, R. H. Busch, G. E. Dagle, R. E. Filipy, and H. A. Ragan. 1983. An overview of the PNL radon experiments with reference to epidemiological data. In Proceedings of the 22nd Hanford Life Sciences Symposium. Richland, Wash.: Batelle, Pacific Northwest Laboratory.
16. Cross, F. T., R. L. Buschbom, G. E. Dagle, R. E. Filipy, P. O. Jackson, S. M. Loscutoff, and R. F. Palmer. 1983. Inhalation hazards to uranium miners. P. 77 in Pacific Northwest Laboratory Annual Report for 1982 to the Department of Energy Assistant Secretary for Environment, Part 1. PNL-4600. Richland, Wash.: Battelle, Pacific Northwest Laboratory.
17. Cross, F. T., R. L. Buschbom, G. E. Dagle, P. O. Jackson, R. F. Palmer, and H. A. Dagan. 1984. Inhalation hazards to uranium miners. P. 41 in Pacific Northwest Laboratory Annual Report for 1983 to the Department of Energy Assistant Secretary for Environment, Part 1. PNL-5000. Richland, Wash.: Battelle, Pacific Northwest Laboratory.

Battelle, Pacific Northwest Laboratory.

18. Cross, F. T., R. F. Palmer, G. E. Dagle, R. E. Busch, and R. L. Buschbom. 1984. Influence of radon daughter exposure rate, unattachment fraction, and disequilibrium on occurrence of lung tumors. Radiat. Prot. Dosimetry 7:381–384.
19. Gray, R. G., J. Lafuma, S. E. Parish, and R. Peto. In press. Radon inhalation by rats: An examination of the dose-response relationship for induced pulmonary tumours and of the exposure parameters and cofactors affecting risk. In Proceedings of the 22nd Hanford Life Sciences Symposium. Richland, Wash.: Battelle, Pacific Northwest Laboratories.
20. Harris, S. J. 1954. Radon levels in mines in New York State. Arch. Ind. Hyg. Occup. Med. 10:54–60.
21. Jackson, M. L. 1940. The Biological Effects of Inhaled Radon. M.S. Thesis. Massachusetts Institute of Technology.
22. Jansen, H., and P. Schultzer. 1926. Experimental investigations into the internal radium emanation therapy. I. Emanatorium experiments with rats. Acta Radiol. 6:631.
23. Kushneva, V. S. 1959. P. 21 in On the Problem of the Long-Term Effects of Combined Injury to Animals of Silicon Dioxide and Radon. TR-4473. Washington, D.C.: U.S. Atomic Energy Commission.
24. Little, J. B., R. B. McGandy, and A. R. Kennedy. 1978. Interactions between polonium-210, α-radiation, benzo(a)pyrene, and 0.9 NaCl solution instillations in the induction of experimental lung cancer. Cancer Res. 38:1929.
25. Morken, D. A. 1955. Acute toxicity of radon. Arch. Ind. Health 12:435.
26. Morken, D. A. 1973. The biological effects of the radioactive noble gases. P. 469 in Noble Gases, R. Stanley and A. A. Moghissi, eds. CONF-730915. U.S. Energy Development and Research Agency, Washington D.C.: National Environmental Research Center.
27. Morken, D. A. 1973. The biological effects of radon on the lung. In Noble Gases, R. Stanley and A. A. Moghissi, eds. CONF-730915. U.S. Energy Development and Research Agency, Washington D.C.: National Environmental Research Center.
28. Morken, D. A., and J. K. Scott. 1966. Effects on Mice of Continual Exposure to Radon and Its Decay Products on Dust. University of Rochester Atomic Energy Project Report UR-669. Springfield, Va.: National Technical Information Service.
29. Read, J., and J. C. Mottram. 1939. The "tolerance concentration" of radon in the atmosphere. Br. J. Radiol. 12:54.
30. Schlesinger, R. B., and M. Lippman. 1978. Selective particle deposition and bronchogenic carcinoma. Environ. Res. 15:424.
31. Shapiro, J. 1954. An Evaluation of the Pulmonary Radiation Dosage from Radon and Its Daughter Products. University of Rochester Atomic Energy Project UR-298. Rochester, N.Y.: University of Rochester.
32. Shami, S. G., L. A. Thibodeau, A. R. Kennedy, and J. B. Little. 1982. Proliferative and morphological changes in the pulmonary epithelium of the Syrian golden hamster during carcinogenesis initiated by ^{210}Po α-Radiation. Cancer Res. 42:1405.
33. Stuart, B. O., R. F. Palmer, R. E. Filipy, and J. Gaven. 1978. Inhaled radon daughters and uranium ore dust in rodents. In Pacific Northwest Laboratory Annual Report for 1977 to the Department of Energy Assistant Secretary for Environment, Part 1. PNL-2500. Richland, Wash.: Battelle, Pacific Northwest Laboratory.

APPENDIX IV
Epidemiological Studies of Persons Exposed to Radon Progeny

INTRODUCTION

The mining of radioactive ores in the Erz Mountains in eastern Europe was the first occupation associated with an increased risk of lung cancer. Metal ores were mined in Schneeberg, on the German side of the mountains, beginning in the fifteenth century, and in Joachimsthal, on what is now the Czechoslovakian side, beginning in the sixteenth century.[26,30] Both areas were later mined for radioactive ores. As early as the sixteenth century, Agricola[1] described exceptionally high mortality from respiratory diseases in miners in this region. The lung-cancer hazard was first recognized by Harting and Hesse[19] and was reported in 1879. Their report provided clinical and autopsy descriptions of intrathoracic neoplasms in miners, which they classified as lymphosarcoma. In a work force of about 650 men, Harting and Hesse counted 150 deaths from "miner's disease" between 1869 and 1877; in retrospect, most of these deaths were probably from lung cancer. During the early twentieth century, histopathological review of a series of cases established that the malignancy prevalent among miners in the Erz Mountains was primary cancer of the lung.[5,49]

The problem was not recognized in the miners on the Czechoslovakian side of the Erz Mountains until 1929, when two cases of lung cancer were reported in Joachimsthal miners. In 1932, Pirchan and Sikl[46] described the autopsy findings in nine miners with lung cancer. These 9 miners were among 19 miners in Joachimsthal who died during 1929–1930. Formal epidemiological studies of the Schneeberg and Joachimsthal miners were not carried out, but published reports documented that about 50% of the miners eventually died from lung cancer.[53] Peller[44] calculated lung-cancer mortality rates for the Schneeberg miners during 1875–1912 and found that they were about 50 times those in Vienna males during 1932–1936.

Many authors offered explanations of the excess cancer in the Schneeberg and Joachimsthal miners (see references 26, 30, and 63 for reviews). Early theories emphasized dust exposure, metals in the ore (particularly arsenic), and increased susceptibility as a result of inbreeding in small mining communities. In 1924, Ludwig and Lorenser[31] reported that radioactivity could be measured in the air and water in the mines of Schneeberg and might contribute to the development of lung cancer. Pirchan and Sikl[46] suggested in 1932 that radioactivity was the most probable cause of the Joachimsthal cancers, on the basis of the finding of radioactivity in both Schneeberg and Joachimsthal mines, the occurrence of lung-cancer in both locations, and the long exposure of underground miners to radioactivity. Teleky's opinion in 1937 was similar.[63] He could find no other satisfactory explanation and concluded that the high level of radioactivity, thought not to be present in other mines, led to the apparently unique lung-cancer problem of Schneeberg and Joachimsthal miners. In 1944, Lorenz[30] argued that radon alone could not be the cause of lung cancer and proposed that genetic susceptibility to lung cancer might be unusually high in the miners. However, during the 1950s and 1960s, as the biological basis of respiratory carcinogenesis became better understood and additional mining groups were studied, it came to be accepted that inhaled radon progeny were the cause of lung cancer in the Schneeberg and Joachimsthal miners and other exposed miners.[26,33,53]

After World War II, several new epidemiological studies were initiated to determine the safety of exposure to radon progeny in mines. Unlike early studies, the newer surveys addressed such important biological questions such as the shape of the dose-response curve, the influence on risk of age at exposure, the effect of dose rate, the temporal expression of risk after exposure, and the interaction of radon daughters with other substances associated with lung cancer. This appendix reviews the epidemiological literature that is now available for addressing these issues.

COLORADO PLATEAU STUDY

Beginning in the late 1940s, the American uranium industry grew rapidly in the Colorado Plateau, a mountainous region of southwestern Colorado and southeastern Utah. In 1949, in response to concerns about the health hazards to workers in this industry and with awareness of the high lung-cancer incidence in European miners in Joachimsthal and Schneeberg,[21] the U.S. Public Health Service (PHS) began to investigate the uranium mines and mills in the Colorado Plateau region. The investigation combined an industrial-hygiene survey with a medical study of the workers. A prospective cohort study of miners and millers was carried out later, first by the PHS and then by the National Institute for Occupational Safety and Health (NIOSH). Until recently, this study offered one of the few epidemiological data bases for estimating the lung-cancer risk associated with exposure to radon progeny.

Field teams from PHS periodically conducted medical surveys of miners and millers between 1950 and 1960.[33] Before 1954, the teams did not attempt to examine all workers, but during 1954–1960, they tried to attain complete coverage. From among the examined miners, a group was assembled for follow-up that included miners who had worked at least a month underground in a uranium mine by January 1, 1964.[33] The number of subjects varied in the reports of this investigation (Table IV-1); in 1971, Lundin et al.[33] provided data on 3,366 white and 780 nonwhite subjects.

An exposure data base was developed from diverse sources: PHS, state agencies, and the mining companies. Holaday (quoted by Lundin et al.[33]) has provided a chronology (Table IV-2). During the period 1951–1968, for which cumulative exposure in working-level months (WLM) was initially calculated, nearly 43,000 measurements of radon-daughter concentrations were made in the approximately 2,500 mines that were worked (Table IV-3).[33] In discussing sources of potential inaccuracy in the working-level (WL) data, Holaday (quoted by Lundin et al.[33]) pointed out that the measurements taken after 1960 were primarily for control purposes and might have led to overestimates of the exposures to miners.

Because coverage was not comprehensive for all mines in all years, several different estimation procedures were used to fill the gaps in the exposure data. These estimation procedures were more important in the earlier years, when exposures to radon daughters were higher and fewer measurements were available.

To make estimates for missing data in the temporal series of WL measurements for a particular mine, the investigators interpolated and extrapolated earlier and later concentrations of radon daughters. When gaps in the data were too wide, area averages by locality, district, and state were used. For 1950 and earlier years, WL values were estimated on the basis of the few available radon measurements and the investigators' knowledge of the mining conditions. Many of the miners worked in other types of hard-rock mines before becoming uranium miners. For exposures to radon daughters in the hard-rock mines, WL values were based on calendar year: 1.0 WL for years before 1935, 0.5 WL for 1935–1939, and 0.3 WL for years 1940 and later.[33]

The arithmetic average of the individual WL measurements made within a mine in a given calendar year was assigned to the mine for that year. Use of the arithmetic average implicitly weighted all measurements equally; error would have been introduced if the numbers of workers

TABLE IV-1 Results of Colorado Plateau Study (Summarized from Principal Reports) of Male Uranium Miners

Followup Cutoff Date	No. of Subjects	No. of Lung-Cancer Deaths (Observed/Expected)	Comment	Reference
1959	2,666	6/3	Increase not statistically significant	2
1959	907	5/1.1[a]	Cohort members with at least 3 yr of experience	2
1962	3,656	15/4.2[a]	Includes 1,156 workers with surface, open-pit, or occasional underground work, respectively, through 1960	65
1963	3,415	22/5.7[b]	Response increases with cumulative WLM	66
1967	3,414	62/10.0[b]	Excess lung cancer in all exposure categories from <120 WLM to 3,720 WLM	32
1968	3,366	70/11.7[b]	Most comprehensive report	33
1974	3,366	144/29.8[b]	Response increases with cumulative WLM in all smoking groups	4
1977	3,362	185/38.4[c]	WLM not considered in analysis	67

[a] $P < 0.05$.
[b] $P < 0.01$.
[c] SMR is 482, 95% lower confidence limit is 425.

TABLE IV-2 Chronology of Radon and Radon-Daughter Measurements in Colorado Plateau Study[a]

Time	Source and Type of Measurement
Before 1950	Few radon measurements; earliest 1949
1950	Few radon samples by the PHS and Colorado State Health Department
1951	Radon and radon-daughter samples by PHS, state agencies, and U.S. Bureau of Mines
1952	Radon and radon-daughter samples taken by PHS in attempt to survey all mines
1953–1954	Scant data collection by PHS and states; Utah tried to survey every mine
1955	Scant PHS coverage; variable among states; mining companies begin measurements
1956–1958	Mine survey work primarily by companies
1959–1960	Colorado, Utah, and New Mexico conducted surveys; company measurements continued
After 1960	State and mining-company programs

[a]Based on data from Lundin et al.[33]

exposed at the concentrations indicated by the measurement were not uniform. The arithmetic average could also be strongly influenced by outlying high values.

To calculate WLM, the WL estimates were combined with work-history information obtained from annual censuses of active miners and from questionnaires. Apparently, a 170-h work month was assumed; and time for vacations, sick leave, or other absences from work was not subtracted from the number of underground hours estimated from the work history.[52] However, cumulative exposures were also not adjusted for time worked beyond 170 h/month, a common practice in the early years of the industry.[52] The investigators did not have enough information to consider work location within a specific mine or job classification, which might have influenced ventilatory demands.

Because WL measurements were sparse in relation to the numbers of mines that were worked, the WLMs accumulated by most miners were based on both measurements and estimates. In fact, WLM totals were calculated solely from measurements on only 10.3% of the white miners. For 36.1% of the white miners, some type of estimation was involved in the calculation of all WLM values; for the remainder, some WLM estimates were based on WL values derived by one of the estimation procedures.[33] In reports published to date, the WLM estimates have extended through September 1969. Information on cigarette smoking was obtained during the survey examinations, at the annual censuses of miners, and from mailed questionnaires.[33,69] As described by Whittemore and McMillan,[69]

TABLE IV-3 Number of Mines Visited and Number of Measurements Made in Colorado Plateau Study

Calendar Year	No. of Mines Visited	No. of Measurements	No. of Measurements/ Mine Visited
1951	5	21	4.2
1952	151	242	1.6
1953	56	474	8.5
1954	33	143	4.3
1955	4	15	3.8
1956	101	1,434	14.2
1957	147	848	5.8
1958	54	475	8.8
1959	281	1,867	6.6
1960	179	1,785	10.0
1961	330	2,952	8.9
1962	336	4,362	13.0
1963	315	2,648	8.4
1964	268	4,196	15.7
1965	268	4,856	18.1
1966	274	5,084	18.6
1967	266	5,696	21.4
1968	259	5,691	22.0
1969	149	1,683	11.3

SOURCE: Dr. Richard W. Hornung, National Institute of Occupational Safety and Health, Cincinnati, Ohio, personal communication.

information on smoking was obtained on one to four occasions between 1950 and 1960, when the surveys were conducted, and at other times between 1963 and 1969.

Mortality in the cohort was determined with follow-up techniques that included records of the Social Security Administration and the Internal Revenue Service, direct contact, and other approaches.[33,67] Only a few subjects could not be traced, and nearly all death certificates were obtained. Most published reports are based on analysis with a modified life-table approach, which is a conventional method for longitudinal studies that compares observed with expected numbers of deaths by cause. More recently, several investigators have applied modeling techniques to the data.[23,24,34,69] In cohort analyses based on an external referent population, expected numbers of deaths were calculated with mortality rates for the western states where the mines were or with the rates for all U.S. white males.

Table IV-1 summarizes the principal reports for the white male miners. At all follow-up intervals, statistically significant excesses of lung-cancer deaths were reported; the standardized mortality ratios (SMRs), which

TABLE IV-4 Lung-Cancer Deaths by Cumulative WLM in White Underground Miners in Colorado Plateau Study[a]

	No. of Lung-Cancer Deaths		
Cumulative WLM	Observed	Expected	Ratio of Observed/Expected
<120	1	1.81	0.55
120–359	12	2.57	4.67[b]
360–839	14	2.95	4.75[b]
840–1,799	12	2.52	4.76[b]
1,800–3,719	21	1.43	14.69[b]
≥3,720	10	0.42	23.81[b]

[a]Based on data from Lundin et al.[33]
[b]$P < 0.01$.

are age- and calendar-year-adjusted ratios of observed to expected deaths, ranged from approximately 4 to 6, without an obvious temporal trend. In several reports, the investigators used stratified analysis to examine the exposure-response relationship of lung-cancer mortality with cumulative WLM by calculating standardized mortality ratios within strata of increasing WLM.[3,4,32,33,66] In one report,[32] the mortality rates were standardized for cigarette smoking; in another,[4] they were stratified by cumulative WLM and smoking. Lundin et al.[33] adjusted the expected numbers of lung-cancer deaths for cigarette smoking. The investigators usually provided tables stratified by the interval after the start of employment in uranium mining.

Lundin et al.[33] compared observed with expected numbers of lung-cancer deaths in six strata of lifetime cumulative WLM (Table IV-4). A statistically significant excess was present in all categories of exposure, except in the category of less than 120 WLM. Archer et al.[4] provided mortality rates by exposure and cigarette smoking but did not include expected numbers of deaths.

Mortality from causes other than lung cancer was also examined. Significant excesses were not observed for cancers at sites other than the respiratory system.[4,33,67] Greater than expected numbers of deaths occurred from tuberculosis, nonmalignant respiratory diseases, accidents, and suicides. The 1981 report by Waxweiler et al.[67] showed a statistically significant excess of deaths (SMR, 262) attributable to the grouping of chronic and unspecified nephritis and renal sclerosis.

The data on the white underground miners have also been analyzed with other statistical approaches. Lundin and coworkers[33,34] developed a descriptive model for the development of lung cancer after radon-daughter exposure; the model was based on the assumption of a time-latency distribution with the same shape and dispersion as that of leukemia incidence after a single radiation exposure. They used the model to examine the effects of latent period, age at exposure, dose rate, and cigarette smoking and to compare absolute- and relative-risk models for the effect of radon-daughter exposure. They found that the relative-risk model was preferable to the absolute-risk model and that a 10-yr latent period gave the best fit. Effects of age at first exposure and of exposure rate on lung-cancer risk were not demonstrated. With regard to cigarette smoking, Lundin et al.[33,34] concluded that nonsmokers had much less radiation-induced lung cancer and that the excess radiation-induced lung cancer in smokers was not heavily influenced by the extent of smoking.

Assuming an exponential form for the relative hazard, Hornung and Samuels[24] used the Cox proportional-hazards model on data accumulated through the 1977 follow-up date. They found that a lag period of 6–11 yr for exposure was most compatible with the data. The modeling also showed that the exposure-response curve was downward at higher doses; that is, lower exposure rates led to greater effects. On a multiplicative scale for assessing the effects of exposures on lung-cancer risk, smoking and radon-daughter exposure had statistically significant effects, but a cross-product term of the two exposures was not statistically significant. These analyses were limited, however, to examination of only the exponential form of the relative risk.

More recently, Hornung and Meinhardt[23] reported on a proportional-hazards analysis of data based on follow-up of the cohort through December 31, 1982. A total of 255 deaths from lung cancer was identified by that date. Hornung and Meinhardt considered exponential, linear, and power-function models of risk and chose the power-function model, because it provided the best fit to the data. The model was developed with a stepwise approach; the data were best fitted by variables for cumulative WLM, cumulative smoking (in packs), and age at initial exposure. In the power-function model, the coefficient for the interaction of radon-daughter exposure and cigarette smoking was negative, although it was of borderline statistical significance ($P = 0.058$). This finding implies a submultiplicative, rather than purely multiplicative, interaction between cigarette smoking and radon-daughter exposure.

Hornung and Meinhardt[23] assessed the effects of several temporal factors: exposure rate, calendar year, age at exposure, and cessation of exposure. They found increasing risk with decreasing exposure rate, greater risk for more recent birth, greater risk for those first exposed at a greater age, and decreasing risk with cessation. The last two effects were thought to suggest a late-stage action of radon daughters, in the context of a multistage model.

Hornung and Meinhardt[23] used their power-function model to develop risk estimates for occupational exposures. Quantitative relative-risk estimates were made for occupational exposure beyond an assumed background exposure rate of 0.4 WLM/yr. For a 30-yr working lifetime, risk estimates were made for exposures of 30–120 WLM (1–4 WLM/yr). The relative risks ranged from 1.42 at 30 WLM to 2.07 at 120 WLM.

Whittemore and McMillan[69] used a case-control approach to examine additive and multiplicative models for the relationship of lung-cancer mortality to radon-daughter exposure and cigarette smoking. The results of their analyses are discussed briefly here and more fully in Appendix VII. A multiplicative linear model, with excess relative risk given by the product of the risk associated with radon-daughter exposure and that associated with cigarette smoking, fitted the data better than an excess-relative-risk model in which excess risks associated with radon and smoking were added. A series of multiplicative relative-risk models was evaluated by the investigators. They found a better fit for a model that incorporated the effects of smoking and WLM on relative risk as simple linear variables than for one that included exponential representations of these factors. Cumulative exposure variables fitted the data better than measures of exposure rate. Risk was not affected by age at the start of underground mining.

The PHS study cohort also includes nonwhite male miners, primarily American Indians. These subjects are of particular interest because of the low incidence of lung cancer in American Indians of the Southwest—a pattern probably attributable to a low prevalence of cigarette smoking.[4,50] Less information has been reported on the nonwhite subjects (Table IV-5). No cases of lung cancer among American Indians were observed initially, but a statistically significant excess was present in the 1974 follow-up.[4] In fact, the expected numbers of cases were probably overestimated because of the use of mortality rates for all nonwhites rather than for American Indians alone. In New Mexico during 1969–1977, for example, the average annual lung-cancer mortality rate in American Indian males was 8.6/100,000, whereas the rate for non-Hispanic white males was 60.8/100,000.[50] Lung-cancer mortality rates for black males have generally been equal to or higher than rates for white males.

Two other reports have addressed lung-cancer risks in American Indians employed in the Colorado Plateau mines. Gottlieb and Husen[18] reported a case series of 17 Navajo males diagnosed as having lung cancer at the Shiprock Indian Health Service Hospital. All but one had worked as a uranium miner, and only two had smoked cigarettes; cumulative WLM ranged from 59 to 2,125. Samet et al.[51] conducted a population-based case-control study to assess the association between uranium mining and lung cancer in Navajo males. Of 32 lung-cancer cases diagnosed between 1969 and 1982, 23 had a documented history of uranium mining. None of

TABLE IV-5 Data on Nonwhite Male Underground Uranium Miners in Colorado Plateau Study

Followup Cutoff Date	No. of Subjects	No. of Lung-Cancer Deaths, Observed/ Expected	Comment	Reference
1959	640	0/[a]	Total of 11 deaths	2
1962	1,103	0/0.8	Comparison rates from state data	65
1974	780	11/2[b]	Comparison with male nonwhite population of Arizona and New Mexico	4

[a]Not reported.
[b]$P < 0.01$.

the 64 matched controls had been uranium miners. The results imply an extremely high relative risk in this nonsmoking population, but individual WLM estimates were not available for all miners, and the data cannot be used for quantitative risk estimation.

The Colorado Plateau study was designed and implemented 35 yr ago. Its strengths include the size of the cohort, the long duration of follow-up, the estimation of WLM for individual subjects, and the availability of cigarette-smoking histories. Application of new techniques to the data set has helped to explore the interaction between cigarette smoking and radon daughters and the effects of time-dependent factors such as dose rate and lag times. Even though investigators have dealt pragmatically with the severely limited number of WL measurements in calculating WLM estimates, the quality of the exposure information must be considered in interpreting the results of the study. Both random error and systematic bias might affect WLM estimates. Much of the exposure occurred before extensive measurement procedures were in place. For example, 36.1% of the total WLM ultimately accumulated by the cohort of white miners occurred before 1956 (Richard W. Hornung, NIOSH, personal communication, 1986). Few measurements were taken during the early years, when exposure rates were highest, so the higher exposures were probably estimated less accurately than later ones. If higher exposures were subject to a greater misclassification, the risk coefficients that have been calculated for the higher WLM values might be artificially low. Bias could also have been introduced by the investigators' decision to rely on measurements taken for control purposes after 1960, in that such measurements can over-represent higher exposures. Finally, the cohort had relatively high exposures and thus provides little information on the results of cumulative exposures of less than 100 WLM.

CZECHOSLOVAKIAN URANIUM MINERS

The retrospective cohort mortality study of the Czechoslovakian uranium miners was initially reported in 1971,[55] and periodic updates have been published.[22,28,29,47,54,56,58] The cohort consisted of miners who began mining uranium ore in 1948–1957. However, the results in the more recent reports are limited to 2,433 miners[55] who began in 1948–1952. The selection criteria for the cohort have not been specifically described. The investigators have not reported whether the study cohort included all eligible miners in a particular geographic area or only a sample, what procedure was used and what records were reviewed to identify the cohort, the total number of miners who died from any causes other than lung cancer, and the distribution of the cohort members by birth year, age, or age when first exposed.

Individual work histories were abstracted from payroll cards for all miners (Langon Swent, personal communication, 1984) from 1948. For each miner, WLM was estimated from radon gas measurements and the number of months of employment at each mine in each calendar year. Since 1948, more than 120,000 radon gas measurements were made by measuring ionization current in an ionization chamber by electrometer. Yearly numbers of radon measurements were not given, but the lowest reported mean number of measurements for a year was 101 ± 8/mine. The range of coefficients of variation of average yearly radon concentrations in mines was 3.5–20.0%. Radon gas concentrations were converted to WL on the basis of ventilation conditions and practices, emanation rates from different types of ores, and after 1959, radon-daughter measurements. Since 1968, each miner's WLM has been determined from individual personal dosimetry cards. Assessment of dosimetry errors was based on the magnitudes of coefficients of variation, which do not provide information on the validity of the dosimetry data.

The cohort was followed with lung-cancer registrations administered by the authors in health facilities, the records of the hygiene service in the uranium industry, and oncology notification cards from throughout the country. The latter two served as independent follow-up sources after 1960. Until 1960, only 12 deaths due to lung cancer occurred. The success of this approach for identifying lung-cancer cases is not established, and the number of persons lost to follow-up is not given in the 1976 report by Sevc et al.[56] Except for a paper on skin cancer, health effects other than lung cancer have not been reported.

In analyses of this cohort, observed lung-cancer mortality was compared with that expected on the basis of age- and calendar-period-specific rates of the male population in Czechoslovakia. In the 1976 report by Sevc et al.,[56] person-years at risk for each subject were classified by the final cumulative WLM category, rather than being distributed across the appropriate WLM categories as they accumulated. This error was corrected in later analyses,[28,29] and only the later analyses are considered here. According to Swent (personal communication, 1984), a miner must have worked at least 4 yr underground to be eligible for inclusion in the cohort. However, person-years at risk were counted from the first date underground, rather than from the date of eligibility, so expected deaths were slightly overestimated.

Cigarette smoking was not assessed for all cohort members individually, but results of studies on a random group of 700 miners indicated that about 70% of the uranium miners were smokers. Data were not given on the amount smoked or the age when smoking started. According to Sevc et al.,[56] the prevalence of smoking in the general male population of Czechoslovakia was comparable with that in the sample of miners. Radon-daughter and other exposures from prior hard-rock mining were not evaluated, because less than 2% of the cohort miners had previously mined nonuranium ores. Other characteristics of this sample have not been reported.

The most recent and thorough analyses were based on follow-up through 1975 of miners who began exposure in 1948–1952. Follow-up averaged 26 yr.[29] In these modified life-table analyses, observed minus expected (based on the male population in Czechoslovakia) lung-cancer deaths were calculated for five categories of cumulative WLM (less than 100, 100–199, 200–399, 400–599, and 600 and over) and, with further stratification, for three categories of duration of exposure (0–7.9 yr; mean, 5.6 yr; 8–11.9 yr; mean, 9.5 yr; and 12 yr or longer; mean, 14 yr) or for three temporal exposure patterns. A temporal pattern of exposure was modeled for each miner individually by the regression, cumulative WLM $= a^b$, where cumulative WLM was calculated for each year of work, t. The cohort was then divided into three groups based on individual members' value of b: group A, b significantly less than 1, implying a high rate followed by a low rate of exposure; group B, b not significantly different from 1, implying a fairly constant rate of exposure; and group C, b significantly greater than 1, implying a low rate followed by a high rate of exposure.

The authors reported two general findings. First, the analyses indicated significant effects of cumulative WLM, duration of exposure, and their interaction. In this study, excess risk is expressed as the excess number of lung cancers per 1,000 miners (Table IV-6) and not per person-year, as is reported in most other cohort studies reviewed here. For those exposed 12 yr or longer, this risk was linearly related to cumulative WLM for miners overall (Figure IV-1), but not for the two shorter exposure periods. Second, excess risk was linearly related (Figure IV-2) to cumulative WLM for miners in groups A and B, but not group C (low followed by

TABLE IV-6 Lung-Cancer Mortality among Czechoslovakian Uranium Miners[a]

Cumulative WLM	Mean WLM[b]	No. of Excess Lung Cancers/1,000 Miners[c]
<100	72 ± 1.8	13.5 (−8.5, 54.3)
100–199	150 ± 1.5	46.6 (28.1, 69.8)
200–399	285 ± 2.6	87.3 (66.9, 109.8)
400–599	570 ± 6.0	116.4 (82.2, 154.2)
≥600		137.3 (89.0, 199.8)
Total	—	80.9 (68.4, 94.1)

[a]Based on data from Kunz et al.[29]
[b]Values are means ± standard deviation.
[c]The 95% confidence limits are given in parentheses.

high exposure rate). Other analyses of the data[29] indicated a significant effect of cumulative WLM on excess risk, but not of exposure pattern or their interaction. From Figures IV-1 and IV-2, it appears that the 95% Poisson-based confidence intervals are wide enough to allow nonlinear interpretations of the relationship between excess risk and cumulative WLM within separate groups of exposure duration or temporal pattern.

An earlier report[28] of follow-up through 1973 is the only report on the cohort of Czechoslovakian uranium miners that provided observed and expected mortality rates per 10,000 person-year and observed to expected lung-cancer mortality ratios, in addition to excess lung-cancer deaths. However, only one independent variable, cumulative WLM (<100, 100–199, 200–399, and ≥400), was reported (Table IV-7).

ONTARIO URANIUM MINERS

A retrospective cohort study of Ontario miners[37–39] engaged in various types of mining included a subcohort of uranium miners who met the following criteria:

- received a miner's physical examination required annually by the company any time in 1955–1977 (uranium mining began in 1955 in Ontario);
- worked at least 1 month as an underground uranium miner; and
- had not worked in a job with any known asbestos exposure, in uranium processing (except in mills), or in any uranium mining in another province as an employee of Eldorado Nuclear.

Radon-daughter exposure was estimated by different methods for 1967 and earlier and for 1968 and later. For 1968 and later, exposure records of WLM maintained by the mining companies were used. For 1957–1967, the

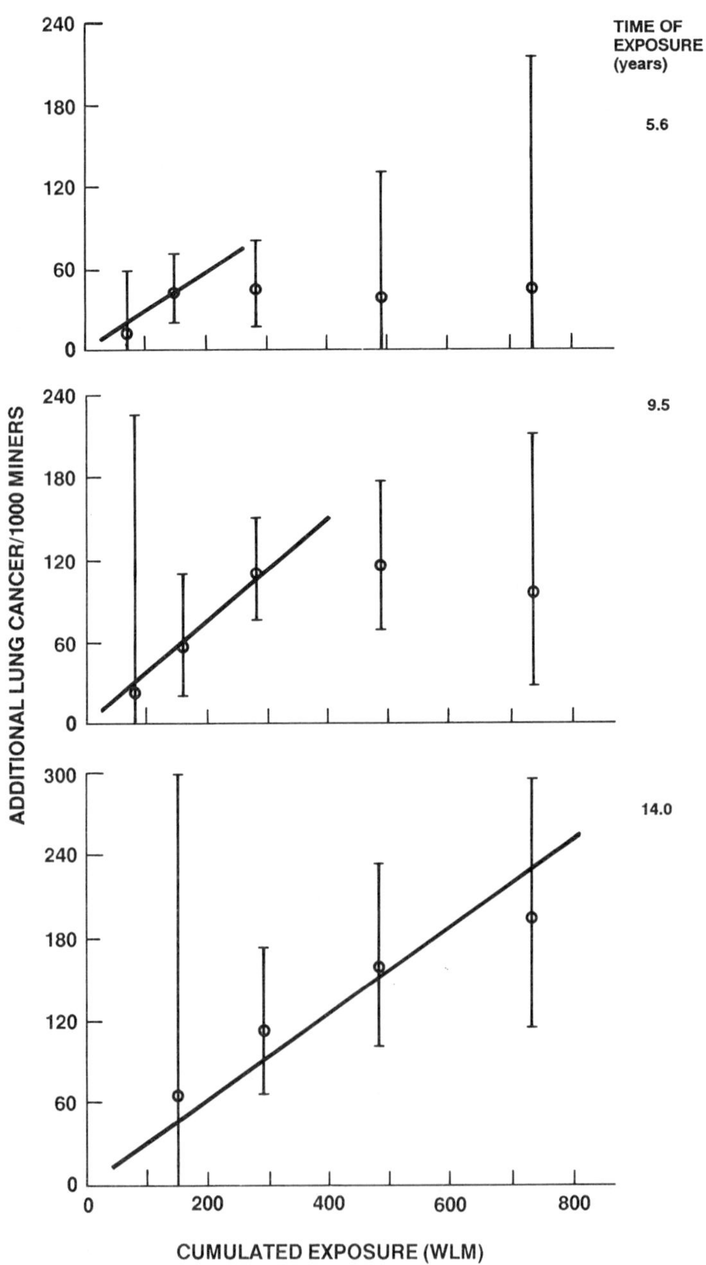

FIGURE IV-1 Relation between additional lung-cancer frequency and cumulative radiation exposure in three groups of Czechoslovakian uranium miners by mean duration of exposure. SOURCE: Kunz et al.[29]

investigators calculated WLM by combining WL information with work histories.[38] Because of the variability of radon-daughter concentrations, the investigators developed two separate sets of WL values for this earlier period. The standard (or lower) WL values were the averages of the four quarterly averages or three 4-month averages for a particular year. To calculate the special (or upper) WL values, the investigators weighted the average of the four highest quarterly measurements or the three highest 4-month measurements in headings, stoops, and raises (a total of 12 or 9 measurements, respectively) by 0.8 and the average of the four highest quarterly or three highest 4-month measurements in travel ways by 0.2. The difference between the standard and special WL values varied with mine and year;[38] for some mines in some years, the special and standard values were equivalent, but the special values were up to 4 times the standard WL estimates in the years and mines for which both were available.[38] The investigators considered that the true exposure of each man lies within this range. During 1958–1967, 13,081 measurements were taken (Table IV-8). For one large mine, WL data for the 4 yr from 1957, when the mine started operating, through 1960 had to be rejected, because they were shown to be unreliable. The values for the missing years were estimated by taking into account tonnage mined, ventilation, and dust concentrations at various times.

Work-history information was obtained primarily from records of pre-employment and yearly examinations carried out by Ontario government agencies.[38] Additional information related to the first 5 yr of employment in the mining industry was collected from work-history cards.

The WLM values for 1955–1967 were calculated by combining the work-history information with a matrix of annual WL values for each mine in each year. Adjustment was made for deviations from normal working hours in a mine, considered to be 2,000 h/yr. No estimates of WL were made for prior gold-mining experience, but persons with such experience were analyzed separately, because Ontario gold miners were at increased risk of lung cancer.[40] It should be noted that the committee's analysis of the Ontario miners, described in Annex 2A, excluded miners with previous gold-mining experience.

Because the WL measurements did not cover the complete working experience of the cohort, some estimation of exposures before 1954 was necessary. These years included the period of highest exposures and, as Muller et al.[38] reported, 22% of the total WLM accumulated by the cohort is based on extrapolation from measured values, with account taken of, for example, ventilation. For one large mine, this percentage includes extrapolation up to 1960. The period of extrapolation weighted by WLM is, however, less than 2 yr.

Follow-up through 1981 was carried out by computer linkage with national mortality data bases combined with manual cross-checking to resolve problems. The investigators did not report on the percentage lost to

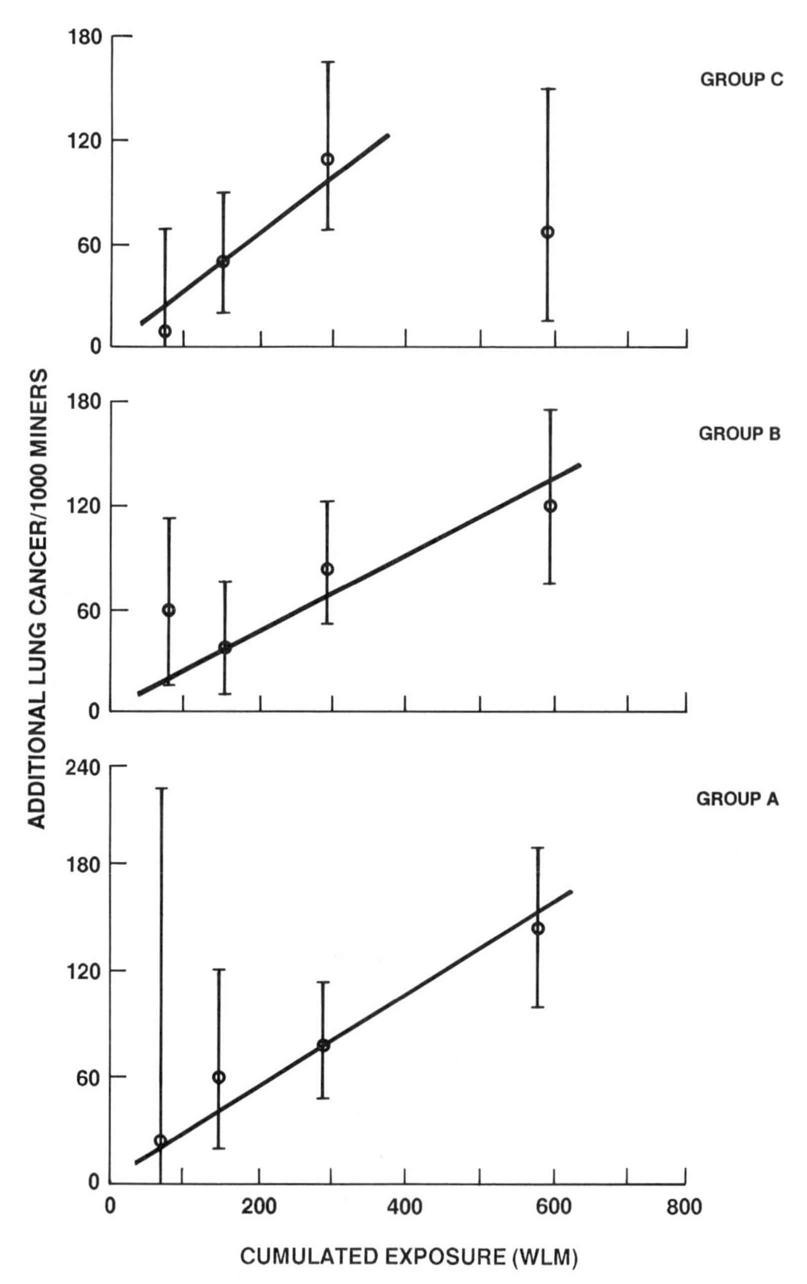

FIGURE IV-2 Relation between additional lung-cancer frequency and cumulative radiation exposure in three groups of Czechoslovakian uranium miners by time course of exposure accumulation (see text). SOURCE: Kunz et al.[29]

follow-up or on the percentage of death certificates not obtained. However, on the basis of a sample of known deaths, 6.3% were not identifiable as deceased with the same follow-up method. Death certificates were the only source of information on cause of death.

Using the modified life-table technique, Muller et al.[38] compared observed with expected mortality (based on the Ontario general male population rates with adjustment for age and calendar period). Results for causes of death other than lung cancer were available only for 1955–1977. The authors did not have information on cigarette-smoking habits of the miners.

The mean cumulative WLM of miners with no previous gold-mining experience was 40 (lower estimate) to 90 (upper estimate). All other descriptions of the cohort included those who had previously mined gold. The median year of birth of the cohort was 1932, and the median year first employed in a mine in Ontario was 1957; thus, the median age at first employment in a mine in Ontario was probably about 25 yr. The median duration of work in a mine was 1.5 yr.

Among uranium miners without any gold-mining experience, Muller et al.[38] found that observed to expected ratios for lung-cancer deaths increased across the six categories of cumulative WLM (Table IV-9). When the upper estimated exposures were used, the first definite excess occurred at a cumulative WLM of 100–170 (mean, 130), with 14 observed and 6.9 expected lung cancers. When the lower estimated exposures were used, there was a definite excess at a cumulative WLM of 40–70 (mean, 53), with 13 observed and 7.0 expected lung cancers. Muller et al.[39] reported that linear regression of the dose-response relationship, weighted by number of person-years at risk (PYAR), showed similar fits for the excess- and relative-risk models.

A 5- and 10-yr exposure lag did not change the slope of the relative-risk model (0.5% excess relative risk per WLM for the upper exposure estimates and 1.3% for the lower ones), but slightly increased the slope of the excess-risk model (from 4.8 to 7.2/million WLM for the lower exposure estimates and from 2.0 to 2.8 for the upper ones). With either model, the use of the upper exposure estimates decreased the dose-response slope by more than 50%. However, dose-response analysis for two age groups of PYAR indicated that the slopes for the relative-risk model were age-independent, whereas the slopes for the excess risk model were not.

ELDORADO URANIUM MINERS

Howe et al.[25] conducted a retrospective cohort mortality study of all 10,945 male employees who had worked at the Eldorado Uranium Mine in Beaverlodge, Saskatchewan, anytime between 1948 (when the mine opened) and December 31, 1980. The cohort was identified from company employment and payroll records. The final study group included

TABLE IV-7 Lung Cancer Among Czechoslovakian Uranium Miners in Relation to Cumulative Radon-Daughter Exposure[a] Based on Modified Life-Table Method[b]

Cumulative WLM	No. of Person-Years at Risk	Calculated No. of Lung-Cancer Deaths/10,000 Person-Years: Observed (95% Confidence Limits)	Expected	Additional (Observed − Expected)	Observed/Expected Ratio
<100	9,380	(2.3)- 6.4-(13.9)	5.5	0.9	1.2
100–199	16,131	(17.7)-24.8-(33.8)	7.6	17.2	3.3
200–399	19,614	(34.0)-42.8-(52.4)	7.7	35.1	5.6
≥400	11,830	(54.8)-69.3-(85.5)	8.4	60.9	8.2
Total	56,955	(34.7)-37.2-(42.5)	7.5	29.7	5.0

[a]Start of exposure was 1948–1952; the cutoff date for analysis was December 31, 1973.
[b]Based on date from Kunz et al.[28]

TABLE IV-8 Numbers of Mines and Measurements in Study of Ontario Uranium Miners[a]

Year	No. of Mines	Total No. of Measurements	No. of Measurements/Mine
1958	15	696	46
1959	14	2,145	153
1960	12	1,879	157
1961	7	1,446	207
1962	6	1,563	260
1963	6	1,170	195
1964	5	776	155
1965	4	985	246
1966	3	1,135	378
1967	4	1,286	322

[a]Personal communication, J. Muller, M.D., 1986.

TABLE IV-9 Observed and Expected Lung-Cancer Deaths by Cumulative WLM among Ontario Uranium Miners with No Gold-Mining Experience[a]

Exposure Group	Mean Cumulative Exposure[b] (WLM)	No. of Lung-Cancer Deaths		Observed/ Expected Ratio	No. of Person-Years at Risk
		Observed	Expected		
Cumulative Special WLM (upper estimates)					
0.1-10	5	14	9.5	1.47	45,055
10.1-40	22	15	17.4	0.86	62,173
40.1-100	64	12	13.2	0.91	47,154
100.1-170	130	14	6.9	2.03	22,041
170.1-340	235	13	6.4	2.03	18,249
340.1+	510	14	3.4	4.1	8,124
Cumulative Standard (lower estimates)					
0.1-6	3	14	11.7	1.20	51,356
6.1-20	12	13	17.2	0.76	61,823
20.1-40	29	15	11.0	1.36	38,751
40.1-70	53	13	7.0	1.86	23,313
70.1-140	98	12	6.0	2.00	17,345
140+	200	15	4.1	3.66	10,208

[a]Based on data from Muller et al.[39]
[b]No exposure lag or minimum latency period was used in estimating the WLM.

8,487 subjects; 1,782 (16%) persons were excluded because of missing or incorrect information, and another 676 (6%) were excluded because they had worked at other company sites. The authors were unable to detect any bias due to these exclusions. Follow-up from 1950 through 1980 was carried out by linkage with a national mortality data base. Only one person was lost to follow-up.

The WLM values for Beaverlodge uranium miners were calculated by Eldorado Resources Ltd., which operated the mine. Different approaches were used for 1966 and earlier years and for 1967 and later years. For 1966 and earlier, the WL estimates were based on all available measurements of radon and radon daughters (Table IV-10). Equilibrium between radon and its daughters was estimated by comparing paired measurements of radon and radon-daughter concentrations. When paired measurements were unavailable for a particular year, the average of the equilibrium factors for adjoining years was interpolated. Because the distribution of measurements was strongly skewed toward higher values, the annual median, rather than the mean, was used to calculate exposure for each year.* For 1967 and later, radon-daughter measurements were generally available. Geometric means or averages of geometric means were used for the calculations. For some locations, adjustments were made on the basis of working conditions.

In calculating the WLM for the work force, the WL values for each year were adjusted for the extent of underground exposure sustained by workers in eight occupational categories. Dates of employment were used to determine the number of weeks worked in each year. Four weeks of holiday time each year were assumed, and adjustments were made for the changing duration of working hours over the study.

Silica exposures to this cohort were always very low, and diesel machinery was never used underground. Potential confounding from other mining exposures was addressed in one analysis by excluding the 540 men who were included in the Ontario miner study[38] and by excluding miners who had reported previous mining experience elsewhere. No measures of cigarette smoking were reported for cohort members individually.

The final cohort consisted of three groups: surface workers only (48%), underground workers only (45%), and both surface and underground workers (7%). The mean years of first exposure for these three groups were 1966, 1966, and 1963, respectively. The mean ages at first employment were 27.7, 28.8, and unreported, respectively. The mean periods of follow-up were 13.9, 13.5, and 17.3 yr, respectively. The mean durations employed

*A recent review of these calculations submitted to the committee, "Beaverlodge Working Level Month Calculations," Draft 4 by S. E. Frost, has suggested possible underestimation of exposures. New calculations for some years indicate that the choice of the median WL value and the method used to determine equilibrium factors might have resulted in bias toward low WLM estimates. For the years reviewed, use of the arithmetic mean, rather than the median, increases the annual WL value.

TABLE IV-10 Numbers of Radon-Daughter and Radon Measurements in Eldorado Beaverlodge Uranium Miner Study, 1954–1966[a]

Year	No. of WL Samples in 1977 Estimates	Total No. of WL Samples Available	No. of Radon Samples Available
1954	—	20[b]	139
1955	—	—	123
1956	38	33	382
1957	—	—	299
1958	—	—	373
1959	17	58	522
1960	—	4	952
1961	122	108	743[c]
1962	179	181	6
1963	160	210	5
1964	171	163	—
1965	286	304	—
1966	459	413	526

[a]Based on data from Beaverlodge Working Level Month Calculations, Draft 4, by S. E. Frost.
[b]Derived from RaA and RaC′ measurements.
[c]Does not include additional 203 shaft Rn measurements.

were 22.2, 15.0, and 43.9 months, respectively. The means of cumulative WLM were 2.8, 16.6, and 28.9, respectively.

A modified life-table analysis was carried out. Comparisons were made with 5-yr, age- and calendar-period-specific mortality rates for the general male population of Canada.

The finding of no lung-cancer excess among those with less than 5 WLM (19 observed versus 18.36 expected) was interpreted as evidence against strong confounding by cigarette smoking in the entire cohort. Furthermore, among those with greater than 5 WLM, no excess of lung cancer was found within the first 5 yr after exposure began. The authors excluded the first 10 yr of follow-up from further analyses, to be consistent with procedures in other studies, although an excess risk of lung cancer was found at higher doses within 5–9 yr after first exposure (6 observed versus 1.54 expected).

The SMRs for lung cancer increased monotonically (Table IV-11) from the lowest to the highest category of cumulative WLM (0–4, 5–24, 25–49, 50–99, 100–149, 150–249, and 250+). The authors used weighted least-squares regression to describe the exposure-response relationship. Exposure within each category was represented by the mean cumulative WLM, and PYAR was used for weighting. The addition of a quadratic term did not significantly improve the fit of the linear model to the data. When the authors multiplied simple linear functions by exponential terms to represent a cell-killing parameter, they were unable to fit a biologically appropriate model to the data. Furthermore, a 5-yr lag of exposures changed the linear-regression coefficients by less than 10%, compared with no lag. Howe et al.[25] also investigated the effects of age at first exposure and age at observation. In both cases, the attributable risk was found to be much more dependent on age than was the relative risk.

The Beaverlodge miners have also been included in a larger study of Eldorado Resources Ltd. employees. Nair et al.[41] conducted a retrospective cohort mortality study of all males employed before 1981 at four major operations: a pitchblende mine at Port Radium from 1932 to 1940 (time period excluded from study) that was later a uranium mine during 1942–1960; a refinery at Port Hope, Ontario, which opened in 1932, refined radium until 1954, and refined uranium and converted it to uranium dioxide and uranium hexaflouride until the present; a uranium mine at Beaverlodge from 1953 until the present; and other sites.

The cohort was assembled from a company employee roll that included full name, sex, place and date of birth, and last year known alive. The Port Radium cohort was divided into those who ever and those who never worked underground. Follow-up was limited to computer linkage with a national mortality data base for 1950–1980. Bias might have been introduced by the rejection of a large percentage of each cohort (Port Hope, 38%; Port Radium, 44%; and Beaverlodge, 13%) because of inadequacy in personal data or loss to follow-up before 1950.

Preliminary findings on lung-cancer mortality were reported only as observed deaths due to lung cancer for each work group versus those expected based on national rates (Table IV-12).

Later examination determined that the Port Radium surface cohort included a number of underground miners. These results are not useful for assessing dose-response relationships, because data on WLM were not available.

FRENCH URANIUM MINERS

Tirmarche et al.[64] carried out a retrospective cohort study of all men who began underground uranium mining during 1947–1972 in any of 12 French mines and worked a minimum of 3 months. For 1947–1955, WLM values were based on a few radon measurements, ventilation conditions, ore characteristics, and working methods. Extensive radon measurements were taken later; there were an average of 20–30 values taken per mine/year during 1957–1970 and twice that during 1971–1980. WL was estimated retrospectively using the current equilibrium factor

TABLE IV-11 Observed and Expected Lung-Cancer Deaths by Cumulative WLM, 1950–1980 (First 10 Years of Followup Excluded) Among Eldorado Beaverlodge Uranium Miners[a]

			No. of Lung-Cancer Deaths			
Cumulative WLM	Mean CWLM[b]	Person-Years	Observed	Expected	RR[c]	AR[d]
0–4	0.9	29,818	14	14.46	0.97 (0.53, 1.62)	−15 (−288, 303)
5–24	11.7	14,815	12	6.48	1.85 (0.96, 3.24)	373 (−19, 978)
25–49	35.6	5,554	5	2.64	1.89 (0.61, 4.42)	425 (−183, 1,625)
50–99	69.8	3,755	6	2.48	2.42 (0.89, 5.26)	937 (−75, 2,817)
100–149	121.1	1,607	7	1.17	5.98 (2.41, 12.35)	3,628 (1,024, 8,248)
150–249	187.4	1,051	6	0.76	7.89 (2.88, 17.10)	4,986 (1,269, 11,705)
250+	294.9	342	4	0.28	14.29 (3.87, 36.35)	10,877 (2,366, 29,165)
Total	20.2	56,942	54	28.27	1.91 (1.43, 2.49)	452 (216, 741)

[a]Based on data from Howe et al.[25] Observed and expected deaths denote the number of deaths based on age-specific and calendar-year-specific Canadian national mortality rates, 1950–1980.
[b]Weighted by person-years at risk.
[c]Relative risk: observed/expected (with 95% confidence limits).
[d]Attributable risk: [(Observed − Expected)/PQ] $\times 10^6$ (with 95% confidence limits).

TABLE IV-12 Standard Mortality Ratios for Lung Cancer among Eldorado Employees[a]

Site	No. of Lung-Cancer Deaths		SMR
	Observed	Expected	
Beaverlodge	112	60.87	184
Port Hope	14	17.16	82
Port Radium (surface)	28	15.97	175
Port Radium (underground)	55	14.67	375
Other sites	5	4.53	110

[a]Based on data from Nair et al.[41]

of 0.22. The only epidemiological results were for lung-cancer mortality: 36 observed versus 18.77 expected among the entire cohort without any lag in exposure or latency considerations. No dose-response results were reported. It appears that PYAR for each miner began inappropriately at the date of first employment and not after 3 months of mining. National mortality rates were used for comparison; however, the mines are all in agricultural areas. The WLM data in this study are potentially limited by the lack of measurements for 1947–1955 and the retrospective estimation of the equilibrium factor. Furthermore, cause of death was not known for 25% of the deceased subjects.

CORNISH TIN MINERS

High concentrations of radon and its daughters have been measured in tin mines in Cornwall, England. Fox et al.[17] conducted a retrospective cohort study of mortality in 1,333 men employed in two tin mines in Cornwall during 1939. In comparison with mortality rates for England and Wales, lung-cancer mortality was increased in the underground miners (SMR, 211), but not in surface workers (SMR, 74) or in workers who were not classifiable into either of these two categories (SMR, 94). WLM were not estimated for the subjects. The authors reported government estimates of 25 WLM and 15 WLM, respectively, annually for the two mines.

CHINESE TIN MINERS

Tin has been mined in the Yunnan region of China for centuries,[62] and the miners in this region are known to have arsenic and radon-daughter exposures. Wang et al.[68] identified a cohort of 12,243 underground miners

TABLE IV-13 Lung-Cancer Mortality by Cumulative Radiation Exposure among Chinese Tin Miners[a]

Cumulative WLM	No. of Person-Years at Risk	No. of Lung-Cancer Deaths		Observed/Expected Ratio
		Observed	Expected[b]	
<140	33,302	31	7.1	4.4
140-279	28,468	44	6.0	7.3
280-559	19,111	106	8.2	13.0
560-839	6,436	92	3.8	24.3
840-1,399	7,045	115	3.7	30.7
≥1,400	1,774	45	1.0	43.6
Total cohort	86,136	433	29.8	14.5

[a]Based on data from Wang et al.[68]
[b]Based on Shanghai population, apparently without adjustment for age, sex, or calendar period.

and followed them from 1975 to 1981 for lung-cancer incidence and mortality. Information has not been reported on the selection of study subjects, their duration of work, latency distribution, smoking distribution, follow-up methods, or losses to follow-up. The age distribution of the cohort was not given, but it has been reported that many persons began underground mine work between the ages of 8 and 14 yr;[62] this practice was phased out around 1949. WLM were calculated from detailed individual work histories and systematic radon-daughter measurements at underground work boxes during 1972–1980. Only natural ventilation was used in the mines in 1953–1972, so exposures were assumed to be constant during this interval. Before 1953, some of the mines were smaller, no wet-mining methods were used, and proportionate adjustments were made. Another adjustment was made for exposures before 1949, when more primitive mining methods, including back-carrying of ore through narrow tunnels, were used.

During the follow-up period, lung-cancer incidence was 515/100,000 (499 cases) among underground miners, 41.3/100,000 (59 cases) among surface workers. The investigators did not report the incidence data by dose, years worked, or latency. Lung-cancer mortality for the underground miners was also compared with that of the Shanghai population (Table IV-13), but apparently without adjustment for age, sex, calendar period, or smoking.

In addition to radon daughters, exposure to arsenic was considered to play an etiological role in the lung-cancer excess. Ore samples contained 1.5–3.5% arsenic; the investigators estimated that a miner's respiratory tract was exposed to 1.97–7.43 mg of arsenic/yr during the years immediately after 1949. Total dust concentrations were estimated at 30 mg/m^3, with peaks during dry drilling of 344 mg/m^3 in the earlier years.

Sun et al.[62] described 929 lung-cancer cases (755 deaths) ascertained during 1954–1978 among workers at three mines of the Geiju Tin Mine Company in Yunnan. Death certificates, histological-cytological reports, and chest x rays were cross-checked to confirm the cases. The little available information on relative risks was based on a crude cohort study that calculated expected deaths using the age distribution of the workers in one of the mines in 1975. The cohort analyses that controlled for duration of mining indicated significant differences in SMRs by age at which the miners began mining. However, for those who began mining before age 14, risk did not increase with duration of mining. The smoking habits of 17,287 miners were recorded. The authors reported that the relative risk for lung cancer in smoking miners was about 20 times higher than that in nonsmokers. There was no significant relationship between latent period and degree of smoking.

The age distribution of the work force in the three mines in 1973 was trimodal, with peaks at 20, 25, and 40 yr, which is a reflection of temporal changes in hiring practices. This unusual age distribution and the emphasis on case ascertainment (i.e., follow-up) during 1971–1978 obscures the relationships among age at which miners started mining, latency period, length of follow-up, and risk. However, of the large number of persons who began work underground before the age of 14, few developed lung cancer before the age of 35.

CANADIAN FLUORSPAR MINERS

The open-pit mining of fluorspar (calcium fluoride) in St. Lawrence, Newfoundland, began in 1933. Underground mining began in 1936 and has been carried out in 12 mines. After the discovery in the 1950s of an unusual excess of lung-cancer deaths among the miners, a retrospective-prospective cohort study was undertaken.[13,14,35,36] The ore itself is not radioactive,[14] but the substantial amounts of water seeping through the mines contain radon gas.

Exposures have been estimated on the basis of occupational histories that include type and place of work and hours of work by year. For years before and to 1960, work hours were converted to working months (167 or 170 h) and used to calculate WLM. WL values were estimated retrospectively from measurements made in only one mine in one year, 1959. Before 1960, the mines were ventilated primarily by natural draft,

TABLE IV-14 Lung-Cancer Mortality by Cumulative Radiation Exposure among Canadian Fluorspar Miners[a]

Cumulative WLM	No. of Person-Years at Risk	No. of Lung-Cancer Deaths		Observed/Expected Ratio	*P* Value
		Observed	Expected		
0	13,657.8	7	7	1.00	—
1-9	3,045.5	3	2.02	1.49	0.50
10-239	9,510.5	13	7.22	1.80	0.09
240-599	5,105.5	10	3.87	2.58	0.06
600-1,979	7,107.0	6	1.71	3.51	0.03
1,980-2,039	2,415.5	25	1.54	16.23	<0.001
≥2,040	1,889.0	40	1.07	37.38	0.001

[a]Based on data from Morrison et al.[36]

occasionally aided by small blowers, and the ventilation varied greatly in each mine, as did the amount of water seepage. Radiation measurements were infrequent during 1960–1968, but were taken daily after 1968.

Recent follow-up of the cohort has been primarily through linkage with the nationwide mortality data base[36] but has also included a small number of deaths certified by local clergy, parish records, hospitals, and relatives, rather than medically. All persons not definitely identified as deceased were assumed to be alive. Initial analyses of lung-cancer mortality in 1952–1960 showed 21 deaths among the miners, compared with 0.7 expected from age-adjusted rates for the remainder of Newfoundland. Recent analyses[35,36] used standard modified life tables of PYAR and age-specific lung-cancer rates among the surface workers for comparison. This analysis is limited by the small number of lung-cancer deaths (seven) in the comparison group and possibly by migration between surface and underground work. Follow-up was accomplished with the use of the company, union, and medical files. Analysis based on follow-up through 1978 (ignoring the first 10 yr of risk after hiring) showed a strong dose-response relationship between lung-cancer risk and cumulative WLM (Table IV-14).[36] Smoking-specific findings are not reviewed because of the lack of adequate ascertainment of smoking status. The authors found that latency periods decreased for men first exposed when older and for men exposed during the earlier years, when exposures were presumably higher.

SWEDISH IRON MINERS: MALMBERGET

A retrospective cohort mortality study by Radford and Renard[48] included miners from two iron mines (in Malmberget and Koskoskulle) owned by one company (LKAB). Selected for study were the 1,415 men born in 1880–1919 who were alive in 1930 and who worked underground for more than 1 calendar year during 1897–1976. The cohort was identified principally from company and union records of active and pensioned miners that dated back to 1900. Additional men were identified from medical surveys and a few were identified from parish records. Time worked underground was determined from company and union records and medical files. Work histories appear to contain data only by year; July 2 was assumed as a starting and stopping date for underground work. For those who stopped and restarted in 1 yr or started and died in 1 yr, April 1 and October 1 were assumed, respectively. The extent to which the cohort covered all employed miners was evaluated for the years 1942–1946 by comparing person-years underground from two sources: the work histories of the cohort and company records.

The WLM values for this analysis were those calculated by Radford and Renard.[48] As described in their 1984 report, radon dissolved in water was assumed to be a major source of radon daughters in the mines. Comparison of radon measurements in water taken in 1915 with data from 1972 and 1975 indicated constant radon concentrations in groundwater. The first measurements of radon in mine air were obtained in 1968. Radon and radon daughters were later measured by the mining company and by the National Radiological Institute. Past concentrations were then reconstructed on the basis of these measurements in combination with information on ventilation conditions. Radford and Renard assumed that ventilation conditions in 1968–1972, when the measurements were made, were not greatly different from those in the past. In support of this assumption, they cited a pattern of natural convection and data on quartz dust concentrations that extended to the 1930s. Radon daughters were found to be at about 70% equilibrium with radon.*

---

*A report submitted to the committee, "Comments to the U.S. Mine Safety Health Administration for the American Mining Congress," by Swent and Chambers, questioned some assumptions underlying the historical reconstruction of the exposures for the years before measurements were taken. Because Radford and Renard[48] assumed that water was the major source of radon in the mines and its strength was constant, Swent and Chambers argued that changing mining practices might alter radon influx into a mine, even in the face of a constant concentration in water. In addition, changing ventilation practices over the years could have influenced exposures. In discussing potential bias in the exposure estimates used by Radford and Renard,[48] Swent and Chambers suggested that the direction of changes in exposures would have been downward. If the exposures were, in fact, underestimated, the estimated risk coefficients would exaggerate the actual risk.

Time worked underground was determined from company and union records and from medical files. Adjustments were made for variation in the average number of hours worked underground in a month. Average yearly WLM were calculated for each 10-yr calendar period from the average number of hours per month underground and from radon-daughter concentration in each area, with weighting by the company data on the numbers of person-hours worked underground in each section of the mines.

Follow-up of the cohort depended on parish records and was thorough (only one person was untraceable through 1976). Of the lung-cancer deaths, 70% had been confirmed by autopsy or thoracotomy, but only death-certificate information was used for comparisons. The expected number of cases was based on age- and calendar-year-specific national mortality rates for males since 1951. Accordingly, PYAR and expected deaths were calculated from the later of two dates: January 1, 1951, and January 1 of the year after a miner began work underground. Induction-latency periods were considered in two ways: by excluding PYAR for each miner for 10.5 yr after mining was begun and by lagging the cumulative WLM by 5 yr.

Information on cigarette smoking was not reported for all cohort members individually, but only for a sample of the responses to a 1972–1973 survey of active miners and surface workers and from a 1977 survey of pensioners in the study cohort. In addition, smoking histories were obtained for each lung-cancer death. The authors estimated smoking-specific lung-cancer SMRs for two categories: smokers combined with recent ex-smokers and all others. They based these SMRs on the ratios between a sample of the responses from the surveyed miners and a national population study of the age-specific proportion of smokers and the amount smoked. Interpretation of the SMRs must be constrained by the lack of information for all cohort members on smoking as presently reported for the surveyed miners (556 of 1,294, or 43%), by differences in the periods associated with the questionnaire data from the miners (1972 and 1977) and from the national population sample (1963–1972), and by the use of information provided by the next of kin for deceased lung-cancer cases.

Other potentially confounding variables for lung cancer were considered. Silicosis, examined in a case-control study nested within the cohort, was found to be equally severe and prevalent in lung-cancer victims (14/50) and in age- and work-period-matched controls (26/100). Diesel equipment, with its exhaust, was not introduced into the mines until the 1960s, by which time 70% of the persons who later developed lung cancer had terminated work. Arsenic, chromium, and nickel—known respiratory carcinogens—were virtually absent in analyses of bedrock. X-ray diffraction of airborne dust samples from the mine showed no identifiable asbestos fibers. Indoor radon concentrations in miners' homes ranged from 0.002 to 0.03 WL, but had been measured in a sample of homes selected because of potentially high concentrations. The lung-cancer rates among nonminers in this region are lower than Swedish national rates.

Of the mining groups exposed to radon daughters, this cohort offers one of the longer follow-up experiences. Over 41% of the cohort (532/1,294) were deceased. The average year first employed underground was 1932, the average age at first employment was 27.8, and the average duration underground was 19.5 yr. The average exposure rate was 4.8 WLM/yr, resulting in an average cumulative WLM of 93.7 (range, 2–300 WLM). Cause-specific and total mortality were assessed with a modified life-table analysis. Excesses of observed deaths were found for total mortality, lung cancer (50 observed versus 12.8 expected), stomach cancer (28 observed versus 15.1 expected), and all causes except cancer combined (393 observed versus 312.6 expected). The latter excess was due to silicosis, occupational accidents, and cardiovascular disease, according to Radford and Renard.[48]

Lung-cancer mortality was studied in detail. Excess risk was not evident until at least 20 yr after the start of underground work. Significantly increased risks were found for both smokers (32 observed versus 11.0 expected; SMR, 291) and nonsmokers (18 observed versus 1.8 expected; SMR, 1,000). The excess-risk coefficient for smokers was 21.8/million person-yr WLM, and for nonsmokers, 16.3. The combined effect of smoking and radon-daughter exposure was reported as nearly additive,[48] although formal statistical testing, as described in Appendix VII, was not carried out. (The rate ratio for smokers versus nonsmokers based on the Swedish population study was estimated by the authors to be 7.4.)

Dose-response relationships were evaluated for five categories of lagged cumulative WLM (0–49, 50–99, 100–149, 150–199, and over 199). An excess of lung cancer was found even in the lowest dose category (8 observed versus 3.4 expected), and the dose-response data were equally consistent with absolute- and relative-risk models, as measured by weighted correlation coefficients. Assessments of effects of age at beginning of work, year of beginning work, and age at risk were undertaken separately and not by multivariate modeling.

SWEDISH IRON MINERS: KIRUNA

A proportionate-mortality-ratio study was carried out in Kiruna, Sweden, to compare cause-specific mortality distributions among underground iron miners from two companies (LKAB and TGA), surface miners and workers (LKAB), and all other male deaths in Kiruna.[27] Selected for study were all deaths registered in Kiruna that occurred in 1950–1970 among men aged 30–74. Because rates of emigration from Kiruna were very small, the authors considered that nearly all deaths among the miners would have been registered there. Lung-cancer deaths were verified from hospital records for 41 of 42 cases, and autopsies were performed in all 13 cases among miners. An additional analysis attempted to calculate SMRs among active employees of the mines (surface and underground combined), but was limited by the absence of data.

After age adjustment, underground miners experienced 13 lung-cancer deaths (12 were after 1957) versus 4.5 expected based on the cause-specific distribution of deaths among all other residents of Kiruna and versus 4.2 based on the cause-specific distribution of deaths among the entire Swedish male population in 1951–1966. Analyses were not presented on dose-response relationships, latency, or interaction of cigarette smoking and exposure to radon daughters.

The iron mine was an open-pit mine until the 1950s, when underground mining began. Diesels were introduced in the late 1950s. Quartz concentrations were around 7% of the particles smaller than 5 μ. The concentration of radon daughters, measured only since 1970, was 10–30 pCi/liter in most places and much higher in some unventilated areas. In 1966, a survey of all employees showed that about two-thirds of both underground and surface workers were smokers. Information from coworkers and next of kin indicated that 12 of the 13 lung-cancer cases among underground miners were smokers (four of these smoked only a pipe).

SWEDISH IRON MINERS: KIRUNA AND GALLIVARE

A case-control study[11] was carried out on 604 lung-cancer victims who died during 1972–1977 in three counties in northern Sweden. These counties contained two major iron mines, which were in two separate municipalities, Kiruna (containing the Kirunavaara mine) and Gallivare (containing the Malmberget mine). The investigators used next-of-kin interviews to determine underground mining and smoking histories. No estimates were made of WLM or duration of mining. The investigators concluded that their data showed that relative risks for smoking and underground iron mining were between additive and multiplicative in their combined effect. However, the wide confidence intervals in their data are consistent with an additive, a multiplicative, or a more extreme interpretation.

A recent extension of this case-control study[12] included 69 deaths during 1972–1982, but was limited to Kiruna and Gallivare. The median age of the subjects was 66. WLM were not estimated, but lung-cancer risks by duration of underground iron mining and lifetime number of cigarettes smoked were found to fit a multiplicative-risk model based on linear logistic regression. Unfortunately, statistical testing of the model was not reported, and the data were limited by the small numbers of nonsmoking miners (four) and nonsmoking nonminers (two) among the cases.

SWEDISH IRON MINERS: GRANGESBERG

Edling[15] carried out a case-control study of all male residents known to have died of lung cancer during 1957–1977 in the iron-mining town of Grangesberg, Sweden. The unmatched controls (897), who all died of other causes, and cases were submitted to the local iron-mining company for identification of history of underground mining. The author found an age-adjusted rate ratio for lung-cancer deaths associated with employment at a mine (16.6) that was significant (95% confidence interval, 7.7–35.3). Strikingly, 42 of the 47 lung-cancer cases had mined underground.

A separate analysis in the same report of the effect of cigarette smoking[15] added cases through 1980, but included only persons who had been underground miners. A new set of controls (individually matched for age, sex, and year of death) who died from causes other than malignancy and had been underground miners was selected (44 pairs). Smoking histories were obtained from next of kin by telephone. A risk ratio of 2.0 (95% confidence interval, 0.7–5.7) was found for smoking and lung cancer; the author interpreted that as not fully consistent with the general experience of at least a 5-fold to 10-fold risk ratio.

A second case-control analysis on the same population[16] used only controls who died during 1966–1977 at ages over 50. The authors found a lower age-standardized lung-cancer death rate ratio than in the previous analysis (relative risk, 11.7; 95% confidence interval, 5.3–26.0). A separate analysis of smoking similar to the one described above resulted in a risk ratio of 1.5 (95% confidence interval, 0.4–5.3) for smoking and lung cancer, on the basis of 28 matched pairs.

Edling and Axelson[16] also estimated a lung-cancer excess risk per million person-years WLM for miners aged 50–64 (26 excess cases) and aged 65 or greater (54 excess cases). These estimates were made by multiplying the number of miner person-years at risk in Grangesberg during 1966–1977, as estimated from town censuses, by the proportion of controls from the case-control study who had previously been underground miners. Cumulative WLM was estimated by multiplying the average number of years worked underground by the product of number of cases and 0.5 WL—an exposure assumed to apply to the entire period, according to 1969 mine measurements. Although these risk estimates were based on extensive assumptions, the authors noted that they were in agreement with estimates in the report by the Committee on the Biological Effects of Ionizing Radiations (BEIR III).[42]

GENERAL SWEDISH MINERS

Snihs[59] presented an epidemiological study of miners in all districts of Sweden. Sweden had 60 underground mines; all mined ferrous and sulfide ores, and none mined uranium. Radon measurements were made in all mines in 1969–1970 with 4.80-liter propane containers and ionization chambers. Radon daughters were sampled with conventional glass-fiber filters and analyzed by the Kusnetz method. In March 1972, exposures were limited by regulation to an annual average of 30 pCi/liter. Snihs[59] reported that air brought into the miners for ventilation, rather than water or rocks in the working areas, was the predominant source of exposure in 17 of 22 mines.

A nationwide cohort of all miners aged 20–64 and employed during 1961–1968 was followed during 1961–1971. The follow-up and analytical methods were not described. It is unclear whether follow-up extended beyond 5 yr after employment ended.

Observed and expected lung-cancer deaths during 1961–1968 were compared among underground miners, aboveground miners, and nonminers in the mining districts. The methods were not fully described, but it appears that the estimated annual number of active miners aged 20–64 during this period was multiplied by district age-specific lung-cancer rates to estimate the expected number of deaths. Observed deaths were included if they occurred within 5 yr of cessation of mining. Annual WL was estimated from measurements only after 1969. Cumulative WLM was estimated from the WL estimates and duration of exposure estimates for all workers based on data on those dying of lung cancer. Limitations of the study noted above make it difficult to judge the validity of the dose-response relationship results.

SWEDISH LEAD-ZINC MINERS: HAMMAR

Axelson and Sundell[7] studied Kiruna (Hammar parish) lead-zinc miners with a case-control study embedded in a crude cohort study. In the case-control study, the 29 cases included all men who died from lung cancer during 1956–1976 in the parish surrounding two physically connected lead-zinc mines. Controls (174) consisted of the first three deaths other than from lung cancer listed before and after the case in the chronologically ordered parish registry, but matching was dropped in the analysis. The authors believed that the registry included fairly complete diagnoses from the death certificates. The local mining company assessed the underground work experience of all subjects.

Smoking habits of the miners were learned from medical files and interviews with two retired foremen. For 2 of 10 subjects on whom smoking information was independently obtained from more than one source, the information was conflicting.

The age-standardized rate ratio for lung cancer among lead-zinc miners was 16.3 (90% confidence interval, 7.8–35.3). Among underground miners, those who had never smoked (nine) appeared to have longer work-related induction latent periods than smokers (nine) (respective medians of 49 versus 37 yr) and to have a greater risk of developing lung cancer.

NORWEGIAN NIOBIUM MINERS

Solli et al.[60] followed a cohort of employees at a niobium-mining company that operated from 1951 to 1965. Niobium itself is not considered to be carcinogenic, but the ore also contained ^{238}U (0.3–2 ppm) and ^{232}Th (50–300 ppm). Exposure estimates for the cohort were of questionable quality. The WLM from both radon and thoronium progeny was calculated for the employees on the basis of measurements of alpha activity during 2 days in 1959. Among the employees, a strong dose-response relationship was found between lung-cancer risk and cumulative WLM (Table IV-15). Poor dosimetry probably resulted in underestimation of exposures by about a factor of 2, according to the authors. Lifetime occupational histories indicated that three of the subjects had been previously exposed to asbestos and one had mined iron. In addition, 75% of the employees were smokers, compared with 60% of the Norwegian population.

FLORIDA PHOSPHATE WORKERS

Some U.S. phosphate ore contains uranium and radium. Workers involved in the mining and the processing of the ore might be exposed to radon and radon daughters. Two retrospective cohort studies of mortality in Florida phosphate workers have been conducted recently; each was performed because of concern raised by apparent clusters of lung cancer.

TABLE IV-15 Lung-Cancer Mortality among Norwegian Niobium Mine Workers[a]

Cumulative WLM (Corrected/Twofold Underestimate)	No. of Person-Years at Risk	No. of Lung-Cancer Deaths		Observed/Expected Ratio
		Observed	Expected	
0	4,622	0	1.73	0
1–38	1,343	3	0.50	6.0
40–158	1,312	4	0.58	6.9
160–238	147	2	0.07	28.6
≥240	169	3	0.08	37.5

[a]Based on data from Solli et al.[60]

Stayner et al.[61] conducted a study of 3,199 workers employed at a phosphate fertilizer plant. Seven samples were taken for radon progeny; the range was 0.00–0.02 WL. Overall respiratory-cancer mortality was not significantly increased (SMR, 113). Further analysis did not show trends of respiratory-cancer mortality with duration of employment or length of follow-up in white men. In black men, respiratory-cancer mortality was significantly increased in those with more than 20 yr of employment. However, only five cases were identified in black men, and two were in the index cluster.

In a larger study, Checkoway et al.[10] examined mortality in 17,601 white and 4,722 nonwhite male employees of the Florida phosphate industry. Lung-cancer mortality was not significantly increased in either group, in comparison with rates for Florida. When mortality from lung cancer was examined in the workers considered to have potential exposure to alpha radiation, a significant excess was apparent (SMR, 1.08).

These studies do not have sufficiently detailed information on exposure for risk estimation. Individual exposures to radon progeny were not estimated, and information on cigarette smoking was not collected. Furthermore, the limited measurements that have been made indicate radon-daughter concentrations only slightly above background concentrations.

RESIDENTIAL EXPOSURE

Within a building, radon-progeny concentrations are determined by the strength of the source and the rate of air exchange with the outside. Most of the radon in buildings enters from the underlying soil and building materials, although water and utility gas can also contribute radon progeny to indoor air.[43] A wide range of radon-daughter concentrations in dwellings has been demonstrated, with different radium concentrations in soil and building materials and different air-exchange rates largely explaining the size of the range.

Epidemiological investigations of domestic radon progeny as a risk factor for lung cancer are still preliminary. Both descriptive and analytical approaches have been used to examine the association between radon-daughter exposure in the home and lung cancer. Techniques for estimating lifetime exposure of people to radon daughters from indoor air are not yet available, and surrogates based on residence type or a few limited measurements have been used in the analytical studies. The available studies are insufficient for the development of quantitative risk estimates for associating exposure to radon progeny in the home and lung cancer.

In the descriptive studies, incidence or mortality rates for lung cancer within geographic units have been correlated with measures of exposure for inhabitants of the units. Edling[15a] compared mortality rates for different Swedish counties with background gamma radiation, described as being correlated with indoor exposure to radon and its daughters. For lung-cancer mortality, the correlation coefficients were 0.46 for males and 0.55 for females. Hess and colleagues[20] performed a similar analysis for lung-cancer mortality during 1950–1969 in the 16 counties of Maine. Using average radon concentrations in water as the measure of exposure, they calculated correlation coefficients of 0.46 for males and 0.65 for females. In a study of 28 Iowa towns served by deep wells, lung-cancer incidence increased with the concentration of ^{226}Ra, a possible surrogate for the radon concentration in the water.[9] These descriptive studies, which did not consider the exposures of people to radon daughters and other agents, provided only suggestive evidence that radon progeny exposure in the home increases lung-cancer risk.

The association has been more directly tested in case-control and cohort studies. Axelson et al.[8] conducted a case-control study in a rural area of Sweden. The investigation included 37 cases and 178 controls. Exposure to radon progeny was inferred from the characteristics of the subjects' residences at the time of death. Those who lived in stone houses were assumed to be most heavily exposed to radon daughters, and those who lived in wooden houses were assumed to be least exposed; other types of dwellings were considered to be sources of intermediate exposure. In spite of the crudeness of this exposure classification, residence in stone houses was associated with a significantly increased odds ratio, in comparison with the reference category of wooden houses (by Mantel-Haenszel method; odds ratio, 5.4; 90% confidence interval, 1.5–19). Data concerning cigarette smoking and residence history were not obtained.

Edling and Axelson[16a] conducted a similar case-control study in a rural area of Sweden. The study subjects were residents of the island of Oeland who died during 1960–1978. The geological characteristics of this island were thought to result in strong differences in background radon concentrations within a small area. Inclusion in the study population required at least 30 yr of residence at the same address before death; 23 lung-cancer cases and 202 controls who died from causes other than lung cancer met this criterion. Most of the dwellings were monitored for radon daughters during 3 months of summer and 1 month of winter. The dwellings were also classified on the basis of structural characteristics, as in the earlier study by Axelson et al.,[8] and cigarette-smoking information was obtained from next of kin. Lung-cancer risk was significantly associated with radon-daughter exposure, as assessed by either the measured concentration or the characteristics of the dwelling, and both crude and smoking-adjusted risk estimates were significantly increased. Logistic analysis yielded smoking-adjusted odds ratio, comparing most with least exposed, of 3.9, and the 90% confidence interval was 1.5–10.0.

Pershagen et al.[45] reported the findings of two small case-control studies in Sweden on domestic radon-daughter exposure, one drawn from a larger study in northern Sweden and the other from a twin registry. The investigators assembled each series with 30 case-control pairs, divided equally between smokers and nonsmokers. Exposure to radon was estimated from information on dwelling type; the investigators attempted to consider all residences lived in by the subjects. In the study group from northern Sweden, imputed radon exposures were significantly higher in smokers than in their smoking controls. Estimated exposures to radon progeny were similar in the nonsmoking cases and controls in the series from northern Sweden and in the smoking and nonsmoking cases and controls in the second series (selected from the twin registry).

In the United States, Simpson and Comstock[57] examined the relationship between lung-cancer incidence and housing characteristics. During a 12-yr period in Washington County, Maryland, lung-cancer incidence was not significantly affected by the type of basement construction or building materials. No measurements of radon or its daughters were made. Rather, dwelling-related variables were assumed to be surrogates for radon-daughter exposure.

SUMMARY

Cause-specific mortality risks for a number of the miner groups discussed above are listed in Table IV-16. Without exception, these studies

TABLE IV-16 Cause-Specific Risks of Mortality among Miners Exposed to Radon Daughters[a]

Study	All Causes			Lung Cancer			Tuberculosis			Other NMRD			Nephritis and Nephrosis		
	Obs.	Exp.	SMR	Obs.	Exp.	SMR	Obs.	Exp.	SMR	Obs.	Exp.	SMR	Obs.	Exp.	SMR
Colorado Plateau Uranium Miners[68]	950	600	158	185	38.4	482	14	3.4	409	83	16.6	499	9	3.7	243
Ontario[38,39c]	1,316	1,113	118	119	65.8	181	4	3.0	132	18	9.5	195			
Elliot Lake Ontario Uranium Miners[38]	999	854	117	81	50	162	1	2.3	44	13	7.2	180			
Bancroft Ontario Uranium Miners[38]	244	203	120	30	12.4	241	2	0.6	333	4	1.8	215			
Eldorado and Ontario[38]	198	116	171	33	7.2	458	1	0.4	250	2	1.0	200			
Eldorado–Port radium[41]															
underground	361	225	160	55	14.7	375	3	1.7	176	3	4.4	68	2	1.0	199
surface	340	259	131	28	16.0	175	9	1.8	0	8	5.6	143	0	1.2	
Eldorado, Beaverlodge															
underground	600	487	123	84	30.0	280	1	2.8	36	6	8.5	71	3	2.1	140
surface[41]	515	529	97	28	30.8	91	3	3.0	100	9	10.4	87	1	2.3	44
uranium miners[25]	604	582	104	65	34.2	190									
Swedish iron miners Malmberget[48]	532	409	130	50	14.6	342									
Norwegian niobium workers[61]	78	67.9	115	12	3.0	405									
Newfoundland fluorspar miners (rate ratios)[35]	244	173	(141)	65	6.5	1,000	24[g]	4.4	550	9[h]	5.4	167	1[i]	2.0	50
Cornish tin miners underground surface[17]	276	—	183	28	—	211	31[j]	—	—	41[k]	—	—			
Colorado Plateau uranium miners[68]	9	6.0	150	1	3.0	33	5	2.3	216	1	0.6	153	9	12.0	75
Ontario[38,39c]	21	16.1	130	6	5.8	104	4	4.9	82	2	1.4	145	26	27.4	95

TABLE IV-16 (*Continued*)

Study	Stomach Obs.	Exp.	151[b] SMR	Kidney 180[b] Obs.	Exp.	SMR	Skin 190, 191[b] Obs.	Exp.	SMR	Bone 196[b] Obs.	Exp.	SMR	Lymph. and Hemo. 200–205 Obs.	Exp.	SMR
Elliot Lake Ontario uranium miners[38]	15	12.2	131	4	4.4	90	2	1.1	187	22	21.1	104			
Bancroft Ontario uranium miners[38]	4	3.1	129	2	1.1	188	1	0.8	122	0	0.2	0	3	4.8	
Eldorado and Ontario[38]	4	1.8	219	0	0.6	0	0	0.5	0	0	0.1	0	2	2.7	74
Eldorado–Port radium[41]															
underground	5	4.2	119	1	1.3	79	1	0.7	147	0	0.3	0	10	9.4	107
surface	2	5.2	39	2	1.4	145	0	0.7	0	0	0.3	0	14	11.6	121
Eldorado, Beaverlodge															
underground	10	8.4	120	2	2.6	76	0	1.6	0	0	0.6	0	9	17.6	51
surface[41]	8	9.5	84	1	2.7	38	2	1.6	127	1	0.7	143	13	21.2	61
uranium miners[25]	[e]														
Swedish iron miners Malmberget[48]	28	15.1	189										7[d]	4.7	149
Norwegian niobium workers[61]	1			1			0			0					
Newfoundland fluorspar miners (rate ratios)[35]	24[f]	12	(200)												
Cornish tin miners underground surface[17]	10	—	200												

[a]Abbreviations: Obs., observed; Exp., expected; NMRD, Non malignant Respiratory Disease.
[b]Numbers are disease identifiers from ICD, 7th revision.
[c]Including previous gold miners, excluding Eldorado uranium miners.
[d]Lymphoma only.
[e]No significant excess.
[f]Digestive system.
[g]Includes silicosis.
[h]Respiratory disease.
[i]Genitourinary disease.
[j]Silico tuberculosis.
[k]Silicosis alone.

indicate an excess probability of death due to lung cancer and, in many cases, other causes of death as well. Continued follow-up of these miner groups will provide additional information on the association of radon-daughter exposure to lung cancer and perhaps other diseases. As discussed in Chapter 2 and Appendix VII, epidemiological information that includes the smoking status of each participant is of paramount value. The committee suggests that every effort be made to collect and report such information for the studies described in this appendix.

REFERENCES

1. Agricola, G. De Re Metallica. Basel, 1556. New York: Dover Publications. English reprint (Hoover translation). 1950, p. 214.
2. Archer, V. E., H. J. Magnuson, D. A. Holaday, and P. A. Lawrence. 1962. Hazards to health in uranium mining and milling. J. Occup. Med. 4:55–60.
3. Archer, V. E., J. K. Wagoner, and F. E. Lundin, Jr. 1973. Lung cancer among uranium miners in the United States. Health Phys. 25:351–371.
4. Archer, V. E., J. D. Gillam, and J. L. Wagoner. 1976. Respiratory disease mortality among uranium miners. Ann. N.Y. Acad. Sci. 271:280–293.
5. Arnstein, A. 1913. Sozialhygienische untersuchungen über die Bergleute in den Schneeberger Kobaltgruben. Wein. Arbeit. Geb. Soz. Med. 5:64–83.
6. Axelson, O., and C. Edling. 1980. Health hazards from radon daughters in dwellings in Sweden, Pp. 79–87 in Health Implications of New Energy Technologies, R. Rom and V. Archer eds. Ann Arbor, Mich.: Ann Arbor Science Publishers, Inc.
7. Axelson, O., and L. Sundell. 1978. Mining, lung cancer and smoking. Scan. J. Work Environ. Health 4:46–52.
8. Axelson, O., C. Edling, and H. Kling. 1979. Lung cancer and residency: A case-referent study on the possible impact of exposure to radon and its daughters in dwellings. Scand. J. Work Environ. Health 5:10–15.
9. Bean, J. A., P. Isacson, R. M. A. Hahne, and J. Kohler. 1982. Drinking water and cancer incidence in Iowa. Am. J. Epidemiol. 116(6):924–932.
10. Checkoway, J., R. M. Mathew, J. L. S. Hickey, C. M. Shy, R. L. Harris, E. W. Hunt, and G. T. Waldman. 1985. Mortality among workers in the Florida phosphate industry. I. Industry-wide cause-specific mortality patterns. J. Occup. Med. 27:885–892.
11. Damber, L., and L. G. Larsson. 1982. Combined effects of mining and smoking in the causation of lung carcinoma-case-control study in northern Sweden. Acta Rad. Oncol. 21:305–313.
12. Damber, L., and L. G. Larsson. 1985. Underground mining, smoking and lung cancer: A case-control study in the iron ore municipalities in northern Sweden. J. Natl. Cancer Inst. 74(6):1207–1213.
13. deVilliers, A. J., and J. P. Windish. 1964. Lung cancer in a fluorspan mining community. Br. J. Ind. Med. 21:94–109.
14. deVilliers, A. J., J. P. Windish, F. de N. Brent, B. Holloywood, C. Walsh, J. W. Fisher, and W. D. Parsons. 1971. Mortality experience of the community and of the fluorspan mining employees at St. Lawrence, Newfoundland. Occup. Health Rev. 22:1–15.

15. Edling, C. 1982. Lung cancer and smoking in a group of iron ore miners. Am. J. Ind. Med. 3:191–199.

15a. Edling, C., P. Combs, O. Axelson, and V. Flodin. 1982. Effects of low-dose radiation—a correlation study. Scand. J. Work Environ. Health 8(Suppl. 1):59–64.

16. Edling, C., and O. Axelson. 1983. Quantitative aspects of radon daughter exposure and lung cancer in underground miners. Br. J. Ind. Med. 40:182–187.

16a. Edling, C., H. Kling, and O. Axelson. 1984. Radon in houses—a possible cause of lung cancer. Scand. J. Work Environ. Health 10:25–34.

17. Fox, A. J., P. Goldblatt, and L. J. Kinlen. 1981. A study of the mortality of Cornish tin miners. Br. J. Ind. Med. 38:378.

18. Gottlieb, L. S., and L. A. Husen. 1982. Lung cancer among Navajo uranium miners. Chest 81:449–452.

19. Harting, F. H., and W. Hesse. 1979. Der lungenkrebs, die Bergkrankheit in den Schneeberger gruben. Vjschr. Gerichtl. Med. Offentl. Gesundheitswesen 31:102–132, 313–337.

20. Hess, C. T., C. V. Weiffenbach, and S. A. Norton. 1983. Environmental radon and cancer correlations in Maine. Health Phys. 45:339–348.

21. Holaday, D. A., W. D. David, and H. N. Doyle. 1952. An Interim Report of a Health Study of the Uranium Mines and Mills by the Federal Security Agency, Public Health Service, Division of Occupational Health, and the Colorado State Department of Public Health. Washington D.C.: U.S. Public Health Service.

22. Horaceck J., V. Placek, and J. Sevc. 1977. Histologic types of bronchogenic cancer in relation to different conditions of radiation exposure. Cancer 40:832–835.

23. Hornung, R. W., and T. J. Meinhardt. 1987. Quantitative risk assessment of lung cancer in U.S. uranium miners. Health Phys. 52:417–430.

24. Hornung, R. W., and S. Samuels. 1981. Survivorship models for lung cancer mortality in uranium miners—is cumulative dose an appropriate measure of exposure? Pp. 363–368 in International Conference, Radiation Hazards in Mining: Control, Measurement, and Medical Aspects, M. Gomez, ed. New York: Society of Mining Engineers of the American Institute of Mining, Metallurgical, and Petroleum Engineers, Inc.

25. Howe, G. R., R. C. Nair, H. B. Newcombe, A. B. Miller, S. E. Frost, and J. D. Abbatt. 1986. Lung cancer mortality (1950–1980) in relation to radon daughter exposure in a cohort of workers at the Eldorado Beaverlodge uranium mine. J. Natl. Cancer 77(2):357–362.

26. Hueper, W. C. 1966. Occupational and Environmental Cancers of the Respiratory System. New York: Springer-Verlag.

27. Jorgensen, H. S. 1973. A study of mortality from lung cancer among miners in Kiruna 1950–1970. Work Environ. Health 10:126–133.

28. Kunz, E., J. Sevc, and V. Placek. 1978. Lung cancer mortality in uranium miners. Health Phys. 35:579–580.

29. Kunz, E., J. Sevc, V. Placek, and J. Horacek. 1979. Lung cancer in man in relation to different time distributions of radiation exposure. Health Phys. 36:699.

30. Lorenz, E. 1944. Radioactivity and lung cancer; a critical review of lung cancer in the miners of Schneeberg and Joachimsthal. J. Natl. Cancer Inst. 5:1–13.

31. Ludwig, P., and E. Lorenser. 1924. Untersuchungen der Grubenluft in den Schneeberger gruben auf den Gehalt and Radiumemanation. Strahlentherapie 19:428–435.

32. Lundin, F. D., Jr., J. W. Lloyd, E. A. Smith, V. E. Archer, and D. A. Holaday. 1969. Mortality of uranium miners in relation to radiation exposure, hardrock mining and cigarette smoking—1950 through September 1967. Health Phys. 16:571–578.
33. Lundin, F. D., Jr., J. K. Wagoner, and V. E. Archer. 1971. Radon Daughter Exposure and Respiratory Cancer, Quantitative and Temporal Aspects. Joint Monograph No. 1. Washington, D.C.: U.S. Public Health Service.
34. Lundin, F. D., Jr., V. E. Archer, and J. K. Wagoner. 1979. An exposure-time-response model for lung cancer mortality in uranium miners—effects of radiation exposure, age and cigarette smoking. Pp. 243–264 in Proceedings of the Work Group at the Second Conference of the Society for Industrial and Applied Mathematics, N. E. Breslow and A. S. Whittemore, eds. Philadelphia: Society for Industrial and Applied Mathematics.
35. Morrison, H. I., D. T. Wigle, H. Stocker, and A. J. deVilliers. 1981. Lung cancer mortality and radiation exposure among the Newfoundland fluorspar miners. Pp. 372–376 in International Conference on Radiation Hazards in Mining: Control, Measurements and Medical Aspects, M. Gomez, ed. New York: Society of Mining Engineers of the American Institute of Mining, Metallurgical, and Petroleum Engineers, Inc.
36. Morrison, H. I., R. M. Semenciw, Y. Mao, D. A. Corkill, A. B. Dory, A. J. deVilliers, H. Stocker, and D. T. Wigle. 1985. Lung cancer mortality and radiation exposure among the Newfoundland fluorspar miners. Pp. 354–364 in Occupational Radiation Safety in Mining, H. Stocker, ed. Proceedings of the International Conference. Toronto: Canadian Nuclear Association.
37. Muller, J., W. C. Wheeler, J. F. Gentleman, G. Suranyi, and R. Kusiak. Pp. 359–362 in The Ontario Miners Mortality Study, General Outline and Progress Report: Radiation Hazards in Mining, M. Gomez, ed. New York: Society of Mining Engineers of the American Institute of Mining, Metallurgical, and Petroleum Engineers, Inc.
38. Muller, J., W. C. Wheeler, J. F. Gentleman, G. Suranyi, and R. A. Kusiak. 1983. Study of Mortality of Ontario Miners, 1955–1977, Part I. Toronto: Ontario Ministry of Labour, Ontario Workers Compensation Board.
39. Muller, J., W. C. Wheeler, J. F. Gentleman, G. Suranyi, and R. Kusiak. 1985. Study of mortality of Ontario miners. Pp. 335–343 in Occupational Radiation Safety in Mining, Proceedings of the International Conference, E. Stocker, ed. Toronto: Canadian Nuclear Association.
40. Muller, J., R. A. Kusiak, G. Suranyi, and A. C. Richie. 1986. Study of Mortality of Ontario Gold Miners, 1955–1957, Part II. Ottawa, Canada: Ontario Ministry of Labour.
41. Nair, R. C., J. D. Abbott, G. R. Howe, H. B. Newcombe, and S. E. Frost. 1985. Mortality experience among workers in the uranium industry. Pp. 154–164 in Occupational Radiation Safety in Mining, H. Stocker, ed. Proceedings of the International Conference. Toronto: Canadian Nuclear Association.
42. National Research Council, Committee on the Biological Effects of Ionizing Radiations (BEIR). 1980. The Effects on Populations of Exposure to Low Levels of Ionizing Radiation. Washington, D.C.: National Academy Press. 524 pp.
43. Nero, A. V. 1983. Airborne radionuclides and radiation in buildings: A review. Health Phys. 45:303–322.
44. Peller, S. 1939. Lung cancer among mine workers in Joachimsthal. Human Biol. 11:130–143.

45. Pershagen, L. D., and R. Falk. 1984. Exposure to radon in dwellings and lung cancer: A pilot study. Pp. 73–78 in Indoor Air, Vol. 2. Radon, Passive Smoking, Particulates and Housing Epidemiology, B. Berglung, T. Lindvall, and J. Sundell, eds. Stockholm: Swedish Council for Building Research.

46. Pirchan, A., and H. Sikl. 1980. Cancer of the lung in the miners of Jachymov (Joachimsthal). Am. J. Cancer 4:681–722, 1932.

47. Placek, V., A. Smid, J. Sevc, L. Tomasek, and P. Vernerova. 1983. Late effects at high and very low exposure levels of the radon daughters. In Proceedings of the 7th International Congress of Radiation Research. Amsterdam: Martinus Nijhoff Publishers

48. Radford, E. P., and K. G. St. Clair Renard. 1984. Lung cancer in Swedish iron miners exposed to low doses of radon daughters. N. Engl. J. Med. 310(23):1485–1494.

49. Rostoski, O., E. Saupe, and G. Schmorl. 1926. Die bergkrankheit der Erzbergleute in Schneeberg in Sachsen ("Schneeberger Lungenkrebs"). Z. Krebforsch. 23:360–384.

50. Samet, J. M., C. R. Key, D. M. Kutvirt, and C. L. Wiggins. 1980. Respiratory disease mortality in New Mexico's American Indians and Hispanics. Am. J. Public Health 70:492–497.

51. Samet, J. M., D. M. Kutvirt, R. J. Waxweiler, and C. R. Key. 1984. Uranium mining and lung cancer in Navajo men. N. Engl. J. Med. 310:1481–1484.

52. Schiager, K. J., and L. W. Hersloff. 1984. Review of radon daughter exposure measurements in U.S. uranium mines—past and present. In Assessment of the Scientific Basis for Existing Federal Limitations on Radiation Exposure to Underground Uranium Miners, Appendix F. Toronto: SENES Consultants Ltd.

53. Seltser, R. 1965. Lung cancer and uranium mining: A critique. Arch. Environ. Health 10:923–935.

54. Sevc, J., and V. Placek. 1973. Radiation induced lung cancer: Relation between lung cancer and long-term exposure to radon daughters. Pp. 305–310 in Proceedings of the 6th Conference on Radiation Hygiene. Jasna pod Chopkom, CSSR.

55. Sevc, J., V. Placek, and J. Jerabek. 1971. Lung cancer risk in relation to long-term radiation exposure in uranium miners. Pp. 315–326 in Proceedings of the 4th Conference on Radiation Hygiene. Parezska, Lhota, CSSR.

56. Sevc, J., E. Kunz, and V. Placek. 1976. Lung cancer mortality in uranium miners and long-term exposure to radon daughter products. Health Phys. 30:433–437.

57. Simpson, S. G., and G. W. Comstock. 1983. Lung cancer and housing characteristics. Arch. Environ. Health 38:248–251.

58. Smid, A., J. Sevc, V. Placek, and E. Kunz. 1983. Lung Cancer in Exposed Human Populations and Dose/Effect Relationship. Presented at the 7th International Congress on Radiation Research. Amsterdam: Martinus Nijhoff Publishers.

59. Snihs, J. O. 1974. The approach to radon problems in non-radium uranium miners in Sweden. Proceedings of the 3rd International Congress of the International Radiation Protection Association. CONF-730907. Oak Ridge, Tenn.: U.S. Atomic Energy Commission.

60. Solli, M., A. Andersen, E. Straden, and S Langand. 1985. Cancer incidence among workers exposed to radon and thoron daughters at a niobium mine. Scan. J. Work Environ. Health 11:7–13.

61. Stayner, L. T., T. Meinhardt, R. Lemen, D. Bayliss, R. Herrick, G. R. Reeve, A. B. Smith, and W. Halperin. 1985. A restrospective cohort mortality study of a phosphate fertilizer production facility. Arch. Environ. Health 40:133–138.

62. Sun, S., X. Yang, Y. Lan, M. Xionyu, L. Shengen, and Y. Zhanyun. 1984. Latent period and temporal aspects of lung cancer among miners. Radiat. Prot. 4(5) (English translation).
63. Teleky, L. 1937. Occupational cancer of the lung. J. Ind. Hyg. Toxicol. 2:73–85.
64. Tirmarche, M., J. Brenot, J. Piechowski, J. Chameaud, and J. Pradel. 1985. The present state of an epidemiological study of uranium miners in France. Pp. 344–349 in Proceedings of the International Conference, Occupational Radiation Safety in Mining, Vol. 1. Canadian Nuclear Association, Toronto, Ontario, Canada.
65. Wagoner, J. K., V. E. Archer, B. E. Carroll. 1964. Cancer mortality patterns among U.S. uranium miners and millers, 1950 through 1962. J. Natl. Cancer Inst. 32:787–801.
66. Wagoner, J. K., V. E. Archer, F. E. Lundin, D. A. Holaday, and J. W. Lloyd. 1965. Radiation as the cause of lung cancer among uranium miners. N. Engl. J. Med. 273:181–188.
67. Waxweiler, R. J., R. J. Roscoe, V. E. Archer, M. J. Thun, J. K. Wagoner, and F. E. Lundin, Jr. 1981. Mortality follow-up through 1977 of the white underground uranium miners cohort examined by the United States Public Health Service. Pp. 823–830 in International Conference, Radiation Hazards in Mining: Control, Measurement and Medical Aspects, M. Gomez, ed. New York: Society of Mining Engineers of the American Institute of Mining, Metallurgical, and Petroleum Engineers, Inc.
68. Wang, X., X. Huang, et al. 1984. Radon and miners lung cancer, Zhonghua Fangshe Yixue yu Fanghu Zazhi 4:10–14 (English translation).
69. Whittemore, A. S., and A. McMillan. 1983. Lung cancer mortality among U.S. uranium miners: A reappraisal. J. Natl. Cancer Inst. 71(3):489–499.

APPENDIX V

Nonmalignant Respiratory and Other Diseases Among Miners Exposed to Radon

Epidemiological evidence on radon progeny as a potential risk factor for nonmalignant respiratory diseases is restricted to uranium miners. As discussed in Appendix III, animal studies are consistent with an association between exposure to radon progeny and nonmalignant respiratory diseases. Animals so exposed develop emphysema and interstitial fibrosis. Pulmonary fibrosis and, to a lesser extent, emphysema are common findings in hamsters, rats, and dogs exposed to radon progeny alone and in mixtures with uranium-ore dust.[7–9,18] These effects are not produced to any appreciable extent in groups of animals until exposures to radon daughters exceed several thousand working-level months (WLM). Thus, the clinical diseases of interest are chronic obstructive pulmonary disease, in which airflow obstruction results from emphysema and airway changes, and interstitial processes such as pulmonary fibrosis.

The occurrence of nonmalignant respiratory diseases has been examined in miners from the Colorado Plateau region and from Ontario, Canada. Several reports from the U.S. Public Health Service study described excess mortality from nonmalignant respiratory diseases.[4,26] Between 1950 and 1977, a fivefold excess of death occurred from nonmalignant respiratory diseases, exclusive of tuberculosis, bronchitis, influenza, and pneumonia.[26] Causes of death were primarily emphysema, fibrosis, and silicosis. The effects of cigarette smoking were not considered.

In the Ontario miners, mortality from influenza, pneumonia, bronchitis, and asthma was not increased. However, mortality from silicosis and chronic interstitial pneumonia was significantly elevated (11 deaths observed, with 2.14 expected).

As part of the U.S. Public Health Service study of Colorado Plateau miners, physical examinations and lung function testing were performed in 1957 and 1960.[2] Data were collected for 2,349 white males, but only 277 participated in both years. Spirometry was carried out at both examinations, and the peak expiratory flow rate was measured in 1960. Three different measures of radiation exposure were used: (1) years of underground uranium mining, (2) a radiation index based on the working level at the time of examination, and (3) a cumulative, ordinal measure of exposure. The analyses were interpreted as showing loss of ventilatory function with increasing cumulative exposure. However, the data were neither collected nor analyzed with techniques that are currently accepted for lung function parameters. Further, the accuracy of the exposure measures that were used is uncertain.

In a later paper, Archer et al.[3] used the same U.S. Public Health Service study data and demonstrated an increasing prevalence of emphysema, as diagnosed by a physician's examination, with increasing WLM. However, the diagnosis of emphysema, a histopathologically identified disease, cannot be established by physical examination. Trapp et al.[23] performed more detailed studies on Colorado Plateau uranium miners and found evidence of pulmonary dysfunction, both restrictive and obstructive. The design of the investigation did not permit assessment of exposure-response relationships with lung function measures.

More recently, Samet et al.[20] surveyed 192 long-term New Mexico uranium miners. The survey procedures included spirometry, completion of a respiratory symptoms questionnaire, physical examination of the chest, and interpretation of chest x rays. Total duration of underground uranium mining, not WLM, was used as the exposure index. The design of the investigation did not permit assessment of the effects of each potentially hazardous agents as radon daughters, silica, and diesel exhaust. With linear multiple-regression analysis that controlled for cigarette smoking, duration of mining was associated with reduction of the forced expiratory volume in 1 s and reduction of the midmaximum expiratory flow rate, but not with reduction of the forced vital capacity. Chest x-ray abnormalities compatible with silicosis were found in 9% of the uranium miners examined for this survey.

These investigations have not separated the effects of radon-daughter exposure from those of other atmospheric contaminants, such as silica, diesel-engine exhaust, and blasting fumes, found in a uranium mine. Given the inadequacies of available exposure data, epidemiological methods cannot assess the individual contributions of all harmful agents to which uranium miners are exposed or are potentially exposed.

OTHER DISEASES AMONG MINERS EXPOSED TO RADON PROGENY

Mortality from selected causes among miners exposed to radon daughters is detailed in Appendix IV, Table IV-16. In addition to the excess risk from lung cancer, a number of the mining cohorts have experienced an excess risk of mortality due to tuberculosis and to "other nonmalignant respiratory diseases," (ICD Code 510-527, 7th revision). Because of the past levels of silica in the mine atmospheres, most of these excesses are believed to be due, in fact, to silicosis, which has been often diagnosed on death certificates as silicotuberculosis, tuberculosis, or other forms of nonmalignant respiratory disease.

Results from many studies of the mining populations have suggested that there is a slight excess risk of stomach cancer, which has an elevated incidence among other mining groups, including gold miners in Ontario without uranium mining experience[15] and coal miners in the United States.[11] Among the Ontario uranium miners, no excess risk existed for those without prior gold mining experience (9 observed versus 9.55 expected).[17] The cases are few, and the risks are low in most studies, but the occurrence of this excess risk in eight different mining populations lends credibility to the causation hypothesis that the excess has resulted from a common occupational risk factor. From the reported analyses, however, it is difficult to determine if that risk factor is radon progeny because exposure-response analyses have not been reported.

Two studies, both based on small numbers, have found nonsignificant excesses of skin cancer. While Sevcova et al.[21] found that basal cell carcinomas predominated, four of the five miners in the Colorado Plateau study who died of skin cancer had melanomas.[26] Since the other mortality studies have not shown any significant skin-cancer excess, it is unlikely that alpha radiation in the mines accounted for these excesses. However, the occurrence of skin cancers, particularly nonmelanomas, cannot readily be evaluated with a mortality study. Most skin cancers other than malignant melanoma can be readily cured and rarely lead to death. No other sites of malignancy appear to be consistently elevated among these mining populations. Despite the fact that airborne radon daughters deposit in the nasal passages, no cases of nasal cancer have been reported in any of the epidemiological studies. Only Muller et al.[17] reported an expected value of 0.8 for nasal cancer mortality among all Ontario uranium miners.

Excess risk of mortality due to nonmalignant renal disease was found in a recent analysis of the Colorado Plateau study.[26] In that survey, chronic and unspecified nephritis was particularly elevated after 10 yr latency (7 observed versus 1.9 expected; standardized mortality ratio, 362). The committee considered that this excess may be either a chance finding or indicative of an occupational risk factor, possibly alpha radiation or uranium. The nephrotoxicity of soluble uranium in animals is well documented in the experimental literature,[24] but most of the uranium in mines occurs as less soluble oxides.

Findings from relevant animal experiments should also be considered in interpreting the epidemiological data. In radon-daughter-exposed animals, lesions observed in organs other than the lung are considered spontaneous, or only indirectly exposure-related, in contrast to the case for most alpha-emitters, which translocate from the lung to irradiate other organs. Because of the extremely short half-life of the radon daughters, their alpha emissions occur before they move to other organs. In animals exposed to high concentrations of uranium-ore dust alone (and presumably to radon daughters and uranium-ore dust mixtures) sufficient long-lived radioactivity from the precursors of radon can concentrate in the kidneys to impair their function. However, direct evidence of renal function impairment from exposure to radon daughters alone is lacking.

CYTOGENETIC STUDIES

The frequency of chromosome aberrations in blood cells has been examined in uranium miners and other underground workers as a marker of injury due to ionizing radiation. Brandom and colleagues[5,6] have reported on chromosome aberrations in uranium miners from the Colorado Plateau region of the United States. In their 1972 report,[5] cytogenetic abnormalities in peripheral lymphocytes from 15 miners were compared with the findings in 15 age-matched nonminer controls; 5 of the miners had lung cancer at the time of the study. Most of the aberrations were more prevalent in the miners, and many of the differences between the two groups attained statistical significance.

A subsequent report by Brandom et al.[6] included 80 underground uranium miners and 20 controls, frequency-matched for age and smoking habits. Again, the various types of chromosomal aberrations were more prevalent in the uranium miners than in the controls. Exposure-response relationships were evident up to a cumulative exposure of 3,000 WLM; however, chromosonal abnormalities were less frequent in those with greater than 3,000 WLM than in those with 1,740–2,890 WLM, the next lowest exposure category.

Badgastein is a spa in Austria with thermal springs that discharge water with a high concentration of radon. At Badgastein, patients are also treated in a former gold mine that has a mean radon concentration of 3,000 pCi/liter. Pohl-Ruling and Fischer[19] evaluated cytogenetic abnormalities in inhabitants of the community, bath attendants, and personnel exposed underground. The investigators estimated blood doses from alpha and gamma radiation and used the dose estimates for assessing dose-response relationships. The analyses did not provide a clear estimate of the effect of alpha radiation, though they concluded that occupational alpha doses flattened the dose-response relationships for radiation.

Relevant data are also provided by a study of cytogenetic abnormalities in persons presumed to be exposed to high concentrations of radon in household water.[22] Chromosome aberrations were evaluated in 18 exposed persons and 9 controls. Dicentrics, chromosome breaks, and cells with chromosome change were significantly more frequent in the 18 exposed subjects. However, exposures to radon daughters were not estimated, and the suitability of the control group was not satisfactorily established.

To date, only the above-mentioned limited data are available on cytogenetic abnormalities in radon-daughter-exposed populations. The study of Colorado Plateau uranium miners indicates exposure-response relationships for chromosome aberrations. However, confirming evidence is not available from other populations, and the biological significance of these observations has not been established.

EFFECTS ON REPRODUCTIVE OUTCOME

Recent and primarily descriptive data have renewed speculation that uranium mining is associated with adverse reproductive outcomes. Muller and colleagues[13,14,16] made the first reports on this subject in a series of papers on Czechoslovakian uranium miners that were published during the 1960s. For 1,000 underground male workers, the numbers of children in relationship to age did not deviate from that expected from nationwide data.[13] However, in this sample and in another with 415 uranium miners,[16] the secondary sex ratio (male to female births) declined following the start of underground employment from 1.08 to 0.85 in the former sample and from 1.18 to 0.99 in the latter.

Potential reproductive effects of uranium mining received little further evaluation until the early 1980s. At that time, descriptive data from New Mexico were interpreted as suggesting the adverse reproductive effects caused by uranium mining, by affecting either uranium miners or those living in the vicinity of mines and mills.[27]

This more recent interest in reproductive effects caused by uranium mining followed reports of high rates of congenital malformations and spontaneous abortion at the Shiprock Indian Health Service Hospital, located in San Juan County, New Mexico, which serves Navajos in the northeastern portion of the Navajo nation. Goodman subsequently examined the secondary sex ratio in New Mexico and Navajo births.[12] His analyses showed a temporal decline in the secondary sex ratio for New Mexico, in comparison with nationwide data, that occurred during the period of extensive uranium mining in the state. The decline in secondary sex ratio was greatest for counties with mining activity; further, the Navajo Area Indian Health Service units with the lowest sex ratios also encompassed areas of mining. A preliminary study of Grants, New Mexico, area miners also suggested effects on the secondary sex ratio, and a study of 11 miners showed distribution of Y bodies in their semen different from that in control populations.[10]

Waxweiler and Roscoe[25] reviewed the results of a 1965 questionnaire survey of Colorado Plateau miners; overall, the secondary sex ratio did not vary with cumulative WLM. When the participants were stratified at the population's median age of 24, the secondary sex ratio was significantly increased in the highest exposure category. This observation could not be readily explained.[1,25]

Two studies were implemented to follow-up on these hypothesis-generating observations: a survey of reproductive outcomes in wives of Navajo uranium miners and a case-control study of births at the Shiprock Hospital. Wiese and Skipper[28] have recently reported preliminary findings of the survey of reproductive outcomes. Questionnaires were distributed to uranium miners in the Grants, New Mexico, area and to potash miners employed in the southeastern portion of New Mexico. The study population included 491 uranium and 226 potash miners. The investigators did not find significant differences between the two groups in the frequency of low-birth-weight infants, sex ratio, miscarriages, or infertility. Birth weights were lower in children born after the men began underground mining, but the effect was present only in those births after 1970, when average exposures to radon daughters were lower in the mines. Findings from the case-control study in Shiprock have not yet been reported. To date, the evidence on the possible reproductive effects of uranium mining is largely descriptive and preliminary. The studies of uranium miners do not show a consistent and readily interpretable pattern of effect. The data related to possible effects of the uranium mining industry on the general population are fragmentary at present.

REFERENCES

1. Archer, V. E. 1981. Discussion of secondary sex ratio of first-born offspring of U.S. uranium miners. In Birth Defects in the Four Corners Area, W. H. Wiese, ed. Transcript of a meeting. Albuquerque, N.M.: University of New Mexico School of Medicine.
2. Archer, V. E., H. P. Brinton, and J. K. Wagoner. 1964. Pulmonary function of uranium miners. Health Phys. 10:1183–1194.
3. Archer, V. E., J. K. Wagoner, and F. E. Lundin, Jr. 1973. Lung cancer among uranium miners in the United States. Health Phys. 25:351–371.
4. Archer, V. E., J. D. Gillam, and J. W. Wagoner. 1976. Respiratory disease mortality among uranium miners. Ann. N.Y. Acad. Sci. 271:280–293.

5. Brandom, W. F., G. Saccomanno, V. E. Archer, P. G. Archer, and M. E. Coors. 1972. Chromosome aberrations in uranium miners occupationally exposed to ^{222}radon. Radiat. Res. 52:204–215.
6. Brandom, W. F., G. Saccomanno, V. E. Archer, P. G. Archer, and D. Bloom. 1978. Chromosome aberrations as a biological dose-response indicator of radiation exposure in uranium miners. Radiat. Res. 76:159–171.
7. Chameaud, J., R. Perraud, J. Chretien, R. Masse, and J. Lafuma. 1982. Lung carcinogenesis during in vivo cigarette smoking and radon daughter exposure in rats. Rec. Results Cancer Res. 82:11–20.
8. Cross, F. T., R. F. Palmer, R. H. Busch, R. E. Filipy, and B. O. Stuart. 1981. Development of lesions in Syrian golden hamsters following exposure to radon daughters and uranium ore dust. Health Phys. 41:135–53.
9. Cross, F. T., R. F. Palmer, R. E. Filipy, G. E. Dagle, and B. O. Stuart. 1982. Carcinogenic effects of radon daughters, uranium ore dust and cigarette smoke in beagle dogs. Health Phys. 42:33–52.
10. Dean, R. G. 1981. Semen analysis among uranium miners. In Birth Defects in the Four Corners Area, W. H. Wiese, ed. Transcript of a meeting. Albuquerque: University of New Mexico School of Medicine.
11. Enterline, P. E. 1972. A review of mortality data for American coal miners. Ann. N.Y. Acad. Sci. 200:260–272.
12. Goodman, A. B. 1986. Sex ratio patterns; a biological indicator of environmental factors. Pp. 14, 36 in Birth Defects in the Four Corners Area, W. H. Wiese, ed. Transcript of a meeting. Albuquerque: University of New Mexico School of Medicine.
13. Muller, C., M. Kubat, and J. Marsalek. 1962. Study on fertility of the miners in Jochimstal. Z. Gynekol. 2:63-68.
14. Muller, C., L. Reiicha, and M. Kubat. 1962. On the question of genetic effects of ionizing rays on the miners of Jochimstal. Z. Gynekol. 15:558-560.
15. Muller, J., R. A. Kusiak, G. Suranyi, and A. C. Richie. 1986. Study of mortality of Ontario Gold Miners, 1955–1957. Part II. Toronto: Ontario Ministry of Labour.
16. Muller, C., L. Ruzicka, and J. Bakstein. 1967. The sex ratio in offspring or uranium miners. Acta Univ. Carolinal Med. 13:549–603.
17. Muller, J., W. C. Wheeler, J. F. Gentlemen, G. Suranyi, and R. A. Kusiak. 1985. Study of mortality of Ontario miners. Pp. 335–343 in Occupational Radiation Safety in Mining, Proceedings of the International Conference. E. Stocker, ed. Toronto: Canadian Nuclear Association.
18. National Council on Radiation Protection and Measurements (NCRP). 1984. Evaluation of Occupational and Environmental Exposures to Radon and Radon Daughters in the United States. Report 78. Washington, D.C.: National Council on Radiation Protection and Measurements.
19. Pohl-Ruling, J., and P. Fischer. 1980. An epidemiological study of chromosome aberrations in a radon spa. In Proceedings of the Specialist Meeting on the Assessment of Radon and Daughter Exposure and Related Biological Effects, G. F. Clemente, A. V. Nero, F. Steinhausler, and M. E. Wrenn, eds. Salt Lake City: R.D. Press.
20. Samet, J. M., R. A. Young, M..V. Morgan, C. G. Humble, G. R. Epler, and T. C. McLoud. 1984. Prevalence survey of respiratory abnormalities in New Mexico uranium miners. Health Phys. 46:361–370.
21. Sevcova, M., J. Sevc, and J. Thomas. 1978. Alpha irradiation of the skin and the possibility of late effects. Health Phys. 5:803–806.

22. Stenstrand, K., M. Annanmäki, and T. Rytömaa. 1979. Cytogenic investigation of people in Finland using household water with high natural radioactivity. Health Phys. 36:441–443.
23. Trapp, E., A. D. Renzetti, Jr., T. Kabayashi, M. M. Mitchell, and A. Bigler. 1970. Cardiopulmonary function in uranium miners. Am. Rev. Respir. Dis. 101:27–43.
24. Voegtlin, C., and H. C. Hodge, ed. 1949. Pharmacology and toxicology of uranium compounds. National Nuclear Energy Series. Division IV. New York: McGraw-Hill.
25. Waxweiler, R. J., and R. J. Roscoe. 1981. Secondary sex ratio of first-born offspring of U.S. uranium miners. In Birth Defects in the Four Corners Area, W. H. Wiese, ed. Transcript of a meeting. Albuquerque, N.M.: University of New Mexico School of Medicine.
26. Waxweiler, R. J., R. J. Roscoe, V. E. Archer, M. J. Thun, J. K. Wagoner, and F. E. Lundin, Jr. 1981. Mortality followup through 1977 of the white underground uranium miners cohort examined by the United States Public Health Service. Pp. 823–830 in Radiation Hazards in Mining: Control, Measurement, and Medical Aspects, M. Gomez, ed. New York: Society of Mining Engineers, American Institute of Mining, Metallurgical, and Petroleum Engineers, Inc.
27. Wiese, W. H., ed. 1981. Birth Defects in the Four Corners Area. Transcript of a meeting. Albuquerque, N.M.: University of New Mexico School of Medicine.
28. Wiese, W. H., and B. J. Skipper. 1986. Survey of reproductive outcomes in uranium and potash mine workers: Results of first analysis. Ann. Am. Conf. Gov. Ind. Hyg. 14:187–192.

APPENDIX VI
Lung-Cancer Histopathology

Correlations between radon-daughter exposures and specific histopathological cell types of lung cancer in humans have been a subject of controversy for many years. Primary cancer of the lung comprises diverse and generally distinct histopathological cell types. The most common are squamous-cell carcinoma, small-cell carcinoma, adenocarcinoma, and large-cell carcinoma, representing, respectively, about 35, 17, 25, and 9% of lung cancers in the male population of the United States.[16] In nonsmokers, adenocarcinoma is the most common cell type and small-cell carcinoma is infrequent, accounting for less than 5%.[24] Clinically, these four cell types of lung cancers differ in their manner of clinical presentation, natural history, and response to therapy. At present, lung cancers are generally classified histologically by conventional light microscopy. Figure VI-1 shows examples of some of the common histological types of lung cancers. Numerous classification schemes have been published, with the most widely used being that developed by the World Health Organization and recently modified.[25] The accuracy of histopathological diagnoses is influenced by the quantity and quality of the tissue available for examination. Observer variability in the classification of lung-cancer histopathology has been well documented and may be of substantial magnitude.[14] Reliance on clinical reports may introduce substantial misclassification. Because of observer variability, classification of lung-cancer cell types for research purposes should incorporate a standardized review of original histological material by a panel of pathologists. However, use of review panels does not ensure comparability between studies.

The distribution of lung-cancer cell types has been examined in uranium and other miners exposed to radon daughters. The principal populations are listed in Table VI-1. Most are from from mining groups in which cigarette smoking was prevalent. Exceptions include early miners in Schneeberg and Joachimsthal, reported to have smoked little, and Navajo miners in the United States.[3,15]

In the 1879 report by Harting and Hesse,[12] the malignant disease in Schneeberg miners was identified as lymphosarcoma, a designation that may have reflected the similarity of cells in small-cell cancers and in some lymphomas when examined with a light microscope. Autopsy specimens from miners in nearby Joachimsthal showed a preponderance of small-cell carcinomas.[17,23] A later investigation of uranium miners in this same area showed that small-cell cancer remained the most common histological type. Horacek et al.[13] compared the histological distribution of 115 cases of lung cancer in Czechoslovakian uranium miners with that in 326 control cases. Diagnoses were made by one pathologist who was not informed about the exposure status of the cases under review. Only a small percentage of the lung-cancer cases in each group were nonsmokers. The percentages of squamous-cell carcinomas were similar in the miners and in the controls, but about 12% more small-cell cancers were observed in miners. The mortality rate of all major cell types was increased beyond that which would have been expected.

In the late 1950s and early 1960s a number of reports from France, which included crude proportionate mortality studies, were published describing the lung-cancer excess among iron miners in Lorraine. Roussel et al.,[18] in a histological review of a series of 225 lung-cancer cases among these miners, found that 44% were anaplastic versus an expected value of 28% for nonminers. Both groups included approximately 1% nonsmokers.

The studies by Saccomanno and colleagues of the Colorado Plateau uranium miners provide the most extensive data concerning lung-cancer cell types and radon-daughter exposure. Findings have been reported periodically since 1964.[2,3,19–21] The case material was derived from miners in the U.S. Public Health Service study and others who lived in the Colorado Plateau area. Review methods varied; most reports were based on a panel's consensus, but some reviews apparently involved only one pathologist. Most of the miners smoked cigarettes, and the total series has included only 14 nonsmokers.[20] Initially, the majority of cases reviewed were small-cell carcinomas. The proportion of this cell type declined from 76% in 1964 to 22% in the late 1970s, while squamous cell carcinomas increased concomitantly. In nonsmokers, eight cases were small-cell carcinomas and the remaining six were of other cell types.

The strong predominance of squamous-cell carcinomas in Newfoundland fluorspar miners appears anomalous (Table VI-1).[26] However, the histological diagnoses were made by sputum cytology, which results in

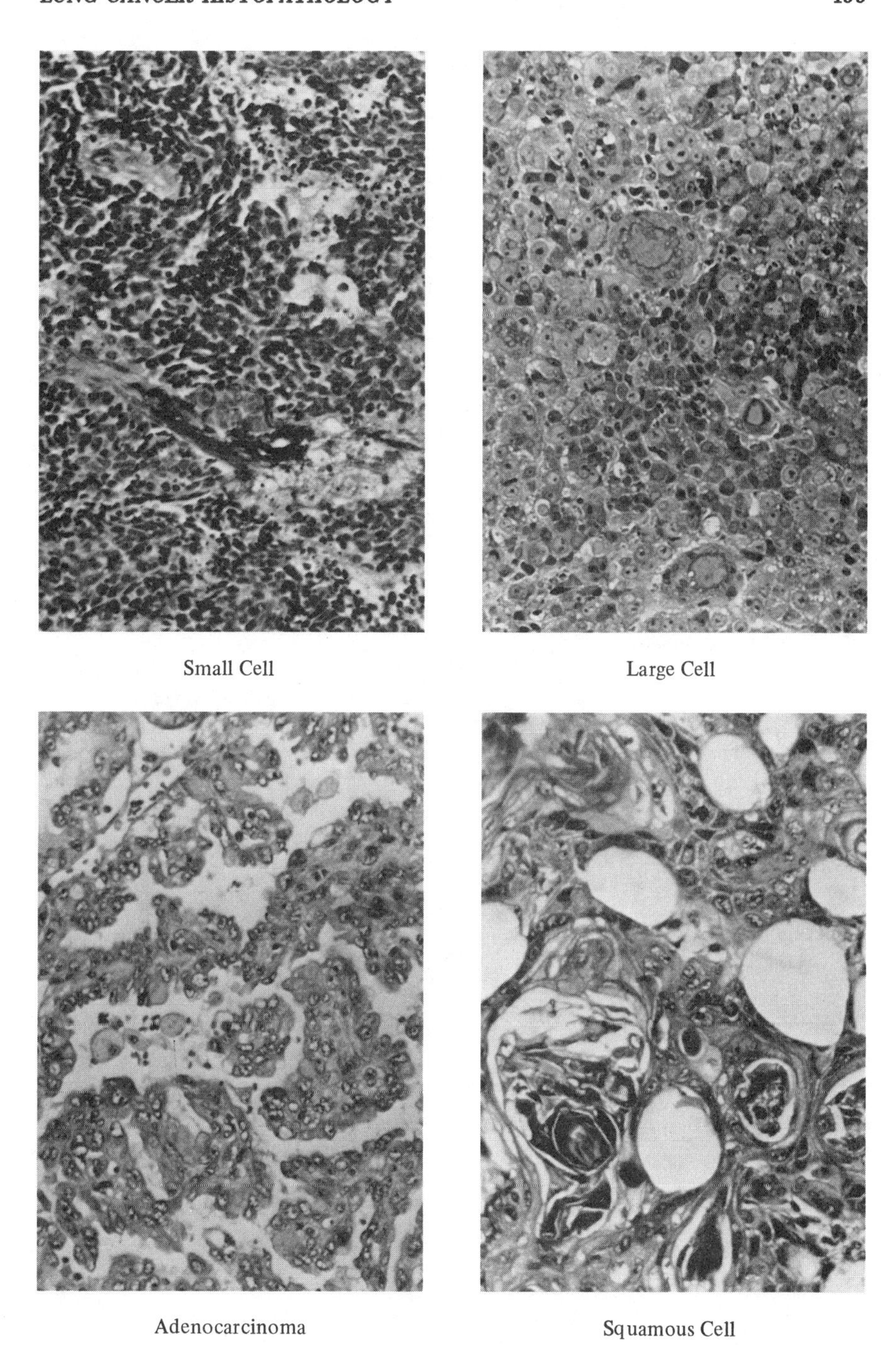

FIGURE VI-1 Histological types of lung cancer. SOURCE: Jonathan Samat, University of New Mexico, personal communication, 1987.

TABLE VI-1 Lung-Cancer Histopathology in Mining Groups Exposed to Radon Daughters

Population	Findings[a]	Comment
Joachimsthal miners[17,23]	28 cases: 16 SCC, 12 squamous	Autopsies 1929–1930 and 1933–1938; one pathologist
Colorado Plateau uranium miners[1,4,19,20,21]	SCC predominance, increasing with WLM	Cases seen at specific hospitals; panel review
Newfoundland fluorspar miners[26]	29 cases: 26 squamous, 2 SCC, 1 adenocarcinoma	All cases diagnosed by sputum cytology
British iron ore miners[5]	69 cases; 27 SCC	Histologic type from pathology report
Swedish iron ore miners[10]	36 cases: 26 SCC, 10 squamous	Histologic type from Swedish cancer registry
Swedish iron ore miners[9]	25 cases: 11 SCC, 11 squamous, 3 adenocarcinoma	Methods not given
Czechoslovakian uranium miners[13]	115 cases: 62 SCC, 40 squamous, 4 adenocarcinoma, 9 other types	One pathologist
Canadian uranium miners[8]	91 cases; 47 SCC	Selected from 134 cases; histologic type from pathology report
Navajo uranium miners[11]	16 cases: 10 SCC, 3 squamous, 3 other types	Methods not given; histologic type from pathology report
Navajo uranium miners[7]	21 cases: 7 SCC, 8 squamous, 4 adenocarcinoma, 2 LCC	Panel review
New Mexico uranium miners[6]	45 cases: 28 SCC, 15 squamous, 1 adenocarcinoma, 1 LCC	Panel review

[a]Abbreviations: SCC, small-cell carcinoma; LCC, large-cell carcinoma; Squamous, squamous cell carcinoma.

the over-representation of centrally located tumors. Both squamous- and small-cell carcinomas tend to be located in the larger airways, but at the time of this investigation the former may have been more readily diagnosed by cytology alone.

Navajo miners who worked in the Colorado Plateau are of interest because only a small proportion smoked cigarettes, with average consumption by the smokers of only a few cigarettes each day.[4,22] Gottlieb and Husen[11] described 16 Navajo miners diagnosed with lung cancer at the Shiprock Indian Health Service Hospital from 1965 through 1979. Based on record review, they reported that 10 of the cases were small-cell carcinomas. Butler et al.[7] reviewed histopathological material for 26 of 32 lung-cancer cases diagnosed among all Navajo males between 1969 and 1982. A panel of three pathologists examined all slides. In contrast with the earlier study of Gottlieb and Husen,[11] small-cell carcinomas did not predominate in the 21 cases of lung cancer occurring in Navajo uranium miners. Seven of these cancers were small-cell carcinoma, eight were squamous-cell carcinoma, four were adenocarcinoma, and two were large-cell carcinoma. While small-cell carcinoma was not the predominant cell type in this series, the proportion with this cell type (33%) is far greater than expected from the distribution of lung-cancer histopathology in nonsmokers. The discrepant findings of these two reports may reflect the use of medical records by Gottlieb and Husen[11] to determine the diagnoses.

Recent reports from Canada and New Mexico document a continued excess incidence of small-cell carcinomas in contemporary uranium miners.[6,8] In iron-ore miners in Great Britain and Sweden, also exposed to radon daughters, small-cell carcinoma has occurred in excess.[5,9,10] The pattern has been consistently observed in populations of miners who smoked cigarettes. Data for nonsmokers are sparse and conflicting. Saccomanno et al. [21] reported that most cases of lung cancer in nonsmokers from the Colorado Plateau region were small-cell carcinomas. Butler et al.[7] found a cell type distribution in Navajo miners comparable to the observed distribution in the general population. Thus, available information does not strongly support the association between uranium mining and small-cell cancers in nonsmokers, although this association in smokers is supported by the available data. This pattern appears to change as miners who smoke age and the interval since cessation of uranium mining exposure lengthens.

REFERENCES

1. Archer, V. E., H. P. Brinton, and J. K. Wagoner. 1964. Pulmonary function of uranium miners. Health Phys. 10:1183–1194.
2. Archer, V. E., J. K. Wagoner, and F. E. Lundin, Jr. 1973. Uranium mining and cigarette smoking effects on man. J. Occup. Med. 15:204–211.
3. Archer, V. E., G. Saccomanno, and J. H. Jones. 1974. Frequency of different histologic types of bronchogenic carcinoma as related to radiation exposure. Cancer 34:2056–2060.

4. Archer, V. E., J. D. Gillam, and J. W. Wagoner. 1976. Respiratory disease mortality among uranium miners. Ann. N.Y. Acad. Sci. 271:280–293.
5. Boyd, J. T., R. Doll, J. S. Faulds, and J. Leiper. 1970. Cancer of the lung in iron ore (haematite) miners. Br. J. Ind. Med. 27:97–195.
6. Butler, C., J. M. Samet, D. M. Kutvirt, C. R. Key, and W. C. Black. 1985. Cigarette smoking and lung cancer cell types in uranium miners. Am. Rev. Respir. Dis. 131(Pt. 2):A176 (abstract).
7. Butler, C., J. M. Samet, W. C. Black, C. R. Key, and D. M. Kutvirt. 1986. Histopathologic findings of lung cancer in Navajo men: Relations to uranium mining. Health Phys. 51:365–368.
8. Chovil, A., and B. Chir. 1981. The epidemiology of primary lung cancer in uranium miners in Ontario. J. Occup. Med. 23:417–421.
9. Damber, L., and L. G. Larsson. 1982. Combined effects of mining and smoking in the causation of lung carcinoma-case-control study in northern Sweden. Acta Rad. Oncol. 21:305–313.
10. Edling, C. 1982. Lung cancer and smoking in a group of iron ore miners. Am. J. Ind. Med. 3:191–199.
11. Gottlieb, L. S., and L. A. Husen. 1982. Lung cancer among Navajo uranium miners. Chest 81:449–452.
12. Harting, F. H., and W. Hesse. 1879. Der Lungenkrebs, die Bergkrankheit in den Schneeberger Gruben. Vierteljahrsschr Gerichtl. Med. Offentl. Gesundheitswesen 30:296–309; 31:102–129, 313–337.
13. Horacek, J., V. Placek, and J. Sevc. 1977. Histologic types of bronchogenic cancer in relation to different conditions of radiation exposure. Cancer 40:832–835.
14. Ives, J. C., P. A. Buffler, and S. D. Greenberg. 1983. Environmental associations and histopathologic patterns of carcinoma of the lung: The challenge and dilemma in epidemiologic studies. Am. Rev. Respir. Dis. 128:195–209.
15. Myers, D. K., C. G. Stewart, and J. R. Johnson. 1981. Review of epidemiological studies on hazards of radon daughters. Pp. 513–524 in Radiation Hazards in Mining: Control, Measurement, and Medical Aspects, M. Gomez, ed. New York: Society of Mining Engineers, American Institute of Mining, Metallurgical, and Petroleum Engineers, Inc.
16. Percy, C., J. W. Horm, and T. E. Goffman. 1983. Trends in histologic types of lung cancer, SEER, 1973–1981. Pp. 153–159 in Lung Cancer: Causes and Prevention, M. Mizell, and P. Correa, eds. Deerfield Beach, Fla.: Verlag Chemie International.
17. Pirchan, A., and H. Sikl. 1932. Cancer of the lung in miners of Jachymov (Joachimsthal). Report of cases observed in 1929–1930. Am. J. Cancer 16:681–722.
18. Roussel, J., C. Pernot, P. Schoumacher, M. Pernot, and Y. Kessler. 1964. Considerations statistiques sur la cancer bronchique du mineur de fer du bassin de la Lorraine. J. Radiol. Electrol. 45:541–546.
19. Saccomanno, G. 1982. The contribution of uranium miners to lung cancer histogenesis. Rec. Results Cancer Res. 82:43-52.
20. Saccomanno, G., V. E. Archer, R. P. Saunders, L. A. James, and P. A. Beckler. 1964. Lung cancer of uranium miners on the Colorado Plateau. Health Phys. 10:1195–1201.
21. Saccomanno, G., V. E. Archer, O. Auerbach, M. Kuschner, R. P. Saunders, and M. G. Klein. 1971. Histologic types of lung cancer among uranium miners. Cancer 27:515–523.

22. Samet, J. M., D. M. Kutvirt, R. J. Waxweiler, and C. R. Key. 1984. Uranium mining and lung cancer in Navajo men. N. Engl. J. Med. 310: 1481–1484.
23. Sikl, H. 1950. The present status of knowledge about the Jachymov disease (cancer of the lungs in the miners of the radium mines). Acta Unio Intl. Contra Cancrum 6:1366–1375.
24. Stayner, L. T., and D. H. Wegman. 1983. Smoking, occupation and histopathology of lung cancer: A case-control study with the use of the Third National Cancer Survey. J. Natl. Cancer Inst. 70:421.
25. World Health Organization. 1982. The World Health Organization Histological Typing of Lung Tumors, 2nd ed. Am. J. Clin. Pathol. 77:123–136.
26. Wright, E. S., and C. M. Couves. 1977. Radiation-induced carcinoma of the lung—the St. Lawrence tragedy. J. Thorac. Cardiovasc. Surg. 74:495–498.

APPENDIX VII
The Combined Effects of Radon Daughters and Cigarette Smoking

Part 1 of this appendix reviews the epidemiological literature on the combined health effects of smoking and radiation. The studies reviewed by the committee are summarized in Table VII-1. Part 2 presents the committee's analyses of lung-cancer occurrence in persons exposed to both carcinogens. Part 3 summarizes the committee's views, including the possible effects of smoking on the validity of dose estimates and the need for further studies of the combined effects of radon daughters and smoking.

PART 1. Epidemiological Studies of Smoking and Radiation

STUDIES AMONG SWEDISH METAL MINERS

Studies of iron-ore miners in Northern Sweden reveal an excess of lung cancer that is related primarily to underground employment and exposure to radon.[17,28] In order to clarify the role of radon exposure combined with tobacco use on the occurrence of lung cancer, Damber and Larsson[10] carried out a case-control study in a three-county area of Northern Sweden, in which cases were ascertained in the years 1972–1977. As all iron mines were found within only two municipalities, succeeding studies were focused on the mining areas Kiruna and Gallivare and were extended to encompass the years 1972–1982.[11] Therefore, only the latest report is considered here.

The case group consisted of 69 lung-cancer cases who were reported to the Swedish cancer registry after 1972 and who were deceased before

July 1982. For each case, one deceased control was drawn from the National Registry for Cause of Death, matched by sex, year of death, age, and municipality; suicides and lung carcinomas were not included in the control group. A second living control for 60 cases aged 80 and under was selected from the Swedish National Population Registry and matched by sex, year of birth, and municipality.

As recognized by the authors, smoking-related risks that are based on deceased controls may be underestimated, since tobacco use was likely to have been greater by the deceased than by the general population. However, use of controls required to be alive until 1982 may overestimate relative risks since their smoking rates may be may have been *less* than those of the general population at risk.[21] Nevertheless, the results presented by Damber and Larrson[10] were generally comparable, regardless of control group.

Interviews were conducted with the index subject or the next of kin to obtain information on smoking practices and work history. Members of the study group who worked underground in an iron mine were considered exposed to radon. Since no accurate measurements of direct radon exposure were available, the surrogate variable, years underground, was used for analysis.

Smoking data consisted of the year tobacco use started the number of cigarettes smoked per day, and the year of cessation of smoking. Smokers were individuals who consumed one cigarette daily for at least 1 yr. For cigar and pipe smokers, 1 g of tobacco was equated to one cigarette.

Results were tabulated by three categories of lifetime tobacco use: nonsmoker, low (<150,000 cigarettes); and heavy (>150,000 cigarettes) consumption. For cases and deceased controls, relative risks rose from the baseline 1.0 for nonsmokers to 2.4 to 8.4 for aboveground workers and from 5.4 to 21.7 to 69.7 among underground miners. Similar results were reported for cases and the combined control group of all living and deceased subjects. Although based on small numbers (23 cases had no or low tobacco consumption, of which only 3 had no underground-mining experience), the results suggest that radon exposure and smoking combine multiplicatively rather than additively on a relative risk scale.

As outlined in Appendix IV, Radford and Renard[27] reported a historical cohort study of 1,415 miners from the Malmberget and Koskoskulle areas of Sweden. Data on current smoking habits were reported from 388 questionnaires administered in 1972–1973 to active miners and surface workers (35% of the contemporary work force) and from 168 pensioners. Pipe smoking was considered equivalent to cigarette smoking. Although pipe smoking has been related to lung-cancer risk, the affect of this assumption is difficult to assess since information on the percentage of pipe smokers and their inhalation patterns was not provided. The authors state

TABLE VII-1 Relative Risks from Selected Studies of Cigarette Use, Radiation Exposure, and Lung-Cancer Risk

Study Area	Design	Results	Comments
Kiruna and Gallivare, Sweden[11]	Cases (69) from death register 1972–1982; two types of controls: alive from general population (60) and deceased from register (67)	Cigarette Use[a] Underground miner / 0 / <150 / >150 No / 1 / 2.4 / 8.4 Yes / 5.4 / 21.7 / 69.7	Smoking data from interviews of subjects or next-of-kin; results consistent with multiplicative, relative risk (RR) model, although formal testing not presented
Malmberget, Sweden[27]	Cohort study of 1,415 miners, with 50 cases of lung cancer	/ Nonsmoker / Smoker Nonminer / 1 / 1 Miner / 10.0 / 2.9	Results suggestive of submultiplicative model for RR, possibly additive; calculation of RRs not precisely described; formal model fitting not presented
Hammar, Sweden[2]	Cases (29) listed in death register 1957–1976; controls (174) also from register, matched on year of death	RR for mining 16.6 (90% confidence interval, 7.8–35.3); RR for smoking among miners 0.5 (90% confidence interval, 0.1–2.2)	Suggestive of a protective effect of smoking among miners; results subject to biases (see text)
Colorado[1]	Cohort study of uranium miners examined through 1960; followup from 1964–1967 with 39 cases of lung cancer	/ Cigarette Use / Lung Cancer Rate × 10^5 / No / Yes Miners / 7.1 / 42.2 Expected[b] / 1.1 / 4.2	Multiplicative combination is suggested; analysis of cases shows shorter latency period for smokers

Colorado[38]

Nested case control study from 3,362 miners followed from 1964–1977, with 194 cases and 776 controls; exposures lagged 10 years

WLM	Cigarette Use (pack yr) 0–10	10–20	30–30	30+
0–21	1	9.1	4.2	7.7
22–119	1.1	13.6	6.5	19.0
120–359	3.6	16.0	8.8	23.1
360–839	7.8	5.2	16.2	46.8
840–1,799	5.2	17.6	27.4	42.7
1,800+	18.2	137.6	52.6	146.8

Analyses formally reject additive RR model; data consistent with multiplicative model

Colorado (Appendix VII, Part 2, this volume)

Cohort study of 3,362 miners followed through 1982, with 256 observed cases of lung cancer, exposures lagged 5 years

WLM	Cigarette Use (no./day) 0–4	5–19	20–30	30+
0–59	1[c]	2.7	7.8	2.9
60–119	0.0	0.0	5.6	26.6
120–239	2.4	9.1	15.3	9.8
240–479	8.4	3.5	14.6	25.8
480–959	17.8	12.6	32.0	34.0
960+	27.6	36.0	63.6	90.3

Data fit well with multiplicative model ($P = 0.53$), while additive was rejected ($P = 0.03$); although not statistically superior to multiplicative model, best fitting power model was submultiplicative

Grand Junction Colorado[32]

Cases (489) and controls (992) drawn from cohort of 9,817 miners followed from 1960–1980, from whom sputum specimens were regularly obtained; cases defined as moderate or worst cell atypia

Yr Underground	Cigarette Use (pack yr) 0	1–20	21+
0	1	0.3	2.9
1–10	7.3	4.1	18.2
11+	9.6	9.8	26.0

Study is of cell atypia; suggests multiplicative effects, although statistical testing not presented

TABLE VII-1 (*Continued*)

Study Area	Design	Results					Comments
New Mexico (Appendix VII, Part 2, this volume)	Cases (52) and controls (222) extracted from cohort of uranium miners		Cigarette Use (no./day)				Both multiplicative and additive RR models consistent with data, although former exhibits better fit
		Yr Mining Underground	<5	5–14	15–24	>25	
		<10	1	5.1	7.0	8.2	
		10–14	1.0	12.0	6.7	6.2	
		15–19	3.7	4.2	17.5	0.0	
		20+	0.0	39.9	24.0	30.1	
Uranium City, Saskatchewan, Canada[4]	Followup for 3 yr of underground miners and controls who participated in lung cancer screening program; cases defined as moderate or worst cell atypia		Cigarette Use				Study is of cell atypia; few events among nonsmokers; data and analysis insufficient to assess interaction of exposures
		WLM	No	Yes			
		0	1	2.7			
		<120	2.6	3.7			
		≥120	1.2	12.6			
Oeland, Sweden[14]	Cases (22) and controls (178) drawn from death registry 1960–1978; smoking habits obtained from next-of-kin using mail questionnaire		Cigarette Use				Data were sparse, and no formal models were fit, but RRs suggest multiplicative interaction, or at least greater than additive
		Housing Type[d]	No	Yes			
		0	1	2.7			
		1	1.3	3.6			
		2	4.4	9.3			
Japan[26]	Cohort study of 40,498 A-bomb survivors for whom smoking data are available; there were 281 lung-cancer deaths		Cigarette Use				Both multiplicative and additive RR models fit data equally well
		Radiation Exposure (rad)	No	Yes			
		<10	1	2.4			
		10–99	1.1	2.4			
		>100	2.3	3.6			

Study	Population	Radiation Exposure (rads)	Cigarette Use (no./day)				Comments
			0	1–10	11–20	>20	
Japan (Appendix VII, Part 2, this volume)	Cases (485) and controls (1,089) identified during 1971–1980 from Life Span Study among A-bomb survivors		*Males*				Both multiplicative and additive RR models fit data equally well
		<10	1	3.7	6.9	26.5	
		10–99	1.3	2.4	6.6	13.2	
		>100	3.3	7.2	10.6	24.8	
			Cigarette Use				
			0	1–10	>10		
			Females				
		<10	1	2.3	4.2		
		10–99	0.7	2.5	2.1		
		>100	5.2	5.2	—		

[a]Lifetime number in thousands.
[b]Incidence based on rates in mountain states.
[c]Baseline category based on 0.7 expected cases compared to 0 observed.
[d]See text for category definitions.

that approximately half the workers who were still living at the time of the study took part in the survey of smoking habits. Smoking histories for lung cancer patients were obtained from next of kin or, in a few instances, from the subject. Evaluation of the quality of tobacco consumption data is not possible, since no attempt was made to compare subjects from whom smoking data were obtained to those for whom data were unavailable. The smoking rate among miners is probably underestimated, since surveys covered only living workers.

The precise method of analysis of smoking is not completely clear in the published report. Among miners, smokers were defined as those who had stopped smoking within 10 yr of the interview or who were currently smoking, while nonsmokers were defined as subjects who stopped smoking 10 yr or more years to the interview or who had never smoked. The authors assumed that risk of lung cancer for smokers relative to that for nonsmokers is constant over age. The smoking status of miners was then compared with a national smoking survey of 25,000 men carried out in 1963.[30] It was determined that the miners had a higher proportion of smokers. Although apparently no adjustment was made for the different time periods of the two surveys, the mortality experience of the national survey[8] was applied to the miners and a relative risk of 7.4 for smokers versus nonsmokers was obtained. The method for deriving the relative risk of 7.4 was not explicitly described. The subsequent relative risks for miners to nonminers were 2.9 for smokers based on 32 lung-cancer cases and 10.0 for nonsmokers based on 18 cases. The authors concluded that mining- and smoking-related risks combine additively. This conclusion seems to go beyond the evidence as presented. Radford and Renard's[27] results, however, do tend to suggest that risks for the two exposures are submultiplicative.

Within the parish of Hammar, Sweden (population, 4,000), Axelson and Sundell[3] compared smoking and mining (zinc and lead) experiences in 29 lung-cancer cases deceased between 1956–1976 with 174 referents who died of causes other than lung cancer and who were matched to cases by time of death. A subject was exposed if he appeared on employee files of the mining company. For workers with mining experience (21 cases and 19 controls), foremen who were contemporaries of the subjects were contacted and queried about the smoking status of the subjects. Smoking status was not determined for nonminers.

Among miners, smoking appeared to be protective for lung cancer, although the 90% confidence interval was large (relative risk, 0.49; 90% confidence interval, 0.1–2.2). The authors explained this finding by suggesting that smokers have a lower radiation-induced risk because of a thickened mucus layer in critical bronchial segments.

Axelson and Sundell[3] did not evaluate the effects of smoking among nonminers or of mining exposure by smoking category. Because of this lack of information on smoking in nonminers and on duration of radon exposure in miners, the study could not address the mode of interaction between radon and smoke exposures. Nevertheless, as noted in Table VII-1, the protective effect of smoking does suggest that an interaction could be additive or subadditive. However, potentially biased exposure assessment procedures (for example, inadequate company files and recall bias by the foremen), inappropriate control selection (inclusion of referents with tobacco-related causes of death), or simply the possibility that nonsmokers spent more time underground than smokers are alternative explanations.

STUDIES AMONG COLORADO PLATEAU MINERS

Several published reports based on the U.S. Public Health Service cohort of uranium miners of the Colorado Plateau have evaluated in detail the roles of radiation and cigarette smoking in the production of lung cancers.[1,22,24,32,38]

The earliest report, by Archer et al.,[1] included 39 cases of lung cancer that arose in a well-defined, physically examined special study group during a 4-yr observation period (1964–1967). Compared to lung-cancer rates among white male residents of mountain states, 1.1 and 4.4/10,000 person-yr for nonsmokers and smokers, respectively, the rates among uranium miners were 7.1 and 42.2/10,000, respectively. These comparisons, which show a 4-fold population-based excess for smoking and a 5.9-fold miner excess, suggest a multiplicative interaction of these agents.

Another analysis by Archer et al.,[1] reported in the same paper, focused on a larger sample of 207 cases, whose ascertainment of health status and population were less clearly defined but which included the 39 special study group cases. This second analysis relied solely on comparisons of age at lung-cancer diagnosis between groups of smoking and nonsmoking miners. Mine-related variables, such as age at start of mining, cumulative working-level months (WLM), and years of other hard-rock mining, were controlled through matching. The induction-latent period was shorter for smokers than for nonsmokers. The authors argue that the agents act synergistically. This analysis is questionable, however, regarding the form of the model, since in a survival model introduction of a second disease-related exposure, that is, radon, increases age-specific hazard rates and thus increases the probability of a tumor appearing earlier.

Lundin et al.[22] evaluated 62 lung-cancer cases that developed in the cohort of 3,366 white uranium miners followed from first medical examination through 1968. Smoking data came from periodic surveys carried out prior to 1970. Lundin et al. used tobacco consumption information from the last examination. Analysis was based on a log-normal model for estimating a yearly effective radiation dose, which weighted exposure in each previous year, in order to account for disease latency. Although no formal statistical testing of hypotheses was carried out, the results suggest that the relative-risk model for WLM exposure, in comparison with the absolute-risk model, is more appropriate. The authors then analyzed the effect of smoking under their assumed effective-dose model and claimed a submultiplicative effect for smoking. The results suggest there is a greater amount of radiation-induced lung-cancer risk among smokers, with slight differences between heavy and light smokers, than in nonsmokers. It is difficult to evaluate conclusions from this analysis because of the lack of formal hypothesis testing.

A detailed study of the Colorado Plateau uranium miners in which 194 lung-cancer cases were used was carried out by Whittemore and McMillan.[38] A nested or synthetic case-control approach was used,[20,21] whereby each lung-cancer case was matched with four controls born within 8 months of the case and alive at the time the case died. Exposure histories for controls were adjusted to reflect values up to the time the case died (minus any lag time). With this type of analysis, relative hazard (or relative risk) is modeled in either a multiplicative or additive way. No direct information on disease rates are obtained, and hence, evaluation of absolute excess risk is not possible. Data were classified by four categories of cigarette pack-years (average cigarette packs smoked per day times duration, in years, of use), accumulated from the start of exposure to a predefined cutoff date, and six categories of WLM. A single 23-parameter relative-risk model was fit to the two-way classification. All subsequent models were then compared for goodness of fit to this saturated model. Several models for the relative risk with combined cigarette and WLM exposure were fit. Multiplicative and additive excess-risk models were fit, as well as other richer variants, for example, mixture models of additive and multiplicative terms for smoking and linear and quadratic terms for radiation; exponential models were also fit. Whittemore and McMillan[38] found substantial support for the multiplicative model, finding that it fit nearly as well as the saturated one. The authors rejected the additive model, which agrees with a preliminary analysis reported by Hornung and Samuels.[15] Further analyses found little improvement when smoking rate was added to the model, although this improvement might have been expected since pack-years incorporate cigarettes smoked per day and the subjects were matched by age. They also reported that the smoking effect did not interact with age.

The joint effect of smoking and radon-daughter exposure in this cohort was also addressed in a National Institute for Occupational Safety and Health (NIOSH) Report to the Mine Safety and Health Administration.[24] Using the Colorado Plateau miner cohort with follow-up through 1982, a synergistic effect between these two factors was reported, that is, combined effects exceeding the sum of the separate effects (as would be predicted by an additive model). However, the data also suggested that the combined effect was less than multiplicative. It is generally difficult to compare these conclusions with other analyses of these data, since the authors relied on a power function relative risk model. Whittemore and McMillan[38] found that linear-relative-risk models for both smoking and radon exposure, individually, were preferrable to power-function relative-risk models.

An analysis of data from another group of Colorado Plateau workers has recently been reported by Saccomanno et al.[32] The cohort included 9,817 miners, underground and open pit, and millers who worked between 1960 and 1980 and who agreed to participate in the study.[31,32] Sputum samples were collected periodically, although irregularly, from 1957. Information on the number of workers lost to follow-up and on the completeness of sputum assessment was not reported. Although not explicitly stated, exposure measurements for radon and cigarette use were likely determined from periodic cohort surveys, as described previously for the other Colorado miner group.

Analyses were based on a selected case-control subsample from cohort members who had at least one sputum specimen taken between 1960–1980 and who had a current exposure history. Cases ($n = 489$) were defined as men who had at least one sputum cytology specimen classified as moderate or worse atypical squamous cell metaplasia. Controls ($n = 992$) were a 11% random sample of the noncase members of the cohort. Variables of interest were age, cumulative WLM, and pack-years. Because of case definition, this is a study of the determinants of moderate cell atypia or worse, and not of lung cancer.

The results suggested a multiplicative association for the combined effects of cigarette use and radon exposure, although formal testing procedures were not described. Based on unmatched analyses, increased age-adjusted relative risks with duration of underground uranium mining were similar within categories of pack-years, as were risks with cigarette consumption for categories of underground duration. The former increases from 1.0 for no underground experience to approximately 10.0 for more than 10 yr of underground experience, while the latter increases from 1.0 to approximately 3.0 for more than 20 pack-years of cigarette use. The authors also present results for logistic model fitting. Their interpretations are not clear, and may be statistically inappropriate.

This study is also subject to several potential biasing factors. Controls were substantially younger than cases (41.8 versus 58.2 yr, respectively) and were more likely to have been lost to follow-up (39% versus 23% respectively) or to have missing WLM data (33% versus 25%, respectively). Although one analysis was matched on age (± 2 yr), the primary analysis was unmatched and relied on age adjustment, with either crude age categories or a single age parameter in a logistic model. The adequacy of this adjustment is hard to assess. Bias is also possible from the method of control selection; controls were selected from all noncase members of the cohort, regardless of length of follow-up, instead of from cohort members at risk at the time of case ascertainment.[21] Controls were therefore likely to be healthier and to have received less exposure to radon and tobacco. Case selection bias, that is, more intense disease evaluation of higher-exposure workers, could have occurred, since workers who were more highly exposed to radon or cigarettes may have been more health conscious and therefore more likely to submit sputum specimens and ultimately categorized as a case. The authors did not give the mean number of specimens evaluated prior to ascertainment for cases or at equivalent follow-up for controls. Sputum specimens were obtained during follow-up and were used to define cases. However, men who were hospitalized or died with suspected lung cancer were apparently also classified as cases, although their atypia status should have been based on evaluated cytology records. Again, this deviation from the case definition criteria may have biased results of this study.

Using data from the 1982 follow-up of the Colorado Plateau cohort initiated by the U.S. PHS, this committee extended the analysis of radon daughters and cigarette use, which was carried out by Whittemore and McMillan.[38] The results of our analysis of 256 lung-cancer deaths are summarized in Table VII-1 and presented in detail in Part 2 of this appendix. They support Whittemore and McMillan's conclusions with some qualifications. The multiplicative relative-risk model fit the data quite well ($P = 0.48$), while the purely additive excess-relative-risk model was rejected ($P = 0.005$). To help clarify these results we studied a larger class of models, which were defined through a mixture of competing models,[34] in which both the additive and multiplicative models were nested. This investigation showed that the best-fitting model was submultiplicative, although it did not provide a statistically significant improvement in fit over the multiplicative model. The fitting of a sequence of models suggested that the data are consistent with a wide range of submultiplicative to supramultiplicative models, and there is no clear a prior reason to accept the multiplicative model, except parsimony.

STUDIES AMONG NEW MEXICO URANIUM MINERS

In the second part of this appendix, an evaluation is presented of the associations of cigarette smoking and duration of underground employment in a uranium mine with lung cancer in case-control data extracted from a cohort of New Mexico uranium miners. The results (Table VII-1) suggest that a multiplicative combination of the two exposures is more compatible with the observed patterns of relative risks than an additive model, although there was not a statistically significant difference between the fit of the additive and relative-risk models.

STUDIES AMONG CANADIAN URANIUM MINERS

Band et al.[4] reported on a study initiated in 1974 of the combined effects of cigarette smoking and radon exposure among a group of miners and nonminers who were residents of Uranium City, a town in northern Saskatchewan, Canada. The miners were employed at the Eldorado Nuclear Mine. All residents were invited to participate in a lung-cancer screening program. Responses were obtained from 80% of the uranium workers (all males) and 50% of the total adult population. The study group consisted of 249 underground miners and 123 male controls. In a manner similar to that of Saccommano et al.,[32] outcome status was determined by degree of cellular atypia, as evaluated from two or three yearly sputum cytology samples. No information was given on whether the evaluations were carried out blindly or if more than one abnormal cytology (moderate or more severe atypia) during the 3-yr study period was required for an individual to be designated as a case.

Information on smoking and an occupational history was obtained by questionnaire. Based on work history, cumulative WLM exposure to radon through 1977 was determined for each underground miner. A nonsmoker was defined as one who never smoked cigarettes.

The results (Table VII-1) show an increasing risk with radon exposure and with cigarette use. However, there were too few nonsmokers with moderate or more severe atypia (three cases) to assess the interaction of these factors. The authors did not provide data on amount smoked by cumulative WLM exposure, but did show results from four separate (miners/nonminers, smokers/nonsmokers) logistic regressions, which included years of smoking and years underground as continuous variables. Although the parameter estimates for the smoking effect were very similar for the miner group and for the controls not exposed to radon (thus suggesting a multiplicative interaction), it would be unwise to infer support for either an additive or multiplicative relative-risk model because of sparse data and the lack of a more formal assessment of model fit.

STUDIES OF THE HOME ENVIRONMENT

Edling et al.[14] reported on a case-control study of the association among exposure to radon in homes, smoking, and lung-cancer mortality. Preliminary results were reported by Axelson.[2] The investigation was carried out on the island of Oeland, Sweden, which is located in the Baltic

Sea, where a narrow strip of uranium-containing alum shale is found on one side of the island. Cases of lung cancer and noncancer referents were obtained from death records between the years 1960 and 1978. In addition, all subjects were aged 40 yr or more and had to have lived for 30 yr or more at their death address. There were 22 cases and 178 controls for whom data were available.

For each subject, investigators classified blindly the type of housing as: wooden without a basement or on normal ground (category 0); wooden with a basement on radiation ground (alum shale) or stone, brick, and plaster with a basement on any ground or without a basement on radiation ground (category 2); or all other types (category 1) (e.g., wooden without basement on radiation ground). Next of kin provided information on smoking status through mail questionnaires.

Among nonsmokers, relative risks by the three categories of housing type were 1.0, 1.3, and 4.4, respectively, while among smokers risks were 2.7, 2.6, and 9.3. Although the data were sparse, a greater than additive interaction is suggested (Table VII-1).

STUDIES AMONG JAPANESE ATOMIC-BOMB SURVIVORS

Studies of lung-cancer mortality among atomic-bomb survivors in Hiroshima and Nagasaki offer information for assessing combined exposure to tobacco smoke and low linear energy transfer (LET) radiation received at a single time point. However, the relevance of such analyses for understanding the combined effect on lung-cancer risk of smoking and protracted, high-LET radiation is uncertain. In Part 2 of this appendix, the committee gives results of its own analysis of combined exposures to the atomic-bomb survivors.

Prentice et al.[26] combined data from several different surveys among atomic-bomb survivors: approximately 20,000 participants from the Adult Health Study who were interviewed in 1963–1964, in 1964–1968, or in 1968–1970;[5] a subset of males from the Life-Span Study cohort who were surveyed by mail in 1965;[18] and a subset of females from the Life-Span Study who were surveyed by mail in 1969–1970. A total of 40,498 subjects were available.

For analysis, Prentice et al.[26] used T65DR dose estimates for total radiation exposure. Information on tobacco use came from the various surveys. Although the questionnaires differed, current smoking pattern at the time of interview could be categorized into nonsmoking; about 5, 10, 20, or 30 cigarettes/day; and 0–4, 5–9, 10–14, 15–19, and 20 or more years of cigarette use. Smoking data, for subjects surveyed more than once, were taken from the earliest interview. To avoid bias resulting from healthy subjects surviving longer and hence having a greater likelihood of interview, follow-up started at the initial interview and continued to death or the end of the study. A total of 281 lung-cancer deaths occurred. The Cox proportional hazards model was used for analysis.[9] Because of the relatively short 15-yr follow-up period, variation in radiation and smoking-induced lung-cancer risk with follow-up time was not evaluated.

Using categorical variables in the proportional hazards regression model and stratifying on city, sex, age at time of bombing, and survey date, relative risk of lung cancer among nonsmokers rose from 1.0 to 1.1 to 2.3 with exposure to <10, 10–99, >100 rad, respectively, while among smokers risks increased from 2.4 to 2.4 to 3.6, respectively (see Table VII-1). Additional analyses incorporating more detailed smoking information revealed no significant departures from either a multiplicative or an additive model, with the maximum likelihood values being nearly identical to each other.

Kopecky et al.[19] considered essentially the same data, but excluded those not in the city at the time of bombing and extended follow-up through 1980. A total of 29,332 subjects were in the study cohort; 351 lung-cancer deaths were observed. The results of Kopecky et al.[19] were similar to those of Prentice et al.[26] The additive-excess-risk model was shown to fit the data quite well, and neither superadditivity nor subadditivity was strongly suggested. However, Kopecky et al.[19] did not fit a multiplicative model so that results could be compared across models. Thus, while some preference for an additive model was suggested, these two analyses of atomic-bomb survivors are inconclusive in favoring a specific model.

In addition to these cohort studies, two case-control studies have been conducted. Lung-cancer cases, which were found among an autopsy series from the Life-Span Study cohort during 1961–1970, were paired with non-lung-cancer autopsy controls, matching on inclusion in the Adult Health Study, city, sex, age at death, and year of death.[16] Interviews with next of kin were conducted to ascertain information on tobacco use and occupation. A total of 180 case-control pairs were analyzed. Risk among lightly exposed (<1 rad) smokers was 3 times that of similarly exposed nonsmokers. Relative to lightly exposed nonsmokers, the risks to heavily exposed (200+ rad) smokers and nonsmokers were 8.6 and 6.2, respectively. Although sample size was small and detailed evaluation was missing, this suggests an additive model for the two effects.

Blot et al.[6] have presented preliminary results of a second case-control study of 582 lung cancers identified during the years 1971–1980 from members of the Life-Span Study cohort. Controls, also from the Life-Span Study cohort, were selected for each case and matched on date of birth, sex, city of participation in the Adult Health Study, and vital status. The 1,306 controls were selected from persons without cancer or chronic respiratory diseases. Interviews were conducted with 485 cases (83%) and

1,089 (83%) controls or next of kin (Table VII-1). Information was obtained on smoking status, passive exposure to tobacco smoke, occupation, and other factors. Among nonsmoking males, relative risks were 1.0, 2.1, and 6.2 for radiation dose categories 0–9, 10–99, and 100 rad or more, respectively, while among smoking males the corresponding risks were 9.7, 7.3, and 14.0, respectively. Among females the corresponding risks were 1.0, 0.7, and 5.3 for nonsmokers and 2.0, 2.1, and 4.5 for smokers, respectively. These patterns, based on the largest case series yet reported, appeared to suggest additive contributions for smoking and radiation, although Blot et al.[6] did not present formal significance tests in this preliminary report. The committee's analyses of related data and discussion of the fits of additive vis à vis multiplicative relative-risk model, are discussed below.

PART 2. The Committee's Analyses of Smoking and Radiation

In this portion of the appendix, we present the results of this committee's analyses of data from three populations, which address the combined effect of radiation exposure and cigarette consumption on the risk of lung cancer. The data sets include case-control studies of New Mexico uranium miners and Japanese atomic-bomb survivors and the cohort study of Colorado Plateau miners with follow-up through 1982.

METHODS

For the case-control data, models were fit using a conditional likelihood for matched data.[7] When data are matched on a time-related variable, such as age, the procedure is similar to a Cox survival time analysis for an entire cohort, and thus, the procedure is closely related to the Poisson methods that the committee employed for risk estimation.

In our modeling, it is assumed that the relative risk (RR) is the same for each matched set, regardless of level of exposure in controls (although variation with age and other matching factors can be evaluated). The RR is then modeled in several ways. Suppose the ranges for number of cigarettes smoked per day and years of mining are divided into categories; then, to estimate the relative risk associated with each variable, ignoring the other, one fits:

$$RR = 1 + \phi(\text{yr}) \text{ or} \qquad \text{(VII-1)}$$

$$RR = 1 + \phi(n/\text{day}), \qquad \text{(VII-2)}$$

where $\phi(\text{yr})$ and $\phi(n/\text{day})$ denote the individual excess *RR* estimates for categories of years of underground mining and number of cigarettes smoked per day, respectively, and are defined so that ϕ takes a value of zero at the baseline category.

More complex models that incorporate more than one variable can be defined. For example, for years of exposure and number of cigarettes per day, one can specify a multiplicative or additive combination of *RR* effects, namely:

$$RR = [1 + \phi(\text{yr})][1 + \phi(n/\text{day})] \text{ or} \tag{VII-3}$$

$$RR = [1 + \phi(\text{yr}) + \phi(n/\text{day})]. \tag{VII-4}$$

These are not nested models, since they involve the same number of parameters. However, each of these formulations can be imbedded in a richer *RR* model and compared to it.

In our evaluation, we also applied the transformation proposed by Thomas.[34] The relative risk for combined exposure is defined as follows:

$$RR = \{[1 + \phi(\text{yr})][1 + \phi(n/\text{day})]\}^{\lambda}[1 + \phi(\text{yr}) + \phi(n/\text{day})]^{1-\lambda}. \tag{VII-5}$$

At $\lambda = 1$, *RR* reduces to the multiplicative model, while at $\lambda = 0$, *RR* reduces to the additive model, as in Equations VII-3 and VII-4. Through the parameter λ, this richer model defines a smooth transformation in *RR*, which incorporates both additive and multiplicative models. Models given in Equations VII-3 to VII-5 can be compared to the saturated model given by:

$$RR = 1 + \phi(\text{yr}, n/\text{day}), \tag{VII-6}$$

were ϕ represents the excess *RR* in each cell of the cross-classification. If there are four categories of each variable, then $\phi(\text{yr}, n/\text{day})$ represents 15 free parameters, with the baseline parameter being fixed at zero.

Analogous to methodology described in Annex 2A, we fit Poisson regression models to the data on smoking and exposure to radiation described below. We assumed that the expected number of events in each cell of a cross-classification is the product of the person-years accrued times the lung-cancer disease rate, which is modeled as an age- and calendar-period-specific rate among nonexposed persons times a relative risk function, namely:

$$\text{person} - \text{yr} \times \alpha_a \times RR, \tag{VII-7}$$

where α_a represents age and year parameters. The models for *RR* are the same as defined above.

CASE-CONTROL STUDY AMONG NEW MEXICO URANIUM MINERS

The committee obtained access to data from an ongoing case-control study of lung cancer in a cohort of New Mexico uranium miners.[23] The cohort includes 4,051 subjects with at least 1 yr of documented underground employment in a New Mexico uranium mine. Large-scale uranium mining did not begin in New Mexico until the late 1950s, and exposures have generally been lower than those to the Colorado Plateau miners.

For assessment of the combined effects of cigarette smoking and uranium-mining exposure, a case-control study was conducted within this cohort. The cases included all Hispanic and non-Hispanic white males diagnosed with lung cancer, regardless of whether cause of death had been coded as lung cancer. The selection date of the case was the earlier of the date of diagnosis or death. For each of the 69 cases, four controls were selected. The controls were also Hispanic or non-Hispanic white males who (1) met the cohort entry criteria before the selection date of the case, (2) were alive and free of lung cancer at the selection date of the case, (3) had some follow-up information, and (4) had a record of a physical examination related to mine employment. From the pool of controls, the four controls closest in age to the case were selected.

Because the computation of cumulative exposures has not been completed for this cohort, the number of documented years of employment was used as the exposure variable. Cigarette-smoking information was available from one or more histories obtained at a pre-employment or annual physical examinations.

Table VII-2 gives the relevant data and shows that risks are elevated with the use of all types of tobacco products and increase with number of cigarettes smoked per day and with years spent in underground mining. The remaining analyses are restricted to nonsmokers and cigarette smokers; those who smoked cigars and/or pipes exclusively were dropped.

Matched *RR* regression models were fit to the 52 cases and 218 controls for whom the committee had data on both variables. Table VII-3 shows the distribution of cases and controls for the cross-classification of years of underground mining and smoking. Based on estimates from fitting a full 15-parameter model, risks are increased with years of underground mining within each cigarette-use category. *RR* estimates from the additive and multiplicative main effects models are also shown. The summary of the model fitting is given at the bottom of Table VII-3 and indicates that neither the multiplicative nor the additive model deviated significantly

TABLE VII-2 Data on Smoking Rate and Radiation Exposure from Case Control Study of New Mexico Uranium Miners by Various Variables[a]

	NS	FS	CS	KS	C+P/S	P/S	P+S	C+P+S	Total
Cases	2	0	33	14	8	1	0	5	63
Controls	49	3	117	56	28	1	3	9	266
RR[b]	1	—	7.4	6.6	7.8	22.6	—	13.7	
	No. of Cigarettes Smoked/day								
	<5	5–14	15–24	25+	Total				
Cases	3	16	28	5	52				
Controls	56	72	77	17	222				
RR[b]	1	5.7	8.3	7.0					
	Years of Undergrounding Mining								
	<10	10–14	15–19	20+	Total				
Cases	28	15	12	14	69				
Controls	151	59	38	24	272				
RR[b]	1	1.5	1.7	4.4					

[a]Abbreviations: NS, nonsmoker; FS, former smoker; CS, current smoker; KS, known to smoke: type, amount, and duration unknown; C+P/S, cigarette smoker who also used pipe or cigars; P/S, pipe or cigar smoker; P+S, smoked both pipe and cigars; C+P+S, smoked cigarettes, pipe, and cigars.
[b]Estimated from matched data.

from the full 15-parameter model. Although based on only 52 cases, the results suggest that the multiplicative model provides a better fit. The committee also fit the model defined by Equation VII-5 by fixing a sequence of λ values. The maximum log-likelihood (MLL) as a function of λ was rather flat, reaching a maximum at $\lambda = 4.0$ with $2 \times \text{MLL} = -127.2$, a value not much different from the simple multiplicative model (cf. Table VII-3).

CASE-CONTROL STUDY FROM THE LIFE-SPAN STUDY COHORT, JAPANESE ATOMIC-BOMB SURVIVORS

The committee's current analysis of radon-exposed miners has revealed substantial differences in the effects of radiation on lung cancer in miners in comparison with the Japanese atomic-bomb survivors. Most notable is the decline in excess risk by time since exposure, whereas there is little evidence of such decline in risk with time since exposure for atomic-bomb survivors. It is important to characterize differences and similarities in exposure effects among different radiation-exposed populations to provide insight into mechanisms of action for the exposures. The committee presents a

TABLE VII-3 Data from Case Control Study of New Mexico Uranium Miners

No. of cigarettes/ day	Years of Underground Mining <10		10–14		15–19		20+	
	No. of Cases	No. of Controls	No. of Cases	No. of Controls	No. of Cases	No. of Controls	No. of Cases	No. of Controls
<5	1	27	1	15	1	7	0	5
5–14	7	40	5	15	2	14	2	1
15–24	7	31	6	21	7	14	8	11
25+	2	8	1	4	0	1	2	4
Total	17	106	13	55	10	36	12	21

	Relative Risks <10	10–14	15–19	20+	RR[a]	RR[b]
<5	1	1.0	3.7	0	1	1
5–14	5.1	12.0	4.2	39.9	6.8	5.7
15–24	7.0	6.7	17.5	24.0	8.6	6.6
25+	8.2	6.2	0.0	30.1	8.2	6.2
RR[a]	1	1.8	3.9	14.6		
RR[b]	1	1.3	1.6	3.8		

Regression Models	No. of Parameters	2×MLL	P-Value
1: $1 + \phi(\text{yr}, n/d)$	15	−121.8	
2: $[1 + \phi(\text{yr})][1 + \phi(n/d)]$	6	−127.6	0.76
3: $1 + \phi(\text{yr}) + \phi(n/d)$	6	−129.6	0.55
4: $1 + \phi(\text{yr})$	3	−135.9	0.29
5: $1 + \phi(n/d)$	3	−133.2	0.50

[a]Relative risks from additive model, Equation VII-2.
[b]Relative risks from multiplicative model, Equation VII-1.

new analysis of radiation exposure and cigarette use on lung-cancer risk, using data from a recent case-control study among atomic-bomb survivors, to formally evaluate their combined effects.

The details of this study have been described by Blot et al.[6] Cases include diagnosed lung cancers from participants of the Life-Span Study cohort (LSS) during 1971–1980. Death certificates were used to identify lung cancers among members of this cohort who resided outside of Hiroshima and Nagasaki, so that the 582 cases do not constitute, precisely, an incident series. Controls were selected from LSS members and matched by date of birth, sex, city, Radiation Effects Research Foundation sample status, and survival status. Two controls were selected for each Hiroshima case, and three were selected for each Nagasaki case. Smoking and other information was obtained by direct interview of index subjects or their next of kin.

Table VII-4 presents the data by sex for categories of number of cigarettes smoked per day and radiation exposure. *RR*s obtained from a matched analysis are given in Table VII-5. Risks generally rise with amount of radiation received and number of cigarettes smoked. (Numbers differ slightly from those of Blot et al.[6] due to exclusion of cases who had suspect diagnoses.)

At first inspection, the data in Table VII-5 appear to support an additive model for these joint exposures. For example, by adding excess risks for males, the estimated relative risk in the highest category for each exposure based on separate exposures, $1 + (26.5 - 1) + (3.3 - 1) = 29.8$, is very close to the observed value of 24.8. Similarly, $1 + (6.9 - 1) + (3.3 - 1) = 9.2$ is close to 10.6. However, a specified model must hold throughout Table VII-5, so that the *RR* for a 20+ cigarette smoker who was exposed to less than 10 rad should be related to the 20+ per day and 10–99 rad category and the 0 per day should be related to the 10–99 rad category. Linking these, we have that $1 + (13.2 - 1) - (1.3 - 1) = 12.9$ should be the approximate risk for a 20+ per day smoker with <10 rad exposure. This is clearly quite different from the observed value of 26.5. Similar discrepancies can be found in other parts of the table. A more appropriate examination of Table VII-5 is provided by comparing the *RR*s with fitted estimates from additive and multiplicative main effects models, respectively:

TABLE VII-4 Data on Smoking Rate and Radiation Exposure from a Case Control Study of Lung Cancer among Japanese A-Bomb Survivors[a]

No. of Cigarettes/ day	Radiation Exposure (rad)							
	<10		10-99		100+		Total	
	Cases	Controls	Cases	Controls	Cases	Controls	Cases	Controls
				Males				
0	6	65	3	19	3	8	12	92
1-10	21	73	7	33	4	7	32	113
11-20	49	111	17	43	7	15	73	169
20+	45	31	11	12	9	4	65	47
				Females				
0	51	151	15	59	16	14	82	224
1-10	16	24	5	9	6	5	27	38
11+	10	9	7	10	2	0	19	19

[a]Based on data from Blot et al.[6]

TABLE VII-5 Relative Risks from Matched Analysis for Radiation Exposure and Number of Cigarettes Smoked per Day by Japanese A-Bomb Survivors

No. of Cigarettes/day	Radiation Exposure (rad)				
	<10	10–99	100+	RR[a]	RR[b]
			Males		
0	1	1.3	3.3	1	1
1–10	3.7	2.4	7.2	3.0	2.7
11–20	6.9	6.6	10.6	6.0	5.5
20+	26.5	13.2	24.8	19.4	17.2
RR[a]	1	0.9	3.5		
RR[b]	1	0.8	1.6		
			Females		
0	1	0.7	5.2	1	1
1–10	2.3	2.5	5.2	2.4	2.2
11+	4.2	2.1	—	3.3	3.7
RR[a]	1	0.6	4.9		
RR[b]	1	0.6	4.0		

[a]Relative risks from additive model, Equation VII-2.
[b]Relative risks from multiplicative model, Equation VII-1.

$$RR = 1 + \phi(\text{rad}) + \phi(n/\text{day}) \text{ and} \qquad \text{(VII-8)}$$

$$RR = [1 + \phi(\text{rad})][1 + \phi(n/\text{day})]. \qquad \text{(VII-9)}$$

These estimates are included in Table VII-5. Visual inspection of the *RR*s based on the fitted estimates does not favor either model.

Tables VII-6 and VII-7 cross-classify data and *RR*s by radiation exposure and duration, in years, of cigarette use. Again, risks increase with both exposures. Formal model comparisons were carried out to assess adequacy relative to the 15 parameter model. Table VII-8 summarizes results and shows that for both sexes the multiplicative model with radiation and number of cigarettes smoked per day fits as well as the additive model. The lower half of Table VII-8 shows little difference between additive and multiplicative models when radiation and duration of cigarette use are included.

Several additional models were fit that included continuous exposures, and similar results were obtained. Because the MLL for the main effects models for females and males were almost the same as that for the full model, power models were not tested.

Prentice et al.[26] suggest that either multiplicative or additive relative-risk models fit the atomic-bomb survivor data for lung cancer, with little

TABLE VII-6 Data on Duration of Cigarette Smoking and Radiation Exposure from a Case Control Study of Lung Cancer among Japanese A-Bomb Survivors[a]

	Radiation Exposure (rads)							
	<10		10–99		100+		Total	
Years of Smoking	Cases	Controls	Cases	Controls	Cases	Controls	Cases	Controls
				Males				
0	6	65	3	19	3	8	12	92
1–34	9	24	47	12	1	3	14	39
35–44	6	38	5	16	4	4	15	58
45+	64	78	16	28	9	4	89	110
				Females				
0	51	151	15	59	16	14	82	224
1–34	5	17	2	8	3	1	10	26
35–44	7	3	4	5	0	1	11	9
45+	9	7	4	4	4	1	17	12

[a]Based on data from Blot et al.[6]

TABLE VII-7 Relative Risks from Matched Analysis for Radiation Exposure and Years of Cigarette Use among Japanese A-Bomb Survivors

	Radiation Exposure (rad)				
Years of Smoking	<10	10–99	100+	RR[a]	RR[b]
			Males		
0	1	1.0	8.8	1	1
1–34	1.8	2.0	1.2	1.6	1.1
35–44	1.6	4.9	9.6	2.0	1.9
45+	18.4	13.3	87.2	14.0	11.3
RR[a]	1	1.4	5.5		
RR[b]	1	1.0	3.4		
			Females		
0	1	0.6	4.6	1	1
1–34	0.9	1.2	8.5	1.0	1.1
35–44	14.9	2.6	0.0	5.6	7.3
45+	5.8	2.9	7.2	4.5	4.4
RR[a]	1	0.7	4.8		
RR[b]	1	0.6	4.1		

[a]Relative risks from additive model, Equation VII-2.
[b]Relative risks from multiplicative model, Equation VII-1.

TABLE VII-8 Results of Fitting Additive and Multiplicative Relative Risk Models to Evaluate Radiation Exposure and Number of Cigarettes Smoked per Day or Duration in Years of Cigarette Use among Japanese A-Bomb Survivors

Model	Males			Females		
	No. of Parameters	2 × MLL	*P* Value	No. of Parameters	2 × MLL	*P* Value
1: $1 + \phi(\text{rad}, n/d)$	11	−273.7		8	−219.1	
2: $[1 + \phi(\text{rad})][1 + \phi(n/d)]$	5	−276.0	0.89	4	−220.8	0.79
3: $1 + \phi(\text{rad}) + \phi(n/d)$	5	−275.9	0.90	4	−219.8	0.95
4: $1 + \phi(\text{rad})$	2	−335.4	<0.01	2	−234.3	0.02
5: $1 + \phi(n/d)$	3	−279.1	0.71	2	−236.7	0.01
1: $1 + \phi(\text{rad}, \text{dur})$	11	−140.8		11	−191.4	0.88
2: $[1 + \phi(\text{rad})][1 + \phi(\text{dur})]$	5	−144.7	0.69	5	−193.8	0.80
3: $1 + \phi(\text{rad}) + \phi(\text{dur})$	5	−146.1	0.51	5	−194.5	0.80
4: $1 + \phi(\text{rad})$	2	−187.9	<0.01	2	−212.5	0.01
5: $1 + \phi(\text{dur})$	3	−150.5	0.29	3	−208.8	0.03

preference. Kopecky et al.[19] fit only the additive model in their analysis. The committee's reanalysis of case-control data extracted from members of the LSS group agrees with the interpretation of Prentice et al.[26] in that the committee found no strong preference for an additive combination of radiation and smoking.

COHORT STUDY OF COLORADO PLATEAU URANIUM MINERS

The committee evaluated the combined effects of cigarette consumption and cumulative WLM exposure, using data from the cohort of Colorado Plateau uranium miners with follow-up through 1982. There were 256 observed lung-cancer cases in over 73,000 person-yr of observation.

Assignment of WLM exposures is described in Appendix IV. A 5-yr lag interval was used to determine exposure. Smoking information was obtained from initial medical examination, periodic surveys, and mail questionnaires. For this analysis, the committee used the mean number of cigarettes smoked per day as calculated from all available sources. Because the last update of cigarette use information was in 1969 and because precise details were not available, the committee did not attempt to evaluate other tobacco-related determinants of risk, such as duration of use, time since cessation of smoking, filter or nonfilter cigarette use, and intensity of inhalation. These factors are strongly related to lung-cancer risk and the inability of the committee to consider them may have an impact on this evaluation of combined effects of tobacco use and radon-daughter exposure. Thus, the results described below should be viewed cautiously, pending a more thorough evaluation.

For this analysis the same Poisson regression techniques were used as for the radon risk estimation. Data on lung-cancer events and person-years were cross-classified by categories of cumulative WLM exposure (<60, 60–119, 120–239, 240–479, 480–959, ≥960), number of cigarettes smoked per day (0–4, 5–19, 20–29, ≥30), age (<55, 55–59, 60–64, 65–69, ≥70), and calendar year (1950–1959, 1960–1964, 1965–1969, 1970–1974, 1975–1982). Table VII-9 shows the number of observed lung cancers, person-years, and crude disease rate by WLM exposure and cigarette use when data were collapsed across age and year categories. The numbers differ slightly from the tables in Annex 2A because of a different cross-classification. Note that cells with means exceeding 2,000 WLM have been excluded.

For these data we fit a more general relative risk model than that given in Equation VII-5, namely,

TABLE VII-9 Observed Lung-Cancer Mortality and Calculated Lung-Cancer Mortality Rate as a Function of Cumulative Exposure and Cigarette Consumption for the Colorado Plateau Miner Cohort[a]

Cumulative WLM	No. of Cigarettes/day	0-4	5-19	20-29	30+	Total
0-59	Observe	0	1	7	1	9
	Rate	12.3[b]	35.8	102.2	39.5	49.9
	P-yr[c]	5,878.8	2,790.5	6,848.3	2,530.5	18,048.0
60-119	Observed	0	0	2	3	5
	Rate	0	0	81.9	404.3	78.8
	P-yr	2,263.0	894.0	2,443.5	742.0	6,342.5
120-239	Observed	1	2	9	2	14
	Rate	34.8	138.9	232.0	157.3	148.0
	P-yr	2,872.0	1,439.0	3,879.0	1,271.5	9,461.5
240-479	Observed	6	1	12	8	27
	Rate	157.5	54.0	229.2	421.7	211.0
	P-yr	3,809.3	1,851.5	5,236.8	1,897.0	2,794.5
480-959	Observed	11	3	29	14	57
	Rate	323.1	216.0	523.8	651.7	456.8
	P-yr	3,404.5	1,389.0	5,536.5	2,148.3	12,478.3
960+	Observed	4	6	10	19	39
	Rate	289.5	554.0	457.5	1,189.0	625.0
	P-yr	1,381.8	1,083.0	2,186.0	1,598.0	6,239.8
Total	Observed	22	13	69	47	151
	Rate	112.2	137.6	264.1	461.8	231.0
	P-yr	19,609.3	9,447.5	26,130.0	10,178.3	65,365.0

[a]Cumulative exposure limited to 2,000 WLM.
[b]Baseline rate per 100,000 computed using expected number of cases, based on U.S. white male mortality rates for lung cancer adjusted to nonsmokers.
[c]Person years.

$$RR = R[w(a), n(a); \lambda] = \{[1 + \phi(w)\gamma(a)][1 + \phi(n)\delta(a)]\}^{\lambda}$$
$$[1 + \phi(w)\gamma(a) + \phi(n)\delta(a)]^{1-\lambda}, \qquad \text{(VII-10)}$$

where, as previously, $\phi(w)$ and $\phi(n)$ represent parameters for categories of cumulative WLM (w) and cigarettes per day (n), and where $\gamma(a)$ and $\delta(a)$ denote modifications of the $\phi(w)$ and $\phi(n)$ effects with age a. We have defined $\gamma(a) = \delta(a) = 1$ for age $a < 65$ yr and $\gamma(a) = \gamma$ and $\delta(a) = \delta$ for $a > 65$ yr. To reduce model complexity, we did not include a parameter for time since exposure. This factor is not significant in the Colorado data.

The marginally saturated model given in Equation VII-6 was generalized to include age as:

$$RR = 1 + \phi(w, n, a), \tag{VII-11}$$

where ϕ denoted 46 parameters, including 23 parameters for the cross-classification of cumulative WLM and cigarettes per day for ages <65 yr and 23 parameters for exposures for ages >65 yr. The baseline rate $r_0(a)$ incorporates nine parameters for multiplicative effects of age and calendar year.

The class of models characterized by Equation VII-10 reduces to Equation VII-5 when $\gamma(a) = \delta(a) = 1$ for all a and includes the multiplicative ($\lambda = 1$) and additive ($\lambda = 0$) models. The inclusion of γ and δ (actually the exponential of each) permits formal likelihood ratio testing of age effects for cumulative WLM exposure and cigarettes per day. As seen in Annex 2A, the effects of cumulative WLM decline with age at risk.

Table VII-10 shows predicted relative risks by age group based on various models (risks relative to the lung-cancer rate in the entire cohort), while Table VII-11 gives results from model fittings. Table VII-11 indicates that age at risk is an important modifier for cumulative WLM exposure ($P = 0.005$), but not for cigarettes per day ($P = 0.8$). The estimate of the effect of age modification for cumulative WLM $\gamma(a)$ is 0.1 for ages 65 or over for both from $R[w(a),n(a);\lambda]$ and $\mathrm{R}[w(a),n;\lambda]$, with $\lambda = 1.0$ and with the maximum likelihood estimates for λ. Note that this confirms the analyses of Annex 2A, in which cigarette use is not included.

Therefore, it is appropriate to consider a reduced form of Equation VII-10, $R[w(a),n;\lambda]$, where $\delta(a) = 1$ for all a. Focusing on models labeled 2 and 3 in Table VII-11, additive effects for WLM and cigarettes per day are rejected relative to the mixture model, while multiplicative effects are not. The maximum likelihood estimates for λ are 0.4 under $R[w(a),n(a);\lambda]$ and 0.6 under $R[w(a),n;\lambda]$. However, the maximized likelihood with $\lambda = 0.4$ (or $\lambda = 0.6$) is very similar to the likelihood with λ fixed at one, indicating a comparable fit. It should be noted that the likelihood in λ was very flat for $\lambda > 0.3$, which precludes precise specification of λ. The value $\lambda =$ 0.4 (or $\lambda = 0.6$) does not indicate that the true model is halfway between additive and multiplicative models. Rather, because of the variability in λ and skewness of the distribution of possible values of λ, one cannot be more precise than to say that the data are consistent with a range of joint-effect models, from submultiplicative to supermultiplicative.

The committee's analyses of the interaction between smoking and cumulative exposure support the conclusions of Whittemore and McMillan.[38] An additive model is rejected in the committee's analysis, while a multiplicative combination of relative risks provides an acceptable fit. However, the committee also found that by embedding the simple models into a larger class of mixture models, a range of submultiplicative to supramultiplicative models was equally compatible with the data.

TABLE VII-10 Relative Risks for Lung Cancer among Colorado Plateau Miner's Cohort Based on Various Models for Cumulative WLM (w) and Cigarettes per day (n); Risks Relative to Rate in Entire Cohort

Cumulative WLM	Model[a]	No. of Cigarettes/day 0-4	5-19	20-29	30+
		Age < 65			
0-59	$[1 + \phi(w, n, a)]$	0.00	0.12	0.10	0.12
	$R[w(a), n; \lambda = 1.0]$	0.04	0.04	0.10	0.12
	$R[w(a), n; \lambda = 0.6]$	0.03	0.03	0.12	0.14
	$R[w(a), n; \lambda = 0.0]$	0.02	0.04	0.20	0.24
60-119	$[1 + \phi(w, n, a)]$	0.00	0.00	0.20	0.58
	$R[w(a), n; \lambda = 1.0]$	0.07	0.07	0.19	0.23
	$R[w(a), n; \lambda = 0.6]$	0.06	0.06	0.18	0.21
	$R[w(a), n; \lambda = 0.0]$	0.02	0.04	0.20	0.24
120-239	$[1 + \phi(w, n, a)]$	0.00	0.17	0.36	0.18
	$R[w(a), n; \lambda = 1.0]$	0.10	0.10	0.26	0.31
	$R[w(a), n; \lambda = 0.6]$	0.09	0.10	0.26	0.31
	$R[w(a), n; \lambda = 0.0]$	0.07	0.09	0.25	0.29
240-479	$[1 + \phi(w, n, a)]$	0.18	0.13	0.41	0.81
	$R[w(a), n; \lambda = 1.0]$	0.17	0.17	0.44	0.53
	$R[w(a), n; \lambda = 0.6]$	0.17	0.17	0.44	0.51
	$R[w(a), n; \lambda = 0.0]$	0.19	0.21	0.37	0.41
480-959	$[1 + \phi(w, n, a)]$	0.48	0.34	0.64	1.04
	$R[w(a), n; \lambda = 1.0]$	0.31	0.32	0.82	0.99
	$R[w(a), n; \lambda = 0.6]$	0.35	0.35	0.82	0.94
	$R[w(a), n; \lambda = 0.0]$	0.48	0.50	0.66	0.70
960+	$[1 + \phi(w, n, a)]$	0.80	1.00	2.20	2.87
	$R[w(a), n; \lambda = 1.0]$	0.79	0.81	2.07	2.51
	$R[w(a), n; \lambda = 0.6]$	0.90	0.91	2.03	2.30
	$R[w(a), n; \lambda = 0.0]$	1.25	1.27	1.44	1.47
		Age ≥ 65			
0-59	$[1 + \phi(w, n, a)]$	0.00	0.00	1.65	0.00
	$R[w(a), n; \lambda = 1.0]$	0.24	0.25	0.88	1.06
	$R[w(a), n; \lambda = 0.6]$	0.33	0.34	0.86	1.04
	$R[w(a), n; \lambda = 0.0]$	0.08	0.17	0.87	1.03
60-119	$[1 + \phi(w, n, a)]$	0.00	0.00	0.00	3.41
	$R[w(a), n; \lambda = 1.0]$	0.26	0.27	0.93	1.13
	$R[w(a), n; \lambda = 0.6]$	0.35	0.36	0.93	1.12
	$R[w(a), n; \lambda = 0.0]$	0.08	0.17	0.87	1.03
120-239	$[1 + \phi(w, n, a)]$	0.50	0.98	1.16	0.99
	$R(w(a), n; \lambda = 1.0]$	0.29	0.30	1.01	1.22
	$R[w(a), n; \lambda = 0.6]$	0.38	0.39	0.99	1.19
	$R[w(a), n; \lambda = 0.0]$	0.13	0.22	0.92	1.08

TABLE VII-10 (*Continued*)

Cumulative WLM	Model[a]	No. of Cigarettes/day			
		0-4	5-19	20-29	30+
240-479	$[1 + \phi(w, n, a)]$	0.60	0.00	0.77	0.74
	$R[w(a), n; \lambda = 1.0]$	0.35	0.36	1.16	1.40
	$R[w(a), n; \lambda = 0.6]$	0.43	0.44	1.13	1.36
	$R[w(a), n; \lambda = 0.0]$	0.24	0.34	1.04	1.20
480-959	$[1 + \phi(w, n, a)]$	1.00	0.78	3.08	1.17
	$R[w(a), n; \lambda = 1.0]$	0.49	0.50	1.50	1.79
	$R[w(a), n; \lambda = 0.6]$	0.54	0.56	1.42	1.72
	$R[w(a), n; \lambda = 0.0]$	0.52	0.62	1.32	1.47
960+	$[1 + \phi(w, n, a)]$	0.55	0.63	0.00	2.48
	$R[w(a), n; \lambda = 1.0]$	0.94	0.96	2.53	2.96
	$R[w(a), n; \lambda = 0.6]$	0.91	0.94	2.39	2.89
	$R[w(a), n; \lambda = 0.0]$	1.27	1.36	2.06	2.22

[a] Age-specific rates are $r_0(a)$ times the relative risk, as defined by the various models. The r_0 term includes nine parameters for age and calendar year. The $\phi(w, n, a)$ term denotes 46 free parameters, including 23 for the exposure cross-classification for age $a < 65$ and 23 for age $a \geq 65$ years. Other models are defined as follows;

$$R[w(a), n(a); \lambda] = \{[1 + \phi(w)\gamma(a)][1 + \phi(n)\delta(a)]\}^{\lambda}[1 + \phi(w)\gamma(a) + \phi(n)\gamma(a)]^{1-\lambda}$$

where $\phi(w)$ and $\phi(n)$ denote parameters for categories of cumulative WLM and cigarettes per day, and $\gamma(a)$ and $\delta(a)$ denote their respective age effects for $a < 65$ and $a \geq 65$. For the values in the table, $\delta(a) = 1$ for all a. At $\lambda = 1$, R specifies multiplicative effects for WLM and cigarettes per day for ages <65 and ≥ 65; and at $\lambda = 0$, R specifies additive effects.

APPLICATION OF THE MULTIPLICATIVE MODEL

In Chapter 2, the committee outlines how it applies the multiplicative model for the combined effects of smoking and exposure to radon progeny. Table 2.4 in Chapter 2 lists the estimated risk to smokers and nonsmokers of both sexes due to lifetime exposure. Exposures for shorter periods of time are also of interest since exposure to elevated levels of radon may occur and end at any age. Tables VII-12 to VII-14 and Tables VII-15 to VII-17 provide for male smokers and nonsmokers, respectively, the ratio of the lifetime risk of lung-cancer mortality due to exposure occurring within stated intervals of age. Included in the tables are two other measures of risk described in Chapter 2 and defined mathematically in Annex 2A. These are R_e, the lifetime risk of lung-cancer mortality, which includes the baseline risk R_0, for exposure between two age intervals; and $L_0 - L_e$, the number of years of life lost due to such exposures. Tables VII-18 to VII-20 and Tables VII-21 to VII-23 provide the same set of results for female smokers and nonsmokers.

TABLE VII-11 Results for Fitting Various Relative Risk Models to Colorado Plateau Miners' Cohort

	Model[a]	2 × MLL	No. of Parameters
1	$[1 + \phi(w, n, a)]$	−338.8	55
2(a)	$R[w(a), n(a); \lambda = 0.4]$	−376.6	20
2(b)	$R[w(a), n(a); \lambda = 1.0]$	−377.4	19
2(c)	$R[w(a), n(a); \lambda = 0.0]$	−382.6	19
3(a)	$R[w(a), n; \lambda = 0.6]$	−377.0	19
3(b)	$R[w(a), n; \lambda = 1.0]$	−377.5	18
3(c)	$R[w(a), n; \lambda = 0.0]$	−384.8	18
4(a)	$R[w, n(a); \lambda = 0.2]$	−381.4	19
4(b)	$R[w, n(a); \lambda = 1.0]$	−385.4	18
4(c)	$R[w, n(a); \lambda = 0.0]$	−384.3	18
5(a)	$R[w, n; \lambda = 0.3]$	−383.4	18
5(b)	$R[w, n; \lambda = 1.0]$	−385.4	17
5(c)	$R[w, n; \lambda = 0.0]$	−388.2	17

	Tests of hypothesis: Chi-Sq (d.f.)	*P* value
Fit of mixture model:		
2(a) vs 1	37.8 (35)	0.343
3(a) vs 1	38.2 (36)	0.370
4(a) vs 1	42.6 (36)	0.208
5(a) vs 1	44.6 (37)	0.183
Age effects for WLM		
4(b) vs 2(b)	8.0 (1)	0.005
5(b) vs 3(b)	7.0 (1)	0.005
Age effects for cigarettes per day:		
3(b) vs 2(b)	0.1 (1)	0.752
5(b) vs 4(b)	0.0 (1)	0.841
Multiplicative fit:		
2(b) vs 2(a)	0.8 (1)	0.371
3(b) vs 3(a)	0.5 (1)	0.480
Additive fit:		
2(c) vs 2(a)	6.0 (1)	0.014
3(c) vs 3(a)	7.8 (1)	0.005

[a]See footnote *a* to Table VII-10.

TABLE VII-12 Ratio of Lifetime Risks (R_e/R_0) by Age Started and Age Exposure Ends for Various Rates of Annual Exposure[a] for Male Smokers[b]

Age (yr)	Age (yr) Exposure Ends								
Started	10	20	30	40	50	60	70	80	110
				Exposure Rate = 0.1 (WLM/yr)					
0	1.011	1.023	1.034	1.046	1.058	1.070	1.079	1.082	1.083
10		1.011	1.023	1.034	1.047	1.059	1.068	1.071	1.071
20			1.011	1.023	1.036	1.048	1.057	1.060	1.060
30				1.012	1.024	1.036	1.045	1.049	1.049
40					1.013	1.025	1.034	1.037	1.037
50						1.012	1.021	1.024	1.025
60							1.009	1.012	1.013
				Exposure Rate = 0.2 (WLM/yr)					
0	1.023	1.045	1.067	1.091	1.115	1.139	1.156	1.163	1.164
10		1.023	1.045	1.068	1.093	1.117	1.134	1.141	1.142
20			1.023	1.046	1.071	1.095	1.112	1.119	1.120
30				1.023	1.048	1.073	1.090	1.096	1.098
40					1.025	1.049	1.067	1.073	1.075
50						1.025	1.042	1.049	1.050
60							1.018	1.024	1.025
				Exposure Rate = 0.5 (WLM/yr)					
0	1.056	1.112	1.167	1.223	1.283	1.340	1.381	1.395	1.398
10		1.056	1.112	1.169	1.229	1.287	1.328	1.343	1.346
20			1.057	1.114	1.175	1.234	1.276	1.291	1.293
30				1.059	1.120	1.180	1.222	1.237	1.240
40					1.062	1.123	1.165	1.181	1.184
50						1.061	1.105	1.120	1.123
60							1.044	1.060	1.063
				Exposure Rate = 1.0 (WLM/yr)					
0	1.112	1.221	1.328	1.436	1.549	1.656	1.729	1.755	1.760
10		1.112	1.221	1.332	1.447	1.557	1.633	1.660	1.664
20			1.112	1.226	1.344	1.456	1.534	1.562	1.567
30				1.116	1.237	1.352	1.432	1.461	1.466
40					1.124	1.242	1.324	1.354	1.359
50						1.121	1.206	1.237	1.242
60							1.088	1.119	1.125
				Exposure Rate = 4.0 (WLM/yr)					
0	1.431	1.823	2.181	2.519	2.848	3.133	3.306	3.357	3.364
10		1.431	1.825	2.196	2.556	2.869	3.060	3.119	3.126
20			1.433	1.841	2.235	2.579	2.792	2.858	2.867
30				1.448	1.881	2.260	2.497	2.572	2.583
40					1.477	1.895	2.161	2.246	2.259
50						1.466	1.765	1.862	1.877
60							1.337	1.449	1.467

TABLE VII-12 (*Continued*)

Age (yr) Started	Age (yr) Exposure Ends 10	20	30	40	50	60	70	80	110
				Exposure Rate = 10.0 (WLM/yr)					
0	2.005	2.799	3.437	3.972	4.433	4.772	4.929	4.961	4.964
10		2.005	2.804	3.466	4.030	4.448	4.648	4.693	4.697
20			2.010	2.837	3.535	4.054	4.311	4.371	4.377
30				2.044	2.915	3.566	3.898	3.980	3.989
40					2.106	2.933	3.367	3.479	3.492
50						2.077	2.655	2.812	2.831
60							1.782	2.004	2.032
				Exposure Rate = 20.0 (WLM/yr)					
0	2.799	3.943	4.704	5.251	5.660	5.896	5.965	5.972	5.973
10		2.800	3.950	4.742	5.314	5.647	5.754	5.767	5.768
20			2.808	4.000	4.826	5.311	5.478	5.502	5.503
30				2.866	4.108	4.836	5.103	5.146	5.149
40					2.965	4.105	4.548	4.627	4.633
50						2.900	3.670	3.822	3.835
60							2.390	2.691	2.719

[a] Estimated with the committee's TSE model (Chapter 2) and a multiplicative interaction between smoking and exposure to radon progeny.
[b] R_0, the calculated lifetime risk for unexposed male smokers, is 0.123.

TABLE VII-13 Lifetime Risk (R_e) by Age Started and Age Exposure Ends for Various Rates of Annual Exposure[a] for Male Smokers

Age (yr) Started	Age (yr) Exposure Ends								
	10	20	30	40	50	60	70	80	110
				Exposure Rate = 0.1 (WLM/yr)					
0	0.124	0.126	0.127	0.128	0.130	0.131	0.133	0.133	0.133
10		0.124	0.126	0.127	0.129	0.130	0.131	0.132	0.132
20			0.124	0.126	0.127	0.129	0.130	0.130	0.130
30				0.124	0.126	0.127	0.128	0.129	0.129
40					0.124	0.126	0.127	0.127	0.127
50						0.124	0.125	0.126	0.126
60							0.124	0.124	0.124
				Exposure Rate = 0.2 (WLM/yr)					
0	0.126	0.128	0.131	0.134	0.137	0.140	0.142	0.143	0.143
10		0.126	0.128	0.131	0.134	0.137	0.139	0.140	0.140
20			0.126	0.129	0.132	0.135	0.137	0.137	0.138
30				0.126	0.129	0.132	0.134	0.135	0.135
40					0.126	0.129	0.131	0.132	0.132
50						0.126	0.128	0.129	0.129
60							0.125	0.126	0.126
				Exposure Rate = 0.5 (WLM/yr)					
0	0.130	0.137	0.143	0.150	0.158	0.165	0.170	0.171	0.172
10		0.130	0.137	0.144	0.151	0.158	0.163	0.165	0.165
20			0.130	0.137	0.144	0.152	0.157	0.159	0.159
30				0.130	0.138	0.145	0.150	0.152	0.152
40					0.131	0.138	0.143	0.145	0.145
50						0.130	0.136	0.138	0.138
60							0.128	0.130	0.131
				Exposure Rate = 1.0 (WLM/yr)					
0	0.137	0.150	0.163	0.176	0.190	0.203	0.213	0.216	0.216
10		0.137	0.150	0.164	0.178	0.191	0.201	0.204	0.205
20			0.137	0.151	0.165	0.179	0.189	0.192	0.193
30				0.137	0.152	0.166	0.176	0.180	0.180
40					0.138	0.153	0.163	0.166	0.167
50						0.138	0.148	0.152	0.153
60							0.134	0.138	0.138
				Exposure Rate = 4.0 (WLM/yr)					
0	0.176	0.224	0.268	0.310	0.350	0.385	0.406	0.413	0.413
10		0.176	0.224	0.270	0.314	0.353	0.376	0.383	0.384
20			0.176	0.226	0.275	0.317	0.343	0.351	0.352
30				0.178	0.231	0.278	0.307	0.316	0.317
40					0.181	0.233	0.266	0.276	0.278
50						0.180	0.217	0.229	0.231
60							0.164	0.178	0.180

TABLE VII-13 (*Continued*)

Age (yr) Started	Age (yr) Exposure Ends								
	10	20	30	40	50	60	70	80	110
	Exposure Rate = 10.0 (WLM/yr)								
0	0.246	0.344	0.422	0.488	0.545	0.586	0.606	0.610	0.610
10		0.246	0.345	0.426	0.495	0.547	0.571	0.577	0.577
20			0.247	0.349	0.434	0.498	0.530	0.537	0.538
30				0.251	0.358	0.438	0.479	0.489	0.490
40					0.259	0.360	0.414	0.428	0.429
50						0.255	0.326	0.346	0.348
60							0.219	0.246	0.250
	Exposure Rate = 20.0 (WLM/yr)								
0	0.344	0.485	0.578	0.645	0.696	0.724	0.733	0.734	0.734
10		0.344	0.485	0.583	0.653	0.694	0.707	0.709	0.709
20			0.345	0.491	0.593	0.653	0.673	0.676	0.676
30				0.352	0.505	0.594	0.627	0.632	0.633
40					0.364	0.504	0.559	0.569	0.569
50						0.356	0.451	0.470	0.471
60							0.294	0.331	0.334

[a] Estimated with the committee's TSE model (Chapter 2) and a multiplicative interaction between smoking and exposure to radon progeny. Note that R_e includes R_0, the calculated lifetime risk for unexposed male smokers, 0.123.

TABLE VII-14 Years of Life Lost, $(L_0 - L_e)$ by Age Started and Age Exposure Ends for Various Rates of Annual Exposure[a] for Male Smokers[b]

Age (yr) Started	Age (yr) Exposure Ends								
	10	20	30	40	50	60	70	80	110
				Exposure Rate = 0.1 (WLM/yr)					
0	0.02	0.04	0.06	0.08	0.10	0.12	0.13	0.13	0.13
10		0.02	0.04	0.06	0.08	0.10	0.11	0.11	0.11
20			0.02	0.04	0.06	0.08	0.09	0.09	0.09
30				0.02	0.04	0.06	0.07	0.07	0.07
40					0.02	0.04	0.05	0.05	0.05
50						0.02	0.03	0.03	0.03
60							0.01	0.01	0.01
				Exposure Rate = 0.2 (WLM/yr)					
0	0.04	0.08	0.12	0.16	0.20	0.24	0.25	0.26	0.26
10		0.04	0.08	0.12	0.16	0.20	0.22	0.22	0.22
20			0.04	0.08	0.12	0.16	0.18	0.18	0.18
30				0.04	0.09	0.12	0.14	0.14	0.14
40					0.04	0.08	0.10	0.10	0.10
50						0.04	0.06	0.06	0.06
60							0.02	0.02	0.02
				Exposure Rate = 0.5 (WLM/yr)					
0	0.10	0.19	0.29	0.39	0.50	0.58	0.63	0.64	0.64
10		0.10	0.19	0.30	0.40	0.49	0.54	0.54	0.54
20			0.10	0.20	0.31	0.40	0.44	0.45	0.45
30				0.10	0.21	0.30	0.35	0.36	0.36
40					0.11	0.20	0.25	0.26	0.25
50						0.09	0.14	0.15	0.15
60							0.05	0.06	0.06
				Exposure Rate = 1.0 (WLM/yr)					
0	0.19	0.38	0.57	0.77	0.97	1.14	1.22	1.24	1.24
10		0.19	0.38	0.59	0.79	0.96	1.05	1.06	1.06
20			0.19	0.40	0.61	0.78	0.87	0.89	0.88
30				0.21	0.42	0.60	0.68	0.70	0.70
40					0.22	0.40	0.49	0.50	0.50
50						0.18	0.27	0.29	0.29
60							0.09	0.11	0.11
				Exposure Rate = 4.0 (WLM/yr)					
0	0.75	1.45	2.13	2.82	3.49	3.98	4.19	4.23	4.23
10		0.75	1.47	2.19	2.90	3.43	3.66	3.70	3.70
20			0.76	1.53	2.28	2.84	3.09	3.14	3.14
30				0.81	1.61	2.22	2.49	2.54	2.54
40					0.85	1.50	1.80	1.86	1.85
50						0.70	1.03	1.10	1.09
60							0.37	0.44	0.43

TABLE VII-14 (*Continued*)

Age (yr) Started	Age (yr) Exposure Ends								
	10	20	30	40	50	60	70	80	110
				Exposure Rate = 10.0 (WLM/yr)					
0	1.79	3.34	4.71	6.01	7.17	7.88	8.10	8.13	8.13
10		1.79	3.37	4.84	6.16	6.99	7.26	7.30	7.30
20			1.82	3.50	5.01	5.98	6.33	6.37	6.37
30				1.94	3.68	4.82	5.25	5.31	5.31
40					2.02	3.39	3.92	4.00	4.00
50						1.66	2.33	2.45	2.45
60							0.86	1.02	1.01
				Exposure Rate = 20.0 (WLM/yr)					
0	3.34	5.86	7.86	9.61	11.01	11.66	11.78	11.79	11.79
10		3.34	5.91	8.08	9.79	10.63	10.81	10.82	10.82
20			3.39	6.15	8.30	9.41	9.67	9.69	9.70
30				3.62	6.40	7.89	8.28	8.32	8.32
40					3.73	5.80	6.41	6.48	6.48
50						3.03	3.99	4.12	4.12
60							1.58	1.80	1.81

[a] Estimated with the committee's TSE model (Chapter 2) and a multiplicative interaction between smoking and exposure to radon progeny.

[b] L_0, the calculated lifetime risk for unexposed male smokers, is 69.0 yr.

TABLE VII-15 Ratio of Lifetime Risks, (R_e/R_0) by Age Started and Age Exposure Ends for Various Rates of Annual Exposure[a] for Male Nonsmokers[b]

Age (yr) Started	Age (yr) Exposure Ends								
	10	20	30	40	50	60	70	80	110
				Exposure Rate = 0.1 (WLM/yr)					
0	1.012	1.025	1.037	1.050	1.064	1.077	1.087	1.091	1.092
10		1.012	1.025	1.038	1.051	1.065	1.075	1.079	1.079
20			1.012	1.025	1.039	1.053	1.062	1.066	1.067
30				1.013	1.027	1.040	1.050	1.054	1.055
40					1.014	1.027	1.037	1.041	1.042
50						1.014	1.023	1.027	1.028
60							1.010	1.014	1.015
				Exposure Rate = 0.2 (WLM/yr)					
0	1.025	1.050	1.074	1.100	1.128	1.155	1.174	1.182	1.183
10		1.025	1.050	1.075	1.103	1.130	1.150	1.157	1.159
20			1.025	1.051	1.078	1.105	1.125	1.133	1.134
30				1.026	1.053	1.080	1.100	1.108	1.109
40					1.027	1.054	1.074	1.082	1.083
50						1.027	1.047	1.055	1.056
60							1.020	1.028	1.029
				Exposure Rate = 0.5 (WLM/yr)					
0	1.062	1.124	1.186	1.250	1.318	1.386	1.435	1.454	1.457
10		1.062	1.124	1.188	1.257	1.324	1.373	1.392	1.396
20			1.062	1.127	1.195	1.262	1.312	1.331	1.334
30				1.064	1.133	1.200	1.250	1.269	1.272
40					1.068	1.136	1.186	1.205	1.208
50						1.068	1.117	1.136	1.140
60							1.050	1.069	1.072
				Exposure Rate = 1.0 (WLM/yr)					
0	1.124	1.247	1.371	1.499	1.635	1.769	1.867	1.904	1.911
10		1.124	1.248	1.376	1.512	1.646	1.744	1.782	1.789
20			1.124	1.253	1.389	1.523	1.622	1.659	1.667
30				1.129	1.265	1.400	1.498	1.536	1.543
40					1.137	1.272	1.371	1.408	1.416
50						1.135	1.234	1.272	1.280
60							1.099	1.137	1.145
				Exposure Rate = 4.0 (WLM/yr)					
0	1.494	1.983	2.471	2.972	3.500	4.016	4.389	4.530	4.557
10		1.494	1.986	2.490	3.023	3.543	3.920	4.063	4.090
20			1.496	2.005	2.542	3.066	3.447	3.591	3.619
30				1.513	2.054	2.583	2.968	3.113	3.141
40					1.546	2.079	2.468	2.615	2.643
50						1.539	1.931	2.080	2.108
60							1.396	1.547	1.576

TABLE VII-15 (*Continued*)

Age (yr) Started	Age (yr) Exposure Ends								
	10	20	30	40	50	60	70	80	110
				Exposure Rate = 10.0 (WLM/yr)					
0	2.227	3.428	4.610	5.808	7.057	8.257	9.108	9.421	9.478
10		2.227	3.434	4.657	5.931	7.156	8.028	8.349	8.407
20			2.233	3.481	4.781	6.033	6.925	7.255	7.315
30				2.274	3.602	4.880	5.793	6.132	6.194
40					2.355	3.662	4.597	4.945	5.010
50						2.337	3.296	3.654	3.722
60							1.984	2.353	2.423
				Exposure Rate = 20.0 (WLM/yr)					
0	3.428	5.757	8.003	10.234	12.511	14.644	16.104	16.616	16.702
10		3.428	5.769	8.092	10.462	12.686	14.214	14.754	14.846
20			3.440	5.860	8.328	10.646	12.246	12.816	12.915
30				3.522	6.092	8.510	10.186	10.787	10.893
40					3.681	6.205	7.963	8.598	8.711
50						3.643	5.491	6.164	6.286
60							2.947	3.661	3.791

[a] Estimated with the committee's TSE model (Chapter 2) and a multiplicative interaction between smoking and exposure to radon progeny.
[b] R_0, the calculated lifetime risk for unexposed male nonsmokers, is 0.0112.

TABLE VII-16 Lifetime Risk (R_e) by Age Started and Age Exposure Ends for Various Rates of Annual Exposure[a] for Male Nonsmokers

Age (yr) Started	Age (yr) Exposure Ends								
	10	20	30	40	50	60	70	80	110
				Exposure Rate = 0.1 (WLM/yr)					
0	0.011	0.011	0.012	0.012	0.012	0.012	0.012	0.012	0.012
10		0.011	0.011	0.012	0.012	0.012	0.012	0.012	0.012
20			0.011	0.011	0.012	0.012	0.012	0.012	0.012
30				0.011	0.012	0.012	0.012	0.012	0.012
40					0.011	0.012	0.012	0.012	0.012
50						0.011	0.011	0.012	0.012
60							0.011	0.011	0.011
				Exposure Rate = 0.2 (WLM/yr)					
0	0.011	0.012	0.012	0.012	0.013	0.013	0.013	0.013	0.013
10		0.011	0.012	0.012	0.012	0.013	0.013	0.013	0.013
20			0.011	0.012	0.012	0.012	0.013	0.013	0.013
30				0.011	0.012	0.012	0.012	0.012	0.012
40					0.012	0.012	0.012	0.012	0.012
50						0.012	0.012	0.012	0.012
60							0.011	0.012	0.012
				Exposure Rate = 0.5 (WLM/yr)					
0	0.012	0.013	0.013	0.014	0.015	0.016	0.016	0.016	0.016
10		0.012	0.013	0.013	0.014	0.015	0.015	0.016	0.016
20			0.012	0.013	0.013	0.014	0.015	0.015	0.015
30				0.012	0.013	0.013	0.014	0.014	0.014
40					0.012	0.013	0.013	0.013	0.014
50						0.012	0.013	0.013	0.013
60							0.012	0.012	0.012
				Exposure Rate = 1.0 (WLM/yr)					
0	0.013	0.014	0.015	0.017	0.018	0.020	0.021	0.021	0.021
10		0.013	0.014	0.015	0.017	0.018	0.020	0.020	0.020
20			0.013	0.014	0.016	0.017	0.018	0.019	0.019
30				0.013	0.014	0.016	0.017	0.017	0.017
40					0.013	0.014	0.015	0.016	0.016
50						0.013	0.014	0.014	0.014
60							0.012	0.013	0.013
				Exposure Rate = 4.0 (WLM/yr)					
0	0.017	0.022	0.028	0.033	0.039	0.045	0.049	0.051	0.051
10		0.017	0.022	0.028	0.034	0.040	0.044	0.046	0.046
20			0.017	0.022	0.028	0.034	0.039	0.040	0.041
30				0.017	0.023	0.029	0.033	0.035	0.035
40					0.017	0.023	0.028	0.029	0.030
50						0.017	0.022	0.023	0.024
60							0.016	0.017	0.018

TABLE VII-16 (*Continued*)

Age (yr) Started	Age (yr) Exposure Ends								
	10	20	30	40	50	60	70	80	110
				Exposure Rate = 10.0 (WLM/yr)					
0	0.025	0.038	0.052	0.065	0.079	0.093	0.102	0.106	0.106
10		0.025	0.038	0.052	0.066	0.080	0.090	0.094	0.094
20			0.025	0.039	0.054	0.068	0.078	0.081	0.082
30				0.025	0.040	0.055	0.065	0.069	0.069
40					0.026	0.041	0.052	0.055	0.056
50						0.026	0.037	0.041	0.042
60							0.022	0.026	0.027
				Exposure Rate = 20.0 (WLM/yr)					
0	0.038	0.065	0.090	0.115	0.140	0.164	0.180	0.186	0.187
10		0.038	0.065	0.091	0.117	0.142	0.159	0.165	0.166
20			0.039	0.066	0.093	0.119	0.137	0.144	0.145
30				0.039	0.068	0.095	0.114	0.121	0.122
40					0.041	0.070	0.089	0.096	0.098
50						0.041	0.062	0.069	0.070
60							0.033	0.041	0.042

[a] Estimated with the committee's TSE model (Chapter 2) and a multiplicative interaction between smoking and exposure to radon progeny. Note that R_e includes R_0, the calculated lifetime for unexposed male nonsmokers, 0.0112.

TABLE VII-17 Years of Life Lost, $(L_0 - L_e)$ by Age Started and Age Exposure Ends for Various Rates of Annual Exposure[a] for Male Nonsmokers[b]

Age (yr) Started	Age (yr) Exposure Ends								
	10	20	30	40	50	60	70	80	110
					Exposure Rate = 0.1 (WLM/yr)				
0	0.00	0.00	0.01	0.01	0.01	0.01	0.01	0.01	0.01
10		0.00	0.00	0.01	0.01	0.01	0.01	0.01	0.01
20			0.00	0.00	0.01	0.01	0.01	0.01	0.01
30				0.00	0.00	0.01	0.01	0.01	0.01
40					0.00	0.00	0.00	0.00	0.00
50						0.00	0.00	0.00	0.00
60							0.00	0.00	0.00
					Exposure Rate = 0.2 (WLM/yr)				
0	0.00	0.01	0.01	0.01	0.02	0.02	0.02	0.02	0.02
10		0.00	0.01	0.01	0.02	0.02	0.02	0.02	0.02
20			0.00	0.01	0.01	0.02	0.02	0.02	0.02
30				0.00	0.01	0.01	0.01	0.01	0.01
40					0.00	0.01	0.01	0.01	0.01
50						0.00	0.01	0.01	0.01
60							0.00	0.00	0.00
					Exposure Rate = 0.5 (WLM/yr)				
0	0.01	0.02	0.03	0.04	0.05	0.06	0.06	0.06	0.06
10		0.01	0.02	0.03	0.04	0.05	0.05	0.05	0.05
20			0.01	0.02	0.03	0.04	0.04	0.04	0.04
30				0.01	0.02	0.03	0.03	0.03	0.03
40					0.01	0.02	0.02	0.02	0.02
50						0.01	0.01	0.01	0.01
60							0.00	0.01	0.01
					Exposure Rate = 1.0 (WLM/yr)				
0	0.02	0.04	0.05	0.07	0.09	0.11	0.12	0.12	0.12
10		0.02	0.04	0.06	0.08	0.09	0.10	0.10	0.10
20			0.02	0.04	0.06	0.08	0.08	0.09	0.09
30				0.02	0.04	0.06	0.07	0.07	0.07
40					0.02	0.04	0.05	0.05	0.05
50						0.02	0.03	0.03	0.03
60							0.01	0.01	0.01
					Exposure Rate = 4.0 (WLM/yr)				
0	0.07	0.14	0.22	0.29	0.37	0.44	0.48	0.48	0.48
10		0.07	0.14	0.22	0.30	0.37	0.41	0.41	0.41
20			0.07	0.15	0.23	0.30	0.34	0.34	0.34
30				0.08	0.16	0.23	0.26	0.27	0.27
40					0.08	0.15	0.19	0.20	0.19
50						0.07	0.11	0.11	0.11
60							0.04	0.05	0.04

TABLE VII-17 (*Continued*)

Age (yr) Started	Age (yr) Exposure Ends								
	10	20	30	40	50	60	70	80	110
				Exposure Rate = 10.0 (WLM/yr)					
0	0.18	0.36	0.53	0.72	0.91	1.07	1.16	1.17	1.17
10		0.18	0.36	0.55	0.74	0.91	0.99	1.01	1.00
20			0.18	0.37	0.57	0.74	0.82	0.84	0.84
30				0.19	0.39	0.56	0.65	0.67	0.66
40					0.20	0.37	0.46	0.48	0.48
50						0.17	0.26	0.28	0.28
60							0.09	0.11	0.11
				Exposure Rate = 20.0 (WLM/yr)					
0	0.36	0.70	1.05	1.40	1.77	2.06	2.21	2.24	2.23
10		0.36	0.71	1.07	1.45	1.75	1.90	1.93	1.93
20			0.36	0.74	1.12	1.43	1.59	1.62	1.62
30				0.38	0.78	1.10	1.26	1.30	1.29
40					0.40	0.74	0.90	0.94	0.93
50						0.34	0.52	0.55	0.55
60							0.18	0.22	0.22

[a] Estimated with the committee's TSE model (Chapter 2) and a multiplicative interaction between smoking and exposure to radon progeny.
[b] L_0, the calculated lifetime for unexposed male nonsmokers, is 70.5 yr.

TABLE VII-18 Ratio of Lifetime Risks (R_e/R_0) by Age Started and Age Exposure Ends for Various Rates of Annual Exposure[a] for Female Smokers[b]

Age (yr)	Age (yr) Exposure Ends								
Started	10	20	30	40	50	60	70	80	110
					Exposure Rate = 0.1 (WLM/yr)				
0	1.012	1.024	1.036	1.049	1.062	1.074	1.083	1.087	1.088
10		1.012	1.024	1.037	1.050	1.063	1.071	1.075	1.076
20			1.012	1.025	1.038	1.051	1.059	1.063	1.064
30				1.013	1.026	1.038	1.047	1.051	1.052
40					1.013	1.026	1.034	1.038	1.040
50						1.012	1.021	1.025	1.026
60							1.009	1.013	1.014
					Exposure Rate = 0.2 (WLM/yr)				
0	1.024	1.048	1.072	1.097	1.124	1.149	1.166	1.173	1.176
10		1.024	1.048	1.074	1.100	1.125	1.142	1.150	1.152
20			1.024	1.050	1.076	1.101	1.118	1.126	1.128
30				1.025	1.052	1.077	1.094	1.102	1.104
40					1.027	1.051	1.069	1.076	1.079
50						1.025	1.042	1.050	1.053
60							1.017	1.025	1.028
					Exposure Rate = 0.5 (WLM/yr)				
0	1.060	1.120	1.180	1.242	1.308	1.368	1.410	1.429	1.435
10		1.060	1.120	1.183	1.249	1.310	1.352	1.371	1.377
20			1.061	1.124	1.189	1.251	1.293	1.312	1.318
30				1.063	1.129	1.191	1.234	1.253	1.259
40					1.066	1.128	1.171	1.190	1.197
50						1.062	1.105	1.124	1.131
60							1.043	1.063	1.069
					Exposure Rate = 1.0 (WLM/yr)				
0	1.120	1.239	1.357	1.480	1.608	1.726	1.807	1.843	1.855
10		1.120	1.240	1.364	1.493	1.612	1.694	1.730	1.742
20			1.121	1.246	1.376	1.496	1.579	1.616	1.628
30				1.126	1.258	1.379	1.463	1.500	1.512
40					1.133	1.255	1.340	1.377	1.390
50						1.124	1.209	1.247	1.260
60							1.086	1.125	1.138
					Exposure Rate = 4.0 (WLM/yr)				
0	1.473	1.929	2.372	2.819	3.270	3.673	3.940	4.053	4.088
10		1.473	1.933	2.396	2.864	3.282	3.560	3.678	3.714
20			1.477	1.957	2.442	2.876	3.165	3.289	3.327
30				1.498	2.001	2.452	2.752	2.882	2.922
40					1.523	1.991	2.304	2.440	2.483
50						1.488	1.815	1.957	2.002
60							1.341	1.490	1.538

TABLE VII-18 (*Continued*)

Age (yr) Started	Age (yr) Exposure Ends								
	10	20	30	40	50	60	70	80	110
					Exposure Rate = 10.0 (WLM/yr)				
0	2.150	3.201	4.168	5.091	5.973	6.712	7.166	7.343	7.389
10		2.151	3.210	4.220	5.183	5.991	6.492	6.688	6.742
20			2.160	3.265	4.319	5.204	5.756	5.974	6.035
30				2.211	3.364	4.335	4.944	5.188	5.257
40					2.268	3.338	4.012	4.286	4.365
50						2.185	2.936	3.243	3.334
60							1.833	2.179	2.283
					Exposure Rate = 20.0 (WLM/yr)				
0	3.200	5.036	6.584	7.942	9.129	10.025	10.512	10.675	10.709
10		3.201	5.052	6.667	8.075	9.140	9.727	9.929	9.973
20			3.219	5.146	6.821	8.093	8.802	9.051	9.108
30				3.311	5.313	6.837	7.697	8.005	8.079
40					3.416	5.261	6.313	6.699	6.795
50						3.259	4.561	5.049	5.176
60							2.603	3.216	3.382

[a] Estimated with the committee's TSE model (Chapter 2) and a multiplicative interaction between smoking and exposure to radon progeny.

[b] R_0, the calculated lifetime risk for unexposed female smokers, is 0.0582.

TABLE VII-19 Lifetime Risk (R_e) by Age Started and Age Exposure Ends for Various Rates of Annual Exposure[a] for Female Smokers

Age (yr) Started	Age (yr) Exposure Ends 10	20	30	40	50	60	70	80	110
					Exposure Rate = 0.1 (WLM/yr)				
0	0.059	0.060	0.060	0.061	0.062	0.063	0.063	0.063	0.063
10		0.059	0.060	0.060	0.061	0.062	0.062	0.063	0.063
20			0.059	0.060	0.060	0.061	0.062	0.062	0.062
30				0.059	0.060	0.060	0.061	0.061	0.061
40					0.059	0.060	0.060	0.060	0.061
50						0.059	0.059	0.060	0.060
60							0.059	0.059	0.059
					Exposure Rate = 0.2 (WLM/yr)				
0	0.060	0.061	0.062	0.064	0.065	0.067	0.068	0.068	0.068
10		0.060	0.061	0.062	0.064	0.065	0.066	0.067	0.067
20			0.060	0.061	0.063	0.064	0.065	0.066	0.066
30				0.060	0.061	0.063	0.064	0.064	0.064
40					0.060	0.061	0.062	0.063	0.063
50						0.060	0.061	0.061	0.061
60							0.059	0.060	0.060
					Exposure Rate = 0.5 (WLM/yr)				
0	0.062	0.065	0.069	0.072	0.076	0.080	0.082	0.083	0.084
10		0.062	0.065	0.069	0.073	0.076	0.079	0.080	0.080
20			0.062	0.065	0.069	0.073	0.075	0.076	0.077
30				0.062	0.066	0.069	0.072	0.073	0.073
40					0.062	0.066	0.068	0.069	0.070
50						0.062	0.064	0.065	0.066
60							0.061	0.062	0.062
					Exposure Rate = 1.0 (WLM/yr)				
0	0.065	0.072	0.079	0.086	0.094	0.100	0.105	0.107	0.108
10		0.065	0.072	0.079	0.087	0.094	0.099	0.101	0.101
20			0.065	0.073	0.080	0.087	0.092	0.094	0.095
30				0.066	0.073	0.080	0.085	0.087	0.088
40					0.066	0.073	0.078	0.080	0.081
50						0.065	0.070	0.073	0.073
60							0.063	0.065	0.066
					Exposure Rate = 4.0 (WLM/yr)				
0	0.086	0.112	0.138	0.164	0.190	0.214	0.229	0.236	0.238
10		0.086	0.113	0.139	0.167	0.191	0.207	0.214	0.216
20			0.086	0.114	0.142	0.167	0.184	0.191	0.194
30				0.087	0.116	0.143	0.160	0.168	0.170
40					0.089	0.116	0.134	0.142	0.145
50						0.087	0.106	0.114	0.117
60							0.078	0.087	0.090

TABLE VII-19 (*Continued*)

Age (yr) Started	Age (yr) Exposure Ends								
	10	20	30	40	50	60	70	80	110
					Exposure Rate = 10.0 (WLM/yr)				
0	0.125	0.186	0.243	0.296	0.348	0.391	0.417	0.427	0.430
10		0.125	0.187	0.246	0.302	0.349	0.378	0.389	0.392
20			0.126	0.190	0.251	0.303	0.335	0.348	0.351
30				0.129	0.196	0.252	0.288	0.302	0.306
40					0.132	0.194	0.234	0.249	0.254
50						0.127	0.171	0.189	0.194
60							0.107	0.127	0.133
					Exposure Rate = 20.0 (WLM/yr)				
0	0.186	0.293	0.383	0.462	0.531	0.584	0.612	0.621	0.623
10		0.186	0.294	0.388	0.470	0.532	0.566	0.578	0.580
20			0.187	0.300	0.397	0.471	0.512	0.527	0.530
30				0.193	0.309	0.398	0.448	0.466	0.470
40					0.199	0.306	0.367	0.390	0.395
50						0.190	0.265	0.294	0.301
60							0.151	0.187	0.197

[a] Estimated with the committee's TSE model (Chapter 2) and a multiplicative interaction between smoking and exposure to radon progeny. Note that R_e includes R_0, the calculated lifetime risk for unexposed female smokers, 0.0582.

TABLE VII-20 Years of Life Lost ($L_0 - L_e$) by Age Started and Age Exposure Ends for Various Rates of Annual Exposure[a] for Female Smokers[b]

Age (yr) Started	Age (yr) Exposure Ends								
	10	20	30	40	50	60	70	80	110
				Exposure Rate = 0.1 (WLM/yr)					
0	0.01	0.02	0.03	0.05	0.06	0.07	0.07	0.07	0.07
10		0.01	0.02	0.03	0.05	0.06	0.06	0.06	0.06
20			0.01	0.02	0.04	0.05	0.05	0.05	0.05
30				0.01	0.02	0.03	0.04	0.04	0.04
40					0.01	0.02	0.03	0.03	0.03
50						0.01	0.01	0.01	0.01
60							0.00	0.01	0.00
				Exposure Rate = 0.2 (WLM/yr)					
0	0.02	0.04	0.07	0.09	0.12	0.13	0.14	0.15	0.14
10		0.02	0.04	0.07	0.09	0.11	0.12	0.12	0.12
20			0.02	0.05	0.07	0.09	0.10	0.10	0.10
30				0.02	0.05	0.07	0.08	0.08	0.08
40					0.02	0.04	0.05	0.05	0.05
50						0.02	0.03	0.03	0.03
60							0.01	0.01	0.01
				Exposure Rate = 0.5 (WLM/yr)					
0	0.06	0.11	0.17	0.23	0.29	0.34	0.36	0.36	0.36
10		0.06	0.11	0.17	0.23	0.28	0.30	0.31	0.30
20			0.06	0.12	0.18	0.23	0.25	0.25	0.25
30				0.06	0.12	0.17	0.19	0.20	0.19
40					0.06	0.11	0.13	0.13	0.13
50						0.05	0.07	0.07	0.07
60							0.02	0.03	0.02
				Exposure Rate = 1.0 (WLM/yr)					
0	0.11	0.22	0.33	0.45	0.57	0.67	0.71	0.71	0.71
10		0.11	0.22	0.34	0.47	0.56	0.60	0.61	0.60
20			0.11	0.23	0.36	0.45	0.49	0.50	0.49
30				0.12	0.25	0.34	0.38	0.39	0.38
40					0.12	0.22	0.26	0.27	0.26
50						0.10	0.14	0.15	0.14
60							0.04	0.05	0.04
				Exposure Rate = 4.0 (WLM/yr)					
0	0.44	0.87	1.30	1.75	2.21	2.54	2.69	2.71	2.69
10		0.44	0.88	1.34	1.80	2.15	2.30	2.32	2.30
20			0.45	0.92	1.39	1.74	1.90	1.93	1.90
30				0.49	0.97	1.33	1.48	1.51	1.48
40					0.49	0.86	1.02	1.05	1.02
50						0.38	0.55	0.57	0.54
60							0.17	0.20	0.17

TABLE VII-20 (*Continued*)

Age (yr) Started	Age (yr) Exposure Ends								
	10	20	30	40	50	60	70	80	110
				Exposure Rate = 10.0 (WLM/yr)					
0	1.08	2.10	3.10	4.12	5.10	5.78	6.06	6.11	6.08
10		1.08	2.13	3.21	4.23	4.95	5.25	5.31	5.28
20			1.10	2.24	3.32	4.09	4.41	4.46	4.43
30				1.19	2.34	3.15	3.49	3.55	3.52
40					1.21	2.08	2.45	2.51	2.46
50						0.93	1.33	1.39	1.34
60							0.43	0.50	0.43
				Exposure Rate = 20.0 (WLM/yr)					
0	2.10	4.00	5.75	7.48	9.02	9.97	10.32	10.38	10.37
10		2.10	4.04	5.95	7.65	8.73	9.13	9.20	9.19
20			2.15	4.25	6.14	7.36	7.82	7.90	7.88
30				2.32	4.42	5.80	6.34	6.43	6.40
40					2.35	3.92	4.54	4.65	4.60
50						1.80	2.53	2.65	2.59
60							0.84	0.98	0.89

[a] Estimated with the committee's TSE model (Chapter 2) and a multiplicative interaction between smoking and exposure to radon progeny.
[b] L_0, the calculated lifetime for unexposed female smokers, is 75.9 yr.

TABLE VII-21 Ratio of Lifetime Risk (R_e/R_0) by Age Started and Age Exposure Ends for Various Rates of Annual Exposure[a] for Female Nonsmokers[b]

Age (yr) Started	Age (yr) Exposure Ends								
	10	20	30	40	50	60	70	80	110
	Exposure Rate = 0.1 (WLM/yr)								
0	1.012	1.025	1.037	1.051	1.064	1.077	1.086	1.090	1.092
10		1.012	1.025	1.038	1.052	1.065	1.074	1.078	1.079
20			1.013	1.026	1.039	1.052	1.061	1.065	1.067
30				1.013	1.027	1.040	1.049	1.053	1.054
40					1.014	1.027	1.036	1.040	1.041
50						1.013	1.022	1.026	1.027
60							1.009	1.013	1.014
	Exposure Rate = 0.2 (WLM/yr)								
0	1.025	1.050	1.075	1.101	1.129	1.154	1.172	1.180	1.183
10		1.025	1.050	1.076	1.104	1.129	1.147	1.156	1.158
20			1.025	1.051	1.079	1.105	1.123	1.131	1.134
30				1.026	1.054	1.079	1.097	1.106	1.108
40					1.028	1.053	1.071	1.079	1.082
50						1.026	1.044	1.052	1.055
60							1.018	1.026	1.029
	Exposure Rate = 0.5 (WLM/yr)								
0	1.062	1.124	1.187	1.253	1.321	1.385	1.430	1.451	1.458
10		1.062	1.125	1.190	1.259	1.323	1.368	1.389	1.396
20			1.063	1.128	1.197	1.261	1.306	1.327	1.333
30				1.066	1.134	1.199	1.244	1.264	1.271
40					1.069	1.133	1.178	1.198	1.205
50						1.064	1.109	1.130	1.137
60							1.045	1.065	1.072
	Exposure Rate = 1.0 (WLM/yr)								
0	1.124	1.249	1.374	1.505	1.642	1.770	1.859	1.900	1.914
10		1.124	1.250	1.381	1.518	1.646	1.736	1.776	1.790
20			1.125	1.256	1.394	1.522	1.612	1.652	1.666
30				1.131	1.269	1.397	1.487	1.527	1.541
40					1.138	1.266	1.356	1.397	1.411
50						1.129	1.219	1.259	1.273
60							1.090	1.131	1.145
	Exposure Rate = 4.0 (WLM/yr)								
0	1.497	1.992	2.490	3.008	3.549	4.053	4.404	4.562	4.615
10		1.497	1.996	2.516	3.059	3.565	3.917	4.076	4.130
20			1.501	2.023	2.568	3.076	3.429	3.589	3.643
30				1.523	2.071	2.580	2.936	3.096	3.150
40					1.550	2.061	2.418	2.579	2.633
50						1.514	1.872	2.033	2.089
60							1.360	1.522	1.578

TABLE VII-21 (*Continued*)

Age (yr) Started	Age (yr) Exposure Ends								
	10	20	30	40	50	60	70	80	110
				Exposure Rate = 10.0 (WLM/yr)					
0	2.239	3.467	4.694	5.963	7.282	8.502	9.346	9.722	9.848
10		2.239	3.477	4.758	6.090	7.321	8.173	8.554	8.681
20			2.250	3.542	4.886	6.129	6.990	7.374	7.504
30				2.305	3.661	4.916	5.785	6.174	6.306
40					2.370	3.637	4.515	4.908	5.042
50						2.280	3.168	3.566	3.701
60							1.897	2.300	2.437
				Exposure Rate = 20.0 (WLM/yr)					
0	3.467	5.889	8.286	10.743	13.273	15.586	17.166	17.861	18.088
10		3.467	5.909	8.412	10.989	13.345	14.957	15.667	15.900
20			3.488	6.037	8.661	11.063	12.707	13.432	13.672
30				3.597	6.271	8.718	10.395	11.137	11.383
40					3.725	6.221	7.934	8.693	8.946
50						3.548	5.298	6.076	6.336
60							2.787	3.583	3.851

[a] Estimated with the committee's TSE model (Chapter 2) and a multiplicative interaction between smoking and exposure to radon progeny.
[b] R_0, the calculated lifetime risk for unexposed female nonsmokers, is 0.00602.

TABLE VII-22 Lifetime Risk (R_e) by Age Started and Age Exposure Ends for Various Rates of Annual Exposure[a] for Female Nonsmokers

Age (yr) Started	Age (yr) Exposure Ends								
	10	20	30	40	50	60	70	80	110
				Exposure Rate = 0.1 (WLM/yr)					
0	0.006	0.006	0.006	0.006	0.006	0.006	0.007	0.007	0.007
10		0.006	0.006	0.006	0.006	0.006	0.006	0.006	0.006
20			0.006	0.006	0.006	0.006	0.006	0.006	0.006
30				0.006	0.006	0.006	0.006	0.006	0.006
40					0.006	0.006	0.006	0.006	0.006
50						0.006	0.006	0.006	0.006
60							0.006	0.006	0.006
				Exposure Rate = 0.2 (WLM/yr)					
0	0.006	0.006	0.006	0.007	0.007	0.007	0.007	0.007	0.007
10		0.006	0.006	0.006	0.007	0.007	0.007	0.007	0.007
20			0.006	0.006	0.006	0.007	0.007	0.007	0.007
30				0.006	0.006	0.007	0.007	0.007	0.007
40					0.006	0.006	0.006	0.006	0.007
50						0.006	0.006	0.006	0.006
60							0.006	0.006	0.006
				Exposure Rate = 0.5 (WLM/yr)					
0	0.006	0.007	0.007	0.008	0.008	0.008	0.009	0.009	0.009
10		0.006	0.007	0.007	0.008	0.008	0.008	0.008	0.008
20			0.006	0.007	0.007	0.008	0.008	0.008	0.008
30				0.006	0.007	0.007	0.007	0.008	0.008
40					0.006	0.007	0.007	0.007	0.007
50						0.006	0.007	0.007	0.007
60							0.006	0.006	0.006
				Exposure Rate = 1.0 (WLM/yr)					
0	0.007	0.008	0.008	0.009	0.010	0.011	0.011	0.011	0.012
10		0.007	0.008	0.008	0.009	0.010	0.010	0.011	0.011
20			0.007	0.008	0.008	0.009	0.010	0.010	0.010
30				0.007	0.008	0.008	0.009	0.009	0.009
40					0.007	0.008	0.008	0.008	0.008
50						0.007	0.007	0.008	0.008
60							0.007	0.007	0.007
				Exposure Rate = 4.0 (WLM/yr)					
0	0.009	0.012	0.015	0.018	0.021	0.024	0.027	0.027	0.028
10		0.009	0.012	0.015	0.018	0.021	0.024	0.025	0.025
20			0.009	0.012	0.015	0.019	0.021	0.022	0.022
30				0.009	0.012	0.016	0.018	0.019	0.019
40					0.009	0.012	0.015	0.016	0.016
50						0.009	0.011	0.012	0.013
60							0.008	0.009	0.010

TABLE VII-22 (*Continued*)

Age (yr) Started	Age (yr) Exposure Ends								
	10	20	30	40	50	60	70	80	110
				Exposure Rate = 10.0 (WLM/yr)					
0	0.013	0.021	0.028	0.036	0.044	0.051	0.056	0.059	0.059
10		0.013	0.021	0.029	0.037	0.044	0.049	0.052	0.052
20			0.014	0.021	0.029	0.037	0.042	0.044	0.045
30				0.014	0.022	0.030	0.035	0.037	0.038
40					0.014	0.022	0.027	0.030	0.030
50						0.014	0.019	0.021	0.022
60							0.011	0.014	0.015
				Exposure Rate = 20.0 (WLM/yr)					
0	0.021	0.035	0.050	0.065	0.080	0.094	0.103	0.108	0.109
10		0.021	0.036	0.051	0.066	0.080	0.090	0.094	0.096
20			0.021	0.036	0.052	0.067	0.077	0.081	0.082
30				0.022	0.038	0.052	0.063	0.067	0.069
40					0.022	0.037	0.048	0.052	0.054
50						0.021	0.032	0.037	0.038
60							0.017	0.022	0.023

[a] Estimated with the committee's TSE model (Chapter 2) and a multiplicative interaction between smoking and exposure to radon progeny. Note that R_e includes R_0, the calculated lifetime risk for unexposed female nonsmokers, 0.00602.

TABLE VII-23 Years of Life Lost ($L_0 - L_e$) by Age Started and Age Exposure Ends for Various Rates of Annual Exposure[a] for Female Nonsmokers[b]

Age (yr) Started	Age (yr) Exposure Ends								
	10	20	30	40	50	60	70	80	110
					Exposure Rate = 0.1 (WLM/yr)				
0	0.00	0.00	0.00	0.00	0.01	0.01	0.01	0.01	0.01
10		0.00	0.00	0.00	0.00	0.01	0.01	0.01	0.01
20			0.00	0.00	0.00	0.00	0.01	0.01	0.01
30				0.00	0.00	0.00	0.00	0.00	0.00
40					0.00	0.00	0.00	0.00	0.00
50						0.00	0.00	0.00	0.00
60							0.00	0.00	0.00
					Exposure Rate = 0.2 (WLM/yr)				
0	0.00	0.00	0.01	0.01	0.01	0.01	0.01	0.02	0.01
10		0.00	0.00	0.01	0.01	0.01	0.01	0.01	0.01
20			0.00	0.00	0.01	0.01	0.01	0.01	0.01
30				0.00	0.01	0.01	0.01	0.01	0.01
40					0.00	0.00	0.01	0.01	0.01
50						0.00	0.00	0.00	0.00
60							0.00	0.00	0.00
					Exposure Rate = 0.5 (WLM/yr)				
0	0.01	0.01	0.02	0.02	0.03	0.04	0.04	0.04	0.04
10		0.01	0.01	0.02	0.02	0.03	0.03	0.03	0.03
20			0.01	0.01	0.02	0.02	0.03	0.03	0.03
30				0.01	0.01	0.02	0.02	0.02	0.02
40					0.01	0.01	0.01	0.01	0.01
50						0.00	0.01	0.01	0.01
60							0.00	0.00	0.00
					Exposure Rate = 1.0 (WLM/yr)				
0	0.01	0.02	0.03	0.05	0.06	0.07	0.07	0.08	0.07
10		0.01	0.02	0.04	0.05	0.06	0.06	0.06	0.06
20			0.01	0.02	0.04	0.05	0.05	0.05	0.05
30				0.01	0.03	0.04	0.04	0.04	0.04
40					0.01	0.02	0.03	0.03	0.03
50						0.01	0.01	0.02	0.01
60							0.00	0.01	0.00
					Exposure Rate = 4.0 (WLM/yr)				
0	0.05	0.09	0.14	0.19	0.24	0.28	0.30	0.30	0.30
10		0.05	0.09	0.14	0.19	0.23	0.25	0.25	0.25
20			0.05	0.10	0.15	0.19	0.21	0.21	0.21
30				0.05	0.10	0.14	0.16	0.16	0.16
40					0.05	0.09	0.11	0.11	0.11
50						0.04	0.06	0.06	0.06
60							0.02	0.02	0.02

TABLE VII-23 (*Continued*)

Age (yr) Started	Age (yr) Exposure Ends								
	10	20	30	40	50	60	70	80	110
	Exposure Rate = 10.0 (WLM/yr)								
0	0.11	0.23	0.34	0.47	0.60	0.69	0.74	0.74	0.73
10		0.11	0.23	0.36	0.48	0.58	0.62	0.63	0.62
20			0.12	0.24	0.37	0.47	0.51	0.52	0.51
30				0.13	0.26	0.35	0.40	0.40	0.40
40					0.13	0.23	0.27	0.28	0.27
50						0.10	0.14	0.15	0.14
60							0.05	0.05	0.04
	Exposure Rate = 20.0 (WLM/yr)								
0	0.23	0.45	0.68	0.93	1.17	1.36	1.45	1.46	1.44
10		0.23	0.46	0.71	0.96	1.14	1.23	1.24	1.23
20			0.23	0.48	0.74	0.93	1.01	1.03	1.01
30				0.25	0.51	0.70	0.79	0.80	0.78
40					0.26	0.45	0.54	0.56	0.54
50						0.20	0.29	0.30	0.28
60							0.09	0.11	0.09

[a] Estimated with the committee's TSE model (Chapter 2) and a multiplicative interaction between smoking and exposure to radon progeny.
[b] R_0, the calculated lifetime for unexposed female nonsmokers, is 76.7 yr.

PART 3. Further Considerations

EFFECTS OF CIGARETTE SMOKING ON THE RESPIRATORY TRACT

Cigarette smoking has well-characterized effects at all levels of the respiratory tract (Table VII-24).[36] Changes in the airways are most relevant for respiratory carcinogenesis. Cigarette smoking produces mucus gland hypertrophy and hyperplasia in the large airways and stimulates mucus production from goblet cells in the small airways. The clinical counterpart of these changes is chronic bronchitis, defined as regular sputum production. The bronchial epithelium develops dysplastic and metaplastic changes in smokers. Certain physiological changes accompany these structural abnormalities. Mucociliary clearance, which removes gases and particles from the large airways, is slowed in cigarette smokers. Increased permeability may facilitate passage of inhaled agents across the epithelium.

Proportionately greater central deposition of particles has been demonstrated in the airways of smokers, in comparison with nonsmokers. This deposition pattern may be a consequence of the abnormal small airway function commonly found in smokers. Impaired lung function can be demonstrated in many smokers, and perhaps 10 to 15% of sustained smokers develop disabling chronic airflow obstruction. The resulting physiological impairment leads to an increased respiratory rate for any particular level of activity.

TABLE VII-24 Histologic and Physiologic Changes in the Respiratory Tract, Other than Malignancy, Associated with Cigarette Smoking[36]

Large airways	Mucous gland hypertrophy and hyperplasia
	Dysplasia and metaplasia of epithelial cells
	Increased epithelial permeability
	Impaired mucociliary transport
	Inflammation
Small airways	Goblet cell metaplasia
	Epithelial cell metaplasia
	Increased mucus production
	Inflammation
Lung parenchyma	Fibrosis
	Increased cell numbers
	Altered cell populations
	Altered function of some cells
	Emphysema

In assessing the consequences of combined exposure to cigarette smoke and radon daughters, consideration must be given to these diverse effects of smoking (Table VII-24), as well as to interaction between the two agents in the process of carcinogenesis itself. Smoking-related changes in the lung's structure and function might alter the dose to target cells at any particular level of exposure. In comparison with nonsmokers, dose might be increased in smokers by the greater central deposition, the increased airways permeability, and the slowed mucociliary transport. Dose might be reduced in smokers by mucosal edema and the increased average mucus thickness due to the heightened mucus production in the airways of smokers. A conclusion concerning the net effect of these smoking-related changes on the dosimetry of radon daughters cannot be reached at present. Nevertheless, the effect of radon daughters in the presence of smoking must be interpreted in the context of the changes in lung structure and function, which can be readily demonstrated in many smokers.[36]

In this regard, several pulmonary disease processes resulting from cigarette smoking have been associated with increased lung-cancer risk: chronic bronchitis and chronic obstructive pulmonary disease. By epidemiological convention, chronic bronchitis refers to chronic sputum production. Clinical diagnosis of chronic obstructive pulmonary disease occurs in patients with disabling and irreversible airflow obstruction. At times, clinical diagnoses such as chronic bronchitis, emphysema, and chronic obstructive pulmonary disease may be applied to persons with irreversible airflow obstruction, regardless of other features.

Nevertheless, epidemiological studies show that these diagnoses are associated with increased risk of lung cancer, even with adjustment for cigarette smoking. In an early case-control study, Doll and Hill[13] found that lung-cancer cases yield a history of chronic bronchitis significantly more often than controls. In two subsequent case-control studies, diagnostic terms applied to patients with chronic airflow obstruction were also associated with lung cancer, even with control for cigarette smoking.[33,37] Davis[12] showed that the incidence of lung cancer in patients with chronic obstructive pulmonary disease was higher than expected in comparison with rates in smokers.

Two studies have demonstrated that mucus hypersecretion, as ascertained by a questionnaire, predicts increased lung-cancer occurrence. Rimington[29] determined lung-cancer incidence in male participants who had given information on their smoking habits and sputum production for a radiological screening program. In all categories of cigarette smoking, lung-cancer incidence was higher in those with a history of daily sputum production for 5 yr at the time of enrollment. Peto et al.[25] examined mortality of 2,518 British men during a 20- to 25-yr follow-up period.

Lung-cancer mortality was higher in those with a lower level of lung function and in those with chronic sputum production. The latter association persisted after adjustment for lung-function level and cigarette smoking. The finding of increased lung cancer in persons with underlying respiratory disease and mucus hypersecretion conflicts with the hypothesis that increased mucus production reduces penetration of alpha particles into the tracheobronchial epithelium and thus protects against cellular damage.[3]

THE ASSOCIATION BETWEEN LUNG CANCER, SMOKING, AND RADIATION

Exposure to radon progeny and cigarette consumption are each associated with lung cancer in a complex way. Because there are only a few studies on the combined effects of radiation exposure and tobacco smoke, the amount of information for their interaction is limited. The committee's analyses described in Chapter 2 and Annex 2A show that cancer risk associated with exposure to radon progeny depends on cumulative dose, age, and time since exposure. The actual biological relationship is undoubtedly more complex than the statistical model that the committee has developed and may be influenced by other factors that cannot be fully evaluated with the available data. These factors might include age at first exposure, dose rate, sex, diet, and genetic predisposition. Moreover, the association of tobacco consumption with lung cancer is also complex and depends on duration and number of cigarettes smoked per day, type of tobacco product, method of inhalation, and years since cessation of use for former smokers.[35] Assessment of the combined effects of cigarette smoking and radon progeny should account for the individual patterns of effect from both insults. Other aspects of the combined exposure may also be important, for example, the effect of the sequencing of exposures and the degree of their overlap in time.

In contrast, the studies of combined exposures, reported in the literature or analyzed by this committee in Part 2 of this appendix, have usually considered only cumulative WLM (or duration of employment or other surrogate) and duration or intensity of cigarette use, and not the effects of the other variables described above. Such assessments of the underlying relationship may be distorted by not accounting for other predictors of risk. Nevertheless, risk models are a useful method for describing patterns in the different data sets. With these complexities in mind, the data currently available on radon daughters and tobacco exposure suggest that risks do not combine additively on the relative-risk scale. Although there is great uncertainty regarding the relative impact of the two exposures, the multiplicative model appears to have greater support in the literature. The analyses by this committee suggest that a submultiplicative model should not be dismissed and may provide a more accurate description of the underlying relationship.

A clear pattern of risk among studies of miner's exposure to radon and tobacco smoke has not yet emerged. A few small studies have shown mixed results, while the largest study of the issue by Whittemore and McMillan[38] indicates a multiplicative interaction. While the committee's analyses of the Colorado Plateau uranium miners in Part 2 of this appendix support this conclusion, the analyses also support submultiplicative and supramultiplicative relationships.

The committee's analysis of the Japanese atomic-bomb survivor data shows that for these data, neither an additive nor a multiplicative model can be rejected on statistical grounds; indeed, their maximum likelihoods are nearly identical. This is consistent with the results of Prentice et al.[26] In summary, the atomic-bomb survivor data appear amenable to either a multiplicative or additive model for the relative risk. The most recent case-control study by Blot et al.[6] based on a large number of lung-cancer cases sustains this interpretation. The relevance of these studies of atomic-bomb survivors to the interaction of radon and smoking in their relationship to lung-cancer induction, however, must still be determined.

Our review suggests that this issue has yet to be resolved. Areas for further study that are needed to clarify the combined effect of these two exposures include the following:

- the impact of smoking rate (cigarettes per day) and smoking duration, as opposed to rate and/or the combined pack-years, on the radiation association with lung cancer;
- implications of low- versus high-LET radiation;
- the role of smoking cessation on the effect of radiation-associated lung cancer;
- the effect on interactions of tobacco use before and after radiation exposure;
- the role of cigarette use on the histological distribution of radiation-associated lung cancer;
- the relationship of smoking to other measures of radiation exposure, for example, working-level rate, cumulative WLM, and duration of exposure; and
- the role of other agents associated with lung diseases, such as asbestos, silica, and arsenic.

REFERENCES

1. Archer, V. E., J. K. Wagoner, and F. E. Lundin. 1973. Uranium mining and cigarette smoking effects on man. J. Occup. Med. 15:204–211.

2. Axelson, O. 1983. Experiences and concerns on lung cancer and radon daughter exposure in mines and dwellings in Sweden. Z. Erkrank. Atm.-Org. 161:232–239.
3. Axelson, O., and L. Sundell. 1978. Mining lung cancer and smoking. Scand. J. Work Environ. Health 4:46–52.
4. Band, P., M. Feldstein, G. Saccomanno, L. Watson, and O. King. 1980. Potentation of cigarette smoking and radiation: Evidence from a sputum cytology survey among uranium miners and controls. Cancer 45:1273–1277.
5. Beebe, W. G., and M. Usagawa. 1968. The major ABCC samples. Pp. 12–68 in Atomic Bomb Casualty Commission Technical Report. Hiroshima, Japan: Atomic Bomb Casualty Commission.
6. Blot, W. J., S. Akiba, and H. Kato. 1984. Ionizing radiation and lung cancer: A review including preliminary results from a case-control study among A-bomb survivors. In Atomic Bomb Survivor Data: Utilization and Analysis, R. L. Prentice and D. J. Thompson, eds. Philadelphia: Society for Industrial and Applied Mathematics.
7. Breslow, N. E., and B. E. Storer. 1985. General relative risk functions for case control studies. Am. J. Epidemiol. 122:149–162.
8. Cederlof, R., L. Friberg, Z. Hrubec, and U. Lorich. 1975. The relationship of smoking and some covariables to mortality and cancer morbidity. Report of the Department of Environmental Hygiene. Stockholm: Karolinska Institute.
9. Cox, D. R. 1972. Regression models and life tables (with discussion). J. R. Stat. Soc. Ser. B 34:187–220.
10. Damber, L., and L. G. Larsson. 1982. Combined effects of mining and smoking in the causation of lung carcinoma. Acta Radiol. Oncol. 21:305–313.
11. Damber, L., and L. G. Larsson. 1985. Underground mining, smoking, and lung cancer: A case-control study in the iron ore municipalities in Northern Sweden. J. Natl. Cancer Inst. 74:1207–1213.
12. Davis, A. L. 1976. Bronchogenic carcinoma in chronic obstructive pulmonary disease. J. Am. Med. Assoc. 235:612–622.
13. Doll, R., and A. B. Hill. 1952. A study of the aetiology of carcinoma of the lung. Br. Med. J. 2:1271–1286.
14. Edling, C., H. Kling, and O. Axelson. 1984. Radon in homes—a possible cause of lung cancer. Scand. J. Work Environ. Health 10:25–34.
15. Hornung, R. W., and S. Samuels. 1981. Survivorship models for lung cancer mortality in uranium miners—is cumulative dose an appropriate measure of exposure? Pp. 363–368 in Radiation Hazards in Mining: Control, Measurement and Medical Aspects, M. Gomez, ed. New York: Society of Mining Engineers of the American Institute of Mining Metallurgical and Petroleum Engineers.
16. Ishimara, T., R. W. Cihak, C. E. Land, A. Steer, and A. Yamada. 1975. Lung cancer at autopsy in A-bomb survivors and controls, Hiroshima and Nagasaki, 1961–1970. II. Smoking, occupation and A-bomb exposure. Cancer 36:1723–1728.
17. Jorgensen, H. S. 1973. A study of mortality from lung cancer among miners in Kiruna 1950–1970. Scand. J. Work. Environ. Health 10:126–133.
18. Kato, H., K. G. Johnson, and K. Yano. 1966. Mail survey of cardiovascular disease study, Hiroshima and Nagasaki. Pp. 19–66 in Atomic Bomb Casualty Commission Technical Report. Hiroshima, Japan: Atomic Bomb Casualty Commission.
19. Kopecky, K. J., T. Yamamoto, T. Fujikura, S. Tokuoka, T. Monzen, I. Nishimari, E. Nakashima, and H. Kato. In press. Lung cancer, radiation exposure and smoking among A-bomb survivors, Hiroshima and Nagasaki, 1950–1980. In press. Techincal Report. Hiroshima, Japan: Radiation Effects Research Foundation.

20. Liddell, F. D. K., J. C. McDonald, and D. C. Thomas. 1977. Methods of cohort analysis: Appraisals by application to asbestos mining. J. R. Stat. Soc. Ser. A 140:469–491.
21. Lubin, J. H., and M. H. Gail. 1984. Biased selection of controls for case-control analyses of cohort studies. Biometrics 40:63–75.
22. Lundin, F. E., Jr., V. E. Archer, and J. K. Wagoner. 1979. An exposure-time-response model for lung cancer mortality in uranium miners—effects of radiation exposure, age and cigarette smoking. Pp. 243–264 in Proceedings of the Work Group at the Second Conference of the Society for Industrial and Applied Mathematics, N. E. Breslow, and A. Whittemore, eds.
23. Morgan, M.V. and J. M. Samet. 1986. Randon daughter exposures of New Mexico U miners, 1967–1982. Health Phys. 50:656–662.
24. National Institute for Occupational Safety and Health (NIOSH). 1986. Evaluation of Epidemiologic Studies Examining the Lung Cancer Mortality of Underground Miners. Division of Standards Development and Technology Transfer. Cincinnati, Ohio: Centers for Disease Control, National Institute for Occupational Safety and Health.
25. Peto, R., F. E. Speizer, A. L. Cochrane, F. Moore, C. M. Fletcher, C. M. Tinker, I. T. Higgins, R. G. Gray, S. M. Richards, J. Gilliland and B. Norman-Smith. 1983. The relevance in adults of air-flow obstruction, but not mucus hypersecretion to mortality from chronic lung disease: Results from 20 years of prospective observation. Am. Rev. Respir. Dis. 128:491–501.
26. Prentice, R. L., Y. Yoshimoto, and M. Mason. 1983. Relationship of cigarette smoking and radiation exposure to cancer mortality in Hiroshima and Nagasaki. J. Natl. Cancer Inst. 70:611–622.
27. Radford, E. P., and K. G. St. Clair Renard. 1984. Lung cancer in Swedish iron ore miners exposed to low doses of radon daughters. N. Engl. J. Med. 310(23):1485–1494.
28. Renard, K. G. St. Clair. 1974. Respiratory cancer mortality in an iron ore mine in Northern Sweden. Ambio. 3:67–69.
29. Rimington, J. 1971. Smoking chronic bronchitis and lung cancer. Br. Med. J. 2:373–375.
30. Rokvanor i Sverige. 1965. Report of the Central Statistical Bureau. Stockholm: Central Statistical Bureau.
31. Saccomanno, G., V. E. Archer, O. Auerbach, R. P. Saunders, and L. M. Brennan. 1974. Development of carcinoma of the lung as reflected in exfoliated cells. Cancer 33:256–270.
32. Saccomanno, G., C. Yale, W. Dixon, O. Auerbach, and G. C. Huth. 1986. An epidemiological analysis of the relationship between exposure to Rn progeny, smoking and bronchogenic carcinoma in the U-mining population of the Colorado Plateau—1960–1980. Health Phys. 50:605–618.
33. Samet J. M., C. G. Humble, and D. R. Pathok. 1986. Personal and family history of respiratory disease and lung cancer risk. Am. Rev. Respir. Dis. 134:466–470.
34. Thomas, D. C. 1981. General relative risk models for survival time in matched case-control analysis. Biometrics 37:673–686.
35. U.S. Department of Health and Human Services, U.S. Public Health Services Office of Smoking and Health. 1972. The Health Consequences of Smoking: Cancer. A report to the Surgeon General. Department of Health and Human Services (PHS) 82-50179 Washington D.C.: U.S. Government Printing Office.
36. U.S. Department of Health and Human Services, U.S. Public Health Service, Office of Smoking and Health. 1985. The Health Consequences of Smoking:

Chronic Obstructive Lung Disease. A Report to the Surgeon General. PHS 85-50207. Washington, D.C.: U.S. Government Printing Office.

37. Van der Wal, A.M., E. Huizingav, N.G. Orie, H.S. Sliuter, and K.de Vries. 1966. Cancer and chronic nonspecific lung disease (C.N.S.L.D.). Scand. J. Respir. Dis. 47:161–172.

38. Whittemore, A. S., and A. McMillan. 1983. Lung cancer mortality among U.S. uranium miners: A reappraisal. J. Natl. Cancer Inst. 71:489–499.

APPENDIX VIII

Previous Estimates of the Risk Due to Radon Progeny

Several expert groups and individual investigators have published estimates of the risk associated with exposure to radon progeny. In this appendix the committee examines some of the more widely cited studies both for their underlying assumptions and for the numerical value of the estimated risk.

Like the committee's lifetime risk estimates developed in Chapter 2, two steps are usually involved in estimating the risks from radon exposure: the development of an appropriate risk coefficient from epidemiological studies, and the projection of risks over a defined exposure and follow-up periods. Table VIII-1 lists risk coefficients developed in a number of epidemiological studies. Two types of risk coefficients are shown; those for absolute excess risk, the number of cases per person-years at risk per working-level month (WLM), and the excess relative risk, the proportional increase per 100 WLM. Estimates from Annex 2A, using a constant relative risk model are included in Table VIII-1 in cases in which the same cohorts were considered by this committee. Except for the Malmberget miners, the results of the Poisson regressions for internal and external controls used in Annex 2 are not too different from those obtained by other investigators using standardized mortality ratios.

As important as the risk coefficients are in estimating the risks associated with radon exposures, the assumptions in the projection models often have a larger numerical impact. The committee examines these assumptions for particular studies in the following sections.

TABLE VIII-1 Published Risk Coefficients for Exposure of Underground Miners to Radon Progeny

Cohort	Study	Attributable Excess Risk Deaths/10^6 Person-Year at Risk/WLM	Excess Relative Risk/100 WLM	Basis of Risk Estimate
Colorado Plateau	BEIR III[11]	3.5	0.45	Group average, 0–3,719 WLM
	Whittemore and McMillan[19]	6.0	0.8	Group Average, 0–360 WLM
			0.31	Regression, nonsmoker
			1.44	Regression, 20-pack year smoker
	NIOSH[4]		1.1	Proportional hazard regression at 120 WLM
	Annex 2A		0.6–0.6	Regression on exposures, $<2{,}000$ WLM[a]
Czechoslovakia	BEIR III[11]	19	1.8	Group average 0–300 WLM
Ontario	Muller[9]	7.2	1.3	Regression, standard WLM
		2.8	0.51	Regression, special WLM[b]
	Annex 2A		1.4–1.2	Regression, standard WLM[a]
Beaverlodge	Howe[3]	20.9	3.28	Regression
	Annex 2A		2.6–2.6	Regression[a]
Malmberget	Radford and Renard[12]	21	—	Group average for smokers
	Annex 2A	16.9	—	Group average for nonsmokers
		19	3.6	Group average
			1.4–1.6	Regression[a]
Newfoundland	BEIR III[11]	17.7	8.0	Group average
	Morrison et al.[7]	5.6	—	Regression

[a]Tables 2A-2 and 2A-3 for internal and external controls, respectively, using a constant relative risk model, not the time since exposure model recommended by this Committee.
[b]Maximum estimated exposure, see Appendix IV.

NCRP REPORT 78

Risk estimation in a 1984 report[10] by the National Council on Radiation Protection and Measurements (NCRP) relies on the Harley and Pasternack Model B of lung-cancer excess due to radon progeny.[2] The following assumptions formed the basis of the model.

- Following a latent period, the tumor rate is an exponentially decreasing function of the time since exposure.
- Disease rate excess associated with a single exposure increases with age at exposure.
- Lung cancer is rare before the age of 40 yr.
- Median age at lung cancer among miners is about 60 yr in nonsmokers and 50 yr or older in smokers.
- The minimal time for tumor growth, from initial cell transformation to clinical detection, is 5 yr.

From these postulated disease patterns, the Harley and Pasternack model specifies a 5-yr latent period for persons first exposed at the age of 35 yr or older and a $(40 - u)$ yr latent period for persons under the age of 35 yr, where u is age at first exposure. For a single annual exposure at age u, the excess radiation-associated risk above background at age $t > u$ (and $t > 40$) is taken to be

$$A(t, u) = Re^{-m(t-u)}S(t)/S(u),$$

where R is the attributable-risk coefficient per WLM, $S(t)$ and $S(u)$ are the probabilities of survival to the designated age, and m is the rate of removal of transformed stem cells due to repair or cell death. For risk projection, the NCRP task group fixed $m = \ln(2)/20\ \text{yr}^{-1}$, corresponding to a 20-yr half-life. For ages within the latent period or before initial exposure, the excess risk is zero. The exponential term allows for the excess risk to decline with time following exposure, and the survival ratio adjusts for competing causes of mortality. Given the parameters of this model, one integrates over t from age 40 to maximal assumed life (age 85) to obtain lifetime risk due to the single exposure at age u, or over years of exposure, $u_1, \ldots, u_n$, to obtain the excess risk at t due to all previous exposure. Lifetime excess risk from all exposures is the integral over t and u.

This model is extremely important, in that it postulates a modified effect with time since exposure. In this way, it is related to the TSE model recommended in Chapter 2 of this report and the latency models of Lundin et al.[6] and Thomas and McNeill,[16] all of which contrast with a relative-risk model constant in age at risk. Indeed, the distinction between a constant-relative-risk model and models that modify risk according to time since exposure is more fundamental than discrimination among the latter types, which offer refinements in basically similar models.

The analysis presented in Annex 2A clearly suggests that risk effects are modulated by time since exposure. This is manifest in the declining parameter estimates of impact of exposures more distant in time. Therefore, the distinction between the Harley and Pasternack model and a relative-risk model that declines with time since exposure is related to the rate of decline in the relative risk. In light of the complexity of risk arising from chronic radiation exposure, substantial data would be required for an adequate evaluation of such subtle patterns of risk. An informal method of considering this issue is to examine additive excess risk after cessation of exposure. This committee's analysis indicated that the relative risk declines with time since cessation of exposure. However, the NCRP risk model requires that this decline be large enough for the attributable risk to decrease.

To test this hypothesis, data on observed and expected cancers and person-years of exposure from the four miner cohorts analyzed in Annex 2A were categorized by age, age at last exposure, and cumulative WLM. Figure VIII-1 presents for each of the four data sets age-specific attributable risks, (observed − expected)/person-years, for three age-at-last-exposure groups. In the figures, the excess risks were smoothed by graphing the mean of the observed excess and two adjacent values and weighting by the inverse variance. Data from these four worker populations do not show a consistent pattern of declining excess risk. In several cohorts, the excess risks generally increase; in others, the excess declines, but only 20 yr or more after the mean age at last exposure. The NCRP model would predict a declining excess shortly after cessation of exposure.

Patterns similar to those shown in Figure VIII-1 were observed after stratification by two categories of cumulative WLM. In addition, Poisson regression models were fit to the observed risk, where the attributable disease rate was postulated to be linear in age at last exposure and cumulative WLM. For each data set, after adjustment for WLM and age at last exposure, there was no significant improvement in model fit with the inclusion of age at risk. Parameter estimates for five age categories tended to increase, as suggested by Figure VIII-1. However, this effect is poorly estimated. Model fit did not improve significantly with inclusion of a continuous age variate, although the coefficients were generally positive.

A difficulty in the application of the NCRP model is the choice of m, the rate of removal of the transformed cell. Harley and Pasternack acknowledge the issue and select a 20-yr half-life as "representative for extrapolation," although they cite no formal data analysis or experimental results. Additional work in this area would be beneficial for refining the model.

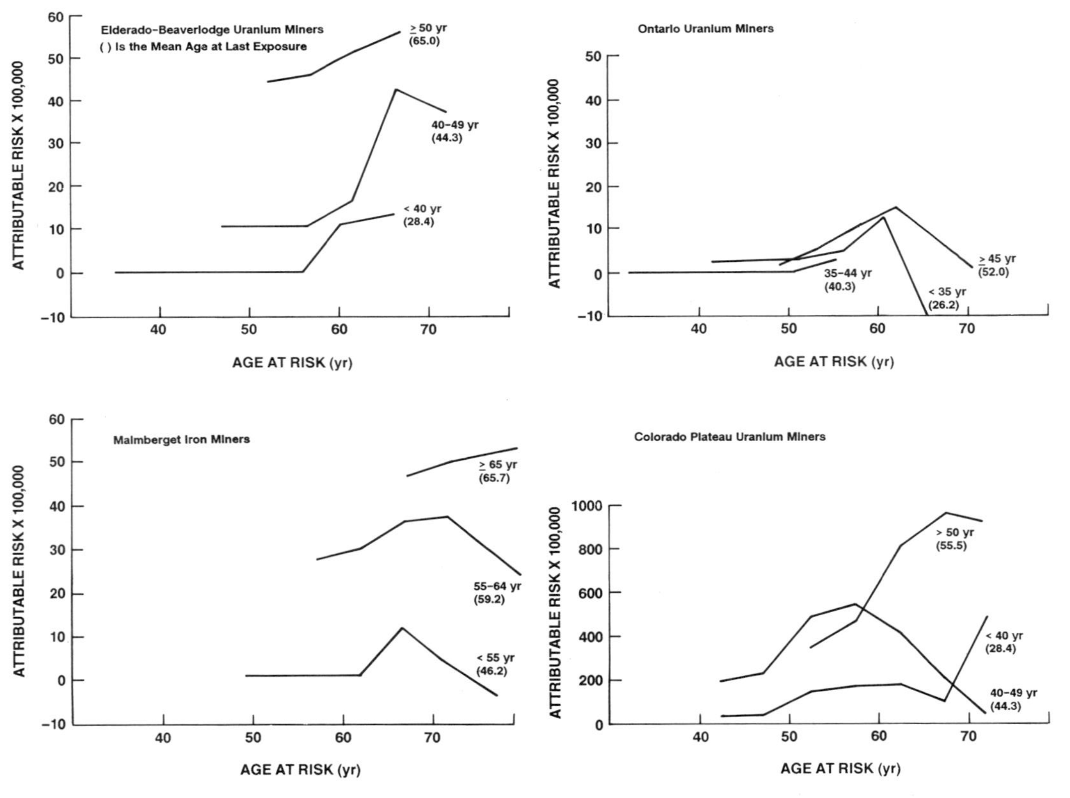

FIGURE VIII-1 Attributable risk of lung cancer at various ages of last exposure as a function of age at risk in four groups of underground miners. The mean age at least exposure is shown in parentheses.

Although it does not have much impact on NCRP lifetime risk estimates, their model limits the occurrence of radiation-induced lung cancer to the age of 40 yr and over, a restriction for which no biological mechanism is readily apparent. In contrast, several studies have observed lung cancers under the age of 40.[7,15] The failure to observe lung cancer in young persons in several other studies could be due to the very low background rates and few person-years. For example, Radford and Renard[12] reported that the mean age at first exposure of the Swedish miners was 28 yr. With a 5-yr latent period, 1,415 miners would accrue a maximum of some 10,000 person-yr by the age of 40, producing 0.5 expected cases if the population lung-cancer mortality rate for ages 35–39 were 5.1×10^{-5}. With this expected value, there is a 0.6 probability that no cases will occur before the age of 40.

A methodological issue concerns the manner in which the attributable risk is adjusted for competing causes of death. As defined, $S(t)$ is the probability that a person who is subject to disease rates of the standard population will survive to age t. For a 1-yr exposure at age $u < t$, the competing-cause adjustment $S(t)/S(u)$—which is the probability of survival of someone in the standard population to t, given survival to age u—does not incorporate the increased lung-cancer risk, and thus decreased survival, of someone exposed. This adjustment error is compounded as risk is integrated over age t and over yearly exposures but is unlikely to be important except at high dose rates.

BEIR III REPORT

The National Research Council's Committee on the Biological Effects of Ionizing Radiations (BEIR III)[11] assumed a linear relationship between exposure in WLM and the additive excess risk of lung cancer. The excess risk was estimated to vary with age at diagnosis, as shown in Table VIII-2. In addition to the minimal age at expression (similar to that in the NCRP model), a minimal latent period of 15–20 yr (for those exposed at age 15–34) or 10 yr (for those exposed above age 34) is assumed. Later risk is independent of latent period.

These risk values were based on the combined estimates from the epidemiology studies of U.S.[6] and Czechoslovakian[13] uranium miners, Swedish iron miners,[12] and Newfoundland fluorspar miners.[7] The techniques for combining the epidemiological data were not described and so cannot be evaluated. It appears that substantial weight was given to the results from the Swedish, Newfoundland, and Czechoslovakian miner surveys. The Colorado Plateau uranium miners had much lower lung-cancer risks, which the committee thought was due to their high dose rates. The Swedish metal miners had a higher risk, even with less prevalent cigarette smoking;

TABLE VIII-2 Excess Risk Estimated To Vary with Age and Diagnosis

Age (yr) at Diagnosis	Excess Cases (per 10^6 person-years at risk per WLM)
<35	0
35–49	10
50–65	20
>65	50

that difference was attributed to longer follow-up. No data are available to indicate whether these risk estimates apply to childhood irradiation.

The BEIR III report[11] discussed, but did not resolve, the effect of cigarette smoking on these radiation risks. The BEIR III report states that if the two exposures are additive, their risk estimates would apply to both smokers and nonsmokers. But if there is a multiplicative interaction (i.e., the lung-cancer risk estimates due to radiation are proportional to the smoking-specific rates), the estimates should be increased by 50% for smokers and reduced by a factor of 6 for nonsmokers.

REPORT OF THOMAS AND MCNEILL

The report of Thomas and McNeill and co-workers[16,17] reviewed epidemiological and animal data on lung cancer, bone and head sarcomas, and some other cancers, with an emphasis on lung cancer from radon progeny. To develop risk estimates, the authors considered data from the Czechoslovakian, Ontarian, and Colorado Plateau uranium miners; the Newfoundland fluorspar miners; the Swedish metal miners; and (for inferences regarding the shape of the dose-response curve, but not the magnitude of risk) the Japanese atomic-bomb survivors. Animal data were used primarily to investigate the effect of modifying factors, as opposed to estimation of magnitude of risk.

The comprehensive report reached qualitative and quantitative conclusions largely in accord with those in Chapter 2. Thomas and McNeill discussed at length the epidemiological and statistical principles underlying selection of a risk model (i.e., relative risk versus additive excess risk), the shape of the dose-response curve, and the role of modifying and confounding factors. We support and have repeated their approach of formally combining evidence from various cohorts. This committee concurs with their argument that simply comparing risk estimates from different cohorts in relation to average exposure of the cohorts is not suitable for studying the shape of the dose-response curve. Thomas and McNeill used a more statistically sound method; that is, they fit a single model to a combination of data sets. They allowed the degree of risk to vary among studies, so that they could adjust for varied confounding factors, but incorporated parameters common to the data sets to model nonlinearities in dose-response relationships. The primary limitation of this analysis, as acknowledged, was the very limited form of the data that could be extracted from published reports concerning the various cohorts.

Thomas and McNeill adopted a model with the relative-risk constant in age and, tentatively, linear in cumulative exposure, except at very high values. Their analysis indicated an estimated value of 2.28/100 WLM for the excess relative risk. In selecting this estimate, they discounted a substantially lower risk among the Colorado Plateau miners; and, to some extent, by using a cell-killing model, they compensated for the lower risks per unit exposure at very high levels of cumulative exposure. Inclusion of an exponential term to represent cell killing resulted in a final model that was nonlinear in dose; however, the decrease in slope caused by this cell-killing term was important only at very high doses. However, this allowance for a decrease in slope at very high exposures was statistically significant. They also considered models in which excess relative risk was proportional to an estimated power of dose; such models provide for a more general nonlinearity in dose. The fitted model, although not providing a statistically significant improvement over a simple linear model, resulted in a convex dose-response function, that is, a generally (but only slightly) decreasing slope of the response with increasing cumulative exposure. As noted above, however, they felt that a linear dose-response relationship at moderate to low doses was adequate for extrapolation, with data from very high doses discounted via the cell-killing model. Their interpretation of the possible curvilinearity was primarily that one should be less confident that low-dose extrapolations are conservative than in the case of low linear energy transfer (LET) radiation, where the curvilinearity is generally held to be of the opposite type (slope increasing with dose).

Although we emphasize that their conclusions are in accord with those drawn in this report, we believe that the adoption of a constant-relative-risk model at all ages for the effect of radon daughters is not well supported. The data available to Thomas and McNeill on this issue were sparse. The most relevant evidence was presented in Section 7.2.1 of their 1982 report,[16] where they argued that, with the meager data available, the additive excess risk increases substantially with age, at a given dose, whereas the relative risk is more stable. In Section 4.2.1.3 of the same report,[16] they attempted to discriminate between the "attributable-risk" (i.e., excess-risk) and relative-risk models, solely on the basis of the total (or average) risk over age (and time). This attempt may have been inappropriate, because information on age-specific risks was not available to Thomas and McNeill.

The present committee was fortunate to have access to much more detailed data on some populations and can confirm to some extent the conclusions drawn by Thomas and McNeill. Their average risk coefficient (2.28/100 WLM) is not very different from that found by this committee (1.5/100 WLM) using external controls and constant-relative-risk model. The difference is largely due to their exclusion of results from the Colorado Plateau cohort, which the committee's analysis includes.

The tentative conclusion of Thomas and McNeill regarding the linearity of the dose-response relationship was supported by the data available to them. We agree with the statistical approach that they used, and for two reasons concur with the tentativeness of their conclusion as to the shape of the dose-response curve. First, at very large doses, there is a suggestion of nonlinearity in specific cohorts, although it is not consistent enough among all the cohorts to be statistically significant. More important, there cannot be enough evidence from epidemiological studies to ascertain the effects at low doses.

On the critical issue of the interaction of cigarette smoking and radiation effects, Thomas and McNeill concluded[16] that the joint effect seemed to be "intermediate between additive and multiplicative, although on balance [they] would favor the multiplicative model." The evidence for this was moderately weak, inasmuch as the effects of other modifying factors—such as age at exposure, exposure rate, and time since cessation of exposure—were not controlled.

In conclusion, the reports of Thomas and McNeill[16,17] provide a strong discussion of principles and methods, but are limited by the data available to them. The present report is complementary in its approach, but more data were accessible to the committee. These were the data from the four cohorts in Eldorado-Beaverlodge, Ontario, Colorado, and Sweden described in Annex 2A. Although we disagree with the claim made in Thomas and McNeill's Appendix J[16] that grouping of doses tends to result in underestimation of risks, the general consistency of conclusions, both qualitative and quantitative, between the two reports is notable.

1981 REPORT OF EVANS ET AL.

In a brief report in the journal *Nature* in 1981, Evans et al.[1] provided an upper bound to the lifetime lung-cancer risk associated with radon-daughter exposure in the general population. The report originated in an international workshop on radiation protection principles for naturally occurring radionuclides. The authors primarily considered the epidemiological evidence in determining the risks of environmental radon. They cited a range of lifetime attributable-risk coefficients, developed by other authors, of 21–54 to 1,000 deaths/10^6 WLM. In their collective judgment, the "most defensible upper bound of the lifetime risk to the general population is 100 lung cancer deaths per 10^6 WLM." This coefficient reflects a reduction in unit exposure for the general population, in comparison with miners, because of differing exposure conditions, smoking habits, and age and sex distributions of the two populations.

Evans et al. acknowledged the informality of their approach for determining a risk coefficient for the general population. They did not use models directly, either to derive a risk coefficient from the miner data or to extrapolate from miners to the general population. They also assumed an attributable-risk model and did not specifically address the effects of cigarette smoking.

1977 UNSCEAR REPORT

The 1977 report of the U.N. Scientific Committee on the Effects of Atomic Radiation (UNSCEAR)[18] provided an attributable-risk coefficient for lung-cancer incidence of 200–450/10^6 WLM, which described a full, for example, 40-yr, expression of the carcinogenic effect on lung tissue of radon and of its daughter products. The report reviewed data from American uranium miners, Swedish underground miners, Newfoundland fluorspar miners, iron-ore miners in the United Kingdom, and Czechoslovakian uranium miners. The upper bound of the attributable-risk range was clearly derived from analysis of the Czechoslovakian data; the derivation of the lower limit is unclear, although the Swedish data reported by Snihs[14] apparently were considered. The Colorado Plateau data do not appear to have been used in setting the range.

The UNSCEAR report emphasized the Czechoslovakian study, because of long latency after the onset of exposure and the availability of appropriate mortality rates. The authors cited the dose-response relationship of excess risk to exposure as 230×10^{-6}/WLM; this coefficient, however, was taken from the 1976 report[13] that was based on an incorrect method of analysis. To obtain the upper bound of 450×10^{-6}/WLM, the authors merely doubled the value reported by Sevc et al.[13] That calculation was justified by assuming that the average follow-up in the Czechoslovakian study (20 yr) represented the median latency for a 40-yr complete expression of the effects of exposure. The report did not provide evidence to support the biological model that is implicit in the doubling of the risk coefficient.

The Swedish data were also characterized as appropriate for consideration, although the original report by Snihs[14] did not provide complete information. The present committee does not regard these data as adequate for risk estimation. For a 40-yr period, Snihs estimated the attributable risk as 140×10^{-6}/WLM, on the basis of the Swedish data. The derivation of the lower bound of 200×10^{-6}/WLM from this value was not described. The report did not make firm statements about the effects of cigarette smoking.

ICRP PUBLICATION 32

Publication 32 by the International Commission on Radiological Protection (ICRP)[5] published in 1981, provided a recommended limit for inhalation of radon progeny by workers. In developing this limit, ICRP considered both the epidemiological evidence and the results of a dosimetric analysis. This committee has focused on ICRP's epidemiological approach.

The ICRP group emphasized the findings of the Colorado Plateau and Czechoslovakian studies. Relying on reports from those studies and on the 1977 UNSCEAR[18] and 1980 BEIR III[11] reviews, it cited a range of attributable risk of 2–20 cases/10^6 person-yr/WLM. Because the effect of exposure was noted to vary with age at exposure, the group considered 5–15 cases/10^6 person-yr/WLM as "the most probable range," on the basis of averaging "over all age periods during occupational work." Over "a mean manifestation period of 30 years," the group translated the attributable-risk range of 5–15 cases/10^6 person-yr/WLM into a total lifetime risk of 1.5–4.5 excess cases/WLM. With adjustment for the higher breathing rate of miners, the excess risks were reduced by about 20%. The ICRP group noted that the risks for miners might be increased by the effects of other exposures and thus tend to overestimate the effects of radon daughters alone.

This committee could not fully critique ICRP's epidemiological approach, because some procedures were not fully described: the derivation of the range of 2–20 cases/10^6 person-yr/WLM, the averaging that reduced this range to 5–15 cases/10^6 person-yr/WLM, and the rationale for the 30-yr period for calculating lifetime risk. As discussed elsewhere, this committee finds a modified relative-risk model to be preferable to the attributable-risk model used by ICRP in 1981.[5]

SUMMARY

The descriptions of risk estimates given above make it clear that a number of approaches have been applied to estimating the risks due to radon-daughter exposure. Some are based largely on expert opinion, while others depend on analyses of limited data on lung-cancer cases associated with exposure to radon progeny. Results vary, as indicated in Table VIII-1 above and Table 2-13 in Chapter 2. There are at least three underlying causes for this lack of agreement between risk estimates.

1. As discussed in Chapter 2 and Appendix IV, there is a fair amount of variability between the results of the individual epidemiological studies. Although these differences are perhaps no greater than would be anticipated on statistical grounds, it is not unreasonable to believe that other factors enter as well. Since some risks estimators put greater weight on one set(s) of observations than another, differences between risk estimates are not surprising.

2. A variety of techniques must be used to project lifetime risk to a general population on the basis of relatively short-term occupational exposures to underground miners, a topic discussed at length in Chapter 2. Foremost among these is the modeling of age-specific lung-cancer risk. Risk projections which use models based on the relative risk depend critically on the age-specific background rates. As discussed in Chapter 2, differences in estimated lifetime risks occur if the relative risk is constant or if it is permitted to vary with time-related factors. Similarly, lifetime risks that are derived from models of additive excess risk depend on the modeling of time-related effects. The different models will produce approximately the same average risk for populations with similar age structure and follow-up such as the underground miners. However, projecting beyond the range of the miner cohort data can produce very different numerical estimates.

3. Finally, several of the risk projections described above seem to depend more on considerations of biological plausibility rather than data analyses by standard methods. Some investigators might perhaps argue that biological plausibility should be the main criteria for risk projections, but others are less sure. Lung cancers observed in the miner studies are largely due to two complete carcinogens, smoking and high-LET radiation, whose joint interaction is not well defined. The committee believes that until underlying processes of carcinogenesis are understood, an objective analysis of observational data is a surer path to valid estimates of radon risks.

REFERENCES

1. Evans, R. D., J. H. Harley, W. Jacobi, H. S. McLean, W. A. Mills, and C. G. Stewart. 1981. Estimate of risk from environmental exposure to Rn-222 and its decay products. Nature 290:98–100.
2. Harley, N. H., and B. S. Pasternack. 1981. A model for predicting lung cancer risks induced by environmental levels of radon daughters. Health Phys. 40:307–316.
3. Howe G. R., R. C. Nair, H. G. Hewcombe, A. B. Miller, and J. D. Abbett. Lung cancer mortality (1950–1980) in relation to radon daughter exposure in a cohort of workers in the Eldorado Beaverlodge uranium mine. J. Natl. Cancer Inst. 77(2):357–362.

4. Hornung, R. W., and T. J. Meinhardt. 1987. Quantitative risk assessment of lung cancer mortality in U.S. uranium miners. Health Phys. 52:417–430.
5. International Commission on Radiological Protection (ICRP). 1981. P. 24 in Limits for Inhalation of Radon Daughters by Workers. ICRP Publication 32. Oxford: Pergamon.
6. Lundin F. E., J. K. Wagoner, and V. E. Archer. Radon Daughter Exposure and Respiratory Cancer Quantitative and Temporal Aspects. 1971. Joint Monograph No. 1. Washington, D.C.: U.S. Public Health Service.
7. Morrision H. I., D. T. Wigle, and A. J. deVilliers. 1981. Lung cancer mortality and radiation exposure among the Newfoundland flurospar miners. Pp. 372–376 in Proceedings of the International Conference on Radiation Hazards in Mining: Control, Measurements, and Medical Aspects, M. Gomez, ed. New York: Society of Mining Engineers of the American Institute of Mining, Metallurgical, and Petroleum Engineers, Inc.
8. Morrison, H. I., R. M. Semenciw, Y. Mao, D. A. Corkill, A. B. Dory, A. J. deVilliers, A. J. Stocker, and D. T. Wigle. 1985. Lung cancer mortality and radiation exposure among the Newfoundland flurospar miners. Pp. 365–368 in Proceedings of the International Conference on Occupational Radiation Safety in Mining, E. Stocker, ed. Toronto: Canadian Nuclear Association.
9. Muller J., W. C. Wheeler, J. F. Gentleman, G. Suranyi, and R. A. Kusiak. 1985. Study of mortality of Ontario miners. Pp. 335–343 in Proceedings of the International Conference on Occupational Radiation Safety in Mining, A. Stocker, ed. Toronto: Canadian Nuclear Association.
10. National Council on Radiation Protection and Measurements (NCRP). 1984. Evaluation of Occupational and Environmental Exposure to Radon and Radon Daughters. NCRP Report 78. Washington, D.C.: National Council on Radiation Protection and Measurements. 204 pp.
11. National Research Council, Committee on the Biological Effects of Ionizing Radiations (BEIR). 1980. The Effects on Populations of Exposure to Low Levels of Ionizing Radiation. Washington, D.C.: National Academy Press. 524 pp.
12. Radford, E. P., and K. G. St. Clair Renard. 1984. Lung cancer in Swedish iron miners exposed to low doses of radon daughters. N. Engl. J. Med. 310:1485–1494.
13. Sevc, J., E. Kung, and V. Placek. 1976. Lung cancer in uranium miners and long-term exposure to radon daughter products. Health Phys. 30:433–437.
14. Snihs, J. O. 1973. The approach to Rn problems in non-uranium mines in Sweden. Pp. 900–911 in Proceedings of Third International Congress of the IRPA. CONF-730907-P2. Technical Information Center, U.S. Department of Energy. Oak Ridge, Tenn.: Oak Ridge Natural Laboratory.
15. Sun S., X. Yang, Y. Lan, M. Xionyu, L. Shengen, and Y. Zhanyun. 1984. Latent period and temporal aspects of lung cancer among miners. Radiat. Prot. 4(5) (English translation).
16. Thomas, D. C., and K. G. McNeill. 1982. P. 23 in Risk Estimates for the Health Effects of Alpha Radiation. INFO-0081. Ottawa, Canada: Atomic Energy Control Board.
17. Thomas, D. C., K. G. McNeill, and C. Dougherty. 1985. Estimates of lifetime lung cancer risks resulting from Rn progeny exposure. Health Phys. 49:825–846.
18. United Nations Scientific Committee on the Effects of Atomic Radiation (UNSCLEAR). 1977. P. 725 in Sources and Effects of Ionizing Radiation. Report E.77.IX.1., New York: United Nations.
19. Whittemore, A. S., and A. McMillan. 1983. Lung cancer mortality among U.S. uranium miners: A reappraisal. J. Natl. Cancer Inst. 71:489–499.

Glossary

Absorbed dose. The mean energy imparted to the irradiated medium, per unit mass, by ionizing radiation. Units: gray (Gy), rad.

Activity. The mean number of decays per unit time of a radioactive nuclide. Units: becquerel (Bq), curie (Ci).

Activity median aerodynamic diameter (AMAD). The diameter of a unit-density sphere with the same terminal settling velocity in air as that of the aerosol particulate whose activity is the median for the entire aerosol.

Additive interaction model. This model is used to find the combined risk for risk factors which have no interaction with each other. For example, the combined mortality risk of cigarette smoking and automobile accidents is the sum of the separate risks.

Adenosarcoma. A mixed tumor which consists of a substance like embroyonic connective tissue together with glandular elements.

Alpha particle. Two neutrons and two protons bound as a single particle that is emitted from the nucleus of certain radioactive isotopes in the process of decay or disintegration.

Aneuploid. Having numbers of chromosomes not equal to exact multiples of the haploid number. Down syndrome is an example.

Background radiation. Radiation arising from radioactive material other than that under consideration; background radiation due to cosmic rays and natural radioactivity is always present; there may also be background radiation due to the presence of radioactive substances in building material.

Bayesian analysis. Analysis in which Bayes' theorem is used to derive posterior probabilities from assumed prior knowledge together with observational data. For example, biological information on the relationship between species and hazardous substances can be combined with data on interspecies dose response to calculate the response of human populations.

Becquerel (Bq). SI unit of activity. (*See* Units.)

Bremsstrahlung. The production of electromagnetic radiation (photons) by the acceleration (positive or negative) that a fast, charged particle (usually an electron) undergoes from the effect of an electric or magnetic field; for instance, from the field of another charged particle (usually a nucleus).

Bronchioles. The small branches of the tracheobronchial tree of the lung.

Cell culture. The growing of cells in vitro, in such a manner that the cells are no longer organized into tissues.

Chromosomal nondisjunction. Either a gain or a loss of chromosomes that occurs when cell division leading to either egg or sperm production goes awry. This results in aneuploidy.

Ciliated mucosa. The mucous membrane in the lung covered with small hairlike structures which serve to move the mucus.

Competing risks. Other causes of death which affect the value of the risk being studied. Persons dying from other causes are not at risk of dying from the factor in question.

Constant-relative-risk model. A risk model which assumes that, after a certain time, the ratio of the risk at a specific dose to the risk in the absence of the dose does not change with time.

Contact inhibition. The cessation of migratory activity and sometimes other functions, including mitosis, when adjacent cells establish firm contact.

Cox proportional hazards model. A relative-risk model that permits the use of internal comparison groups as controls for confounding variables such as cigarette smoking and age.

Curie (Ci). A unit of activity equal to 3.7×10^{10} disintegrations/s. (*See* Units.)

Daughter product. An isotope formed as a result of radioactive decay. One daughter atom is formed for each particle emitted.

Decay chain or decay series. A sequence of radioactive decays of the same nucleus. An initial nucleus, the parent, decays into a daughter nucleus that differs from the first by whatever particles were emitted during the decay. If further decays take place, the subsequent nuclei are also usually called daughters. Sometimes, to distinguish the sequence, the daughter of the first daughter is called the granddaughter, etc.; ordinarily, however, this quickly becomes too complicated.

Diffusion. The random path followed by very small particles due to the impact of surrounding molecules. This Brownian motion governs the transport of submicrometer-size particles in air.

Dominant mutation. The mutation is dominant if it produces its effect in the presence of an equivalent normal gene from the other parent.

Dose-distribution factor. A factor which accounts for modification of the dose effectiveness in cases in which the radionuclide distribution is nonuniform.

Dose equivalent. A quantity that expresses, for the purposes of radiation protection and control, the assumed effectiveness of dose on a common scale for all kinds of ionizing radiation. SI unit is the Sievert. (*See* Units.)

Doubling dose. The amount of radiation needed to double the natural incidence of a genetic or somatic anomaly.

Electron volt (eV). A unit of energy = 1.6×10^{-12} ergs = 1.6×10^{-19} J; 1 eV is equivalent to the energy gained by an electron in passing through a potential difference of 1 V; 1 keV = 1,000 eV; 1 MeV = 1,000,000 eV.

Epithelium. A membranous cellular tissue that covers the surface of some organ or part of the body.

Equilibrium fraction. In equilibrium, the parents and daughters have equal radioactivity, that is, as many decay into a specific nuclide as decay out. When fresh radon enters a volume, the daughter products have not yet accumulated, and there is disequilibrium. The working-level definition of radon does not take into account the amount of equilibrium.

Equilibrium, radioactive secular. The condition in which the activities of a parent and daughter in a radioactive decay chain are (very nearly) equal.

Equilibrium, radiation. The condition in a radiation field where the energy of the radiations entering a volume equals the energy of the radiations leaving that volume.

Euploid. Having uniform exact multiples of the haploid number of chromosomes.

Gamma ray. Short-wavelength electromagnetic radiation of nuclear origin (range of energy, 10 keV to 9 MeV).

Gray (Gy). SI unit of absorbed dose. (*See* Units.)

Half-life, biologic. Time required for the body to eliminate half of an administered dose of any substance by regular processes of elimination; it is approximately the same for both stable and radioactive isotopes of a particular element.

Half-life, radioactive. Time required for a radioactive substance to lose 50% of its activity by decay.

Impaction. As air is taken into the lung, it follows a tortuous path, changing direction many times. At each change of direction, the momentum of the particles carried in the airstream causes them to impact on the bifurcations of the lung. The force on the particle causing it to move and impact on the lung surface is the Stokes force, which is proportional to the velocity of the air moving with respect to the particle. Impaction is important for particles with large aerodynamic diameters.

Incidence. The rate of occurrence of a disease within a specified period; usually expressed in number of cases per 100,000 persons per year.

Ionization. The process by which a neutral atom or molecule acquires a positive or negative charge.

Ionization density. Number of ion pairs per unit volume.

Ionization path (track). The trail of ion pairs produced by ionizing radiation in its passage through matter.

Isotopes. Nuclides that have the same number of protons in their nuclei, and hence the same atomic number, but that differ in the number of enutrons, and therefore in the mass number; chemical properties of osotopes of a particular element are almost identical. The term should not beused as a synonym for nuclide.

Latent period. The period of time between exposure and expression of the disease. After exposure to a dose of radiation, there is a delay of several years (the latent period) before any cancers are seen.

Life-span study (LSS). Life-span study of the Japanese atomic-bomb survivors; the sample consists of 120,000 persons, of whom 82,000 were exposed to the bombs, mostly at low doses.

Lifetime risk. The lifetime probability of dying of a specific disease.

Lifetime risk ratio. The ratio of the lifetime risk (R_e) of an exposed person to the lifetime risk of an unexposed person (R_0). This number minus 1 is the proportional increased risk associated with exposure ($R_e - R_0$).

Linear dose model. This model postulates that the excess risk is linearly proportional to the dose.

Linear energy transfer (LET). Average amount of energy lost per unit track length.

Low LET. Radiation characteristic of electrons, x rays, and gamma rays; the distance between ionizing events is large on the scale of a cellular nucleus.

High LET. Radiation characteristic of protons and fast neutrons; the distance between ionizing events is small on the scale of a cellular nucleus. Average LET is specified to even out the effect of a particle that is slowing down near the end of its path and to allow for the fact that secondary particles are not all of the same energy.

Lymphosarcoma. A sarcoma of the lymphoid tissue. This does not include Hodgkin's disease.

Minute volume. The amount of air moving through the lung per minute; the product of the breathing rate times the volume of air per breath.

Multiplicative interaction model. A model in which independent risk factors interact so that the combined risk is the product of the relative risks due to each factor alone.

Neoplasms. Any new and abnormal growth, such as a tumor; *neoplastic disease* refers to any disease that forms tumors, whether malignant or benign.

Nonstochastic. Describes effects whose severity is a function of dose; for these, a threshold may occur; some nonstochastic somatic effects are cataract induction, nonmalignant damage to skin, hematological deficiencies, and impairment of fertility.

Nuclide. A species of atom characterized by the constitution of its nucleus, which is specified by its atomic mass and atomic number (Z), or by its number of protons (Z), number of neutrons (N), and energy content.

Oncogenes. Genes which carry the potential for cancer.

Person-gray. Unit of population exposure obtained by summing individual dose-equivalent values for all people in the exposed population. Thus, the number of person-grays contributed by 1 person exposed to 1 Gy is equal to that contributed by 100,000 people each exposed to 10 μGy.

Person-years-at-risk (PYAR). The number of persons exposed times the number of years after exposure minus some lag period during which the dose is assumed to be unexpressed (latent period).

Prevalence. The number of cases of a disease in existence at a given time per unit population, usually 100,000 persons.

Progeny. The decay products resulting after a series of radioactive decays. Progeny can also be radioactive, and the chain continues until a stable nuclide is formed.

Quadratic-dose model. A model which assumes that the excess risk is proportional to the square of the dose.

Quality factor (Q). A linear energy transfer dependent factor by which absorbed doses are multiplied to obtain (for radiation-protection purposes) a quantity which corresponds more closely to the degree of biological effect produced by x or low-energy gamma rays.

Rad. A unit of absorbed dose. Replaced by the gray in SI units. (*See* Units.)

Radioactivity. The property of some nuclides of spontaneously emitting particles or gamma radiation, emitting x radiation after orbital electron capture, or undergoing spontaneous fission.

Artificial radioactivity. Man-made radioactivity produced by fission, fusion, particle bombardment, or electromagnetic irradiation.

Natural radioactivity. The property of radioactivity exhibited by more than 50 naturally occurring radionuclides.

Radioisotopes. A radioactive atomic species of an element with the same atomic number and usually identical chemical properties.

Radionuclide. A radioactive species of an atom characterized by the constitution of its nucleus.

Radiosensitivity. Relative susceptibility of cells, tissues, organs, and organisms to the injurious action of radiation; radiosensitivity and its antonym, radioresistance, are used in a comparative sense rather than an absolute one.

Recessive gene disorder. This requires that a pair of genes, one from each parent, be present in order for the disease to be manifest. An example is cystic fibrosis.

Relative biological effectiveness (RBE). Biological potency of one radiation as compared with another to produce the same biological endpoint. It is numerically equal to the inverse of the ratio of absorbed doses of the two radiations required to produce equal biological effect. The reference radiation is often 200-kV x rays.

Relative mutation risk. The ratio of the risk of a genetic mutation among the exposed population to that in the absence of exposure.

Risk coefficient. The increase in the annual incidence or mortality rate per unit dose: (1) absolute risk coefficient is the observed minus the expected number of cases per person year at risk for a unit dose; (2) the relative-risk coefficient is the fractional increase in the baseline incidence or mortality rate for a unit dose.

Risk estimate. The number of cases (or deaths) that are projected to occur in a specified exposed population per unit dose for a defined exposure regime and expression period: number of cases per person-Gray or, for radon, the number of cases per person cumulative working-level month.

Rem. A unit of dose equivalent. Replaced by the sievert. (*See* Units.)

Sedimentation. The gravitational force on a particle is partially balanced by the viscous force of the air. The resultant velocity toward the earth is the sedimentation velocity. Important for particles with intermediate aerodynamic diameters.

Sex-linked mutation (or X-linked). A mutation associated with the X chromosome. It will usually only manifests its effect in males (who have only a single X chromosome).

SI units. The International System of Units as defined by the General Conference of Weights and Measures in 1960. These units are generally based on the meter/kilogram/second units, with special quantities for radiation including the becquerel, gray, and sievert.

Sievert. The SI unit of radiation dose equivalent. It is equal to dose in grays times a quality factor times other modifying factors, for example, a distribution factor; 1 sievert equals 100 rem.

Specific activity. Total activity of a given nuclide per gram of a compound, element, or radioactive nuclide.

Specific energy. The actual energy per unit mass deposited per unit volume in a given event. This is a stochastic quantity as opposed to the average value over a large number of instances (i.e., the absorbed dose).

Squamous cell carcinoma. A cancer composed of cells that are scaly or platelike.

Standard mortality ratio (SMR). Standard mortality ratio is the ratio of the disease or accident mortality rate in a certain specific population compared with that in a standard population. The ratio is based on 100 for the standard so that an SMR of 200 means that the test population has twice the mortality from that particular cause of death.

Stochastic. Describes random events leading to effects whose probability of occurrence in an exposed population (rather than severity in an affected individual) is a direct function of dose; these effects are commonly regarded as having no threshold; hereditary effects are regarded as being stochastic; some somatic effects, especially carcinogenesis, are regarded as being stochastic.

Stopping power. The average rate of energy loss of a charged particle per unit thickness of a material or per unit mass of material traversed.

Straggling. The statistical variation in the range of a particle caused by the large number of interactions and scatterings within the material being traversed.

Surface-seeking radionuclide. An internal emitter that is deposited and remains on the surface of bone for a long period of time. This contrasts with a volume seeker, which deposits more uniformly throughout the bone volume.

Target theory (hit theory). A theory explaining some biological effects of radiation on the basis that ionization, which occurrs in a discrete volume (the target) within the cell, directly causes a lesion that later results in a physiological response to the damage at that location; one, two, or more hits (ionizing events within the target) may be necessary to elicit the response.

Threshold hypothesis. The assumption that no radiation injury occurs below a specified dose.

Time-since-exposure (TSE) model. A model in which the relative risk is not constant but varies with the time after exposure.

Transformed cells. Tissue culture cells changed in vitro from growing in an orderly pattern and exhibiting contact inhibition to growing in a pattern more like that of cancer cells, resulting in the loss of contact inhibition.

Translocation. A chromosome aberration resulting from chromosome breakage and subsequent structural rearrangement of the parts between the same or different chromosomes.

Tumorigenicity. Ability of cells to proliferate into tumors when inoculated into a specified host organism under specified conditions.

Unattached fraction. That fraction of the radon daughters, usually ^{218}Po (Radium A), which has not yet attached to a particle. As a free atom, it has a high probability of being retained within the lung and depositing alpha energy when it decays.

Units[a]	Conversion Factors
Becquerel (SI)	1 disintegration/s = 2.7×10^{-11} Ci
Curie	3.7×10^{10} disintegrations/s = 3.7×10^{10} Bq
Gray (SI)	1 J/kg = 100 rad
Rad	100 erg/g = 0.01 Gy
Rem	0.01 Sievert
Sievert (SI)	100 rem

[a]International Units are designated (SI).

Working level (WL). Any combination of short lived radon daughters in 1 liter of air that will result in the ultimate emission 1.3×10^5 MeV of potential alpha energy. This number was chosen because it is approximately the alpha energy released from the decay of daughters in equilibrium with 100 picocuries of ^{222}Ra.

Working-level month (WLM). Exposure resulting from inhalation of air with a concentration of 1 working level of radon daughters for 170 working hours.

Years of life lost. The expected years of life for a nonexposed person minus the expected lifetime of an exposed person.

Index

B

E

M

N

S

T

Z

RANDALL LIBRARY-UNCW
3 0490 0078822 0